RECENT DEVELOPMENTS IN APPLIED MICROBIOLOGY AND BIOCHEMISTRY

RECENT DEVELOPMENTS IN APPLIED MICROBIOLOGY AND BIOCHEMISTRY

Edited by

VISWANATH BUDDOLLA
Department of Bionanotechnology, Gachon University, South Korea

ELSEVIER

ACADEMIC PRESS
An imprint of Elsevier

Academic Press is an imprint of Elsevier
125 London Wall, London EC2Y 5AS, United Kingdom
525 B Street, Suite 1650, San Diego, CA 92101, United States
50 Hampshire Street, 5th Floor, Cambridge, MA 02139, United States
The Boulevard, Langford Lane, Kidlington, Oxford OX5 1GB, United Kingdom

Notices

Knowledge and best practice in this field are constantly changing. As new research and experience broaden our understanding, changes in research methods, professional practices, or medical treatment may become necessary.

Practitioners and researchers must always rely on their own experience and knowledge in evaluating and using any information, methods, compounds, or experiments described herein. In using such information or methods they should be mindful of their own safety and the safety of others, including parties for whom they have a professional responsibility.

To the fullest extent of the law, neither the Publisher nor the authors, contributors, or editors, assume any liability for any injury and/or damage to persons or property as a matter of products liability, negligence or otherwise, or from any use or operation of any methods, products, instructions, or ideas contained in the material herein.

Library of Congress Cataloging-in-Publication Data
A catalog record for this book is available from the Library of Congress

British Library Cataloguing-in-Publication Data
A catalogue record for this book is available from the British Library

ISBN 978-0-12-816328-3

For information on all Academic Press publications
visit our website at https://www.elsevier.com/books-and-journals

Publisher: John Fedor
Acquisition Editor: Linda Versteeg-buschman
Editorial Project Manager: Timothy Bennett
Production Project Manager: Poulouse Joseph
Cover Designer: Victoria Pearson

Typeset by SPi Global, India

Dedication

This book is dedicated to Professor Bontha Rajasekhar Reddy who has been a constant source of knowledge and inspiration.

Contents

20. Development in the Treatment of Municipal and Industrial Wastewater by Microorganism

MEKAPOGU MADAKKA, NADIMIKERI JAYARAJU, NAMBI RAJESH, MUNI RAMANNA GARI SUBHOSH CHANDRA

21. Emerging Trends of Microorganism in the Production of Alternative Energy

GOLLA RAMANJANEYULU, BONTHA RAJASEKHAR REDDY

22. Present Status and Future Perspectives of Marine Actinobacterial Metabolites

GURUSHANKARA HUNASANAHALLY PUTTASWAMYGOWDA, SHILPA OLAKKARAN, ANET ANTONY, ANUPAMA KIZHAKKE PURAYIL

5
METHODS IN MICROBIOLOGY

23. Status and Recent Developments in Analytical Methods for the Detection of Foodborne Microorganisms

SESHADRI REDDY ANKIREDDY, JONGSUNG KIM

24. The Impact of Bioinformatics Tools in the Development of Antimicrobial Drugs and Other Agents

SWETHA KUMARI KODURU

25. Recent Trends in the Development of Paper-Based Diagnostic Chips for the Detection of Human Viruses

SARAVANAN GOVINDARAJU, MIN HO LEE, KYUSIK YUN

Contributors

Seshadri Reddy Ankireddy Department of Chemical and Biological Engineering, Gachon University, Seongnam, South Korea

Anet Antony Department of Animal Science, School of Biological Sciences, Tejaswini Hills, Periye, Kasaragod, India

Meriga Balaji Department of Biochemistry, Sri Venkateswara University, Tirupati, India

Hari Balaji Department of Biochemistry, Sri Venkateswara University, Tirupati, India

Maram Bhargavi Department of Applied Microbiology, Sri Padmavati Mahila Visvavidyalayam, Tirupati, India

Muni Ramanna Gari Subhosh Chandra Department of Microbiology, Yogi Vemana University, Kadapa, India

P.B.B.N. Charyulu Department of Microbiology, Sri Krishnadevaraya University, Anantapuramu, India

Narayanaiah Cheedarla Department of HIV/AIDS, National Institute for Research in Tuberculosis-ICMR, Chennai, India

Subramanyam Dasari Medical Sciences Program, Indiana University School of Medicine, Bloomington, IN, United States

Routhu Gyana Deepika Aurobindo Pharma Ltd, APL Research Center-2, Sanga Reddy Mandal, India

Velluri Deepika Department of Applied Microbiology, Sri Padmavati Mahila Visvavidyalayam, Tirupati, India

Yaramolu Divya Department of Applied Microbiology, Sri Padmavati Mahila Visvavidyalayam, Tirupati, India

Muni Swamy Ganjayi Department of Biochemistry, Sri Venkateswara University, Tirupati, India

Saravanan Govindaraju Department of Bionanotechnology, Gachon University, Seongnam, Republic of Korea

Krishnamoorthy Gowthami Department of Applied Microbiology, Sri Padmavati Mahila Visvavidyalayam, Tirupati, India

Dilip Reddy Gunturu Department of Veterinary Pharmacology and Toxicology, College of Veterinary Science, Tirupati, India

Luke Elizabeth Hanna Department of HIV/AIDS, National Institute for Research in Tuberculosis-ICMR, Chennai, India

M. Hema Department of Virology, Sri Venkateswara University, Tirupati, India

Avilala Janardhan Jiangsu Key Laboratory for Molecular and Medical Biotechnology, College of Life Sciences, Nanjing Normal University, Nanjing, People's Republic of China

Nadimikeri Jayaraju Department of Geology, Yogi Vemana University, Kadapa, India

Reddy Sankaran Karunakaran Department of Biochemistry, Sri Venkateswara University, Tirupati, India

Chandrasekhar Kathera Jiangsu Key Laboratory for Molecular and Medical Biotechnology, College of Life Sciences, Nanjing Normal University, Nanjing, People's Republic of China

Jongsung Kim Department of Chemical and Biological Engineering, Gachon University, Seongnam, South Korea

Anupama Kizhakke Purayil Department of Animal Science, School of Biological Sciences, Tejaswini Hills, Periye, Kasaragod, India

Swetha Kumari Koduru Molecular Cloning Lab, Department of Biosciences and Sericulture, Sri Padmavati Women's University, Tirupati, India

M.S. Krishnaveni Department of Microbiology, Sri Krishnadevaraya University, Anantapuramu, India

Bestha Lakshmi Department of Zoology, Sri Krishnadevaraya University, Anantapur, India

Min Ho Lee Chung Ang University, School of Integrative Engineering, Seoul, Republic of Korea

Mekapogu Madakka Department of Biotechnology & Bioinformatics, Yogi Vemana University, Kadapa, India

Ravuri Jaya Madhuri Department of Applied Microbiology, Sri Padmavati Mahila Visvavidyalayam, Tirupati, India

Nimgampalle Mallikarjuna Department of Zoology, Sri Venkateswara University, Tirupati, India

Sravanthi Mannem Department of Veterinary Pharmacology and Toxicology, College of Veterinary Science, Tirupati, India

Alpha Raj Mekapogu Department of Veterinary Pharmacology and Toxicology, College of Veterinary Science, Tirupati, India

M.R.N. Murthy Molecular Biophysics Unit, Indian Institute of Science, Bangalore, India

Shilpa Olakkaran Department of Animal Science, School of Biological Sciences, Tejaswini Hills, Periye, Kasaragod, India

Pandeeti Emmanuel Vijay Paul Aurobindo Pharma Ltd, APL Research Center-2, Sanga Reddy Mandal, India

Gurushankara Hunasanahally Puttaswamygowda Department of Animal Science, School of Biological Sciences, Tejaswini Hills, Periye, Kasaragod, India

Mokula Mohammed Rafi Department of Natural Resource Management, College of Agriculture and Rural Transformation, University of Gondar, Gondar, Ethiopia;Department of Microbiology, Sri Krishnadevaraya University, Anantapuramu, India

Nambi Rajesh Department of Biotechnology & Bioinformatics, Yogi Vemana University, Kadapa, India

Golla Ramanjaneyulu Department of Microbiology, Sri Krishnadevaraya University, Anantapur; Council of Scientific and Industrial Research—Central Institute of Medicinal and Aromatic Plants Research Centre, Bangalore, India

Bontha Rajasekhar Reddy Department of Microbiology, Sri Krishnadevaraya University, Anantapur, India

Veeraiah Sangeetha Aurobindo Pharma Ltd, APL Research Center-2, Sanga Reddy Mandal, India

Maddu Saraswathi Department of Applied Microbiology, Sri Padmavati Mahila Visvavidyalayam, Tirupati, India

Marella Saritha Nanotechnology Laboratory, Institute of Frontier Technology, Regional Agricultural Research Station, Acharya N.G. Ranga Agricultural University, Tirupati, India

H.S. Savithri Department of Biochemistry, Indian Institute of Science, Bangalore, India

Anshul Sharma Department of Food Science and Biotechnology, Gachon University, Seongnam, Republic of Korea; Department of Biotechnology, Shoolini University of Biotechnology and Management Sciences, Solan, India

Dasari Sreenivasulu Department of Biochemistry, Sri Venkateswara University, Tirupati, India

Myla Sunil Kumar Department of Virology, Sri Venkateswara University, Tirupati; NextBio Research Pvt., Ltd., Bengaluru, India

Shameer Syed Nanotechnology Laboratory, Institute of Frontier Technology, Regional Agricultural Research Station, Acharya N.G. Ranga Agricultural University, Tirupati, India

Hamsa D. Tadepally Scientific Curator (Consultant), Cleveland, OH, United States

N.V.K.V. Prasad Tollamadugu Nanotechnology Laboratory, Institute of Frontier Technology, Regional Agricultural Research Station, Acharya N.G. Ranga Agricultural University, Tirupati, India

G.P. Vishnu Vardhan Department of Biochemistry, Indian Institute of Science, Bangalore, India

Viswanath Buddolla Department of Bionanotechnology, Gachon University, South Korea

Muralidhar Yegireddy Department of Veterinary Pharmacology and Toxicology, College of Veterinary Science, Tirupati, India

Kuna Yellamma Department of Zoology, Sri Venkateswara University, Tirupati, India

Kyusik Yun Department of Bionanotechnology, Gachon University, Seongnam, Republic of Korea

Foreword

This book, *Recent Developments in Applied Microbiology and Biochemistry*, is published as a fitting tribute to Professor Rajasekhar Reddy, Sri Krishnadevaraya University, Anantapur, India, on the occasion of his superannuation. Prof. Reddy is internationally recognized and respected as a teacher, researcher, and academician. I recall his enthusiasm and meticulous approach in his research when he was working for his doctorate in our laboratory at the Central (now National) Rice Research Institute, Cuttack, India. During the short tenure (1981–84) with us, he contributed to research in the area of microbiology of rice soils in a big way. I take pride in the fact that he continued active research in his area of specialization later at Bhabha Atomic Research Centre, Mumbai, and then at Sri Krishnadevaraya University, Anantapur, after leaving us. Prof. Reddy has since been involved in research and teaching and has served as a source of inspiration to many students and colleagues.

The present book is organized into five sections—medical microbiology and biochemistry, microbial biotechnology, soil and agriculture microbiology, environmental microbiology, and methods in microbiology—which together contains 30 chapters contributed by acknowledged experts from various parts of the world. All the five sections are aimed at reviewing the key concepts in applied microbiology and biochemistry. Dr. Viswanath Buddolla, one of the foremost students of Prof. Reddy and the editor of this book, has done a fine job of bringing together a diverse range of topics to effectively highlight some of the important contributions to the topics.

The book chapters cover both traditional and recent developments combined with discussions and critical analysis of topics. I congratulate the contributors of the chapters of the book for their diligent work, and I have no hesitation in recommending this book as a valuable resource to the multidisciplinary researchers, academicians, and scholars in the field of microbiology and biochemistry.

Nambrattil Sethunathan
Fellow: Indian Academy of Sciences,
Indian National Science Academy
Formerly Principal Scientist and Head at
Division of Soil Science and Microbiology
Specialization: Environmental
Microbiology
Central Rice Research Institute,
Cuttack, India

Preface

Recent Developments in Applied Microbiology and Biochemistry is a book with diverse chapters related to medical microbiology and biochemistry, microbial biotechnology, soil and agriculture microbiology, environmental microbiology, and methods in microbiology. The book is being published on the occasion of superannuation of an inspiring teacher, Prof. Bontha Rajasekhar Reddy.

(*Prof. B.R. Reddy, a man with a great vision*: 30 years of dedicated efforts as a teacher and as a researcher in enhancing student learning)

Prof. B.R. Reddy was born on May 16, 1956, in Andhra Pradesh, India. Dr. Reddy worked at the Central (now National) Rice Research Institute, which was affiliated to Utkal University, as a doctoral researcher under the guidance of Dr. N. Sethunathan and obtained his PhD in 1985. Thereafter, in June 1985, he joined the Bhabha Atomic Research Centre, India, as a research associate and worked for 4 years. Since March 1989, he has been serving as a faculty of Department of Microbiology, Sri Krishnadevaraya University, India. In 1991, he received the Commonwealth Academic Staff Fellowship, and in 1993, he was the recipient of University Grants Commission Career award for young teachers. He has also mentored a number of master's and doctoral students, the majority of whom have become successful independent investigators and are also well respected in their own right. I feel privileged to be his mentee. The royalties from this book are going to be used toward an award in the name of the Prof. B.R. Reddy award of excellence, to encourage students who demonstrate academic achievement, strong character, and leadership skills from the Department of Microbiology, Sri Krishnadevaraya University, India.

There is no doubt that the fields of microbiology and biochemistry have seen wonderful developments by virtue of keen observations and discoveries made by several great scientists. This book pools knowledge of microbiological and biochemical aspects to provide an integrated version of the most recent developments in the fundamental and applied aspects of microbiology and biochemistry. The book has 25 chapters, which have been classified into five sections. All the chapters have included the most important achievements that have taken place in recent years with state-of-the-art information. The unifying and interdisciplinary approaches in the book may be beneficial for bridging the gaps among diverse knowledge domains. I'm hopeful that this book will be useful to students, teachers, and researchers interested in the area of applied microbiology and biochemistry as this book covers topics from basic concepts to novel methodologies. I would be delighted if the book will be found more often on the bench top than in the book shelf.

Viswanath Buddolla
Gachon University, South Korea
May 2018

Acknowledgments

I thank Dr. N. Sethunathan for all his blessings and guidance in the completion of this book. I would like to take this opportunity to thank all the authors for their contributions in this book. Their efforts have improved the quality of the book considerably. I am grateful to Mrs. Linda Versteeg, Mr. Timothy Bennett, and the entire team of Elsevier/Academic Press for all their incentive, patience, and support in all cases that needed specific consideration and in producing this book.

SECTION 1

MEDICAL MICROBIOLOGY AND BIOCHEMISTRY

CHAPTER

1

Recent Findings of *Lactobacillus* Diversity and Their Functional Role in Vaginal Ecosystems

Subramanyam Dasari

Medical Sciences Program, Indiana University School of Medicine, Bloomington, IN, United States

OUTLINE

1.1 INTRODUCTION

The microbial communities have evolved their relation and association with us over thousands of years and are critical to our survival. For example, we rely on some bacteria for the absorption of nutrients and for conditioning the innate immune response; however, now, we are beginning to understand the complex mechanisms of these mutually beneficial interactions. On the other hand, disturbances in that microbiome may result in disease as well, and thus, improved knowledge of this association is critical to a better understanding of both

https://doi.org/10.1016/B978-0-12-816328-3.00001-5

human health and illness (Martin, 2012). An enormous number of microorganisms, especially bacterial species, are known to colonize and form complex communities, or microbiota, at various sites within and on the human body (Ling et al., 2010). Growing evidence suggests that the composition and function of the microbiota in different human body habitats play a pivotal role in human development, physiology, immunity, and nutrition (Dethlefsen et al., 2007). Female genital tract is created by a succession of cavities such as the uterine cavity, endocervix, and vagina with the exterior through the vulvar cleft. These cavities allow the exteriorization of the menstrual flow; on the other hand, it is normally colonized by a complex mixture of indigenous microbial flora. In the female —the vagina—is important microbial habitats that harbor different species of bacteria, which are known to have important effects on health (Ling et al., 2010). Experimental and clinical studies have been devoted to elucidating the microbiota of the vagina and, to a lesser extent, the cervix and vulva (Dasari et al., 2016).

In 1892, Döderlein isolated an organism from vaginal specimen of normal pregnant woman, called Döderlein's bacillus, which later he named *Lactobacillus* (Linhares et al., 2010). It is well established that these *Lactobacillus* species inhabit the vaginal tract and play an important role in the maintenance of health by significantly inhibiting the growth of other organisms through the production of various defense mechanisms (Cribby et al., 2008). Production of lactic acid, bacteriocins, and hydrogen peroxide has led to an abundance of research involving the ability of *Lactobacillus* to inhibit the growth of bacterial and fungal pathogens that commonly cause urogenital tract infections (Westbroek et al., 2010). Hence, we try to discuss more about the recent findings of *Lactobacillus* diversity and their functional role (antimicrobial, antiviral, anticancer, and immunomodulatory properties of lactic acid) in maintaining the women's health management.

1.2 VAGINAL (BIRTH CANAL) MICROBIAL COMMUNITY

The lower female reproductive tract (the vagina and ectocervix) is composed of a stratified squamous epithelium and glandular cells that make mucus, which is bathed in mucus and colonized with a large diversity of microbial flora that can have either commensal (benefit of one of the organisms, without effect on the other), mutualistic (both organisms benefit), or parasitic (microbes profit at the expense of the host) nature (Tachedjian et al., 2017). The vaginal microbial ecosystem undergoes significant structural changes at various growth stages in woman's life that are directly associated with hormone levels (estrogen) in the body (Hickey et al., 2012). First microbial colonization occurs during birth, when the infant is first exposed to her mother's birth canal. At the age of 2–4 weeks, maternal hormones mediate the lowering of the vaginal pH through the production of glycogen, which is fermented by indigenous bacteria in vaginal epithelial cells. Due to this low pH, until several weeks after birth, *Lactobacillus* is the predominant flora in vagina of a newborn girl. Nearly neutral pH was maintained during childhood and colonized by a diverse group of aerobic and strictly anaerobic bacteria and is highly susceptible to a variety of bacterial pathogens. During the menstrual process, follicular development causes hormone (estrogen) production and is accompanied by a thickening of the vaginal epithelium and intracellular production of glycogen. *Lactobacillus* again become prominent in this condition and are capable of fermenting glycogen to lactic acid, and the associated acidification of the vaginal environment is characteristic of women reproductive age. Apart from *Lactobacillus*, a smaller number of yeasts and other bacterial species are also present and are more resistant to infection (Dasari et al., 2016; Hickey et al., 2012).

1.3 LACTOBACILLUS-DOMINATED VAGINAL MICROBIOTA

Lactobacillus is the keystone species present in the microbiota of the birth canal. It is well established that the high number of *Lactobacillus* in the vagina is often considered as "healthy" or "normal" and low or absence considered as "abnormal" condition. *Lactobacillus* is popularized as lactic acid-producing bacteria by fermenting glycogen secreted from follicular cells of the vaginal epithelium, resulting in an enhanced acidic environment (pH $<$ 4.5) (Linhares et al., 2010). The acidic environment of a healthy vagina prevents the growth of many potential pathogens. The most frequently detected *Lactobacillus* species are *Lactobacillus iners*, *L. crispatus*, and *L. gasseri*. Other species as *L. jensenii*, *L. gallinarum*, and *L. vaginalis* have also been identified in some women and approximately 10^8 lactobacillus cells can be found per milliliter of vaginal fluid (Fredricks et al., 2005; Zhou et al., 2004). *Lactobacillus* forms a massive growth of biofilm, which tightly attaches to the vaginal epithelia surface, creating the first line of defense mechanism against potential pathogens. The vaginal fluid secreted from the stratified squamous epithelium contains specific organic acids, peroxides, and polypeptides that act as antibacterial agents against pathogenic or abnormal microbes (Dover et al., 2008). Vagina's acid environment is an important defense mechanism against the proliferation of pathogens. When the flora of predominating *Lactobacillus* is reduced in the vaginal environment due to any reason, other bacteria (*Megasphaera* and *Leptotrichia*) start occupying and are capable of producing metabolites characterized by an unpleasant smell. Fig. 1.1 shows that the protective role of vaginal *Lactobacillus* through the production of various inhibitory substances and

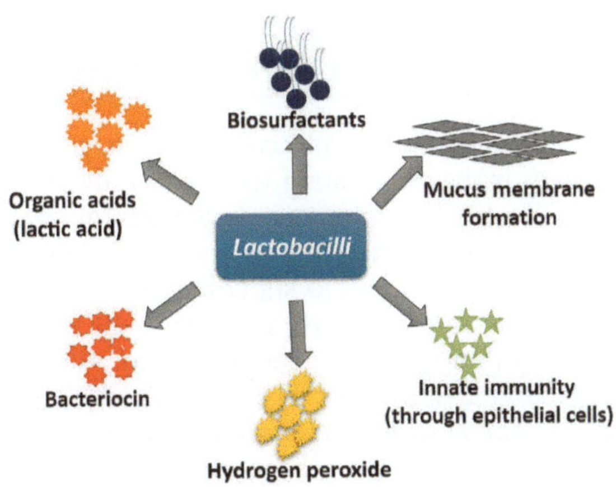

FIG. 1.1 Mode of action of vaginal *Lactobacillus* species: *Lactobacillus* are potentially inhibiting the pathogens in a direct manner by producing various compounds such as lactic acid, biosurfactants, hydrogen peroxide, and bacteriocin. In addition, they promote the integrity of the epithelium through stimulating mucus secretion and regulate the immune cells through inhibiting the bacterial and viral pathogens.

TABLE 1.1 Functional Role of Various Vaginal *Lactobacillus*

S. No.	*Lactobacillus* Species	Functional Role	Reference
1	*Lactobacillus crispatus*	Inhibition of hyphal formation	Wang et al. (2017)
2	*Lactobacillus acidophilus* and *L. gasseri*	Adhesion	Boris et al. (1998)
3	*Lactobacillus crispatus*	Adhesion through S-layer proteins	Mousavi et al. (2017)
4	*Lactobacillus crispatus*	Regulate the innate immunity	Doerflinger et al. (2014)
5	*Lactobacillus crispatus*	Matrix metalloproteinase inducer	Witkin et al. (2013)
6	*Lactobacillus crispatus*	Regulating the toll-like receptors, interleukin 8 (IL-8), and human β-defensin 2	Petrova et al. (2015)
7	*Lactobacillus acidophilus*	Virucidal effect against HIV-1	Botić et al. (2007)
8	*Lactobacillus crispatus*	Inactivates the HSV	Zabihollahi et al. (2012)
9	*Lactobacillus crispatus*	Antiviral-HSV-2	Mousavi et al. (2017)
10	*Lactobacillus* and *Bifidobacterium* species	Antiviral on rotavirus	Galán et al. (2016)
11	*Lactobacillus gasseri*	Anticancer—cervical cancer	Sungur et al. (2017)
12	*Lactobacillus crispatus* and *L. gasser*	Cytotoxic effects on cervical tumor cells	Wang et al. (2017)

compounds blocks the adhesion sites and degradation of toxin receptors (Dasari et al., 2016). Table 1.1 summarizes the major findings on variations in composition of *Lactobacillus* species from female genital tract and their functional diversity. In the next sections, we are discussing more about the functional role of *Lactobacillus* in the management of female genital tract.

1.4 GEOGRAPHICAL DIVERSITY OF VAGINAL *LACTOBACILLUS*

In concern to the women's health, the microbiome of vagina has not been extensively studied. Historically, *L. acidophilus* is the dominating species in human vagina, which motivates the scientific community to identify more therapeutic options for treatment of many diseases including cancer. It is an established fact that the composition of microbial flora is associated with various diseases; hence, it is important to know the diversity of *Lactobacillus* geographically and functionally. Previous studies clearly indicated that there is a global association between human microbial flora and different geographic locations for a better assessment of disease diagnosis and relative treatment of that population (Gupta et al., 2017). Table 1.2 summarizes the diversity of vaginal *Lactobacillus* in various countries.

Lactobacillus species are known to be the predominant component of the healthy vaginal microbial flora, but some studies indicated that other species such as *Gardnerella, Atopobium, Prevotella, Pseudomonas,* or *Streptococcus* are predominant in some healthy women instead of *Lactobacillus*. Verstraelen et al. (2016) reported that vaginal microbiome of North American-Asian, North American white, and North American black women was dominated by

TABLE 1.2 Geographic Diversity of Vaginal *Lactobacillus* Around the Global

S. No	Geographic/Ethnicity Diversity	Predominant Vaginal Microbial Flora	Reference
1	North American-Asian, North American white, and North American black women	*Lactobacillus* (51%–96%)	Ravel et al. (2011) and Verstraelen et al. (2016)
2	Belgium population	Bacteroides (34%)	Ravel et al. (2011) and Verstraelen et al. (2016)
3	South Africa	*Lactobacillus crispatus*, *L. iners*, *L. jensenii*, *L. gasseri*, and *L. vaginalis*	Pendharkar et al. (2013)
4	Japanese women	*Lactobacillus crispatus* and *L. gasseri*	Pavlova et al. (2002)
5	United States	*Lactobacillus gasseri* and *L. fermentum*	Pavlova et al. (2002)
6	Brazil	*Lactobacillus jensenii*, *L. gasseri*, and *L. mucosae*	Pavlova et al. (2002)
7	India	*Lactobacillus gasseri* and *L. jensenii*	Pavlova et al. (2002)
8	Caucasian women	*Atopobium*	Zhou et al. (2007)
9	North American women	*Lactobacillus crispatus*, *L. gasseri*, *L. iners*, or *L. jensenii*	Ravel et al. (2011)

Lactobacillus (51%–96%), but in Belgium population, vaginal microbiome was dominated by *Bacteroides* (34%) (Ravel et al., 2011; Verstraelen et al., 2016). However, a cohort study showed that vaginal *Lactobacillus* species isolated from South African women are similar to those identified in European populations, in which *L. crispatus* is related to a normal vaginal microbiota (Pendharkar et al., 2013). In another study, Zhou et al. demonstrated that *Lactobacillus* species were found to be more common in Caucasian women, while *Atopobium* is the dominated species in the communities of North American black women of reproductive age (Zhou et al., 2004). A vaginal microbial survey based on 16S rRNA in North American women from four ethnic groups (white, black, Hispanic, and Asian) elucidated the existence of five distinct groups of vaginal microbiome profiles: four were dominated by *Lactobacillus* species (*L. crispatus*, *L. gasseri*, *L. iners*, or *L. jensenii*), while the fifth contained lower proportions of lactic acid bacteria and higher proportions of obligate anaerobic bacteria (Ravel et al., 2011). Similar type of divergences in vaginal microbial flora has also been noticed between Belgian and Canadian women (Verstraelen et al., 2016) with *L. crispatus*. *L. iners*, and *Prevotella prevailing* in Belgian population and *L. iners* and *L. jensenii* in Canadian population.

1.5 FUNCTIONAL DIVERSITY IN VAGINAL *LACTOBACILLUS*

1.5.1 Production of Inhibitory Substances

Lactobacillus are known to produce various inhibitory substances and antimicrobial compounds to prevent vaginal infection caused by abnormal microbial flora (Dasari et al., 2014a).

The antimicrobial compounds, such as bacteriocin-like substances, biosurfactants, and hydrogen peroxide, have been acting as an antagonistic effect of probiotic *Lactobacillus* strains against a variety of vaginal pathogens. *Lactobacillus* isolated from healthy vaginal swabs shows in vitro antimicrobial activity against cervicovaginal pathogens (*Salmonella, Gardnerella, Enterococcus,* and *Neisseria*) by the production of antimicrobial substances. The isolated *Lactobacillus* species are exhibiting probiotic characteristic and are more active under acidic conditions (Dasari et al., 2014b). The antimicrobial activity of *Lactobacillus* sp. is mainly due to the reduction in transmembrane potential ($\Delta\psi$) and pH gradient produced by host cell immunity. *Lactobacillus* sp. produced antimicrobial peptides (bacteriocins), which are important defense mechanisms against competitive microorganisms. Bacteriocins are extracellular substances and exert their inhibitory activity through adsorption to specific receptors located on the surface of sensitive bacteria, followed by metabolic, biological, and morphological changes resulting in the killing of such bacteria (Daw and Falkiner, 1996). Vulvovaginal candidiasis, caused by *Candida albicans*, affects women at reproductive stage and is associated with the disruption of the healthy vaginal microbiota (Achkar and Fries, 2010). Recently, Wang et al. (2017) isolated nearly 18 *Lactobacillus* strains from healthy vagina and are tested to inhibit the growth of and hyphal formation of *C. albicans*. Interestingly, they found that the *L. crispatus* inhibited hyphal formation by downregulating the hypha-specific genes *ALS3*, *HWP1*, and *ECE1* possibly through the overexpression of transcriptional repressor gene NRG1.

In another study, Dover et al. reported that the antimicrobials derived from vaginal probiotics were nontoxic to both animal and human vaginal tissue (Hillier et al., 1993). The production of lactic acid by vaginal *Lactobacillus* may be essential for the maintenance of a healthy ecosystem. Resulting acid pH prevents the excessive proliferation of potentially pathogenic microorganisms. It was confirmed by the O'Hanlon et al. (2013) that women with *Lactobacillus*-dominated vaginal microbiota have significantly more protection against infection than women with less vaginal *Lactobacillus* (O'Hanlon et al., 2013). Juárez Tomás et al. (2003) reported that *L. acidophilus* CRL 1259 isolated from the vagina inhibits the growth of uropathogenic *Escherichia coli*. Most of these *Lactobacillus* produce hydrogen peroxide, which possesses a defense mechanism against pathogens and seems to be less frequent in African-American women (Antonio et al., 1999).

1.5.2 Blocking of Adhesion Sites

Adherence of bacteria to epithelial cells has been shown to be an important factor in the colonization of mucous membranes. Competitive inhibition for bacterial adhesion sites on vaginal epithelial surfaces is another mechanism of action for *Lactobacillus*. Glycoproteins and carbohydrates produced by *L. acidophilus* and *L. gasseri* are responsible for adherence to the epithelial cells, which presumably were the targets of the competition observed between the *Lactobacillus* and the pathogenic microbes (Boris et al., 1998). Adhesion of human *Lactobacillus* species to mucin through various binding factors has been studied. Collagen-binding A precursor and aggregation-promoting factor-like proteins are suggested to participate on adhesion to HT-29, Caco-2, and HeLa cells. Interestingly, the extracellular proteins of *L. plantarum* Li69 and of *L. salivarius* Lv72 significantly inhibited the adhesion of *Actinomyces neuii* to HeLa cells (Martín et al., 2012). *L. crispatus* strains encode adhesion components called S-layer proteins that play a significant role on adherence of bacteria on HeLa cells.

The potentiality of *Lactobacillus* to attach to the cells depends on various factors like the cell wall components of bacteria, cell receptors, and soluble proteins secreted by the host cells. However, more studies should be conducted on normal vaginal cell lines to elucidate the adhesion ability of *Lactobacillus* on epithelial cells and protection against bacterial and viral infections (Mousavi et al., 2017).

1.5.3 *Lactobacillus* Induced Immunity

The indigenous vaginal *Lactobacillus* of the genital tract are also involved in the prevention of microbial infection through innate and acquired immunity. Vaginal epithelial cells play a crucial role in the regulation of microbial infection by host-microbe interaction. In fact, epithelial cell secretions have some soluble factors with innate immunity (defensins, complement components, TLRs, nitric oxide, and phagocytic cells). Epithelial cells having TLRs recognize the pathogenic microbial components and trigger the development of antigen-specific acquired immunity by various signal transduction pathways (Dasari et al., 2016). A recent study using in vitro colonization of vaginal epithelial cell monolayers with common vaginal bacteria such as *L. crispatus*, *Prevotella bivia*, and *Atopobium* showed that these bacteria regulate the epithelial innate immunity in a species-specific manner (Doerflinger et al., 2014). Witkin et al. (2013) reported that D-lactic acid, present in vaginal secretions of women with *L. crispatus*-dominated microbiota, can inhibit expression of extracellular matrix metalloproteinase inducer (EMMPRIN) production. EMMPRIN induces matrix metalloproteinase 8 (MMP-8), which acts as a vaginal barrier more prone to upper genital infections associated to preterm birth (Witkin et al., 2013). Therefore, D-lactic acid production is an important factor in the protection against preterm delivery. Moreover, *L. crispatus* can also have beneficial effects via immunomodulation. For example, *L. crispatus* ATCC 33820 was shown to inhibit *C. albicans* in vitro by regulating the toll-like receptors (TLRs) 2/4, interleukin 8 (IL-8), and human β-defensin 2 and 3 expression in epithelial cells (HeLa) (Petrova et al., 2015).

1.5.4 *Lactobacillus*—Antiviral Activity

In vitro and in vivo experimental studies suggest that reduction in vaginal *Lactobacillus* content was identified as a predisposing factor for viral infections such as human papillomavirus (HPV) and herpes simplex virus (HSV) type. Recent prospective investigations demonstrated that there is a direct association between vaginal flora alterations and the occurrence of human immunodeficiency virus (HIV) infection (Zabihollahi et al., 2012). On the other hand, genital herpes and HPV infections are major risk factors for initiation and transmission of HIV via sexual contact (Atashili et al., 2008). Moreover, the HIV copy number in female genital tract discharge inversely correlates with vaginal *Lactobacillus* counts in bacterial flora (Sha et al., 2005). Therefore, *Lactobacillus* exhibit an important role both in preventing viral infections in the female and reducing the risk of infectious transmission to a healthy man. In vitro studies have revealed that hydrogen peroxide production by *L. acidophilus* strain shows a virucidal effect against HIV-1. Additionally, it has been reported that preincubation with different lactobacillus strains reduces the infection titer and thereby reduces the rate of infection of vesicular stomatitis virus (VSV) (Botić et al., 2007). Zabihollahi et al. (2012) evaluated that *L. crispatus* inactivated the 74% of HSV and HIV virion in coculture condition, suggesting that *Lactobacillus* may secrete a sort of molecules with viral neutralization activity (Zabihollahi et al., 2012). Recently, Mousavi et al. (2017) reported that *L. crispatus* (LC) shows potential antiviral activity against HSV-2 infection by either formation of LC microcolonies in the cell surface that could block HSV-2 receptor binding and prevent viral entry to cells during the initial stage of infection or entrapment of viral particles by direct interaction of *L. crispatus* with HSV-2 envelope (Mousavi et al., 2017). Galán and group determined the antiviral activity of *Lactobacillus* and *Bifidobacterium* species against rotavirus infection through the intracellular NSP4 protein production and Ca^{2+} release (Galán et al., 2016).

1.5.5 *Lactobacillus* in Cancer Treatment and Management

Several in vitro and in vivo experimental studies suggest that the consumption of *Lactobacillus* sp. reduces the risk of cancers such as colon and cervical cancer. Recent reports demonstrated that there is an inverse association between cancer risk and yogurt and other fermented milk containing the probiotic *Lactobacillus*. Since most of the vaginal *Lactobacillus* species exhibit probiotic characters and are potentially used in the prevention and treatment of cancer through various mechanisms (Dasari et al., 2017). Sungur et al. (2017) reported that lyophilized EPS (L-EPS) produced by two vaginal *Lactobacillus* (*L. gasseri*) have anti-inflammatory effect on human cervical cancer cells (HeLa cells), and they inhibit the cell proliferation, by decreasing the production of TNF-α and increasing the IL-10 production (Sungur et al., 2017). Several in vitro studies demonstrated that common vaginal *Lactobacillus* (*L. crispatus* and *L. gasser*) exert cytotoxic effects on cervical tumor cells (HeLa cells), but not on normal cells, through the regulation of autophagy, cell cycle, and expression of oncogenes (Wang et al., 2017).

1.6 CONCLUSION

One of the most rapidly emerging researches is the management of vaginally acquired infections through its own normal flora—*Lactobacillus*. Several *Lactobacillus* species inhibit the growth of vaginally acquired pathogens by producing various compounds or reduce the adherence of pathogens to urogenital epithelial cells in culture and prevent pathogen colonization of the urogenital tract. Obviously, our knowledge about the vaginal microbiota dominating *Lactobacillus* and its functional output would contribute to an understanding of what truly constitutes "normal" and "abnormal." It is expected that the application of this study will translate into better diagnostics and improved personalized treatments in female genital tract infections.

References

Achkar, J.M., Fries, B.C., 2010. Candida infections of the genitourinary tract. Clin. Microbiol. Rev. 23, 253–273.

Antonio, M.A., Hawes, S.E., Hillier, S.L., 1999. The identification of vaginal Lactobacillus species and the demographic and microbiologic characteristics of women colonized by these species. J. Infect. Dis. 180, 1950–1956.

Atashili, J., Poole, C., Ndumbe, P.M., Adimora, A.A., Smith, J.S., 2008. Bacterial vaginosis and HIV acquisition: a meta-analysis of published studies. AIDS (London, England) 22, 1493–1501.

Boris, S., Suárez, J.E., Vázquez, F., Barbés, C., 1998. Adherence of human vaginal lactobacilli to vaginal epithelial cells and interaction with uropathogens. Infect. Immun. 66, 1985–1989.

Botić, T., Klingberg, T.D., Weingartl, H., Cencic, A., 2007. A novel eukaryotic cell culture model to study antiviral activity of potential probiotic bacteria. Int. J. Food Microbiol. 115, 227–234.

Cribby, S., Taylor, M., Reid, G., 2008. Vaginal microbiota and the use of probiotics. Interdiscip. Perspect. Infect. Dis. 2008. https://doi.org/10.1155/2008/256490.

Dasari, S., Rajendra, W., Valluru, L., 2014a. Evaluation of microbial enzymes in normal and abnormal cervicovaginal fluids of cervical dysplasia: a case control study. Biomed. Res. Int. 2014, 716346. https://doi.org/10.1155/2014/716346.

Dasari, S., Shouri, R.N.D., Wudayagiri, R., Valluru, L., 2014b. Antimicrobial activity of *Lactobacillus* against microbial flora of cervicovaginal infections. Asian Pac. J. Trop. Dis. 4, 18–24.

Dasari, S., Anandan, S.K., Rajendra, W., Valluru, L., 2016. Role of microbial flora in female genital tract: a comprehensive review. Asian Pac. J. Trop. Dis. 6, 909–917.

Dasari, S., Kathera, C., Janardhan, A., Praveen Kumar, A., Viswanath, B., 2017. Surfacing role of probiotics in cancer prophylaxis and therapy: a systematic review. Clin. Nutr. 36, 1465–1472.

Daw, M.A., Falkiner, F.R., 1996. Bacteriocins: nature, function and structure. Micron 27, 467–479.

Dethlefsen, L., McFall-Ngai, M., Relman, D.A., 2007. An ecological and evolutionary perspective on human-microbe mutualism and disease. Nature 449, 811–818.

Doerflinger, S.Y., Throop, A.L., Herbst-Kralovetz, M.M., 2014. Bacteria in the vaginal microbiome alter the innate immune response and barrier properties of the human vaginal epithelia in a species-specific manner. J. Infect. Dis. 209, 1989–1999.

Dover, S.E., Aroutcheva, A.A., Faro, S., Chikindas, M.L., 2008. Natural antimicrobials and their role in vaginal health: a short review. Int. J. Probiotics Prebiotics 3, 219–230.

Fredricks, D.N., Fiedler, T.L., Marrazzo, J.M., 2005. Molecular identification of bacteria associated with bacterial vaginosis. N. Engl. J. Med. 353, 1899–1911.

Galán, N.N.O., Rubiano, J.C.U., Reyes, F.A.V., Duarte, K.P.F., Cárdenas, S.P.S., Fernandez, M.F.G., 2016. In vitro antiviral activity of *Lactobacillus casei* and *Bifidobacterium adolescentis* against rotavirus infection monitored by NSP4 protein production. J. Appl. Microbiol. 120, 1041–1051.

Gupta, V.K., Paul, S., Dutta, C., 2017. Geography, ethnicity or subsistence-specific variations in human microbiome composition and diversity. Front. Microbiol. 8. https://doi.org/10.3389/fmicb.2017.01162.

Hickey, R.J., Zhou, X., Pierson, J.D., Ravel, J., Forney, L.J., 2012. Understanding vaginal microbiome complexity from an ecological perspective. Transl. Res. J. Lab. Clin. Med. 160, 267–282.

Hillier, S.L., Krohn, M.A., Rabe, L.K., Klebanoff, S.J., Eschenbach, D.A., 1993. The normal vaginal flora, H_2O_2-producing lactobacilli, and bacterial vaginosis in pregnant women. Clin. Infect. Dis. 16, S273–S281.

Juárez Tomás, M.S., Ocaña, V.S., Wiese, B., Nader-Macías, M.E., 2003. Growth and lactic acid production by vaginal *Lactobacillus acidophilus* CRL 1259, and inhibition of uropathogenic *Escherichia coli*. J. Med. Microbiol. 52, 1117–1124.

Ling, Z., Kong, J., Liu, F., Zhu, H., Chen, X., Wang, Y., Li, L., Nelson, K.E., Xia, Y., Xiang, C., 2010. Molecular analysis of the diversity of vaginal microbiota associated with bacterial vaginosis. BMC Genomics 11, 488. https://doi.org/10.1186/1471-2164-11-488.

Linhares, I.M., Giraldo, P.C., Baracat, E.C., 2010. New findings about vaginal bacterial flora. Rev. Assoc. Med. Bras. (1992) 56, 370–374.

Martin, D.H., 2012. The microbiota of the vagina and its influence on women's health and disease. Am J Med Sci 343, 2–9.

Martín, R., Sánchez, B., Suárez, J.E., Urdaci, M.C., 2012. Characterization of the adherence properties of human Lactobacilli strains to be used as vaginal probiotics. FEMS Microbiol. Lett. 328, 166–173.

Mousavi, E., Makvandi, M., Teimoori, A., Ataei, A., Ghafari, S., Samarbaf-Zadeh, A., 2017. Antiviral effects of *Lactobacillus crispatus* against HSV-2 in mammalian cell lines. J. Chin. Med. Assoc. 81, 262–267.

O'Hanlon, D.E., Moench, T.R., Cone, R.A., 2013. Vaginal pH and microbicidal lactic acid when *Lactobacilli* dominate the microbiota. PLoS One 8, e80074.

Pavlova, S.I., Kilic, A.O., Kilic, S.S., So, J.S., Nader-Macias, M.E., Simoes, J.A., Tao, L., 2002. Genetic diversity of vaginal lactobacilli from women in different countries based on 16S rRNA gene sequences. J. Appl. Microbiol. 92 (3), 451–459.

Pendharkar, S., Magopane, T., Larsson, P.G., de Bruyn, G., Gray, G.E., Hammarström, L., Marcotte, H., 2013. Identification and characterisation of vaginal lactobacilli from south African women. BMC Infect. Dis. 13, 43. https://doi.org/10.1186/1471-2334-13-43.

Petrova, M.I., Lievens, E., Malik, S., Imholz, N., Lebeer, S., 2015. *Lactobacillus* species as biomarkers and agents that can promote various aspects of vaginal health. Front. Physiol. 6, 81. https://doi.org/10.3389/fphys.2015.00081.

Ravel, J., Gajer, P., Abdo, Z., Schneider, G.M., Koenig, S.S.K., McCulle, S.L., Karlebach, S., Gorle, R., Russell, J., Tacket, C.O., Brotman, R.M., Davis, C.C., Ault, K., Peralta, L., Forney, L.J., 2011. Vaginal microbiome of reproductive-age women. Proc. Natl. Acad. Sci. U. S. A. 108, S4680–S4687.

Sha, B.E., Zariffard, M.R., Wang, Q.J., Chen, H.Y., Bremer, J., Cohen, M.H., Spear, G.T., 2005. Female genital-tract HIV load correlates inversely with Lactobacillus species but positively with bacterial vaginosis and *Mycoplasma hominis*. J. Infect. Dis. 191, 25–32.

Sungur, T., Aslim, B., Karaaslan, C., Aktas, B., 2017. Impact of exopolysaccharides (EPSs) of *Lactobacillus gasseri* strains isolated from human vagina on cervical tumor cells (HeLa). Anaerobe 47, 137–144.

Tachedjian, G., Aldunate, M., Bradshaw, C.S., Cone, R.A., 2017. The role of lactic acid production by probiotic Lactobacillus species in vaginal health. Res. Microbiol. 168 (9–10), 782–792.

Verstraelen, H., Vilchez-Vargas, R., Desimpel, F., Jauregui, R., Vankeirsbilck, N., Weyers, S., Verhelst, R., De Sutter, P., Pieper, D.H., Van De Wiele, T., 2016. Characterisation of the human uterine microbiome in non-pregnant women through deep sequencing of the V1-2 region of the 16S rRNA gene. PeerJ 4, e1602.

Wang, K.D., Xu, D.J., Wang, B.Y., Yan, D.H., Lv, Z., Su, J.R., 2017. Inhibitory effect of vaginal lactobacillus supernatants on cervical cancer cells. Probiotics Antimicrob. Proteins. https://doi.org/10.1007/s12602-017-9339-x.

Westbroek, M.L., Davis, C.L., Fawson, L.S., Price, T.M., 2010. Interactions of lactobacilli with pathogenic *Streptococcus pyogenes*. Infect. Dis. Obstet. Gynecol. 2010, 289743.

Witkin, S.S., Mendes-Soares, H., Linhares, I.M., Jayaram, A., Ledger, W.J., Forney, L.J., 2013. Influence of vaginal bacteria and D- and L-lactic acid isomers on vaginal extracellular matrix metalloproteinase inducer: implications for protection against upper genital tract infections. MBio 4 (4). pii: e00460-13.

Zabihollahi, R., Motevaseli, E., Sadat, S.M., Azizi-Saraji, A.R., Asaadi-Dalaie, S., Modarressi, M.H., 2012. Inhibition of HIV and HSV infection by vaginal lactobacilli in vitro and in vivo. Daru 20 (1), 53. https://doi.org/10.1186/2008-2231-20-53.

Zhou, X., Bent, S.J., Schneider, M.G., Davis, C.C., Islam, M.R., Forney, L.J., 2004. Characterization of vaginal microbial communities in adult healthy women using cultivation-independent methods. Microbiology 150, 2565–2573.

Zhou, X., Brown, C.J., Abdo, Z., Davis, C.C., Hansmann, M.A., Joyce, P., Foster, J.A., Forney, L.J., 2007. Differences in the composition of vaginal microbial communities found in healthy Caucasian and black women. ISME J. 1 (2), 121–133.

CHAPTER

2

Role of Maternal Microorganisms in the Development of the Immune System

Avilala Janardhan

Jiangsu Key Laboratory for Molecular and Medical Biotechnology, College of Life Sciences, Nanjing Normal University, Nanjing, People's Republic of China

OUTLINE

2.1 INTRODUCTION

The human microbiome thought to have some predictable changes during pregnancy and offspring development. These changes must be integrated to protect the health of both the mother and the offspring (Penders et al., 2006; Biasucci et al., 2008). During or after delivery, the neonate is exposed to different types of microbes, many of them are from the mother during and after the passage through the birth canal (Zhou et al., 2007). The microbiota of neonates plays an important role in the postnatal development of the immune system. So, the maternal microbes during pregnancy impact neonatal and infant health, including

Recent Developments in Applied Microbiology and Biochemistry
https://doi.org/10.1016/B978-0-12-816328-3.00002-7

preterm birth, cardiometabolic complications of pregnancy such as preeclampsia and gestational hypertension, and gestational weight gain (Dunlop et al., 2015). The maintenance of a low pH in the vagina is by the microbial production of lactic acid, an important defense against infectious disease in reproductive age women. Due to cesarean section (C-section), babies are not exposed to vagina microbes during birth. The difference in this mode of birth can influence the intestinal microbiota of babies (Biasucci et al., 2008). So, the mode of birth, by either vaginal or cesarean, also has effects on the infant's initial microbiota, which shows the effect on infant nutrition for the first 2 years. In the first week of life, the neonatal gut is largely colonized by Actinobacteria (*Bifidobacterium*), Proteobacteria, and up to some extent Firmicutes (*Lactobacillus* species) (Stout et al., 2013). It is known that initial microbial exposure is important to develop immunity and may fight with pathogenic bacteria early in life. The presence or absence of some pathogenic and beneficial microbes in early childhood plays an important role in the initiation of early or preterm birth (Bäckhed et al., 2004). For instance, some microbes involved in the development of asthma (Samuel et al., 2008), allergy (Bäckhed et al., 2007), autism (Nicholson et al., 2005), and other immunologic deficiencies (Swann et al., 2009). Immunologic tolerance of babies is mainly caused by the T lymphocytes, which are generated from the mother (Mold et al., 2008). Different studies on neonate gut microbiome show that there are fluctuations in the microbiome till the child becomes 3 years of age (Azad et al., 2013). It is difficult to understand the microbiota in pregnancy, at which stage the fetus encounters microbes, and how the microbiome begins (Nuriel-Ohayon et al., 2016). The researchers and physicians raise the doubt on this hypothesis and finally conclude that the bacteria is already present in the fetoplacental unit (Aagaard et al., 2014). Although some research has started about the microbial changes that occur during pregnancy, it is important to understand the influence of the maternal microbiome in the development of the immune system. In this chapter, we explained about the microbial changes that occur in the pregnant female, as well as the role of maternal microorganisms in the development of the immune system.

2.2 HUMAN MICROBIOTA

The human body consists of different types of microorganisms, which are called human microbiota. The differences in microbial composition vary with body sites; some environmental conditions such as pH, levels of oxygen, availability of nutrients, humidity, and temperature also influence microbial population (Ursell et al., 2012). These microbial populations play a key role on host health, by affecting host metabolism (Turnbaugh et al., 2006), immunity (Naik et al., 2012), and hormones (Neuman et al., 2015). Some hundreds of bacterial species majorly reside in the gut (Lozupone et al., 2012), among them are the dominant bacterial phyla Firmicutes and Bacteroidetes (Rajilić-Stojanović et al., 2009). In addition to this, the other parts like the skin and vagina have important niches for different bacterial populations, which can fight against the potential pathogens (Naik et al., 2012). There are some natural changes inside the body that occur in the microbiota that lead to host development from infancy to early childhood, from young to older adults, and during pregnancy. These microbial populations were considered as beneficial microbiota. Microbiotic population differ according to different parts, ages, sexes, races, and diets of the host (Hollister et al., 2014). The commensal bacteria start to grow immediately after birth and slowly develop their community. Some symbiotic bacteria supply essential nutrients, and some bacteria defend against the pathogens. Some beneficiary microbiota in the intestinal microbiota helps in the digestion of foods that cannot be digested by the small intestine (Wang et al., 2017). The gut bacteria can also stimulate the development of humoral and mucosal immune systems. The metabolites and some signals of these microorganisms can be sensed by hematopoietic and nonhematopoietic cells of the innate immune system and translated into physiological responses (Thaiss et al., 2016). The human microbiota can also influence the infectious diseases in the host. Infection is caused by dysbiosis of the microbiota and infectious disease, and its treatment has impact on the human microbiota, which are responsible for the outcome of the disease in the host. Offending pathogens colonize the intestinal mucosa, which results in the influence of the inflammatory response, followed by the translocation of the intestinal bacteria (Hand et al., 2012).

2.3 EARLY MICROBIOTA IN INFANTS

Microorganisms that transfer from the mother to the offspring can develop the infant immune system by modifying the composition of the infant gut microbiota. As the neonates have less capacity to initiate an immune response, the microorganisms involved in the neonate play an important role on postnatal immune responses that result in microbiota disruption during adult life. So, the microbiota in neonatal period is considered as critical developmental window. In the first 3 years of life, many biological systems will develop, and the microbiota in these systems was stimulated by some environmental conditions. Many studies suggest that nutrient deficiency or nutrient excess in utero or in early infancy increases susceptibility to metabolic disease later in life (Friedman, 2015). Early microbiome changes in the infant can lead to some diseases like asthma, allergies, obesity, and other inflammatory diseases (Russell et al., 2012; Cho et al., 2012; Arrieta et al., 2015). The early microbiomes can also lead to some liver diseases and to infection of other body organs. Maternal diet and poor health impact the disease conditions in the offspring; these changes can show effect on infant microbiome. Maternal nutrition show influence on the offspring immune system. The infants born to obese persons or those who take food with high-fat diet (HFD) that affects the fetal liver, microbiome, and offspring immunity have increased nonalcoholic fatty liver disease (NAFLD) and obesity risks (Vijay-Kumar et al., 2010).

2.4 MATERNAL IMMUNE SYSTEM DURING PREGNANCY

Pregnant women have susceptibility to some infectious diseases due to their microbiota and special immunological condition caused by pregnancy. The function of the immune system is to protect the host from pathogenic microorganisms or foreign particles. The capacity of the immune system mainly depends on the innate immune system. Pregnancy is the most important period for women; the immune system plays a key role to protect the mother and the offspring. The immune system protects the mother and the fetus from certain environmental conditions. During pregnancy, the maternal immune system is defined by a reinforced network of recognition, communication, trafficking, and repair, which is able to give signals

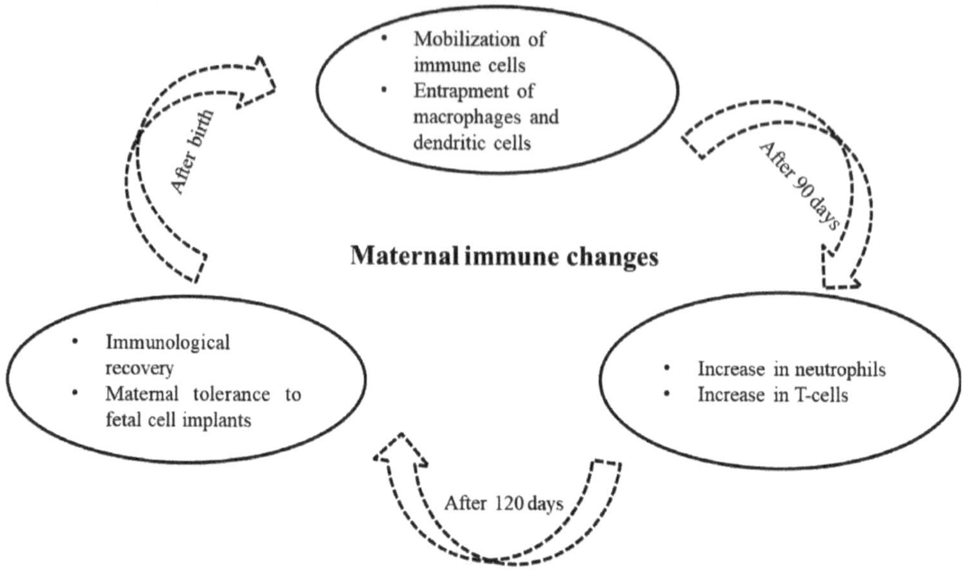

FIG. 2.1 Maternal immune system changes in women that occur during pregnancy. During mid- or late pregnancy, the leukocytes (neutrophils), natural killer cells, macrophages, T cells, and a small number of dendritic cells were increased.

to protect the mother and the fetus. The fetus also develops active immune system that modifies the mother's response due to environmental factors; this shows the uniqueness of the immune system during pregnancy.

The immune cells and molecules are the key players in the development and function of the placenta and the fetus. During pregnancy, the decidua, the vaginal mucosal membrane, has 70% of leukocytes or natural killer cells followed by 20%–25% of macrophages, up to 10% of T cells, and a small number of dendritic and B cells (Morelli et al., 2015). The maternal immune changes in different stages of pregnancy were represented in Fig. 2.1. The absence of natural killer cells and trophoblast cells in the uterus vascular membrane leads to termination of the pregnancy. These studies indicate that natural killer cells are important for trophoblast invasion in the uterus. Similarly, exhaust of dendritic cells prevented blastocyst implantation and formation of the decidua (Moffett-King, 2002).

2.5 MICROBIAL CHANGES DURING PREGNANCY AND BIRTH

Many of the metabolic and immune changes occur during pregnancy by the microbiota. In the first 3 months of pregnancy, the gut microbial population is similar to that of healthy, nonpregnant women. After 3 months, the gut microbiota population changes in dramatic manner. These changes majorly include the members of the Actinobacteria and Proteobacteria phyla. The levels of *Faecalibacterium* are decreased after 6 months of pregnancy (Koren et al., 2012). Like in the gut, the human vaginal microbiome undergoes changes during pregnancy; vaginal microbiota plays an important role in the defense system against some microbial and viral infections (Turovskiy et al., 2011). The vaginal microbial population consists of major *Lactobacillus* species followed by Clostridiales, Bacteriodales, and Actinomycetales (Aagaard et al., 2012). The changes in vaginal microbiome lead to decrease in the vaginal pH and an increase in vaginal secretions (Prince et al., 2014). Vaginal microbiome differs based on gestational age, but during last stages of pregnancy, the vaginal microbiome is same as in nonpregnant vaginal microbiome (Aagaard et al., 2012).

While coming to microbial population in the placenta, it is very difficult to understand because some studies explained that the placenta is germ-free and produces a sterile environment for fetus from infections. However, other studies suggest that the placental infection originates from the lower genital tract. Some evidence proves that the normal placenta itself has bacteria (Goldenberg et al., 2000). Some other studies explained that the placenta contains a unique microbiome, which shows similarity with oral microbiota at species levels. This similarity between the oral and placental microbiota suggests that the bacteria may pass from the oral cavity to the placenta (Aagaard et al., 2014).

2.6 MATERNAL MICROBIOME INFLUENCE ON DEVELOPMENT OF NEONATAL IMMUNE SYSTEM

The microbial exposure of human fetus describes the development of immune tolerance. There are several reports suggesting that the transfer of foreign antigens including proteins, parasites, and even cells from the mother to the fetus is common, but the mechanism and response of fetal immune system in recognizing to such antigens are unclear (Remington et al., 2006). Peripheral lymphoid tissues in the human fetus are populated by T cells especially CD4+ and CD8+ at the end of the third month (Romano-Keeler and Weitkamp, 2015). Mononuclear blood cells from the mother carry bacteria or their DNA at high frequency during pregnancy. These bacterial signs are seen in the infant feces that represent the transfer of maternal microorganisms or antigens during fetal development. This is interesting particularly in certain known beneficial bacterial species like bifidobacterium and lactobacilli. The genetic material from these two beneficial bacteria has also been detected in human placenta. This development that creates a balanced immune response to commensal organisms in the fetus and immune dysregulation was also observed in the uterus after exposure to pathogenic bacteria (Perez et al., 2007; Vlasova et al., 2013). The mucosal membrane of the fetus may alter with the colonization of *Ureaplasma* species and has a risk factor for adverse postnatal outcomes like necrotizing enterocolitis, a common and often fatal complication of prematurity (Okogbule-Wonodi et al., 2011).

Molecular signals from the maternal microbiota during pregnancy can reach the offspring during in utero development via the placenta and after birth through breast feeding; these signals are involved in the development of offspring immune system (Fig. 2.2). The transfer of microorganisms from the mother to the fetus is common in the animals, but the exact mechanism in transfer of materno-fetal bacteria is not clear. Some evidences in human beings suggest that the maternal fecal microbiome appears like the microbiota in the fetal intestine. So, the maternal gastrointestinal tract is mostly the source for the transfer of microbiota (Thum et al., 2012; Hu et al., 2013).

The translocation of bacteria from the intestine to maternal blood stream and from there to the other organs is increased during pregnancy and lactation of the mother (Perez et al., 2007).

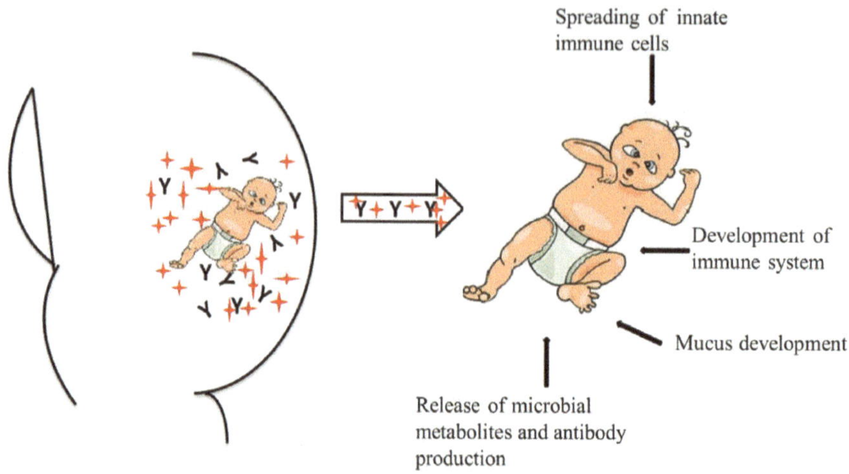

FIG. 2.2 Transfer of maternal microbiota and antibodies to the neonate. This microbiota has a key role in the development of the neonate immune system.

But it is unknown by which mechanism the transfer of bacteria may occur. But some studies revealed that the transfer is by some of these mechanisms like migration of luminal bacteria through the small intestinal epithelium (Macpherson and Uhr, 2004; Lelouard et al., 2012). The diet, medications, genetic predisposition, and other environmental influences also shape the maternal microbiome, which is responsible for fetal microbiome in preparation of perinatal and postnatal bacterial colonization. Injecting, or as a diet, of some probiotic bacteria like *L. rhamnosus* and *B. lactis* during pregnancy can change the fetal immune response and also enhances the immune protective components in the breast milk. The intact between the fetal intestine to commensal organisms or their products is an important activity in gut maturation (Prescott et al., 2008). Gluten-free diet during pregnancy may also change the gut microbiota in the mother and the offspring, which results in immunologic milieu in the gut and pancreas (Hansen et al., 2014).

2.7 CONCLUSION AND FUTURE PROSPECTIVE

The maternal microorganism serves as the source for the infant microbiome and plays a key role in the health of the neonate and infant. In this chapter, we tried to explain about the influence of maternal microorganisms on the host immune system. During pregnancy of women, the innate immune response was upregulated, and the cell-mediated response was decreased. The lymphoid cells in the decidual membrane are actively involved in the development of the placenta. Though the existing evidences explained about the maternal microbiota and function of immune cells during pregnancy, many questions have to be answered about the underlying mechanisms about the transfer of microbiome to improve health of neonate, and the proliferation and function by endocrine and other factors also still remain to be found out clearly.

References

Aagaard, K., Riehle, K., Ma, J., Segata, N., Mistretta, T.-A., Coarfa, C., et al., 2012. A metagenomic approach to characterization of the vaginal microbiome signature in pregnancy. PLoS ONE 7, e36466.

Aagaard, K., Ma, J., Antony, K.M., Ganu, R., Petrosino, J., Versalovic, J., 2014. The placenta harbors a unique microbiome. Sci. Transl. Med. 6, 237ra65.

Arrieta, M.-C., Stiemsma, L.T., Dimitriu, P.A., Thorson, L., Russell, S., Yurist-Doutsch, S., et al., 2015. Early infancy microbial and metabolic alterations affect risk of childhood asthma. Sci. Transl. Med. 7, 307ra152.

Azad, M.B., Konya, T., Maughan, H., Guttman, D.S., Field, C.J., Chari, R.S., et al., 2013. Gut microbiota of healthy Canadian infants: profiles by mode of delivery and infant diet at 4 months. CMAJ 185, 385–394.

Bäckhed, F., Ding, H., Wang, T., Hooper, L.V., Koh, G.Y., Nagy, A., Semenkovich, C.F., Gordon, J.I., 2004. The gut microbiota as an environmental factor that regulates fat storage. Proc. Natl. Acad. Sci. U. S. A. 101, 15718–15723.

Bäckhed, F., Manchester, J.K., Semenkovich, C.F., Gordon, J.I., 2007. Mechanisms underlying the resistance to diet-induced obesity in germ-free mice. Proc. Natl. Acad. Sci. U. S. A. 104, 979–984.

Biasucci, G., Benenati, B., Morelli, L., Bessi, E., Boehm, G., 2008. Cesarean delivery may affect the early biodiversity of intestinal bacteria. J. Nutr. 138, 1796S–1800S.

Cho, I., Yamanishi, S., Cox, L., Methé, B.A., Zavadil, J., Li, K., et al., 2012. Antibiotics in early life alter the murine colonic microbiome and adiposity. Nature 488, 621–626.

Dunlop, A.L., Mulle, J.G., Ferranti, E.P., Edwards, S., Dunn, A.B., Corwin, E.J., 2015. Maternal microbiome and pregnancy outcomes that impact infant health: a review. Adv. Neonatal Care 15, 377–385.

Friedman, J.E., 2015. Obesity and gestational diabetes mellitus pathways for programming in mouse, monkey, and man—where do we go next? The 2014 Norbert Freinkel award lecture. Diabetes Care 38, 1402–1411.

Goldenberg, R.L., Hauth, J.C., Andrews, W.W., 2000. Intrauterine infection and preterm delivery (FH Epstein, Ed). N. Engl. J. Med. 342, 1500–1507.

Hand, T.W., Dos Santos, L.M., Bouladoux, N., Molloy, M.J., Pagan, A.J., Pepper, M., Maynard, C.L., Elson, C.O., Belkaid, Y., 2012. Acute gastrointestinal infection induces long-lived microbiota-specific T cell responses. Science 337, 1553–1556.

Hansen, C.H.F., Krych, L., Buschard, K., Metzdorff, S.B., Nellemann, C., Hansen, L.H., et al., 2014. A maternal gluten-free diet reduces inflammation and diabetes incidence in the offspring of NOD mice. Diabetes 63, 2821–2832.

Hollister, E.B., Gao, C., Versalovic, J., 2014. Compositional and functional features of the gastrointestinal microbiome and their effects on human health. Gastroenterology 146, 1449–1458.

Hu, J., Nomura, Y., Bashir, A., Fernandez-Hernandez, H., Itzkowitz, S., Pei, Z., Stone, J., Loudon, H., Peter, I., 2013. Diversified microbiota of meconium is affected by maternal diabetes status. PLoS ONE 8, e78257.

Koren, O., Goodrich, J.K., Cullender, T.C., Spor, A., Laitinen, K., Bäckhed, H.K., et al., 2012. Host remodeling of the gut microbiome and metabolic changes during pregnancy. Cell 150, 470–480.

Lelouard, H., Fallet, M., de Bovis, B., Méresse, S., Gorvel, J., 2012. Peyer's patch dendritic cells sample antigens by extending dendrites through M cell-specific transcellular pores. Gastroenterology 142, 592–601.e3.

Lozupone, C.A., Stombaugh, J.I., Gordon, J.I., Jansson, J.K., Knight, R., 2012. Diversity, stability and resilience of the human gut microbiota. Nature 489, 220–230.

Macpherson, A.J., Uhr, T., 2004. Induction of protective IgA by intestinal dendritic cells carrying commensal bacteria. Science 303, 1662–1665.

Moffett-King, A., 2002. Erratum: natural killer cells and pregnancy. Nat. Rev. Immunol. 2, 656–663.

Mold, J.E., Michaëlsson, J., Burt, T.D., Muench, M.O., Beckerman, K.P., Busch, M.P., Lee, T.-H., Nixon, D.F., McCune, J.M., 2008. Maternal alloantigens promote the development of tolerogenic fetal regulatory T cells in utero. Science (New York, N.Y.) 322, 1562–1565.

Morelli, S., Mandal, M., Goldsmith, L.T., Kashani, B.N., Ponzio, N.M., 2015. The maternal immune system during pregnancy and its influence on fetal development. Res. Rep. Biol. 6, 171.

Naik, S., Bouladoux, N., Wilhelm, C., Molloy, M.J., Salcedo, R., Kastenmuller, W., et al., 2012. Compartmentalized control of skin immunity by resident commensals. Science (New York, N.Y.) 337, 1115–1119.

Neuman, H., Debelius, J.W., Knight, R., Koren, O., 2015. Microbial endocrinology: the interplay between the microbiota and the endocrine system (E Banin, Ed). FEMS Microbiol. Rev. 39, 509–521.

Nicholson, J.K., Holmes, E., Wilson, I.D., 2005. Gut microorganisms, mammalian metabolism and personalized health care. Nat. Rev. Microbiol. 3, 431–438.

Nuriel-Ohayon, M., Neuman, H., Koren, O., 2016. Microbial changes during pregnancy, birth, and infancy. Front. Microbiol. 7, 1031.

Okogbule-Wonodi, A.C., Gross, G.W., Sun, C.-C.J., Agthe, A.G., Xiao, L., Waites, K.B., Viscardi, R.M., 2011. Necrotizing enterocolitis is associated with ureaplasma colonization in preterm infants. Pediatr. Res. 69, 442–447.

Penders, J., Thijs, C., Vink, C., Stelma, F.F., Snijders, B., Kummeling, I., van den Brandt, P.A., Stobberingh, E.E., 2006. Factors influencing the composition of the intestinal microbiota in early infancy. Pediatrics 118, 511–521.

Perez, P.F., Dore, J., Leclerc, M., Levenez, F., Benyacoub, J., Serrant, P., Segura-Roggero, I., Schiffrin, E.J., Donnet-Hughes, A., 2007. Bacterial imprinting of the neonatal immune system: lessons from maternal cells? Pediatrics 119, e724–e732.

Prescott, S.L., Wickens, K., Westcott, L., Jung, W., Currie, H., Black, P.N., et al., 2008. Supplementation with Lactobacillus rhamnosus or Bifidobacterium lactis probiotics in pregnancy increases cord blood interferon-γ and breast milk transforming growth factor-β and immunoglobin A detection. Clin. Exp. Allergy 38, 1606–1614.

Prince, A., Antony, K., Ma, J., Aagaard, K., 2014. The microbiome and development: a mother's perspective. Semin. Reprod. Med. 32, 014–022.

Rajilić-Stojanović, M., Heilig, H.G.H.J., Molenaar, D., Kajander, K., Surakka, A., Smidt, H., de Vos, W.M., 2009. Development and application of the human intestinal tract chip, a phylogenetic microarray: analysis of universally conserved phylotypes in the abundant microbiota of young and elderly adults. Environ. Microbiol. 11, 1736–1751.

Remington, J.S., Klein, J.O., Wilson, C.B., Baker, C.J., 2006. Dedication. In: Infectious Diseases of the Fetus and Newborn Infant. Elsevier, Amsterdam.

Romano-Keeler, J., Weitkamp, J.-H., 2015. Maternal influences on fetal microbial colonization and immune development. Pediatr. Res. 77, 189–195.

Russell, S.L., Gold, M.J., Hartmann, M., Willing, B.P., Thorson, L., Wlodarska, M., et al., 2012. Early life antibiotic-driven changes in microbiota enhance susceptibility to allergic asthma. EMBO Rep. 13, 440–447.

Samuel, B.S., Shaito, A., Motoike, T., Rey, F.E., Backhed, F., Manchester, J.K., et al., 2008. Effects of the gut microbiota on host adiposity are modulated by the short-chain fatty-acid binding G protein-coupled receptor, Gpr41. Proc. Natl. Acad. Sci. U. S. A. 105, 16767–16772.

Stout, M.J., Conlon, B., Landeau, M., Lee, I., Bower, C., Zhao, Q., et al., 2013. Identification of intracellular bacteria in the basal plate of the human placenta in term and preterm gestations. Am. J. Obstet. Gynecol. 208, 226.e1–226.e7.

Swann, J., Wang, Y., Abecia, L., Costabile, A., Tuohy, K., Gibson, G., et al., 2009. Gut microbiome modulates the toxicity of hydrazine: a metabonomic study. Mol. BioSyst. 5, 351.

Thaiss, C.A., Zmora, N., Levy, M., Elinav, E., 2016. The microbiome and innate immunity. Nature 535, 65–74.

Thum, C., Cookson, A.L., Otter, D.E., McNabb, W.C., Hodgkinson, A.J., Dyer, J., Roy, N.C., 2012. Can nutritional modulation of maternal intestinal microbiota influence the development of the infant gastrointestinal tract? J. Nutr. 142, 1921–1928.

Turnbaugh, P.J., Ley, R.E., Mahowald, M.A., Magrini, V., Mardis, E.R., Gordon, J.I., 2006. An obesity-associated gut microbiome with increased capacity for energy harvest. Nature 444, 1027–1031.

Turovskiy, Y., Sutyak Noll, K., Chikindas, M.L., 2011. The aetiology of bacterial vaginosis. J. Appl. Microbiol. 110, 1105–1128.

Ursell, L.K., Clemente, J.C., Rideout, J.R., Gevers, D., Caporaso, J.G., Knight, R., 2012. The interpersonal and intrapersonal diversity of human-associated microbiota in key body sites. J. Allergy Clin. Immunol. 129, 1204–1208.

Vijay-Kumar, M., Aitken, J.D., Carvalho, F.A., Cullender, T.C., Mwangi, S., Srinivasan, S., et al., 2010. Metabolic syndrome and altered gut microbiota in mice lacking toll-like receptor 5. Science (New York, N.Y.) 328, 228–231.

Vlasova, A.N., Chattha, K.S., Kandasamy, S., Liu, Z., Esseili, M., Shao, L., Rajashekara, G., Saif, L.J., 2013. Lactobacilli and Bifidobacteria promote immune homeostasis by modulating innate immune responses to human rotavirus in neonatal gnotobiotic pigs (E Subbiah, Ed). PLoS ONE 8, e76962.

Wang, B., Yao, M., Lv, L., Ling, Z., Li, L., 2017. The human microbiota in health and disease. Engineering 3, 71–82.

Zhou, X., Brown, C.J., Abdo, Z., Davis, C.C., Hansmann, M.A., Joyce, P., Foster, J.A., Forney, L.J., 2007. Differences in the composition of vaginal microbial communities found in healthy Caucasian and black women. ISME J. 1, 121–133.

CHAPTER

3

Microbial Proteomics and Their Importance in Medical Microbiology

Chandrasekhar Kathera

Jiangsu Key Laboratory for Molecular and Medical Biotechnology, College of Life Sciences, Nanjing Normal University, Nanjing, People's Republic of China

OUTLINE

3.1 INTRODUCTION

Microbial infection is a leading cause of death around the world. Most of the infectious diseases are caused by drug-resistant microbes; this may lead to a delay in the administration of microbiologically effective therapy (Chen et al., 2017; Del Chierico et al., 2014). Therefore, exhaustive understanding of microbial physiologies, infection and defense systems, and survival strategies is of great interest in order to actively defeat microbial infection. Microbial proteomics provides complete information of microbial physiology and expression and function of the proteins that are involved in infection and also gives a clue in clinical diagnosis and antimicrobial therapy (Pérez-Llarena and Bou, 2016; Vranakis et al., 2014). Microbial proteomics helps to identify the proteins associated with microbial activity, microbial host-pathogen

Recent Developments in Applied Microbiology and Biochemistry
https://doi.org/10.1016/B978-0-12-816328-3.00003-9

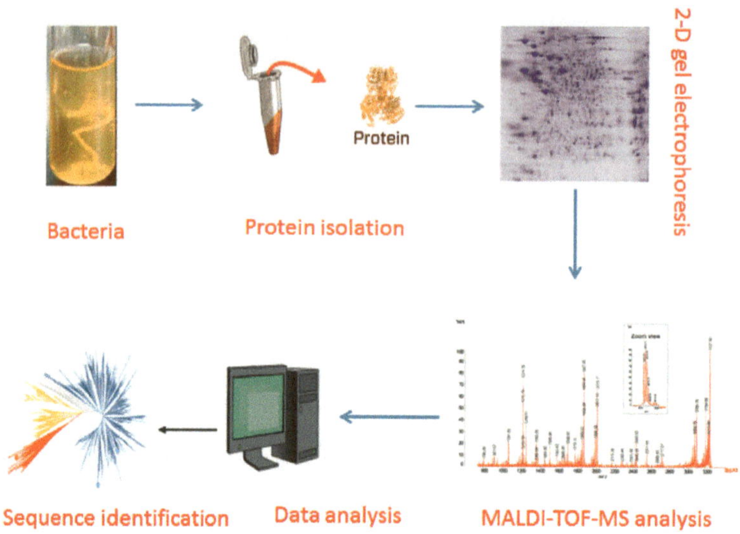

FIG. 3.1 Shotgun proteomic analysis of bacterial sample. Isolation of protein from bacteria, purification of isolated protein sample, separation of the proteins based on isoelectric point and molecular weight of the expressed protein—the new proteins have to be digested with trypsin, before the sample mixed with a suitable matrix for MALDI analysis—and identification with the protein sequence through the database.

interactions, and antimicrobial resistant mechanism. Microbial activity of pathogens can be confirmed by using the 2-D gel-based and gel-free method with the combination of MALDI-TOF-LC-MS/MS. Proteomic analysis of microbial host-pathogen interaction reveals valuable information about the virulence of the pathogen and its resistance; it helps in better understanding of the infection and for developing strategies against microbial infections (Cheng et al., 2016). Fig. 3.1 schematically illustrates the proteomic analysis of the bacterial samples.

3.2 UNDERSTANDING HOST-PATHOGEN INTERACTIONS AND INFECTIOUS DISEASES

Proteomic study of microbial pathogenicity provided invaluable information about the interaction between host and pathogen, captured a clear understanding of infectious process, and identified the proteins expressed by pathogens during infection (Walters and Mobley, 2010). Pathogens affect various molecules for their adhesion and invasion to host cells; infection of neighbor cells, spreading into host systemic circulation; and evasion of host defense mechanisms. Proteomic profiling provides new insights of virulence proteins like proteoglycans; it mediates host-pathogen interactions to affect invasion, progression, and outcome of infection. Host-pathogen interactions of microbes will affect posttranslational modifications, like phosphorylation, glycosylation, and acetylation, which can be rapidly identified by MS (Jean Beltran et al., 2017; Ravikumar et al., 2015).

Many proteomic techniques are used to study the host-pathogen interactions; it may lead to the discovery of important biological insights. Different kinds of methods available for detecting host-pathogen interactions, that is, yeast 2 hybrid, affinity purification-mass spectrometry, chemical cross-linking mass spectrometry, proximity-dependent labeling strategies-mass spectrometry, and protein microarray-based technologies and protein correlation profiling, have its own advantage and disadvantages (Table 3.1) (Nicod et al., 2017). Viral and bacterial proteins interact with host hub proteins or central proteins and alter the cellular processes. Viral pathogen seems to unavoidably disturb cellular processes as they rely on the transcriptional machinery, whereas bacteria tend to mesh with the immune response to prevent their clearance. Also, the manipulation of the host ubiquitin pathways by viruses and bacterial effector proteins is a recurrent finding. By controlling protein degradation and cell signaling, ubiquitination is a critical regulator of various cellular processes such as inflammatory responses, vesicular trafficking, and cell cycle, altogether making it an ideal target to hijack for bacterial and viral pathogenicity (Martinez-Martin, 2017; Yang et al., 2015). These infection-involved proteins inevitably contribute to the elucidation of the path physiology of pathogen-host interactions. Table 3.1 gives an overview of the main advantages and disadvantages for the commonly used host-pathogen protein-protein interaction (HP-PPI) detection methods.

TABLE 3.1 An Overview of the Main Advantages and Disadvantages for the Commonly Used HP-PPI Detection Methods

S.No.	Technique	Merits	Demerits
1.	Chemical cross-linking mass spectrometry	Provides information on interacting protein domains, residue-to-residue resolution, information on the topology and structural arrangement of protein complexes; can be applied on infected tissues directly	Need for large amounts of purified protein complexes, complex data acquisition and analysis, low sensitivity/resolution
2.	Proximity-dependent labeling strategies-mass spectrometry	Appropriate for weak and transient interactions, sensitive, adequate for resolving the spatial organizations of the tagged proteins	Long reaction times, not suitable for time course experiments, hard to distinguish direct from indirect/ proximal interactions
3.	Yeast 2 hybrid (Y2H)	High throughput, existing human and pathogen ORFeome collections, universality—any cDNA from any protein is testable	Need for exhaustive screens, nonphysiological experimental conditions, detects only binary interactions, no PTM information
4.	Affinity purification-mass spectrometry (AP-MS)	High throughput, sensitive, detects entire protein complexes, PTM sensitive, sing antibodies against the bait of interest, can be applied from infected tissues directly	Need for transgenic cell lines, needs additional experimental data to distinguish direct from indirect interactors, identification of PPI depends on the biochemical extraction conditions
5.	Protein correlation profiling	Whole proteome studies, unbiased, stoichiometric and quantitative information readily available	Dynamic range of protein abundances between host and pathogen might be too important, low sensitivity, hard to detect kiss and run interactions

3.3 CLINICAL MICROBIOLOGY AND CURRENT PROBLEMS

Microbial infections are the most common cause of disease development, and it causes millions of deaths around the world. Recurrence and antibiotic resistance of microbial infections demand advanced techniques in medical and diagnostic fields. Timely diagnosis is the most effective approach to prevent diseases, but it requires an accurate and fast identification and characterization of the infecting pathogens, such as bacteria, fungi, viruses, and parasites. New developments in diagnostic techniques not only are for identification purpose but also illuminate bacterial infection mechanisms and bacterial adaptation to the human host. These inputs will help the discovery of new therapeutic approaches to control or eradicate microbial-associated infections.

Clinical diagnosis of microbes is based on, culturing on solid media, molecular and new emerging techniques. Culture-based diagnosis of clinical samples on solid selective enrichment medium is very sensitive, easily contaminated, labor-dependent, and a time-consuming process; there's a chance for error to occur in these diagnosis reports. Currently, some automated identification systems give reliable results in shorter periods of time and also antimicrobial susceptibility testing (van Belkum et al., 2013). Molecular methods are more accurate, specific, sensitive, and faster than traditional culture-based methods, because they use more stable genotypic characteristics than reversible phenotypic characteristics. Molecular methods are of two types, such as immunologic and nucleic acid-based techniques. Immunologic techniques are based on the binding between diagnostic antibodies and specific antigenic determinants of the target. The human body produces specific antibodies in response to microbial infection; it is possible to use those antibodies to detect an acute or past infection. ELISA is the most commonly used technique in clinical diagnosis due to their speed, capacity, easy handling, and relatively low cost. Molecular methods also have some limitations due to the difficulty to generate selective antibodies and the requirement of large amounts of the respective antigen to quantify bacteria. The development of antibodies with the required degree of specificity is difficult for complex organisms (Gehring et al., 2015; Mangal et al., 2016).

Nucleic acid techniques are DNA- and RNA-based techniques; DNA-based techniques include DNA microarray; PCR- and RNA-based techniques are RT-PCR and TMA. PCR is one of the greatest inventions discovered by Kary Banks Mullis; it provides great improvements in microbial diagnosis, and the use of species-specific sequences to detect and identify pathogens requires the prediction of the species presence in the clinical samples. Limitations of PCR technique are as follows: There is a limited number of PCR assays that can be performed per clinical sample, infections or diseases are not monocultures being frequently caused by species-species interactions or even strain-strain interactions, and it requires proper standards and controls for valid interpretation (Maurer, 2011). Nucleic acid-based technologies worked based on entire DNA. The presence of extracellular DNA might influence the DNA-based signals detected; significant amounts of extracellular DNA released by bacterial cells destroyed by immune defenses have been detected. Therefore, DNA-based signals obtained during diagnosis will be influenced by the presence of extracellular DNA that might lead to misidentification (Maurer, 2011).

3.4 ROLE OF MALDI-TOF- IN DIAGNOSIS

Clinical diagnosis requires rapid, reliable, and cost-effective methods for the identification of potential pathogens in clinical samples so that appropriate antimicrobial therapy may be initiated early. MALDI-TOF MS can be used for early identification of bacteria in blood, urinary tract infections, and other clinical samples. MS-based proteomic approaches have been used in clinical diagnostic procedures, including the characterization, identification, and classification of microorganisms. Matrix-assisted laser desorption/ionization time-of-flight MS (MALDI-TOF MS) has been broadly adopted by many clinical microbiology laboratories over the past decade (Chen et al., 2017). MALDI-TOF MS is the most emerging tool used to identify microorganisms at species level, even strain differentiation based on molecular signatures (Wieser et al., 2012). The successful implementation of this technique in the routine of laboratories is due to its fast, easy, and high-throughput characteristics. Moreover, it generates simple and easily interpretable spectra. Bacterial identification by MALDI-TOF MS process involves different steps, i.e., protein separation, 2-D gel analysis, and gel digestion and analysis. Samples were mixed with suitable matrix and target plate today. Samples were exposed to short laser pulses; it (laser energy) vaporizes the protein sample and leads to ionization; these ions reached the detector at the time-of-flight (TOF) tube. Based on this TOF information, a characteristic spectrum is recorded constituting a specific sample fingerprint, which is unique for a given species. Increased mass and ion charge ratio (m/z) corresponds to lower speed and longer time needs to reach the detector (Tracz et al., 2018). MALDI-TOF MS is a very reproducible and reliable tool for microbial identification and can identify bacterial isolates in a few minutes and with low costs, with high efficiency from both a diagnostic and a cost-per-analysis point of view (Singhal et al., 2015).

Proteomics provides valuable data in the field of microbiology; proteomics characterizes the molecular model systems to investigate specific aspects of microbiological gene expression and metabolism. In medical microbiology, these data help to analyze the disease mechanism and how the pathogen interact with their hosts; these investigations are largely focused on the development of novel approaches for treatment and prevention of infectious diseases (Herschend et al., 2017; Wang et al., 2016).

3.5 ROLE OF PROTEOMICS IN MEDICAL MICROBIOLOGICAL RESEARCH AND DIAGNOSIS

Proteomic tools can be used in different ways to diagnose diseases. Proteomic tactics are being used to improve the screening and early detection of cancer; this is attained by identifying proteins whose expression is affected by the disease process. An individual protein is called a biomarker, whereas a set of proteins with altered expression levels are called protein signature (Kim and Hwang, 2016; Kuppusamy et al., 2017). Applying proteomic technologies may improve the understanding of disease processes, develop new biomarkers for diagnosis and early detection of disease, and accelerate drug development. Emerging techniques create numerous opportunities and challenges to meet the needs for high sensitivity and high throughput required for disease-related probes (Gregorich and Ge, 2014; Mehan et al., 2012).

Biomarker is to be useful as a candidate for early screening and detection of cancer. The in vitro estimation of two pathological proteins (rRv3369 and rRv3874) (Ag) through proteomic technology is a high sensitivity method that diagnosed the tuberculosis in clinical isolates (Geyer et al., 2017; Thomas et al., 2016). Pathogenesis of severe acute respiratory syndrome (SARS) is not well unstated, and a specific diagnostic method is critical for the management and control of this disease. Proteomic analysis of sera from patients with SARS has identified potential protein markers shortened from α(1)-antitrypsin that was consistently found in higher concentrations in the sera of SARS patients related to healthy controls. These markers may be useful as diagnostic tools and therapeutic and potential vaccine targets (Clay et al., 2012; McBride and Fielding, 2012). Proteomics is also used to identify differentially expressed cancer cells, which may reflect differences in insensitivity and tendency in developing resistance to treatment (Mohammad et al., 2015). Proteomics has identified proteins involved in metastasis and multidrug resistance that are highly expressed in glioblastoma multiforme compared with brain tissue. These proteins have potential as diagnostic, drug resistance, and invasiveness markers for glioblastoma tumors. Drug resistance is a major clinical problem in the treatment of many infectious diseases (Chowdhury et al., 2018; Hanif et al., 2017; Olar and Aldape, 2014). Proteomics shows solution for drug development against the resistance. Chloroquine is the most successful drug to treat malaria but has been reduced almost ineffective in many parts of the world by the widespread emergence of chloroquine resistance. Through the help of proteomic technology, the potential therapeutic targets, host-pathogen interaction, and protein-drug interactions in *Plasmodium* species are identified (Bhattacharyya and Chakrabarti, 2015; Hossain et al., 2016; Suazo et al., 2016).

In clinical microbiology, the combination of MALDI-TOF MS plays a key role in bacterial identification, susceptibility testing, and nucleic acid targeting and facilitated a better workflow for the timely management of positive blood cultures and other samples. Detection of proteomic changes in resistant versus susceptible bacterial species and can be performed in the classical MALDI-TOF MS mode (Rahi et al., 2016). Identification and characterization of a bacterial isolate from patient within a transient period of diagnosis have reduced the morbidity and mortality value. Early detection is the key to successful treatment of most diseases; clinical test result is based on chemical composition and man power, so there is a chance of false/error to occur and being late. It's time to reduce errors and improve the experimental design with improved outcomes (Sung et al., 2018). The MS-based biomarker technology enables early detection of cancer and cardiovascular diseases. The field of biomarker research is mainly focused on genomics, transcriptomics, and proteomics and understanding the gene regulation, functions, and how are they involved in disease development. Expectations for the ideal MS-based biomarker should reflect the significant recent and future improvements in its technological basis (Crutchfield et al., 2016). Fig. 3.2 illustrates the research areas in medical microbiology that utilize proteomic technologies.

3.6 IMPORTANCE OF MALDI-TOF MS IN IDENTIFICATION OF PATHOGENIC MICROORGANISMS

In normal methods, the identification of bacteria is based on phenotypic characteristics and biochemical reactions; all of these methods can achieve high accuracies, and it takes longer

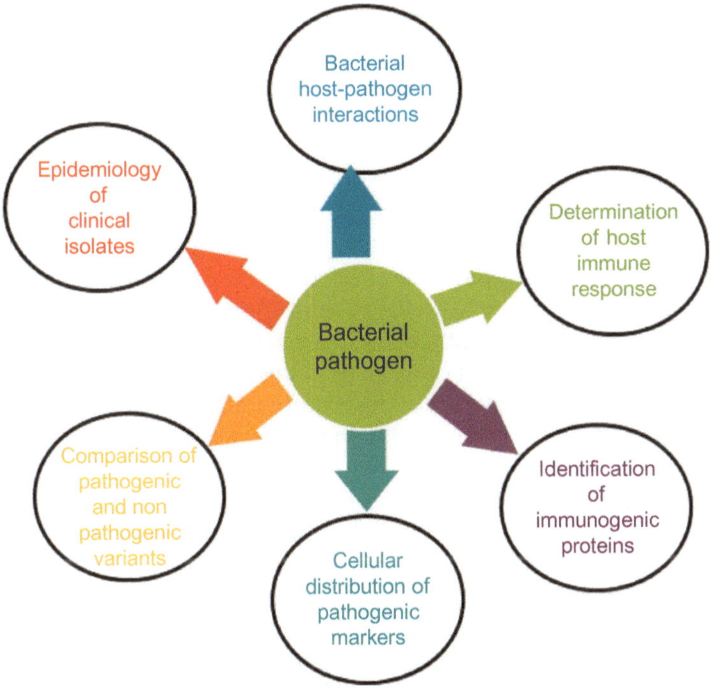

FIG. 3.2 Research areas in medical microbiology that utilize proteomic technologies.

period to total identification process. MALDI-TOF MS is a technical revolution in the microbial field; it is a faster and easier technique for microbial identification of pathogenic microorganisms. Apart from protein extraction from whole-cell bacteria lysates prepared with chemical treatments, direct bacterial analysis is usually adopted for bacterial identification by MALDI-TOF MS; it plays a key role in clinical diagnostic and prognostic workup of biofilms formation and control. Protein profiles can be obtained from a single colony of bacteria directly deposited on the MALDI-TOF MS target plate and overlaid with a matrix solution. MALDI-TOF MS was used to identify microorganisms not only at the species level but also at the subspecies and strain levels and also identify some bacterial toxins (Crutchfield et al., 2016). MALDI-TOF MS is a rapid, easy, relatively inexpensive, and high-throughput method for identifying clinically relevant bacteria and fungi. MALDI-TOF MS methods are implemented in identifying common microorganisms (Cobo, 2013; van Belkum et al., 2017). Characterization of pathogen, based on MALDI-TOF MS fingerprinting profiling, is now a routine in diagnostic microbiology; the entire process is completed in four steps, i.e., detection, ID, susceptibility testing, and epidemiology. The MALDI-TOF MS protein fingerprint employs a pattern-matching algorithm to compare microbial cell extracted peak lists with a reference database, hence creating a consensus "biomarker" peak list, characterized by averaged m/z and intensity. MALDI-TOF MS also has the potential to identify pathogens in direct samples, such as urine and blood cultures. Furthermore, the potential for ID at the serotype or strain level and antibiotic resistance profiling within minutes makes MALDI-TOF MS an ongoing revolution in the clinical microbiology laboratory (Emonet et al., 2010; Suttisunhakul et al., 2017).

3.7 ROLE OF PROTEOMICS IN DEVELOPMENT OF THERAPY AND VACCINE

Proteomics plays a key role in the development of novel therapeutic strategies through the identification of vaccine and antibiotic targets. Antigenic proteins can be identified by an immune response against the infection; these proteins can be used in the development of vaccine. Specific proteins can be selectively identified to optimize the vaccine development. In order to develop vaccine, the bacterial cells were treated with stress or bacterial infection conditions. Then, proteins were analyzed by 2-D electrophoresis, which described the newly expressed proteins with the support of a bioinformatic approach to the identification of vaccine candidate proteins, which took advantage of the extensive gene sequence data now available (Bilgic et al., 2016). Drug resistance of pathogens is the major problem to recurrence of infection or disease. Proteomics can determine the antimicrobial resistance mechanisms through the analysis of microbial proteins; it explains that drug resistance development will lead to improvements in extending the efficacy of current antimicrobial (Li et al., 2010). *Pseudomonas aeruginosa* develops resistant against beta-lactam antibiotics; proteomics can identify the resistance mechanism; it reduces the expression of outer membrane protein against imipenem, ceftazidime, and ampicillin treatment. *Streptococcus pneumoniae* developed resistance against penicillin and erythromycin also, two phenotypes of *S. pneumoniae* specifically the MLS and M phenotypes. Proteomics has been investigating these resistance mechanisms; phenotype M is resistant through the expression of mefE, which encodes a membrane transporter protein that reduces the intracellular levels of erythromycin; MLS phenotype owes their resistance to the methylation of rRNA. Antibiotic action can be investigated by locating proteins that show differential expression patterns when bacteria are grown in the presence or absence of antibiotics. All these data gives a deep boost for the development of novel therapeutic compounds (Editorial Mini Focus Issue: Systems Biology, 2014; Li et al., 2010; Wang et al., 2015; Yari et al., 2016).

3.8 FUTURE PERSPECTIVES

Proteomics has made an important impact in medical biology, and it provides valuable data related to host-pathogen interactions and posttranslational modifications and is helpful for better understanding of microbial community functions and microbial physiology. Pathogenic microbial identification and characterization are very difficult before automation; to overcome these difficulties, MALDI-TOF comparative sequencing is very helpful. Therefore, MALDO-TOF MS has become a powerful tool for routine pathogen identifications in clinical laboratories. Before MALDI-TOF, bacterial identification and sequencing approaches were time-consuming and cost-effective process, but now, MALDI-TOF analysis is routine in microbiological laboratories. Proteomics analyzes the host-pathogen resistance mechanism that can be investigated based on the data. Proteomics may serve neither as the primary screening method for antigen detection nor as a complementary method to confirm bioinformatic predictions of immunogenic analysis (Chiang et al., 2018). The ability to rapidly identify pathogens in specific clinical specimens will help us to study the clinical burden resulting from the emergence of these species as human pathogens, definitely suggesting

MALDI-TOF MS as an alternative to molecular methods in clinical laboratories and as a revolutionized bacterial identification in the field of medical microbiology in coming future.

Acknowledgments

The authors are indebted to all the researchers whom we cited in this review for their significant and valuable research. No funding was received to perform this review.

Conflicts of Interests

The authors declare no relevant competing financial interests to disclose.

References

Bhattacharyya, M., Chakrabarti, S., 2015. Identification of important interacting proteins (IIPs) in *Plasmodium falciparum* using large-scale interaction network analysis and in-silico knock-out studies. Malar. J. 14, 70.

Bilgic, H.B., Karagenc, T., Bakırcı, S., Shiels, B., Tait, A., Kinnaird, J., Eren, H., Weir, W., 2016. Identification and analysis of immunodominant antigens for ELISA-based detection of *Theileria annulata*. PLoS ONE 11, e0156645.

Chen, B., Zhang, D., Wang, X., Ma, W., Deng, S., Zhang, P., Zhu, H., Xu, N., Liang, S., 2017. Proteomics progresses in microbial physiology and clinical antimicrobial therapy. Eur. J. Clin. Microbiol. Infect. Dis. 36, 403–413.

Cheng, K., Chui, H., Domish, L., Hernandez, D., Wang, G., 2016. Recent development of mass spectrometry and proteomics applications in identification and typing of bacteria. Proteomics Clin. Appl. 10, 346–357.

Chiang, C.-Y., Uzoma, I., Moore, R.T., Gilbert, M., Duplantier, A.J., Panchal, R.G., 2018. Mitigating the impact of antibacterial drug resistance through host-directed therapies: current progress, outlook, and challenges. MBio 9, e01932-17.

Chowdhury, F.A., Hossain, M.K., Mostofa, A.G.M., Akbor, M.M., Bin Sayeed, M.S., 2018. Therapeutic potential of thymoquinone in glioblastoma treatment: targeting major gliomagenesis signaling pathways. Biomed. Res. Int. 2018, 1–15.

Clay, C., Donart, N., Fomukong, N., Knight, J.B., Lei, W., Price, L., Hahn, F., Van Westrienen, J., Harrod, K.S., 2012. Primary severe acute respiratory syndrome coronavirus infection limits replication but not lung inflammation upon homologous rechallenge. J. Virol. 86, 4234–4244.

Cobo, F., 2013. Application of maldi-tof mass spectrometry in clinical virology: a review. Open Virol. J. 7, 84–90.

Crutchfield, C.A., Thomas, S.N., Sokoll, L.J., Chan, D.W., 2016. Advances in mass spectrometry-based clinical biomarker discovery. Clin. Proteomics 13, 1.

Del Chierico, F., Petrucca, A., Vernocchi, P., Bracaglia, G., Fiscarelli, E., Bernaschi, P., Muraca, M., Urbani, A., Putignani, L., 2014. Proteomics boosts translational and clinical microbiology. J. Proteome 97, 69–87.

Editorial Mini Focus Issue: Systems Biology, 2014. Bioanalysis 6, 2519–2521.

Emonet, S., Shah, H.N., Cherkaoui, A., Schrenzel, J., 2010. Application and use of various mass spectrometry methods in clinical microbiology. Clin. Microbiol. Infect. 16, 1604–1613.

Gehring, A., Brewster, J., He, Y., Irwin, P., Paoli, G., Simons, T., Tu, S.-I., Uknalis, J., 2015. Antibody microarray for *E. coli* O157:H7 and Shiga toxin in microtiter plates. Sensors 15, 30429–30442.

Geyer, P.E., Holdt, L.M., Teupser, D., Mann, M., 2017. Revisiting biomarker discovery by plasma proteomics. Mol. Syst. Biol. 13, 942.

Gregorich, Z.R., Ge, Y., 2014. Top-down proteomics in health and disease: challenges and opportunities. Proteomics 14, 1195–1210.

Hanif, F., Muzaffar, K., Perveen, K., Malhi, S.M., Simjee, S.U., 2017. Glioblastoma multiforme: a review of its epidemiology and pathogenesis through clinical presentation and treatment. Asian Pac. J. Cancer Prev. 18, 3–9.

Herschend, J., Damholt, Z.B.V., Marquard, A.M., Svensson, B., Sørensen, S.J., Hägglund, P., Burmølle, M., 2017. A meta-proteomics approach to study the interspecies interactions affecting microbial biofilm development in a model community. Sci. Rep. 7, 16483.

Hossain, M.U., Khan, M.A., Hashem, A., Islam, M.M., Morshed, M.N., Keya, C.A., Salimullah, M., 2016. Finding potential therapeutic targets against *Shigella flexneri* through proteome exploration. Front. Microbiol. 7, 1817.

Jean Beltran, P.M., Federspiel, J.D., Sheng, X., Cristea, I.M., 2017. Proteomics and integrative omic approaches for understanding host–pathogen interactions and infectious diseases. Mol. Syst. Biol. 13, 922.

Kim, M., Hwang, D., 2016. Network-based protein biomarker discovery platforms. Genomics Inform. 14, 2.

Kuppusamy, P., Govindan, N., Yusoff, M.M., Ichwan, S.J.A., 2017. Proteins are potent biomarkers to detect colon cancer progression. Saudi J. Biol. Sci. 24, 1212–1221.

Li, S.-L., Ye, F., Cai, W.-J., Hu, H.-D., Hu, P., Ren, H., Zhu, F.-F., Zhang, D.-Z., 2010. Quantitative proteome analysis of multidrug resistance in human ovarian cancer cell line. J. Cell. Biochem. 109, 625–633.

Mangal, M., Bansal, S., Sharma, S.K., Gupta, R.K., 2016. Molecular detection of foodborne pathogens: a rapid and accurate answer to food safety. Crit. Rev. Food Sci. Nutr. 56, 1568–1584.

Martinez-Martin, N., 2017. Technologies for proteome-wide discovery of extracellular host-pathogen interactions. J Immunol Res 2017, 1–18.

Maurer, J.J., 2011. Rapid detection and limitations of molecular techniques. Annu. Rev. Food Sci. Technol. 2, 259–279.

McBride, R., Fielding, B.C., 2012. The role of severe acute respiratory syndrome (SARS)-coronavirus accessory proteins in virus pathogenesis. Viruses 4, 2902–2923.

Mehan, M.R., Ayers, D., Thirstrup, D., Xiong, W., Ostroff, R.M., Brody, E.N., et al., 2012. Protein signature of lung cancer tissues (R Rota, Ed). PLoS ONE 7, e35157.

Mohammad, R.M., Muqbil, I., Lowe, L., Yedjou, C., Hsu, H.-Y., Lin, L.-T., et al., 2015. Broad targeting of resistance to apoptosis in cancer. Semin. Cancer Biol. 35, S78–S103.

Nicod, C., Banaei-Esfahani, A., Collins, B.C., 2017. Elucidation of host-pathogen protein-protein interactions to uncover mechanisms of host cell rewiring. Curr. Opin. Microbiol. 39, 7–15.

Olar, A., Aldape, K.D., 2014. Using the molecular classification of glioblastoma to inform personalized treatment. J. Pathol. 232, 165–177.

Pérez-Llarena, F.J., Bou, G., 2016. Proteomics as a tool for studying bacterial virulence and antimicrobial resistance. Front. Microbiol. 7, 410.

Rahi, P., Prakash, O., Shouche, Y.S., 2016. Matrix-assisted laser desorption/ionization time-of-flight mass-spectrometry (MALDI-TOF MS) based microbial identifications: challenges and scopes for microbial ecologists. Front. Microbiol. 7, 1359.

Ravikumar, V., Jers, C., Mijakovic, I., 2015. Elucidating host-pathogen interactions based on post-translational modifications using proteomics approaches. Front. Microbiol. 6, 1313.

Singhal, N., Kumar, M., Kanaujia, P.K., Virdi, J.S., 2015. MALDI-TOF mass spectrometry: an emerging technology for microbial identification and diagnosis. Front. Microbiol. 6, 791.

Suazo, K.F., Schaber, C., Palsuledesai, C.C., Odom John, A.R., Distefano, M.D., 2016. Global proteomic analysis of prenylated proteins in *Plasmodium falciparum* using an alkyne-modified isoprenoid analogue. Sci. Rep. 6, 38615.

Sung, J.Y., Hwang, Y., Shin, M.H., Park, M.S., Lee, S.H., Yong, D., Lee, K., 2018. Utility of conventional culture and MALDI-TOF MS for identification of microbial communities in bronchoalveolar lavage fluid in comparison with the GS junior next generation sequencing system. Ann. Lab. Med. 38, 110.

Suttisunhakul, V., Pumpuang, A., Ekchariyawat, P., Wuthiekanun, V., Elrod, M.G., Turner, P., et al., 2017. Matrix-assisted laser desorption/ionization time-of-flight mass spectrometry for the identification of *Burkholderia pseudomallei* from Asia and Australia and differentiation between *Burkholderia* species. PLoS ONE 12, e0175294.

Thomas, S., Hao, L., Ricke, W.A., Li, L., 2016. Biomarker discovery in mass spectrometry-based urinary proteomics. Proteomics Clin. Appl. 10, 358–370.

Tracz, D.M., Tober, A.D., Antonation, K.S., Corbett, C.R., 2018. MALDI-TOF mass spectrometry and high-consequence bacteria: safety and stability of biothreat bacterial sample testing in clinical diagnostic laboratories. J. Med. Microbiol. 67, 341–346.

van Belkum, A., Durand, G., Peyret, M., Chatellier, S., Zambardi, G., Schrenzel, J., Shortridge, D., Engelhardt, A., Dunne, W.M., 2013. Rapid clinical bacteriology and its future impact. Ann. Lab. Med. 33, 14.

van Belkum, A., Welker, M., Pincus, D., Charrier, J.P., Girard, V., 2017. Matrix-assisted laser desorption ionization time-of-flight mass spectrometry in clinical microbiology: what are the current issues? Ann. Lab. Med. 37, 475–483.

Vranakis, I., Goniotakis, I., Psaroulaki, A., Sandalakis, V., Tselentis, Y., Gevaert, K., Tsiotis, G., 2014. Proteome studies of bacterial antibiotic resistance mechanisms. J. Proteome 97, 88–99.

Walters, M.S., Mobley, H.L., 2010. Bacterial proteomics and identification of potential vaccine targets. Expert Rev. Proteomics 7, 181–184.

Wang, Z., Liang, S., Lian, X., Liu, L., Zhao, S., Xuan, Q., et al., 2015. Identification of proteins responsible for adriamycin resistance in breast cancer cells using proteomics analysis. Sci. Rep. 5, 9301.

Wang, D.-Z., Kong, L.-F., Li, Y.-Y., Xie, Z.-X., 2016. Environmental microbial community proteomics: status, challenges and perspectives. Int. J. Mol. Sci. 17. pii:E1275.
Wieser, A., Schneider, L., Jung, J., Schubert, S., 2012. MALDI-TOF MS in microbiological diagnostics—identification of microorganisms and beyond (mini review). Appl. Microbiol. Biotechnol. 93, 965–974.
Yang, Y., Hu, M., Yu, K., Zeng, X., Liu, X., 2015. Mass spectrometry-based proteomic approaches to study pathogenic bacteria-host interactions. Protein Cell 6, 265–274.
Yari, S., Tasbiti, A.H., Ghanei, M., Shokrgozar, M.A., Vaziri, B., Mahdian, R., Yari, F., Bahrmand, A., 2016. Proteomic analysis of sensitive and multi drug resistant Mycobacterium tuberculosis strains. Microbiology 85, 350–358.

CHAPTER

4

Importance of Probiotics in Cancer Prevention and Treatment

Anshul Sharma

Department of Food Science and Biotechnology, Gachon University, Seongnam, Republic of Korea; Department of Biotechnology, Shoolini University of Biotechnology and Management Sciences, Solan, India

OUTLINE

4.1 INTRODUCTION

For the last few decades, scientific and health care communities are giving serious attention to probiotics. The main focus is to restore and maintain the beneficial microflora as preventive and therapeutic tools for the management of several diseases including cancer. Worldwide, regardless of improvements in diagnosis and treatment, cancer remains to be a principal cause of mortality (13% of all deaths) (Bray et al., 2015). According to a recent report by the World Health Organization (WHO), approximately 70% of deaths from cancer occur in low- and middle-income countries, and around one-third deaths from cancer are because of high body mass index, the lack of physical activity, low fruit and vegetable intake, and tobacco and alcohol use (WHO, 2018). It has been reported that 90%–95% of the cases are related to the external factors, while 5%–10% of the cases are due to genetic defects (alterations in the

https://doi.org/10.1016/B978-0-12-816328-3.00004-0

normal cell processes fundamental to the regulation of the cell cycle, cell signaling, and differentiation); therefore, it can be said that cancer is a multifactorial disease (Tian et al., 2011). The human intestine harbors about 10^{14} species of microorganisms (bacteria, viruses, and some eukaryotes) that colonize the digestive tract soon after birth (Blaser, 2014). The bacterial, fungal, and viral intestinal communities are generally referred to as the gut microbiota, and altogether, their genomes are discussed as the gut microbiome. The research declares a symbiotic association between microbiota and humans for mutual benefits, which results in the maintenance of normal metabolic, immunologic, and motor functions and correct nutrient digestion and absorption (Blaser, 2014; Goulet, 2015). On the other hand, dysbiosis of the microbial community has been shown to be linked with a diseased state. Dysbiosis has also been found to be associated with carcinogenesis, chemotherapeutic metabolism (Garrett, 2015), and cancer-related symptoms (Kelly et al., 2016). There is a bidirectional relationship between the gut microbiome and cancer; cancer-related changes in the microbiome may happen as a result of the disease and can contribute to cancer development (Zitvogel et al., 2017).

The probiotics may be defined as "live microorganisms which, when administered in adequate amounts, confer a health benefit on the host" (Hill et al., 2014). Recently, the fame of probiotics has expanded exponentially, because of their health-promoting effects and their abilities to prevent or treat different diseases, including some types of cancers (de Moreno de LeBlanc, 2014). Several genera of bacteria and yeast have been reported as probiotics. By considering the importance of the subject, this chapter will summarize some recent studies correlating various strains of probiotics mainly *Lactobacillus*, *Lactococcus*, and *Bifidobacterium* and their cancer preventing and treatment potential.

4.2 PROBIOTICS AND POSSIBLE MECHANISMS IN CANCER PREVENTION AND MANAGEMENT

It has been reported that 90% of gut microbiota is obligatory anaerobes, producing mainly lactic acid (*Bifidobacterium* and *Lactobacillus*) and volatile fatty acids (Bacteroidaceae and *Eubacterium*). Genus *Enterococcus* makes up <1% of the intestinal microbiota. Moreover, the genera *Clostridium*, *Proteus*, *Candida*, *Saccharomyces*, *Pseudomonas*, *Aspergillus*, *Penicillium*, and *Staphylococcus* represent 0.01% of the intestinal microflora (Kverka and Tlaskalova-Hogenova, 2017). Among lactic acid bacteria (LAB), *Lactobacillus* and *Bifidobacterium* are prominent probiotic organisms, and several reports had demonstrated that these strains may mediate anticancer response (Zitvogel et al., 2017). It is well documented that probiotics can have multiple interactions with the host as they colonize host tissues, the gastrointestinal tract, the skin, and other organ systems and contribute to the refurbishment of the healthy microbiota. The mechanisms of probiotics include the intricate interaction between diet, gut microbiota, and host energy metabolism, which persuades several effects, namely, (1) neutralization of carcinogens; (2) support in the process of reconstruction of intestinal barrier and also improving its function by increasing the production of mucin, defensins, and immunoglobulin A (IgA) and by altering the pro-inflammatory cytokine and chemokine's response; (3) synthesis of vitamins and short-chain fatty acids (SCFAs), nutrients, and growth signals for the intestinal epithelium that could play a role in colon and other cancer prevention; (4) increased production of cytokines (IL-2 and IL-12), antioxidants, and anti-angiogenic factors; (5) lowering of intestinal pH; (6) modulation and enhancement of the host's innate and adaptive immune response through the secretion of anti-inflammatory molecules and affecting the helper T-cell response and regulation; (7) immune checkpoint blockades; (8) alteration of the intestinal microflora; (9) antiproliferative effects via regulation of apoptosis and cell differentiation; (10) inhibition of tyrosine kinase and other signaling pathways; and (11) convention of linoleic acid to conjugated linoleic acid (CLA), a polyunsaturated fatty acid with anti-inflammatory effects that helps in decreasing cancer cell viability and inducing apoptosis. A number of studies have been published on the effect of probiotics on various types of cancer. The following section describes a couple of recent studies about the use of different strains of common probiotic bacteria (*Lactobacillus*, *Lactococcus*, and *Bifidobacterium*) for the prevention and treatment of cancer (Table 4.1).

TABLE 4.1 Examples of In Vitro and In Vivo Effect of Probiotics Against Different Cancer Cells

Probiotic Strain	Cancer Type	Biological Effects	References
Lactobacillus			
Lactobacillus acidophilus and *Lactobacillus plantarum*	Glioblastoma and breast cancer	Cytotoxic effects	Bharti et al. (2015)
Lactobacillus lactis	Breast cancer	Inhibition of cancer growth	Aragón et al. (2015)
Lactobacillus plantarum and *L. rhamnosus*	Murine adenocarcinoma	Antitumor response	Hu et al. (2015)
Lactobacillus acidophilus	Ehrlich ascites carcinoma (EAC) cells	Reduced tumor volume	Abd El Ghany et al. (2015)
Lactobacillus acidophilus	Lung cancer	Reduction in tumor size and survival rates increased	Gui et al. (2015)
Lactobacillus plantarum and *L. rhamnosus*	Colon cancer	Reduction of tumor incidence, suppression of COX-2 expression	Walia et al. (2015)
Lactobacillus salivarius	Colorectal cancer	Decreased cancer incidence	Zhang et al. (2015)
Lactobacillus crispatus and *L. rhamnosus*	Cervical and colon cancer	Cytotoxic effects	Nouri et al. (2016)
Lactobacillus crispatus and *L. rhamnosus*	Cervical cancer	Antiproliferative effects	Motevaseli et al. (2016)
Lactobacillus crispatus and *L. rhamnosus*	Breast cancer	Cytotoxic effects	Esfandiary et al. (2016)
Lactobacillus casei BL23	Colorectal cancer	Modulation of regulatory T cells	Lenoir et al. (2016)
Prohep[a]	Hepatocellular carcinoma (HCC)	Reduction in tumor size and growth	Li et al. (2016)
Lactobacillus plantarum LS/07	Breast cancer	Antiproliferative and immunomodulatory	Kassayová et al. (2016)
Lactobacillus plantarum	Sarcoma	Antiproliferative and immunomodulatory	Shin et al. (2016)

Continued

TABLE 4.1 Examples of In Vitro and In Vivo Effect of Probiotics Against Different Cancer Cells—cont'd

Probiotic Strain	Cancer Type	Biological Effects	References
Lactobacillus acidophilus, B. bifidum and *B. infantum*	Colorectal cancer	Antiproliferative	Kuugbee et al. (2016)
Lactobacillus rhamnosus	Colorectal cancer	Reduction of tumor incidence	Gamallat et al. (2016)
Lactobacillus casei	Colon cancer	Apoptosis via JNK signaling pathway	Konishi et al. (2016)
Lactobacillus casei	Colon carcinogenesis	Antimutagenic	Irecta-Nájera et al. (2017)
Lactobacillus fermentum and *Lactobacillus acidophilus*	Colorectal cancer	Anticancer	Kahouli et al. (2017)
Lactobacillus casei	Colorectal cancer	Antiproliferative and immunomodulation	Jacouton et al. (2017)
Lactobacillus johnsonii with other strains	Colon cancer	Anticancer	Chen et al. (2017)
Lactococcus			
Lactococcus lactis subsp. *lactis*	Stomach cancer	G0/G1 cell cycle arrest and apoptosis	Kim et al. (2009)
Lactococcus lactis subsp. lactis	Colon, cervical, gastric, and breast cancer	Apoptosis	Haghshenas et al. (2014)
Lactococcus lactis	Lung, breast, and colon carcinoma	Decreased cell proliferation	Lee et al. (2015)
Lactococcus lactis	Lung, breast, and colon carcinoma	Antiproliferative	Han et al. (2015)
Lactococcus lactis subsp. lactis	Breast cancer	Antitumor	Wu et al. (2016)
Bifidobacterium			
Bifidobacterium lactis	Colorectal cancer	Antiproliferative	Le Leu et al. (2010)
Recombinant *B. infantis*	Lung cancer	Antitumor effect and survival rates increased	Zhu et al. (2011)
Recombinant *B. infantis*	Lung cancer	Increased necrosis rate of tumor and survival rates increased	Li et al. (2012)
Bifidobacterium cocktail	Melanoma	Antitumor	Sivan et al. (2015)

[a]*A mixture of different probiotic strains used in the study.*

4.2.1 *Lactobacillus*

The genus *Lactobacillus* is one of the most prominent probiotic members present in the gastrointestinal tract and vagina. Few strains are commonly used for the production of yogurt, cheese, beer, wine, and other fermented foods. Numerous in vivo and in vitro studies have shown that different isolates of *L casei*, *L. plantarum*, *L. crispatus*, *L. rhamnosus* GG, *L. salivarius* Ren (Ren), and *L. acidophilus* may mediate anticancer effects through various mechanisms

(Zitvogel et al., 2017). Recently published review article has nicely summarized the preclinical results of *L. rhamnosus* GG on cancer cell proliferation and tumor invasion and the potential use of LGG for the prevention and management of diarrhea associated with cancer-treated patients (Banna et al., 2017). Bharti et al. (2015) have described the comparative cytotoxic effects of live whole cell (LWC), heat-killed cell (HKC), and cell-free extract (CFE) of *L. acidophilus* KP942831 and *L. plantarum* KP894100 strains toward U-87 (glioblastoma cell line) and MCF-7 (breast cancer cells). Glioblastoma, also known as glioblastoma multiforme (GBM), is the most common and aggressive type of cancer that starts within the brain. The authors reported that both the strains have the potential to inhibit the viability of MCF7 and U-87 cells. However, *L. plantarum* showed the strongest cytotoxic effect on both the cell lines.

The study performed by Nouri et al. (2016) has described antimetastatic and antiproliferative activity of *L. crispatus* supernatant (LCS) and *L. rhamnosus* supernatant (LRS) on HeLa (cervical cell line) and HT-29 (colon adenocarcinoma cell lines). The results revealed cytotoxic effects of two lactobacilli against HeLa cells; however, LRS treatment was only effective against HT-29 cell line. And also, the *CASP3* gene expression was reduced after treatment with lactobacilli supernatants. According to the study, LRS and LCS are efficacious in the prevention of metastasis potency in HeLa cells with decreased expression of *MMP2* and *MMP9* and increased expression of their inhibitors. In the case of HT-29 cells, only LRS showed this effect. The authors mentioned the requirement of further assessment of these strains on other bacterial cells for better conclusion. Another research demonstrated that the treatment with the culture supernatant of *L. crispatus* and *L. rhamnosus* on cervical cancer cells resulted in the antiproliferative effect accompanied by downregulation of autophagy genes (*ATG14* and *BECN1*) and oncogene (*HPV E6*). Here, the researchers have confirmed that the key changes occurring during cervical carcinogenesis in cell machinery can be reversed by the suppression of HPV oncogenes. Therefore, downregulation of *HPV E6* by lactobacilli may have therapeutic potential for cervical cancer (Motevaseli et al., 2016). Similarly, the effect of both above cited strains has also been demonstrated by analyzing the expression of hypoxia-inducible factor (HIF) pathway in breast cancer cells (MDA-MB-231). Hypoxia-inducible factor (HIF)-1, a transcription factor, regulates the expression of several genes implicated in homeostasis, vascularization, anaerobic metabolism, and immunologic responses. Results showed cytotoxic effects toward cancer cells and alongside suppression of hypoxia-inducible factor (HIF)-1-pathway-mediated oncogenes (Esfandiary et al., 2016).

Recently, a report revealed the type of molecule isolated from *L. casei* ATCC334 with a strong tumor-suppressive effect. *Lactobacillus casei* probiotic-derived ferrichrome may exert a cancer-specific tumor-suppressive effect (colon cancer cells) through the apoptosis-inducing c-jun N-terminal kinase (JNK) signaling pathway (Konishi et al., 2016). Moreover, the effect of ferrichrome was compared with drugs such as cisplatin and 5-fluorouracil and likewise on noncancerous intestinal cells. A significant tumor-suppressive effect was reported on cancer cells compared with drugs. In addition, less effect was observed on noncancerous intestinal cells compared with agents. However, the stability and delivery of ferrichrome in vivo remain major problems in relation to its use in the clinical setting (Konishi et al., 2016). In a recent study, the effect of seven strains of *Lactobacillus*, i.e., *L. johnsonii* BCRC17010, *L. delbrueckii* subsp. *bulgaricus* BCRC10696, *L. salivarius* BCRC14759, *L. reuteri* BCRC14625, *L. brevis* PM150, *L. plantarum* PM153, and L. *brevis* PM177, has been described on HT-29 cells. According to the authors, *L. johnsonii* BCRC17010 strain has revealed the best probiotic potential for the inhibition of HT-29 growth among seven strains. The increased lactate dehydrogenase (LDH) activity and good abilities of the adhesion of BCRC17010 strain were responsible for the inhibitory effect (Chen et al., 2017). So et al. (2017) have described the role of various probiotic bacterial strains against gastrointestinal cancer in vitro: *L. reuteri* NCIMB 701359 and *L. fermentum* strains against colorectal cancer (SW-480) and epithelial colorectal cancer adenocarcinoma (Caco-2) cells; *L. paracasei* subsp. *paracasei* × 12, *L. crispatus* and *L. rhamnosus* strains against colon cancer (HT-29); *L. reuteri*, *L. acidophilus* and *L. casei* strains against colorectal cancer (Caco-2); *L. casei* ATCC 393 against murine (CT26) and human (HT-29) colon carcinoma cell lines; *L. plantarum* A7, *L. pentosus* B281, and *L. plantarum* B282 against colon cancer cells (Caco-2 and HT-29); and *L. brevis* SBL8803 against colorectal cancer (SW620).

4.2.1.1 A Couple of In Vivo Studies

Aragón et al. (2015) have represented the effect of fermented milk by *L. casei* CRL 431 strain on breast cancer. The BALB/c mouse was administered with fermented milk and resulted in the suppression of the tumor growth with less tumor vascularity and extravasation of tumor cells. The findings suggested it was due to the immune response such as decreasing the infiltration of macrophages in the tumor and the lungs and significantly increasing $CD4^+/CD8^+$ ratio. Several lines of evidences suggest that part of the mechanism of action of *L. casei* CRL431 toward breast cancer would be the lowering of IL-6 in blood serum and in the mammary glands. The anticancer and immunomodulatory effect of another probiotic strain, i.e., *L. plantarum* LS/07, on the breast cancer rat model has been described. SD rats were administered with *L. plantarum* LS/07 (probiotic) alone and also with prebiotics (oligofructose-enriched inulin) or with both prebiotics and melatonin to mice for 16 weeks, starting from 14 days before giving injection of a carcinogen, i.e., *N*-nitroso-*N*-methylurea (NMU). Results showed no effect on tumor growth; however, there was a reduction in the cell proliferation marker. In addition, it was observed that combined treatments (probiotic + prebiotic + melatonin) increased T-cell infiltration to a tumor and the Treg-cell level in tumor (Kassayová et al., 2016).

A comparative study investigated by Hu et al. (2015) has shown the antitumor potential of immune responses induced by *L. plantarum* and *L. rhamnosus* strains on CT26 murine adenocarcinoma cells. Results showed that compared with *L. rhamnosus*, oral administration with *L. plantarum* inhibited CT26 cell growth in BALB/c mice and prolonged the survival time of tumor-bearing mice. In addition, an increase in natural killer (NK) cell migration into tumor tissue and effector functions of CD8+ cells was observed. Moreover, upregulated production was reported for IFN-γ, not for IL-4 or IL-17. Another research has presented the protective effects of *L. casei* BL23 on colorectal cancer (CRC) in mice. It appeared that oral treatment with this probiotic bacterium reduces 1,2-dimethylhydrazine (DMH)-associated colorectal cancer by modulation of host immune response. This protection may be associated with the significant increase in the production of Th17 cells accompanied by the higher expression of IL-6, IL-17, IL-10, and TGF-β (Lenoir et al., 2016).

Likewise, in protective effect of oral *L. casei* BL23, a probiotic strain has also been described against colorectal cancer in mice C57BL6. The strain protected the mice by a reduction in tumor development. This antiproliferative effect was demonstrated due to the upregulation of caspase-7, caspase-9, and Bik genes involved in apoptosis. Moreover, an immunomodulation effect, i.e., downregulation of the IL-22 cytokine, was also reported (Jacouton et al., 2017).

Similarly, the prophylactic measure of another probiotic strain *L. salivarius* Ren (Ren) has also been demonstrated in the colorectal cancer mouse model. F344 rats were injected with DMH for 10 weeks, and probiotic was administered orally for 2 weeks before and until the end of the experiment. Results showed a significant decrease in cancer incidence (87.5%–25%). These findings suggested that Ren is a potential agent for colon cancer prevention (Zhang et al., 2015). Exopolysaccharides (EPS) from probiotic bacteria have health-promoting effects, including antitumor effect. One recent in vivo study has described the potential role of *L. acidophilus* (LA1); isolated breast-fed Egyptian infants (15–90 days) and its exopolysaccharides (EPS), with regard to antitumor activity, were evaluated against Ehrlich ascites carcinoma (EAC) cells. Swiss albino mice injected with tumor cells were treated with probiotics or its exopolysaccharides. The treatment reduced tumor volume, improved liver function, and decreased lactate dehydrogenase (LDH) and alkaline phosphatase (ALP) enzymes when compared with positive control. The strain also revealed antioxidant properties by suppression of malondialdehyde and nitric oxide serum levels (Abd El Ghany et al., 2015).

Another in vivo research showed the effect of probiotics on Lewis lung cancer (LLC)-bearing mice (C57BL/6J) as tumor models. Treatments were used in the three groups, i.e., cisplatin, cisplatin/ABX (an antibiotic cocktail of vancomycin, ampicillin, and neomycin, which disturbs intestinal microflora homeostasis), and cisplatin/*L. acidophilus* (probiotic). Results showed that there was a reduction in the tumor size and survival rates were increased. In cisplatin and cisplatin/ABX groups, lesser survival rates were observed, as compared with cisplatin/*L. acidophilus* group, which showed longer survival rates in mice. Moreover, the increased antitumor response in *Lactobacillus*-cotreated mice was correlated with interferon (IFN)-γ, Gzmb, and Prf1mRNA expression (Gui et al., 2015). Li et al. (2016) have demonstrated the role of novel probiotic mixture "Prohep" on hepatocellular carcinoma (HCC) cells. Prohep was composed of *L. rhamnosus* GG (LGG), viable nonpathogenic *Escherichia coli* Nissle 1917 (EcN), and heat-inactivated VSL#3 (1:1:1). VSL#3 was containing *Streptococcus thermophilus, B. breve, B. longum, B. infantis, L. acidophilus, L. plantarum, L. paracasei*, and *L. delbrueckii* subsp. C57BL6/N mice are fed with probiotic-containing diet 1 week before or on the same day as tumor injection. The results showed a reduction in tumor size and downregulated angiogenic factors in probiotic-treated mice. In another study, effect of *L. plantarum* BF-LP284 was demonstrated on sarcoma. BALB/c mice were injected with Meth-A tumor cells and treated with live or heat-killed bacteria. The results showed suppression of tumor growth, together with modulation of immune responses including cytokine production and increased population of antitumor immune cells, i.e., $CD3^+$ (Shin et al., 2016).

Another research has demonstrated the effect of LBB supplemented with oligofructose and maltodextrin on the composition of gut microbiota and colorectal cancer. The LBB includes *L. acidophilus, B. bifidum*, and *B. infantum*. Sprague-Dawley (SD) rats (different sets) were treated with ampicillin, carcinogen 1,2-dimethylhydrazine dihydrochloride (DMH), and probiotics for 23 weeks. Administration of LBB modulates the gut microbiota and reduces colon cancer development by decreasing tumor incidence (80% in GPC (LBB+ DMH) vs 100% in GCA, (DMH only—cancer control) $P=0.0001$), tumor volume (GPC 84.23 (42.75–188.4) mm (3) vs. GCA 243 (175.5–344.5) mm (3), $P<0.0001$) and multiplicity/count (GPC 2.92 ± 0.26 vs GCA 6.27 ± 0.41; $P<0.0001$). Such suppression of apoptosis and inflammation was associated with the improvement of intestinal integrity and mediated by toll-like receptor (TLR) 2 (Kuugbee et al., 2016). A study published by Gamallat et al. (2016) has described the effect of probiotic strain *L. rhamnosus* GG CGMCC 1.2134 (LGG) on colorectal cancer. The effect was revealed on nuclear factor kappa B (NFκB) inflammatory pathway and apoptosis. Sixty-four Sprague-Dawley (SD) rats were grouped into four as follows: group 1 (healthy control), group 2 (LGG), group 3 (dimethylhydrazine (DMH) cancer control), and group 4 (LGG+DMH). LGG was orally administered by groups 2 and 4. The results displayed the reduction of tumor incidence, multiplicity, and volume. In addition, with *L. rhamnosus* GG treatment, a decreased expression of β-catenin; the inflammatory proteins NFκB-p65, COX-2, and TNFα; and Bcl-2 (antiapoptotic protein) was observed. Moreover, the expression of the proapoptotic proteins Bax, p53, and casp3 was found to be increased as compared with cancer control group.

Similarly, another study has demonstrated the effects of potential probiotics, namely, *L. plantarum* (AdF10) and *L. rhamnosus* GG in colon carcinogenesis in rats. The Sprague-Dawley (SD) rats were divided into six groups, namely, normal control, *L. plantarum* (AdF10) treated, *L. rhamnosus* GG (LGG) treated, DMH treated (cancer control), AdF10+DMH treated, and LGG+DMH treated. Results showed a reduction of tumor incidence, multiplicity, and volume. In addition, cyclooxygenase (COX)-2 protein expression and total sialic acid (TSA) level were found to be decreased after probiotic treatment. Therefore, treatment with probiotics has the potential of providing protection against colon cancer by suppressing the COX-2 expression as one of the protective mechanisms (Walia et al., 2015). Irecta-Nájera et al. (2017) have described the effect of *L. casei* ATCC 393 on colon carcinogenesis in mouse. BALB/c mice were administered 1,2-dimethylhydrazine dihydrochloride (DMH) to induce colon cancer. *Lactobacillus casei* was given orally (10^6 CFU, twice a week), 2 weeks before DMH administration. The experimental data showed that the protective action of the *L. casei* and its antimutagenic activity may be associated with the maintenance of polyamine metabolism. Recently, a study for the first time described the in vitro and in vivo effect of probiotic combinations using live bacterial cells of *L. acidophilus* ATCC 314 mixed with the *L. fermentum* NCIMB 5221. The superior synergistic activity of the two probiotic strains has been reported against $Apc^{Min/+}$ colorectal cancer cell (CRC) mouse model through the downregulation of important tumor proliferation markers. In addition, anticancer activity of La-Lf coculture was significantly enhanced in vitro with reduced proliferation (38.8±6.9%, $P=0.009$) and increased apoptosis (413 RUL, $P<0.001$) toward cancer cells and significant protection of normal colon cell growth from toxic treatment (18.6±9.8%, $P=0.001$) (Kahouli et al., 2017).

4.2.2 *Lactococcus*

Lactococcus lactis is a food bacterium that has been used for centuries in the fermentation of food, especially yogurt, cheese, and sauerkraut. It is also used for the production of industrially important products with potentially great bioeconomic value. Similar to this, antiproliferative effects of the cytoplasmic fraction of *L. lactis* ssp. *lactis* (L. lac CF) were demonstrated on the SNU-1 human stomach cancer cell line. The results showed a reduction in the proliferation of SNU-1 cancer cell line, and it attributed to the induction of G0/G1 cell cycle arrest and apoptosis (Kim et al., 2009). Another study demonstrated the anticancer effects of *L. lactis* KC24 on various cancer cell lines, i.e., lung carcinoma (SK-MES-1), colon carcinoma (HT-29 and LoVo), and breast carcinoma (AGS and MCF-7). Results demonstrated antiproliferative effect by means of 10^6 CFU/well of *L. lactis* KC24 against all cancer cell lines using 3-(4,5-dimethylthiazol-2-yl)-2,5-diphenyltetrazolium bromide (MTT) assay. Cell line proliferation of

MCF-7, AGS, SK-MES-1, HT-29, and LoVo was inhibited by 91.89%, 90.12%, 86.53%, 68.30%, and 67.27%, respectively (Lee et al., 2015).

Another probiotic strain, whose effect has been evaluated against various cell lines, namely, SK-MES-1 (human lung carcinoma cell line; KCLB 30058), DLD-1 (human colon adenocarcinoma cell line; KCLB 30058), HT-29 (human colon adenocarcinoma cell line; KCLB 30038), LoVo (human colon adenocarcinoma cell line; KCLB 10229), AGS (human stomach adenocarcinoma cell line; KCLB 21739), and MCF-7 (human breast adenocarcinoma cell line; KCLB 30022), was *L. lactis* NK34. The results revealed the anticancer and anti-inflammatory activity of this strain. Cytotoxicity of the probiotic strain was observed in normal and cancer cells using MTT assay and also through morphology observation. In this study, 10^6 CFU/well of *L. lactis* NK34 was used for the treatment of cancer cells that resulted in strong inhibition of proliferation, i.e., MCF-7 (97.99%), LoVo (97.64%), HT-29 (97.05%), SK-MES-1 (96.71%), AGS (82.07%), and DLD-1(77.23%). In addition, the anti-inflammatory effect of *L. lactis* NK34 revealed a decrease in NO production and pro-inflammatory cytokines (Han et al., 2015). One research demonstrated the isolation, molecular identification, and probiotic characterization of *Lactobacillus* and *Lactococcus* strains from traditional Iranian dairy products and their cytotoxic effects against human cancer cell lines. Suspension of bacterial cell cultures *L. lactis* subsp. *lactis* 44Lac *and L. plantarum* 15HN was used for the cytotoxicity assessments against different cancer cell lines, i.e., colon cancer (HT29), cervical cancer (HeLa), gastric cancer (AGS), and breast cancer (MCF-7), and a normal umbilical vein endothelial cell (i.e., HUVEC). The authors reported that *L. plantarum* exhibited no significant anticancer effects on human cancer cells, whereas *L. lactis* subsp. *lactis* 44Lac has a significant anticancer activity ($P \leq 0.05$) against all of the examined cell lines. It has been stated that apoptosis is the main cytotoxic mechanism for secreted metabolites of *L. lactis* subsp. *lactis* 44Lac (Haghshenas et al., 2014). In further support, recent research has shown that exopolysaccharide (EPS) from *L. lactis* subsp. *lactis* affected the production of inflammatory cytokines. Results also showed significantly increased production of tumor necrosis factor-alpha and inducible NO synthase release in MCF-7 cells (breast cancer cells). This supports the use of *L. lactis* as potential anticancer agent (Wu et al., 2016).

4.2.3 *Bifidobacterium*

Generally, *Bifidobacterium* constitutes the dominant population of the GI tract of human, but in breast-fed infants, they constitute 90% of the total bacterial population (Hidalgo-Cantabrana et al., 2017). Bifidobacteria belong to the symbionts particularly important to the human and animal organism. They are some of the first inhabitants of the digestive tract of the newborn mammals. One study revealed the effect of *B. lactis*, carbohydrate "resistant starch" (RS), and their combination (synbiotic) on their ability to protect against colorectal cancer (CRC). For the research, Sprague-Dawley (SD)-azoxymethane rats were used, and results showed that rats fed with RS in combination with *B. lactis* showed significantly lower incidence and multiplicity of colonic neoplasms ($P < 0.01$) by >50% compared with the control group. The effect was due to apoptosis of the tumor cells at the time of the beginning of the cancer (Le Leu et al., 2010).

Sivan et al. (2015) have demonstrated the antitumor effect of the oral administration of *Bifidobacterium* cocktail to a mice bearing melanoma with and without programmed cell death ligand-1 (PD-L1) blockade immunotherapy (check point blockade). The authors reported a decrease in tumor size when treated with probiotic alone, compared with immunotherapy. In addition, two treatments together almost stopped the tumor outgrowth. With *Bifidobacterium* treatment, improvement was seen in immune responses, including CD8+ T-cell activation and costimulation, cytokine-cytokine receptor interaction, augmented dendritic cell (DC) function, and the chemokine-mediated recruitment of immune cells to the tumor microenvironment.

Li et al. (2012) have demonstrated the use of the recombinant *B. infantis*. For this research, *B. infantis*-mediated soluble kinase insert domain receptor (sKDR) prokaryotic expression system was constructed, and three LLC mice model groups were used. Group a is treated with saline, group b with recombinant *B. infantis* containing pTRKH2-PsT plasmid, and group c with recombinant *B. infantis* containing pTRKH2-PsT/sKDR plasmid. The better quality of life was recorded in mice group c than in the other two groups. The recombinant *B. infantis* containing pTRKH2-PsT/sKDR plasmid improved the efficacy of tumor growth suppression by increasing the necrosis rate of the tumor and prolongation of survival time of the LLC C57BL/6 mice. The anti-angiogenesis effect was also evaluated by MTT assay in vitro.

In another study, *B. infantis*-mediated sFlt-1 gene transferring system was constructed using electroporation, and antitumor effect was investigated on Lewis lung cancer (LLC) in mice (Zhu et al., 2011). Soluble fms-like tyrosine kinase receptor (sFlt-1) is a soluble form of extra membrane part of vascular endothelial growth factor receptor-1 (VEGFR-1) that has antitumor effects. The gene transferring system successfully expressed sFlt-1 at the levels of gene and protein. This system significantly inhibited the growth of human umbilical vein endothelial cells induced by VEGF in vitro. In addition, tumor growth was inhibited, and prolonged survival time of LLC C57BL/6 mice was recorded. These findings suggested that *B. infantis*-mediated sFlt-1 gene transferring system presented a promising therapeutic approach for the treatment of lung cancer (Zhu et al., 2011). Furthermore, anticancer effects of some different probiotic strains, i.e., *Enterococcus lactis* IW5, *Bacillus polyfermenticus* KU3, *Kluyveromyces marxianus* YAS, *Pediococcus pentosaceus* GS4, *E. faecalis*, and *S. hominis*, have been described suitably in recent literature (So et al., 2017). Recently, one research has established the proapoptotic and antiproliferative effects of *Leuconostoc mesenteroides*-derived anticancer pharmaceuticals on colon adenocarcinoma cells (HT-29) for the first time (Zununi et al., 2017).

4.3 NUTRITION, DIET, AND RISK ASSESSMENT

It is well established that diet and nutrition constituents affect the incidence, progression, and prevention of cancer. Certainly, multiple reasons, namely, imbalanced diets, consumption of tobacco and alcohol, decreased physical activity and exposure to environmental toxicants, poor education, low income, and socioeconomic status, are considered for the cancer incidence and progression. Dietary intervention tools may play an emerging and safe strategy in cancer prevention. Management of the gut microbiota by diet alteration appears to be the most effective and economical way to reduce the cancer risk. However, this area of knowledge is in the early stages, and there are still many questions to answer. Recently, a systematic review reported the available data on the efficacy and safety of people with cancer using probiotics (Redman et al., 2014). Another article has thrown light on the safety concerns related to probiotics and cancer (Dasari et al., 2016). In addition, a survey was conducted at the outpatient department of the National Cancer Institute in Slovakia about probiotic use and the association with patient tumor characteristics in cancer-treated patients. In total, 499 patients (323 women, 64.7%, and 176 men, 35.3%) were included in the presented survey. 91.6% were undertaking chemotherapy (2.6% together with radiotherapy), and 8.4% had no anticancer therapy. The most common cancers included in the study were gastrointestinal tumors (26.1%), breast cancer (19.1%), and lymphomas (14.4%). The evaluation was performed by questionnaire form, including the length and method of use relative to anticancer therapy, expectations, side-effect experiences, understanding of the possible risks, dietary supplement use, and others regarding their overall experience with probiotics. The pertinent data were statistically evaluated. However, as mentioned, 86.6% of patients declared no knowledge of possible risks (Ciernikova et al., 2017). Therefore, more research and human clinical trials are required, including sophisticated immunomonitoring approaches for better conclusion.

4.4 CONCLUSION AND FUTURE PERSPECTIVES

A couple of studies described above corroborate that the role of specific probiotic bacteria and their metabolites as therapeutic agents against cancer opens up an entirely new approach to the understanding and treatment of cancer disease. Alterations in the microbiota may affect oncogenesis and tumor progression at multiple levels. In future, the possibility of engineered probiotic microorganisms may offer new prospects for their prophylactic and therapeutic use in the prevention and treatment of cancer and other diseases. Therefore, productive research on genetic manipulation is prerequisite to improve the existing features of particular probiotic strains and to prepare novel engineered probiotic organisms with specific properties for treatment and prophylaxis of various cancers.

References

Abd El Ghany, K., Hamouda, R., Abd Elhafez, E., Mahrous, H., Salem-Bekhit, M., Hamza, H.A., 2015. A potential role of *Lactobacillus acidophilus* LA1 and its exopolysaccharides on cancer cells in male albino mice. Biotechnol. Biotechnol. Equip. 29, 977–983.

Aragón, F., Carino, S., Perdigón, G., de Moreno de LeBlanc, A., 2015. Inhibition of growth and metastasis of breast cancer in mice by milk fermented with *Lactobacillus casei* CRL 431. J. Immunother. 38, 185–196.

Banna, G.L., Torino, F., Marletta, F., Santagati, M., Salemi, R., et al., 2017. *Lactobacillus rhamnosus* GG: an overview to explore the rationale of its use in cancer. Front. Pharmacol. 8, 603.

Bharti, V., Mehta, A.S.S., Jain, N., Ahirwal, L., 2015. Cytotoxicity of live whole cell, heat killed cell and cell free extract of *Lactobacillus* strain in U-87 human glioblastoma cell line and MCF-7 breast cancer cell line. Int. J. Probiotics Prebiotics 10, 153–158.

Blaser, M.J., 2014. The microbiome revolution. J. Clin. Invest. 124, 4162–4165.

Bray, F., Znaor, A., Cueva, P., Korir, A., Swaminathan, R., et al., 2015. Planning and Developing Population-Based Cancer Registration in Low- and Middle Income Settings. IARC Technical Publication No. 43. IARC, Lyon.

Chen, Z.-Y., Hsieh, Y.-M., Huang, C.-C., Tsai, C.-C., 2017. Inhibitory effects of probiotic *Lactobacillus* on the growth of human colonic carcinoma cell line HT-29. Molecules 22, 107.

Ciernikova, S., Mego, M., Semanova, M., Wachsmannova, L., Adamcikova, Z., et al., 2017. Probiotic survey in cancer patients treated in the outpatient Department in a Comprehensive Cancer Center. Integr. Cancer Ther. 16 (2), 188–195.

Dasari, S., Kathera, C., Janardhan, A., Kumar, A.P., Viswanath, B., 2016. Surfacing role of probiotics in cancer prophylaxis and therapy: a systematic review. Clin. Nutr. 2016, 1–6.

de Moreno de LeBlanc, A., 2014. The administration of probiotics and fermented products containing lactic acid bacteria exert beneficial effects against intestinal and non-intestinal cancers. J. Food Nutr. Disord., S1–S005.

Esfandiary, A., Taherian-Esfahani, Z., Abedin-Do, A., Mirfakhraie, R., Shirzad, M., Ghafouri-Fard, S., Motevaseli, E., 2016. Lactobacilli modulate hypoxia-inducible factor (HIF)-1 regulatory pathway in triple negative breast cancer cell line. Cell J. 18 (2), 237–244.

Gamallat, Y., Meyiah, A., Kuugbee, E.D., Hago, A.M., Chiwala, G., et al., 2016. *Lactobacillus rhamnosus* induced epithelial cell apoptosis, ameliorates inflammation and prevents colon cancer development in an animal model. Biomed Pharmacother 83, 536–541.

Garrett, W.S., 2015. Cancer and the microbiota. Science 348 (6230), 80–86.

Goulet, O., 2015. Potential role of the intestinal microbiota in programming health and disease. Nutr. Rev. 73, 32–40.

Gui, Q.-F., Lu, H.-F., Zhang, C.-X., Xu, Z.-R., Yang, Y.-M., 2015. Well-balanced commensal microbiota contributes to anti-cancer response in a lung cancer mouse model. Genet. Mol. Res. 14 (2), 5642–5651.

Haghshenas, B., Abdullah, N., Nami, Y., Radiah, D., Rosli, R., et al., 2014. Different effects of two newly-isolated probiotic *Lactobacillus plantarum* 15HN and *Lactococcus lactis* subsp. *lactis* 44Lac strains from traditional dairy products on cancer cell lines. Anaerobe 30, 51–59.

Han, K., Lee, N., Park, H., Paik, H., 2015. Anticancer and anti-inflammatory activity of probiotic *Lactococcus lactis* NK34. J. Microbiol. Biotechnol. 25, 1697–1701.

Hidalgo-Cantabrana, C., Delgado, S., Ruiz, L., Ruas-Madiedo, P., Sánchez, B., Margolles, A., 2017. Bifidobacteria and their health-promoting effects. Microbiol. Spectr. 5 (3). BAD-0010-2016.

Hill, C., Guarner, F., Reid, G., Gibson, G.R., Merenstein, D.J., et al., 2014. Expert consensus document: the International Scientific Association for Probiotics and Prebiotics consensus statement on the scope and appropriate use of the term probiotic. Nat. Rev. Gastroenterol. Hepatol. 11, 506–514.

Hu, J., Wang, C., Ye, L., Yang, W., Huang, H., et al., 2015. Anti-tumour immune effect of oral administration of *Lactobacillus plantarum* to CT26 tumour-bearing mice. J. Biosci. 40, 269–279.

Irecta-Nájera, C.A., del Rosario Huizar-López, M., Casas-Solís, J., Castro-Félix, P., Santerre, A., 2017. Protective effect of *Lactobacillus casei* on DMH-induced colon carcinogenesis in mice. Probiotics Antimicrob. Proteins 9 (2), 163–171.

Jacouton, E., Chain, F., Sokol, H., Langella, P., Bermúdez-Humarán, L.G., 2017. Probiotic strain *Lactobacillus casei* BL23 prevents colitis-associated colorectal cancer. Front. Immunol. 8, 1553.

Kahouli, I., Malhotra, M., Westfall, S., Alaoui-Jamali Moulay, A., Satya, P., 2017. Design and validation of an orally administrated active *L. fermentum-L. acidophilus* probiotic formulation using colorectal cancer Apc Min/+ mouse model. Appl. Microbiol. Biotechnol. 101, 1999–2019.

Kassayová, M., Bobrov, N., Strojny´, L., Orendáš, P., Demečková, V., et al., 2016. Anticancer and immunomodulatory effects of *Lactobacillus plantarum* LS/07, inulin and melatonin in NMU induced rat model of breast cancer. Anticancer Res. 36, 2719–2728.

Kelly, D.L., Lyon, D.E., Yoon, S.L., Horgas, A.L., 2016. The microbiome and cancer: implications for oncology nursing science. Cancer Nurs. 39 (3), 56–62.

Kim, S.Y., Kim, J.E., Lee, K.W., Lee, H.J., 2009. *Lactococcus lactis* ssp. *lactis* inhibits the proliferation of SNU-1 human stomach cancer cells through induction of G0/G1 cell cycle arrest and apoptosis via p53 and p21 expression. Ann. N. Y. Acad. Sci. 1171, 270–275.

Konishi, H., Fujiya, M., Tanaka, H., Ueno, N., Moriichi, K., et al., 2016. Probiotic-derived ferrichrome inhibits colon cancer progression via JNK mediated apoptosis. Nat. Commun. 7, 12365.

Kuugbee, E.D., Shang, X., Gamallat, Y., Bamba, D., Awadasseid, A., et al., 2016. Structural change in microbiota by a probiotic cocktail enhances the gut barrier and reduces cancer via TLR2 signaling in a rat model of colon cancer. Dig. Dis. Sci. 61, 2908–2920.

Kverka, M., Tlaskalova-Hogenova, H., 2017. Intestinal microbiota: facts and fiction. Dig. Dis. 35, 139–147.

Le Leu, R.K., Hu, Y., Brown, I.L., Woodman, R.J., Young, G.P., 2010. Synbiotic intervention of *Bifidobacterium lactis* and resistant starch protects against colorectal cancer development in rats. Carcinogenesis 31, 246–251.

Lee, N.-K., Han, K.J., Son, S.-H., Eom, S.J., Lee, S.-K., Paik, H.-D., 2015. Multifunctional effect of probiotic *Lactococcus lactis* KC24 isolated from kimchi. LWT Food Sci. Technol. 64, 1036–1041.

Lenoir, M., del Carmen, S., Cortes-Perez, N.G., Lozano-Ojalvo, D., Muñoz-Provencio, D., et al., 2016. *Lactobacillus casei* BL23 regulates Treg and Th17 T-cell populations and reduces DMH-associated colorectal cancer. J. Gastroenterol. 51 (9), 862–873.

Li, Z.-J., Zhu, H., Ma, B.-Y., Zhao, F., Mao, S.-H., et al., 2012. Inhibitory effect of *Bifidobacterium infantis* mediated sKDR prokaryotic expression system on angiogenesis and growth of Lewis lung cancer in mice. BMC Cancer 12, 155.

Li, J., Sung, C.Y.J., Lee, N., Ni, Y., Pihlajamäki, J., Panagiotou, G., El-Nezami, H., 2016. Probiotics modulated gut microbiota suppresses hepatocellular carcinoma growth in mice. Proc. Natl. Acad. Sci. U. S. A. 113 (9), 1306–1315.

Motevaseli, E., Azam, R., Akrami, S.M., Mazlomy, M., Saffari, M., Modarressi, M.H., et al., 2016. The effect of *Lactobacillus crispatus* and *Lactobacillus* rhamnosus culture supernatants on expression of autophagy genes and HPV *E6* and *E7* oncogenes in the HeLa cell line. Cell J. (Yakhteh) 17 (4), 601–607.

Nouri, Z., Karami, F., Neyazi, N., Modarressi, M.H., Karimi, R., et al., 2016. Dual anti-metastatic and anti-proliferative activity assessment of two probiotics on HeLa and HT-29 cell lines. Cell J. (Yakhteh) 18 (2), 127–134.

Redman, M.G., Ward, E.J., Phillips, R.S., 2014. The efficacy and safety of probiotics in people with cancer: a systematic review. Ann. Oncol. 25, 1919–1929.

Shin, R., Itoh, Y., Kataoka, M., Iino-Miura, S., Miura, R., et al., 2016. Antitumor activity of heat-killed *Lactobacillus plantarum* BF-LP284 on meth-a tumor cells in BALB/c mice. Int. J. Food Sci. Nutr. 67, 641–649.

Sivan, A., Corrales, L., Hubert, N., et al., 2015. Commensal *Bifidobacterium* promotes antitumor immunity and facilitates anti PD-L1 efficacy. Science 350, 1084–1089.

So, S.S., Wan, M.L., El-Nezami, H., 2017. Probiotics-mediated suppression of cancer. Curr. Opin. Oncol. 29 (1), 62–72.

Tian, T., Olson, S., Whitacre, J.M., Harding, A., 2011. The origins of cancer robustness and evolvability. Integr. Biol. 3, 17–30.

Walia, S., Kamal, R., Kanwar, S., Dhawan, D., 2015. Cyclooxygenase as a target in chemoprevention by probiotics during 1,2-dimethylhydrazine induced colon carcinogenesis in rats. Nutr. Cancer 67, 603–611.

World Health Organization (WHO), 2018. Cancer. . (Accessed 9 March 2018).

Wu, Z., Wang, G., Pan, D., Guo, Y., Zeng, X., et al., 2016. Inflammation-related pro-apoptotic activity of exopolysaccharides isolated from *Lactococcus lactis* subsp. *lactis*. Benefic. Microbes 7 (5), 761–768.

Zhang, M., Fan, X., Fang, B., Zhu, C., Zhu, J., Ren, F., 2015. Effects of *Lactobacillus salivarius* Ren on cancer prevention and intestinal microbiota in 1, 2-dimethylhydrazine-induced rat model. J. Microbiol. 53, 398–405.

Zhu, H., Li, Z., Mao, S., Ma, B., Zhou, S., et al., 2011. Antitumor effect of sFlt-1 gene therapy system mediated by *Bifidobacterium Infantis* on Lewis lung cancer in mice. Cancer Gene Ther. 18, 884–896.

Zitvogel, L., Pietrocola, F., Kroemer, G., 2017. Nutrition, inflammation and cancer. Nat. Immunol. 18 (8), 843–850.

Zununi, V.S., Barzegari, A., Rahbar Saadat, Y., Goreyshi, A., Omidi, Y., 2017. *Leuconostoc mesenteroides*-derived anticancer pharmaceuticals hinder inflammation and cell survival in colon cancer cells by modulating NF-κB/AKT/PTEN/MAPK pathways. Biomed Pharmacother 94, 1094–1100.

CHAPTER

5

Recent Developments in the Prevention of Obesity by Using Microorganisms

Muni Swamy Ganjayi, Meriga Balaji, Dasari Sreenivasulu, Hari Balaji, Reddy Sankaran Karunakaran

Department of Biochemistry, Sri Venkateswara University, Tirupati, India

OUTLINE

5.1 INTRODUCTION

5.1.1 Human Microbial World

The human cells of the body and skin are outnumbered by microorganisms present in/on our body. Microbial genome, acquired beginning at birth, evolved with humans helping for normal growth and development and molding the immune system. According to

https://doi.org/10.1016/B978-0-12-816328-3.00005-2

the hologenome concept, evolution in humans can be linked to its symbiotic microbiota. The human gut microbiota is a complex ecosystem that includes majorly bacteria (over 10^{14}, about 1.5 kg, >1000 species, and predominantly anaerobes) and other microbes such as fungi, parasites, viruses, and archaea (Sommer and Backhed, 2013). More convincing evidences suggest that modern trends, which include cesarean deliveries, food habits, obsession with cleanliness, antibiotics, and work habits, are disrupting this delicate balance, contributing to some of the most severe ailments like allergies, asthma, diabetes, obesity, autoimmune diseases, and cancer. Apart from several human diseases caused by microbes, human gut microbiota perform many beneficial functions to the host, which include regulating gut motility, producing vitamins, transforming bile acid and steroids, metabolizing xenobiotics, absorbing minerals, and activating toxins. According to Human Microbiome Project led by Lita M. Proctor at the National Institutes of Health, the "human supraorganism" is one term coined to describe the human host and all the attendant microorganisms. Researchers are trying to identify beneficial organisms as "probiotics" that people could take in or substances that people might ingest to nurture the beneficial microbes to keep up good health. Some drugs may mimic the helpful compounds that these organisms produce and contribute to health.

The concept that some people may be prone to some diseases could be because they may be missing certain microbes. Studies indicate that gut dwellers secrete messengers to cells lining the digestive tract to modulate key hormones, such as leptin and ghrelin, which are players in regulating hunger, satiety, and metabolism. "In recent decades, we may have a generation of children growing up without the proper bacteria for various reasons to regulate their leptin, ghrelin, or some other adipocytokines." Research reports show that manipulating gut microbiomes of mice influences their anxiety and activity (Heijtz et al., 2011). Clues also are emerging about how the gut microbes influence brain chemistry and alterations in corresponding behaviors such as stress, anxiety, and depression, via the vagus nerve. Studies on obese people's gut revealed a distinctive mix of digestive bacteria that make them prone to weight gain. Interestingly, thin mice get fatter when their microbiota is replaced with the microbes present in obese animals. Our ancient microbiome is gradually losing the equilibrium it used to have with the hosts and that has tremendous physiological changes (Sekirov et al., 2010). The major microbiota present in the gut (bacterial phyla) in the order of numerical importance is Firmicutes, Bacteroidetes, Actinobacteria, Proteobacteria, Verrucomicrobia, and Fusobacteria. Lactic acid bacteria and bifidobacteria are two important types of gut bacteria, which are present right from birth or acquired through digested food. *Lactobacillus* and *Leuconostoc* species are the main lactic acid bacteria found in the human intestine. *Bifidobacterium* species are the predominant bacteria among the first colonizers found in newborns, and in adults, their levels are low.

5.1.2 Obesity Scenario

Obesity is a global health problem, being considered as one of the most serious and prevalent noncommunicable diseases of the 21st century. Once considered a problem only in high-income countries, overweight and obesity are now dramatically on the rise in low- and middle-income countries, particularly in urban locales (WHO, 2018). Overweight/obesity is a gateway to many diseases and disabilities and even life-threatening conditions like type

2 diabetes, cardiovascular diseases, and several types of cancer (Swinburn et al., 2015). The worldwide prevalence of obesity nearly tripled between 1975 and 2016, and consequently, WHO has declared obesity as a current global epidemic. In 2016, about 1.9 billion adults were deemed to be overweight. This number corresponds to about 39% of the world population older than 18 years. About 13% of worldwide adult population (more women than men) corresponding to 600 million people is considered obese. The most distressing fact is that childhood obesity has increased exponentially during the last two decades, having increased from 32 million in 1990 to 41 million in 2016 globally (WHO, 2018). Considering the huge clinical impact and associated health-care cost, obesity treatment requires a multidisciplinary approach (Duranti et al., 2017).

Lifestyle intervention is the basis for the treatment of overweight and obesity, but it is difficult to implement for long years for various reasons. Obesity treatment appears a thankless job because the patient always requires for the stable result in a very short time period or is awaiting for the "miracle" of minus 20 kg in 2 weeks, and the lifelong change of the lifestyle is not his/her aim. However, there is always a group of "manageable" patients who change their lifestyle and can follow the "eat-smart strategies" lifelong. Weight-loss objectives should always be realistic, individualized, and aimed at the long term. The transdisciplinary approaches, including dietary, exercise, and behavior modifications, while also considering psychosocial characteristics of patients are of great help (Meriga et al., 2017; WHO, 2018). One should always remember that obesity is a chronic disease and a follow-up and continued supervision is necessary to prevent weight regain, to monitor disease risks, and to treat comorbidities.

5.1.3 Existing AntiObesity Drugs

The option of bariatric surgery is not widely followed owing to high risk and high costs. Although a few drugs approved by the FDA are available in the market, their usage is limited by significant side effects. Table 5.1 describes the existing antiobesity drugs, their mechanism of action, their adverse effects, and their current status (Meriga et al., 2017). Some of the recently developed drugs are in various stages of preclinical and clinical trials.

5.1.4 Basis of Obesity

Obesity is a multifactorial disease caused by genetic, metabolic, behavioral, and cultural factors. Obesity is no longer considered a single disease but is rather a condition that is associated with many diseases, collectively referred as metabolic syndrome (Balaji et al., 2016). As per the previous studies, "overeating of junk foods" is the major contributing factor (54%) for obesity. Other reasons cited are "physical inactivity" (31.7%), stress (8%), medication (3.65%), and belief in genetics (4.67%) (Fig. 5.1). In the past one decade, changed circadian patterns due to day and night work shifts, increased photoperiod and snacking frequency, shifting of dinners, lesser exposure to sunlight, and decreased physical activity due to increased comfort and electronic devices have collectively brought about various physiological changes that contributed to the development of obesity and metabolic syndrome (Meriga et al., 2017).

TABLE 5.1 Drugs That Have Been Used for Obesity Therapy and Their Current Status

Drug	Mechanism of Action	Adverse Effects	Current Status
Dinitrophenol	Uncouples oxidative phosphorylation in mitochondria	Cataracts, neuropathy, sensation of warmth, sweating	Withdrawn
Amphetamine	Increases the neurotransmitters dopamine, noradrenaline, and serotonin in the brain	Addiction, headache, nausea, dry mouth, nervousness, anxiety, hypertension, tachycardia	Banned as an antiobesity drug
Phentermine	Sympathomimetic amine	Headache, insomnia, irritability, palpitations, nervousness, and increased blood pressure	Short-term use (<1 week)
Diethylpropion	Sympathomimetic amine	Dry mouth, insomnia, palpitation, and headache	Short-term use (< 12 weeks)
Aminoxaphen (aminorex)	Indirect sympathomimetic action, pulmonary	Hypertension	Withdrawn
Phenylpropanol amine	Central α1-adrenergic receptor agonist	Hemorrhagic stroke, psychosis	Banned in the United States
Fenfluramine	Dexfenfluramine selective serotonin reuptake inhibitor	Pulmonary hypertension valvulopathy	Withdrawn
Sibutramine	Norepinephrine and serotonin reuptake inhibitor	Headache, insomnia, dry mouth, and constipation	Withdrawn
Rimonabant	Cannabinoid receptor antagonist	Depression, nausea, dizziness, arthralgia, and diarrhea	Withdrawn from Europe
Orlistat	Lipase inhibitor	Diarrhea, flatulence, bloating, abdominal pain, and dyspepsia	Marketed
Lorcaserin	Selective 5HT2C receptor agonist	Headache, dizziness, and nausea; difficult breathing; swelling of the face, lips, tongue, or throat	Initially approved, awaiting final approval
Qnexa (topiramate and phentermine)	Topiramate blocks voltage-dependent sodium channels, glutamate receptors, and carbonic anhydrase and augments GABA activity; phentermine is sympathomimetic	Diarrhea, possible harm to fetus, increased heart rate	Initially approved, awaiting final confirmation, marketed
Tensofesine	Norepinephrine, dopamine, and serotonin reuptake inhibition	Dry mouth, nausea, increased anger, and hostility	Not approved
Cetilistat	Lipase inhibitor	Diarrhea, flatulence, bloating, abdominal pain, fatty stool	Not approved
Contrave	Bupropion increases activity of proopiomelanocortin (POMC) neurons. Naltrexone blocks opioid receptors on the POMC neurons preventing feedback inhibition and increasing POMC activity	Nausea, constipation, headache, vomiting, dizziness, insomnia, dry mouth, and diarrhea	Still under investigation
Metformin	Works by suppressing glucose production in the liver, thereby decreasing blood glucose levels. It is thought to work in weight loss because of its appetite-suppressing properties	Abdominal or stomach discomfort, cough or hoarseness, diarrhea	Still under investigation

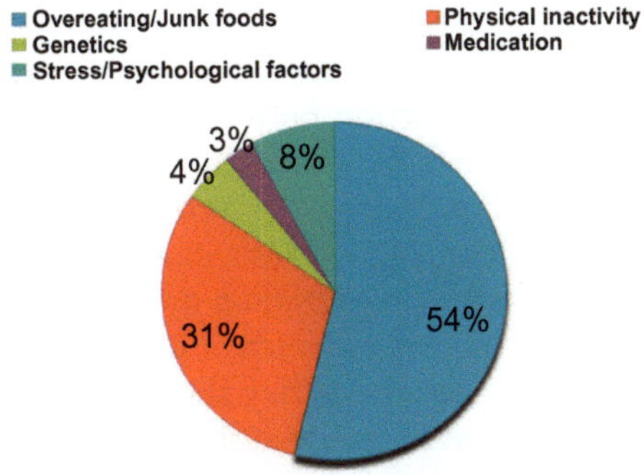

FIG. 5.1 Major factors causing obesity/overweight.

5.2 DIGESTION OF FAT FOODS

Dietary fat contributes a significant caloric value to our diet. Dietary lipids supply 45%–55% of the energy requirements in breastfed human infants (Meek et al., 2010) and 40%–55% of the calories in the Western diet (Meek et al., 2010). Recent studies indicate that fat digestion is a prerequisite for the effects of fat on gastric emptying, gastrointestinal hormone secretion, appetite, and energy intake (Little et al., 2007). An increasing number of gastrointestinal enzymes involved in nutrient digestion are being identified and characterized, representing a rich pool of potential therapeutic targets for obesity and other metabolic disorders (Birari and Bhutani, 2007). More specifically, enzymes such as lingual and gastric lipases, pancreatic lipase (PL), cholesterol-ester lipase, and bile-salt-stimulated lipase involved in the digestion of dietary fats were studied (Armand, 2007). Most of the dietary fat is ingested as triglycerides, and their hydrolysis starts in the mouth by lingual lipase followed by an acid-stable gastric lipase in the stomach and then in the duodenum through the synergistic actions of gastric and pancreatic lipases (PL), leading to the formation of free fatty acids (FFAs) and monoglycerides. These FFAs are absorbed by the enterocytes to synthesize new triglyceride molecules, so as to be transported to different organs via lipoproteins, especially chylomicrons (Armand, 2007). An improved understanding of factors controlling the dietary FA absorption and transport could lead to new approaches for decreasing energy intake in the context of obesity and increasing energy efficiency in the context of malnutrition.

5.3 GUT MICROBES: ROLE IN HUMAN HEALTH

Right from birth, the digestive tract is colonized by complex assemblages of microorganisms called gut microbiota that greatly influence intestinal and extraintestinal physiology of humans (Sekirov et al., 2010). The ability of the gut microbiota to modify dietary nutrient metabolism has emerged as a key feature of host-microbe relationships in the gut (Musso et al., 2011). Intestinal bacteria can digest dietary carbohydrates to produce more short-chain fatty acids (SCFAs) such as acetate, propionate, butyrate, and lactate and thus help in energy homeostasis in the host organism (Rios-Covian et al., 2016). Furthermore, such SCFAs play a role as metabolic regulators that can influence the expression of host genes in the intestine and facilitate the absorption of nutrients and differentiation of adipose tissue (Morrison and Preston, 2016). Of late, the research studies have concentrated on the consequences of disturbances in the flora or in the interaction between the flora and the host. For example, when many components of the normal flora are eliminated or suppressed by a course of broad-spectrum antibiotics or for other reasons, the stage is set for other organisms to step in and cause disorder (Rios et al., 2016). The classic example of this is antibiotic-associated diarrhea and its deadliest manifestation, *Clostridium difficile* colitis. Similar disturbances in the flora are thought to be involved in a severe form of intestinal inflammation in children. In other situations, bacteria may simply be where they should not be. For instance, if motility of the bowel is impaired or acid secretion from the stomach is much reduced, an environment favors the growth of organisms in the small intestine that are otherwise confined to the colon. As a result, the syndrome of small bowel bacterial overgrowth occurs. In other situations, damage to the intestinal epithelium renders the gut wall leaky and permits bacteria to gain access to the submucosal compartments or even to the systemic circulation and cause sepsis. This mechanism is thought to account for many of the infections that occur in critically ill patients in the intensive care unit.

Fecal microbiota transplant (FMT) is a new approach that involves administration of fecal microbiota collected from the stools of a healthy donor into the intestine of diseased recipient with an alteration in their normal gut microbiota through infusion (via enema, colonoscopy, or upper gastrointestinal tract) (Kelly et al., 2015). The ultimate objective of FMT is to treat the disease by restoring the microbial communities more typical of a healthy person. In recent years, quite a number of studies have focused on the comprehension and evaluation of the host-gut microbiota interactions, for greater understanding of the pathophysiology of several complex diseases. The concomitant implementation of metagenomic technology to analyze and characterize the composition of the intestinal microbiota has reinforced this field of investigation, with the hope that some form of modulation of the intestinal microbiota may represent a potential therapeutic basis for the disease. Thus, the gut microbiota composition is a contributing determining factor for an individual to the development of these disorders; therefore, altering a person's microbiota suitably may ameliorate disease.

5.3.1 Alterations in Gut Microbiota: Obesity

Basically, obesity results from an imbalance in the ratio between energy intake and energy consumption that eventually leads to storage of excess energy as triglycerides in white adipose tissue (WAT) (Balaji et al., 2016). In healthy, lean individuals, the typical metabolic tissues like WAT, liver and skeletal muscles are specially equipped to handle storage nutrients like triglycerides and glycogen. However, their storage capacity is limited, and when surpassed, excess nutrients will be channeled into nonmetabolic tissues and cells unable to handle them efficiently. The consequence will be that both metabolic and nonmetabolic tissues are exposed to nonphysiological levels of metabolic substrates that will affect a number of cell-intrinsic and cell-extrinsic processes. The human gut is considered as a bioreactor with a huge diversity of bacterial taxa, predominantly belonging to the Firmicutes and Bacteroidetes phyla (Ley et al., 2006), and shaped by different environmental parameters. Recent studies of obesity have focused on the role of intestinal symbiotic bacteria (Aggarwal et al., 2013). The human gastrointestinal tract contains at least 1013–1014 microbial cells consisting of up to

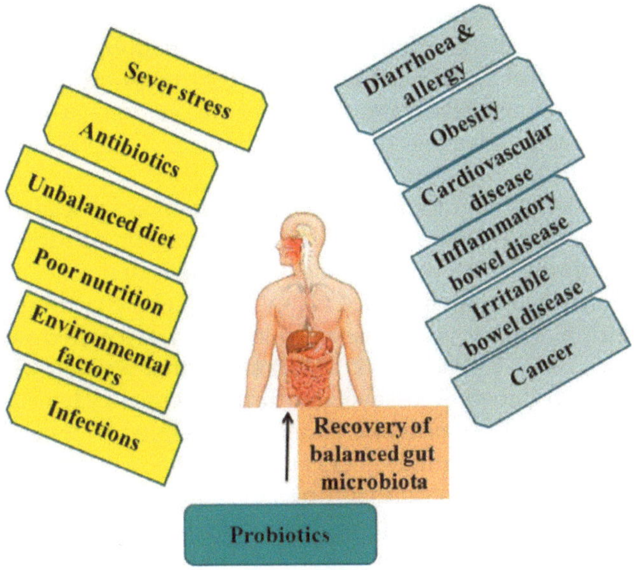

FIG. 5.2 Factors influencing gut microbiota and consequential diseases.

2000 species of bacteria, which is referred to as the gut microbiota. The genome size of the gut microbiota is >100 times greater than that of the human genome and encodes genes for many metabolic functions that are unique to microbes and are physiologically essential for human health (Fig. 5.2) (Xu et al., 2007).

The microbiota has been shown to stimulate hepatic triglyceride production through suppression of the lipoprotein lipase (LPL) inhibitor, fasting-induced adipose factor (also known as angiopoietin-like 4), thereby leading to continued expression of LPL, a key regulator of fatty acid release from triglycerides in the liver. The gut microbiota also can modulate systemic lipid metabolism through the modification of bile acid metabolic patterns, also impacting directly on the emulsification and absorption properties of bile acids and, thus, indirectly the storage of fatty acids in the liver. In the obese host, such cellular dysfunction leads to the activation of stress responses that can be monitored by the activation of signaling pathways that include the MAP kinases (extracellular signal-regulated kinases (ERKs) and c-Jun N-terminal kinase (JNK)) and inhibition of NF-kB kinase beta (IKKb) involved in the activation of the NF-kB pathway, the Ser/Thr kinase inositol-requiring protein-1 (IRE-1), and the mammalian target of rapamycin (mTOR) (Requena et al., 2013). Previous studies using experimental models like transgenic, knockout, and gnotobiotic animals and human studies provide evidence for the pivotal role of the intestinal microbiota in energy harvest and consequently obesity. More precisely, they show a potential link between obesity and alterations in the gut flora (Ley et al., 2006), the specific gut microbiome being associated with the obese phenotype. The gut microbiota facilitate the extraction of calories from ingested dietary foods, help to store these calories in host adipose tissue to be used later, and provide energy and nutrients for microbial growth and proliferation. Individual differences in energy usage and recovery may provide a physiological basis as to why some obese patients do not seem to overeat. Indeed, it has been suggested that a person's gut microbiota has a specific metabolic efficiency and that certain characteristics of the microbiota composition might predispose to obesity (Backhed et al., 2005). The composition and duration of diet intake directly infulences the composition and structure of gut microbiota. Recent studies with ketogenic diet reveled increased bacteriodes and decreased fermicutes, a condition seen in non obese subjects (David et al., 2014).

One of the major shifts in the gut microbiota composition is associated with the introduction of solid foods in infancy (Koenig et al., 2011). Transmission of bacteria from mothers to babies can take place through intrauterine and breast milk delivery (Jimenez et al., 2008). Therefore, breast milk acts as a source of bacteria and of oligosaccharides that further promote the colonization of bifidobacteria, which has been inversely associated with weight gain and inflammation (Isolauri, 2012). Collado et al. (2010) found that women who were overweight prior to pregnancy or who gained excess weight during pregnancy had lower concentrations of *Bifidobacterium*. Likewise, *Bifidobacterium* counts were higher in infants of normal-weight mothers than in infants of overweight mothers and in normal-weight children in comparison with those becoming obese. These studies address the concept that alterations in the gut microbiota at early-life stages could lead to long-term vulnerability to obesity.

5.4 THERAPIES FOR PREVENTION AND TREATMENT OF OBESITY

Unfortunately, current medical and lifestyle treatments for obesity have largely failed to achieve long-term success due to side effects and high cost. A major challenge is the need for lifelong adherence to maintain modest effects in the face of evolutionarily driven compensatory responses to weight loss induced by voluntary caloric restriction. Thus, alternative and novel strategies are urgently needed for long-term prevention and treatment of obesity. Nature represents an enormous reservoir of biologically active compounds to treat various ailments from times immemorial (Balaji et al., 2016). In view of the side effects encountered with the long-time usage of synthetic drugs and due to stringent guidelines to be fulfilled during drug approvals, plant-herbal drugs have gained much attention as a reliable option to clinical remedy, and the claim for these herbal remedies has greatly increased recently. A variety of phytochemicals such as polyphones, alkaloids, terpenoids, flavonoids, tannins, saponins, glycosides, steroids, and proteins present in plants and their products are key factors in the treatment of several disorders (Balaji et al., 2016). Our group has recently reported that natural compounds play crucial role in treating obesity and associated ailments. One such alternative is suggested by studies showing a reciprocal relationship between the composition of the gut microbiota and metabolic disorders (Balaji et al., 2017). In recent years, there is growing interest on the potential role of the gut microbiota in obesity research (Sonnenburg and Backhed, 2016). A discussion on available probiotics/microbes and their antiobesity activity is also presented in this text (Fig. 5.3). An obvious implication from the potential link between the gut microbiota and weight gain is that selective alterations in the gut microbiota might have a positive impact on the control of weight gain.

The microbiological options to bring about such alterations are by increasing selected microbial populations through the intake of probiotics and/or prebiotics and removing

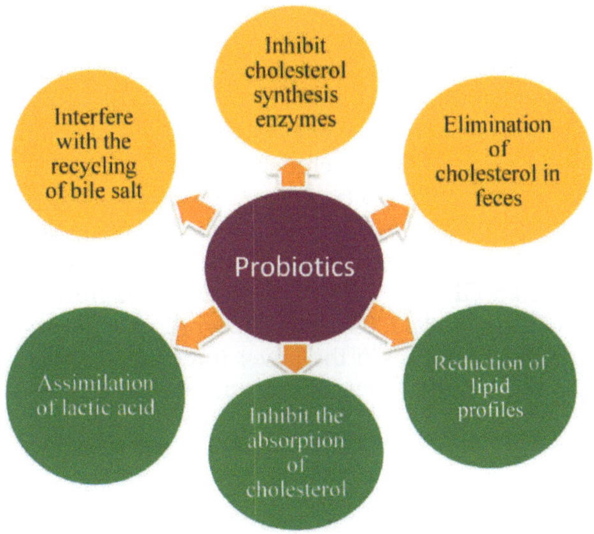

FIG. 5.3 Possible mechanism action of probiotics in obesity.

undesirable microbial populations by administering specific antimicrobial agents. In the latter case, antibiotics and bacteriocins have been investigated, while other options such as phage may ultimately be considered. It is important to note, however, that studies to date in which broad-spectrum antibiotics have been used to alter the composition of the obesity-associated gut microbiota are primarily "proof-of-concept" exercises and that, ultimately, the identification of clearer targets and the use of highly specific (narrow-spectrum) antimicrobials will be needed to minimize collateral damage and to prevent the spread of resistance to clinically important drugs (Requena et al., 2013).

As a consequence of the mentioned association between high proportions of Firmicutes and weight gain, the approach taken by Murphy et al. (2013) was to assess the relative impact of antimicrobials with anti-Firmicutes activity, that is, the antibiotic vancomycin or the bacteriocin-producing probiotic *Lactobacillus salivarius* UCC118, on weight gain using the murine model of diet-induced obesity (DIO). It was established that, over an 8-week intervention period, vancomycin administration did reduce weight gain in animals in receipt of a high-fat diet despite not impacting on caloric intake. Mice in receipt of vancomycin also had lower fasting blood glucose, plasma TNFα, and triglyceride levels compared with controls. In contrast, the bacteriocin-producing probiotic only briefly impacted on weight gain, when compared with the non-bacteriocin-producing control, and this difference was no longer significant by the end of the study (Murphy et al., 2013). It was also apparent that antimicrobial administration, of vancomycin in particular, had an impact on the gut microbiota of these animals (as assessed by high-throughput sequencing) that surpassed that predicted by traditional culture-based experiments. Thus, beneficial effects cannot be attributed specifically to anti-Firmicutes activities (Requena et al., 2013). In recent years, modulation of gut microbiota through the use of prebiotics and probiotics gathered much attention as a possible strategy for the prevention of weight gain in obese-prone subjects and a noninvasive treatment for severely obese individuals.

5.4.1 Role of Prebiotics

Prebiotics are a special form of dietary fiber that acts as a fertilizer for the good bacteria in your gut. Prebiotics are food components that confer health benefits to the host through the stimulation of growth and/or metabolism of beneficial bacteria including specific members of the gut microbiota. Prebiotics usually include carbohydrate forms that are not accessible to the enzymes produced by the human body (e.g., fructooligosaccharides (FOS), glucooligosaccharides, galactooligosaccharides (GOS), inulin, xylooligosaccharides, resistant starch, arabinoxylan and arabinogalactan, lactulose, and raffinose), which means that they reach the distal sections of human GIT still intact where they constitute fermentable substrates for intestinal bacteria (Gibson et al., 2004). In other words, prebiotics work as a nutritional tool to promote bacterial proliferation in the lower intestine; to induce modification of gut microbiota; and, thus, to counteract the fat mass accumulation and obesity ailments (Fig. 5.3) (Dewulf et al., 2011).

5.4.2 Role of Probiotics

Probiotics are health-promoting microorganisms, which, according to FAO/WHO definition, are "live microorganisms that after ingestion provide health benefits to the host maintaining the gut microbiota in correct equilibrium" (FAO/WHO), a definition recently updated as follows: "live microorganisms that, when administered in adequate amounts, confer a health benefit on the host" (Hill et al., 2014). The health-promoting effects exerted by probiotic bacteria are mediated by the interaction with other gut-associated microorganisms and with the host (Rowland et al., 2010). Reported health benefits encompass modulation of the immune response, maintenance of the intestinal barrier, antagonism of pathogen adhesion to host tissue, and production of different metabolites such as vitamins, SCFAs, and molecules that act as neurotransmitters involved in gut-brain axis communication (Turroni et al., 2014; Sanchez et al., 2017).

Currently, the large majority of commercially used probiotic bacteria belong to two genera, that is, *Bifidobacterium* and *Lactobacillus*, both of which are typical inhabitants of the human intestine. In a small number of cases, the role of particular *Lactobacillus* species on the modification of the body weight, both in animals and human beings, has been assessed (Million et al., 2012). However, the resulting data are rather contradictory and variable depending on the species and the strains involved. In this context, probiotic therapy based on L. *rhamnosus* GG (Kim et al., 2016) and L. *reuteri* (Fak and Backhed, 2012) was shown to result in weight increase perhaps because of their pro-inflammatory effects, while other *Lactobacillus* species such as L. *fermentum* and L. *gasseri* (Jung et al., 2013) were associated with weight loss. Finally, certain species, such as *L. sporogenes*, showed no significant effects on weight (Million et al., 2012). In a similar manner, species belonging to the *Bifidobacterium* genus, which represent microorganisms that are common inhabitants of mammalian GIT and are associated with conferring beneficial effects on the host, were shown to have antiobesity effects. For example, *Bifidobacterium pseudocatenulatum*, *B. longum*, and *B. adolescentis* have been shown to reduce body weight and blood serum levels (i.e., total cholesterol, HDL cholesterol, LDL cholesterol, triglyceride, glucose, leptin, and lipase levels) (An et al., 2011; Chen et al., 2012), and some of the probiotics and their metabolic activity are presented in Table 5.2.

TABLE 5.2 Probiotic Strains and Metabolic Alterations in Experimental Models

Probiotic Strain and Intervention	Experimental Models	Metabolic Observation	Reference
Lactobacillus plantarum DK211	Male rats	↓ Fat composition and serum lipids	Hong et al. (2015)
Streptococcus thermophilus, *L. delbrueckii* ssp. *bulgaricus*, and *L. gasseri* SBT2055 (1011 CFU/d), 4 weeks	Humans	↓ Postprandial serum, nonesterified fatty acid, and triacylglycerol levels	Ogawa et al. (2014)
Streptococcus thermophilus, *L. delbrueckii* ssp. *bulgaricus*, and *L. gasseri* SBT2055 (1011 CFU/d), 12 weeks	Humans	↓ Weight and waist-hip circumferences	Kadooka et al. (2013)
L. gasseri BNR17 (6×1010 CFU/d)	Humans	↓ Weight and waist-hip circumferences No change in glucose, insulin, and lipid profile	Jung et al. (2013)
Yogurt + *L. amylovorus* (1.39×10^9 CFU/d), yogurt + *L. fermentum* (1.08×10^9 CFU/d), 43 days	Humans	↓ Total fat mass with *L. amylovorus* ↓ Abundance of Clostridial cluster IV with *L. amylovorus*	Omar et al. (2013)
Lactobacillus paracasei ssp. *paracasei*	Humans	No change in glucose, insulin, lipid profile, hepatic parameters	Karlsson Videhult et al. (2015)
Lactobacillus plantarum	Humans	↓ Total cholesterol, glucose, and IL-6 levels	Guarner and Malangelada (2003)
Lactobacillus curvatus HY7601 and *L. plantarum* KY1032	HFD/placebo or normal chow for 8 weeks	↓ Body weight gain and fat accumulation plasma insulin, leptin, total cholesterol, and liver toxicity biomarkers ↓ Pro-inflammatory genes (TNFα, IL6, IL1β, and MCP1) in adipose tissue ↓ Fatty acid oxidation-related genes (PGC1α, CPT1, CPT2, and ACOX1) in the liver	Park et al. (2013)
Fourteen alive probiotic strains (*Lactobacillus*, *Lactococcus*, *Bifidobacterium*, *Propionibacterium*, *Acetobacter*)	Monosodium glutamate (MSG)-induced obesity/ placebo (vehicle)	↓ Body weight gain and visceral fat accumulation ↓ Liver lipid contents (protection from NAFLD development) ↓ Serum cholesterol, triglyceride, glucose, insulin, and leptin levels ↑ Insulin sensitivity (decreased HOMA-IR, increased adipocytokine)	Savcheniuk et al. (2014)
Lactobacillus paracasei CNCM I-4270, *L. rhamnosus* I-3690, or *Bifidobacterium animalis* subsp. *lactis* I-2494	HFD/placebo or normal chow	↓Body weight gain macrophage infiltration into epididymal adipose tissue ↓ Hepatic steatosis ↑ Glucose-insulin homeostasis Strain-specific attenuation of obesity comorbidities through impacts on MS-associated phylotypes of gut microbiota in mice	Wang et al. (2015)
Lactobacillus acidophilus, *L. bulgaricus*, *L. bifidum*, and *L. casei*	Randomized, single-blinded, placebo-controlled trial	↓ Triglyceride ↓ Malondialdehyde (MDA) ↓ IL-6 and ↓ insulin resistance All changes not statistically significant	Mazloom et al. (2013)

5.5 CONCLUSION

In order to overcome obesity and metabolic disorders, many dietary strategies have been put forwarded for effective therapeutics. However, most of these fail to maintain a long-term effect in reducing body weight and regulation of metabolic pathways. Alteration of microbiota has been suggested to have a significant impact on this. The use of probiotics for the manipulation of the gut microbiota as a therapeutic option for obesity/overweight is likely to be more complex than anticipated and may require long-term treatment and multiple follow-up evaluations over time. The effective dose and feasibility of using probiotics as an antiobesity treatment should be determined in future experimental trials. With the wide range of fermentable substrates that can be used to produce products containing different strains, the future development of new complementary dietetic probiotics will likely represent a crucial contribution to efforts to stem the worldwide increase in obesity. Research into the development of genetically modified gut microbial flora may pave the way for discovering more effective means to treat obesity ailments.

References

Aggarwal, J., Swami, G., Kumar, M., 2013. Probiotics and their effects on metabolic diseases: an update. J. Clin. Diagn. Res. 7, 173–177.

An, H.M., Park, S.Y., Lee, D.K., Kim, J.R., Cha, M.K., et al., 2011. Antiobesity and lipid-lowering effects of Bifidobacterium spp. in high fat diet-induced obese rats. Lipids Health Dis. 10, 116.

Armand, M., 2007. Lipases and lipolysis in the human digestive tract: where do we stand? Curr. Opin. Clin. Nutr. Metab. Care 10, 156–164.

Backhed, F., Ley, R.E., Sonnenburg, J.L., Peterson, D.A., Gordon, J.I., 2005. Host bacterial mutualism in the human intestine. Science 307 (5717), 1915–1920.

Balaji, M., Ganjayi, M.S., Kumar, G.E., Parim, B.N., Mopuri, R., Dasari, S., 2016. A review on possible therapeutic targets to contain obesity: the role of phytochemicals. Obes. Res. Clin. Pract. 10 (4), 363–380.

Birari, R., Bhutani, K., 2007. Pancreatic lipase inhibitors from natural sources: unexplored potential. Drug Discov. Today 12, 879–889.

Chen, J., Wang, R., Li, X.F., Wang, R.L., 2012. *Bifidobacterium adolescentis* supplementation ameliorates visceral fat accumulation and insulin sensitivity in an experimental model of the metabolic syndrome. Br. J. Nutr. 107 (10), 1429–1434.

Collado, M.C., Isolauri, E., Laitinen, K., Salminen, S., 2010. Effect of mother's weight on infant's microbiota acquisition, composition, and activity during early infancy: a prospective follow-up study initiated in early pregnancy. Am. J. Clin. Nutr. 92, 1023–1030.

David, L.A., Maurice, C.F., Carmody, R.N., Gootenberg, D.B., Button, J.E., Wolfe, B.E., et al., 2014. Diet rapidly and reproducibly alters the human gut microbiome. Nature 505 (7484), 559.

Dewulf, E.M., Cani, P.D., Neyrinck, A.M., Possemiers, S., Van Holle, A., et al., 2011. Inulin-type fructans with prebiotic properties counteract GPR43 overexpression and PPAR gamma-related adipogenesis in the white adipose tissue of high-fat diet-fed mice. J. Nutr. Biochem. 22 (8), 712–722.

Duranti, S., Ferrario, C., van Sinderen, D., Ventura, M., Turroni, F., 2017. Obesity and microbiota: an example of an intricate relationship. Genes Nutr. 12 (1), 18.

Fak, F., Backhed, F., 2012. *Lactobacillus reuteri* prevents diet-induced obesity, but not atherosclerosis, in a strain dependent fashion in Apoe-/- mice. PLoS One 7 (10), e46837.

Gibson, G.R., Probert, H.M., Loo, J.V., Rastall, R.A., Roberfroid, M.B., 2004. Dietary modulation of the human colonic microbiota: updating the concept of prebiotics. Nutr. Res. Rev. 17 (2), 259–275.

Guarner, F., Malangelada, J.R., 2003. Gut flora in health and disease. Lancet 361, 512–519.

Heijtz, R.D., Wang, S., Anuar, F., Qian, Y., Björkholm, B., et al., 2011. Normal gut microbiota modulates brain development and behavior. Proc. Natl. Acad. Sci. 108, 3047–3052.

Hill, C., Guarner, F., Reid, G., Gibson, G.R., Merenstein, D.J., Pot, B., et al., 2014. The International Scientific Association for probiotics and prebiotics consensus statement on the scope and appropriate use of the term probiotic. Nat. Rev. Gastroenterol. Hepatol. 11 (8), 506–514.

Hong, S.M., Chung, E.C., Kim, C.H., 2015. Anti-obesity effect of fermented whey beverage using lactic acid bacteria in diet-induced obese rats. Korean J. Food Sci. Anim. Res. 35, 653–659.

Isolauri, E., 2012. Development of healthy gut microbiota early in life. J. Paediatr. Child Health 48, 1e6.

Jimenez, E., Marín, M.L., Martín, R., Odriozola, J.M., et al., 2008. Is meconium from healthy newborns actually sterile? Res. Microbiol. 159, 187e193.

Jung, S.P., Lee, K.M., Kang, J.H., et al., 2013. Effect of *Lactobacillus gasseri* BNR17 on overweight and obese adults: a randomized, double-blind clinical trial. Korean J. Fam. Med. 34, 80–89.

Kadooka, Y., Sato, M., Ogawa, A., et al., 2013. Effect *of Lactobacillus gasseri* SBT2055 in fermented milk on abdominal adiposity in adults in a randomised controlled trial. Br. J. Nutr. 110, 1696–1703.

Karlsson Videhult, F., Ohlund, I., Stenlund, H., Hernell, O., West, C.E., 2015. Probiotics during weaning: a follow-up study on effects on body composition and metabolic markers at school age. Eur. J. Nutr. 54, 355–363.

Kelly, C.R., Kahn, S., Kashyap, P., Laine, L., Rubin, D., et al., 2015. Update on fecal microbiota transplantation: indications, methodologies, mechanisms, and outlook. Gastroenterology 149 (1), 223–237.

Kim, B., Park, K.Y., Ji, Y., Park, S., Holzapfel, W., et al., 2016. Protective effects of Lactobacillus rhamnosus GG against dyslipidemia in high-fat diet-induced obese mice. Biochem. Biophys. Res. Commun. 473 (2), 530–536.

Koenig, J.E., Spor, A., Scalfone, N., Fricker, A.D., Stombaugh, J., et al., 2011. Succession of microbial consortia in the developing infant gut microbiome. Proc. Natl. Acad. Sci. 108, 4578–4585.

Ley, R.E., Turnbaugh, P.J., Klein, S., Gordon, J.I., 2006. Microbial ecology: human gut microbes associated with obesity. Nature 444 (7122), 1022–1023.

Little, T., Horowitz, M., Feinle-Bisset, C., 2007. Modulation by high-fat diets of gastrointestinal function and hormones associated with the regulation of energy intake: implications for the pathophysiology of obesity. Am. J. Clin. Nutr. 86, 531–541.

Mazloom, Z., Yousefinejad, A., Dabbaghmanesh, M.H., 2013. Effect of probiotics on lipid profile, glycemic control, insulin action, oxidative stress, and inflammatory markers in patients with type 2 diabetes: a clinical trial. Iran J Med Sci. 38, 38–43.

Meek, T.H., Eisenmann, J.C., Garland Jr., T., 2010. Western diet increases wheel running in mice selectively bred for high voluntary wheel running. Int. J. Obes. 34 (6), 960.

Meriga, B., Ganjayi, M.S., Parim, B.N., 2017. Phytocompounds as potential agents to treat obesity-cardiovascular ailments. Cardiovasc. Hematol. Agents Med. Chem. https://doi.org/10.2174/1871525715666170830131217. (Epub ahead of print).

Million, M., Angelakis, E., Paul, M., Armougom, F., Leibovici, L., Raoult, D., 2012. Comparative meta-analysis of the effect of Lactobacillus species on weight gain in humans and animals. Microb. Pathog. 53 (2), 100–108.

Morrison, D.J., Preston, T., 2016. Formation of short chain fatty acids by the gut microbiota and their impact on human metabolism. Gut Microbes 7 (3), 189–200.

Murphy, E.F., Cotter, P.D., Hogan, A., O'Sullivan, O., Joyce, A., et al., 2013. Divergent metabolic outcomes arising from targeted manipulation of the gut microbiota in diet-induced obesity. Gut 62, 220–226.

Musso, G., Gambino, R., Cassader, M., 2011. Interactions between gut microbiota and host metabolism predisposing to obesity and diabetes. Annu. Rev. Med. 62, 361–380.

Ogawa, A., Kadooka, Y., Kato, K., Shirouchi, B., Sato, M., 2014. *Lactobacillus gasseri* SBT2055 reduces postprandial and fasting serum non-esterified fatty acid levels in Japanese hypertriacylglycerolemic subjects. Lipids Health Dis. 13, 36.

Omar, J.M., Chan, Y.M., Jones, M.L., Prakash, S., Jones, P.J.H., 2013. *Lactobacillus fermentum* and *Lactobacillus amylovorus* as probiotics alter body adiposity and gut microflora in healthy persons. J. Funct. Foods 5, 116–123.

Park, D.Y., Ahn, Y.T., Park, S.H., Huh, C.S., Yoo, S.R., Yu, R., et al., 2013. Supplementation of *Lactobacillus curvatus* HY7601 and *Lactobacillus plantarum* KY1032 in diet-induced obese mice is associated with gut microbial changes and reduction in obesity. PLoS One 8, e59470.

Requena, T., Cotter, P., Shahar, D.R., Kleivel, C.R., Martinez-Cuesta, M.C., Pelaez, C., Lea, T., 2013. Interactions between gut microbiota, food and the obese host. Trends Food Sci. Technol. 34, 44–53.

Rios-Covian, D., Ruas-Madiedo, P., Margolles, A., Gueimonde, M., de los Reyes-Gavilán, C.G., Salazar, N., 2016. Intestinal short chain fatty acids and their link with diet and human health. Front. Microbiol. 7, 185.

Rowland, I., Capurso, L., Collins, K., Cummings, J., Delzenne, N., et al., 2010. Current level of consensus on probiotic science—report of an expert meeting—London, 2009. Gut Microbes 1, 436–439.
Sanchez, B., Delgado, S., Blanco-Míguez, A., Lourenço, A., Gueimonde, M., Margolles, A., 2017. Probiotics, gut microbiota, and their influence on host health and disease. Mol. Nutr. Food Res. 61 (1). https://doi.org/10.1002/mnfr.201600240.
Savcheniuk, O.A., Virchenko, O.V., Falalyeyeva, T.M., Beregova, T.V., Babenko, L.P., et al., 2014. The efficacy of probiotics for monosodium glutamate-induced obesity: dietology concerns and opportunities for prevention. EPMA J. 5, 2.
Sekirov, I., Russell, S.L., Antunes, L.C., Finlay, B.B., 2010. Gut microbiota in health and disease. Physiol. Rev. 90, 859–904.
Sommer, F., Backhed, F., 2013. The gut microbiota – masters of host development and physiology. Nat. Rev. Microbiol. 11, 227–238.
Sonnenburg, J.L., Backhed, F., 2016. Diet-microbiota interactions as moderators of human metabolism. Nature 535, 56–64.
Swinburn, B., Dietz, W., Kleinert, S., 2015. A lancet commission on obesity. Lancet 386, 1716–1717.
Turroni, F., Ventura, M., Butto, L.F., Duranti, S., O'Toole, P.W., et al., 2014. Molecular dialogue between the human gut microbiota and the host: a Lactobacillus and Bifidobacterium perspective. Cell. Mol. Life Sci. 71 (2), 183–203.
Wang, J., Tang, H., Zhang, C., Zhao, Y., Derrien, M., Rocher, E., et al., 2015. Modulation of gut microbiota during probiotic-mediated attenuation of metabolic syndrome in high fat diet-fed mice. ISME J 9, 1–15.
WHO, 2018. Obesity Fact Sheet. http://www.who.int/mediacentre/factsheets/fs311/en/.
Xu, J., Mahowald, M.A., Ley, R.E., Lozupone, C.A., Hamady, M., et al., 2007. Evolution of symbiotic bacteria in the distal human intestine. PLoS Biol. 5, e156.

Further Reading

FAO/WHO JFWEC, 2002. Guidelines for the evaluation of probiotics in food.
Kadooka, Y., Sato, M., Imaizumi, K., et al., 2010. Regulation of abdominal adiposity by probiotics (*Lactobacillus gasseri* SBT2055) in adults with obese tendencies in a randomized controlled trial. Eur. J. Clin. Nutr. 64, 636–643.
Robson, A.A., 2009. Preventing diet induced disease: bioavailable nutrient-rich, low-energy-dense diets. Nutr. Health 20 (2), 135–166.

CHAPTER

6

Emerging Trends in the Development of Plant Virus-Based Nanoparticles and Their Biomedical Applications

M. Hema, G.P. Vishnu Vardhan†, H.S. Savithri†, M.R.N. Murthy‡*

***Department of Virology, Sri Venkateswara University, Tirupati, India †Department of Biochemistry, Indian Institute of Science, Bangalore, India ‡Molecular Biophysics Unit, Indian Institute of Science, Bangalore, India**

OUTLINE

6.1 INTRODUCTION

Viral nanoparticles (VNPs) are nucleoprotein-based, biologically evolved supramolecular ensembles that could be used to develop bioinspired nanomaterials and nanocarriers due to their simple and inexpensive production, well-defined structural features, unique shapes and sizes, genetic programmability, and robust chemistries (Ma et al., 2012; Koudelka et al., 2015;

Recent Developments in Applied Microbiology and Biochemistry
https://doi.org/10.1016/B978-0-12-816328-3.00006-4

Mateu, 2016). Virus-like particles (VLPs) are a subset of VNPs that lack the viral genome and assemble spontaneously from virus structural proteins to noninfectious protein cagelike structures either naturally in hosts or when expressed in standard heterologous expression systems such as *Escherichia coli*, yeasts, plants, and mammalian or insect cells (Zeltins, 2013; Kushnir et al., 2012; Shirbaghaee and Bolhassani, 2015). Such VLPs tend to mimic the wild-type virus particles structurally/morphologically and demonstrate similar cell tropism, uptake, and intracellular trafficking. In certain cases, VLPs can also be produced directly from VNPs by pH-induced swelling followed by alkaline hydrolysis of the released nucleic acids. Alternatively, intact VNPs can be disassembled into their individual coat protein (CP) subunits and reassembled into VLPs under appropriate conditions including protein concentration, temperature, pH, ionic strength, and the presence of macromolecular crowding agents (Liu et al., 2012; Mateu, 2016).

VNPs/VLPs do not have intrinsic functionalities such as drugs, imaging agents, targeting ligands, and other molecular moieties that are required for potential biomedical applications. However, three distinct interfaces (internal, external, and intersubunit) of VNPs and VLPs could be modified to include functionalities: by direct chemical conjugation (Strable and Finn, 2009; Pokorski and Steinmetz, 2011; Mateu, 2011) and/or by protein engineering using genetic techniques (Mateu, 2011; Koudelka et al., 2015; Narayanan and Han, 2017c). Encapsulation of materials in the interior surface of the VNPs/VLPs is also possible by in vitro assembly, mineralization, and infusion (reviewed in Uchida et al., 2007; Koudelka and Manchester, 2010; Yildiz et al., 2011; Ma et al., 2012; Kan Davelaar et al., 2014; Schwarz and Douglas, 2015; Lee et al., 2016b; Jeong and Seong, 2017; Schwarz et al., 2017). Some of the bacterial and animal virus-based nanoparticles have been modified and demonstrated to be effective nanoparticles for a variety of applications in biomedical and materials science (Singh et al., 2006; Steinmetz, 2010; Koudelka et al., 2015; Rohovie et al., 2017; Zdanowicz and Chroboczek, 2016; Wen and Steinmetz, 2016). Plant virus-based nanoparticles (VNPs and VLPs derived from plant viruses, termed as PVNs) have been successfully used as powerful expression vectors and scaffolds to present proteins or epitopes for the development of vaccines (reviewed in Scholthof et al., 1996; Palucha et al., 2005; Yusibov et al., 2006; Canizares et al., 2005; Lomonossoff and Evans, 2011; Culver et al., 2015; Lebel et al., 2016; Steele et al., 2017; Hefferon, 2017; and Pichova et al., 2017).

During the last decade, utilization of functionalized PVNs is increasingly explored and found to be particularly advantageous for biomedical applications, because of their noninfectious, biocompatible, and biodegradable nature in humans and other mammals (Kan Davelaar et al., 2014; Czapar and Steinmetz, 2017). The other fascinating properties of PVNs that make them excellent candidates as nanocarriers are their size range, high degree of symmetry, polyvalency, monodispersity in shape and size (batch-to-batch consistency), stability over a wide range of pH and temperatures, efficient production and scalability (high yields from plants by mechanical inoculation or agroinfection), and stable large exterior and interior surface areas for controlled display of functional molecules. Extensive molecular and structural (high-resolution atomic structure) studies have facilitated researchers to accomplish structure-based engineering of these scaffolds using genetic and chemical procedures, which has allowed the introduction of deterministic, precise, and reproducible modifications along with the sophisticated rational design of ligand orientation and stoichiometry in order to maximize target recognition (Strable and Finn, 2009; Koudelka and Manchester, 2010; Mateu, 2011, 2016; Pokorski and Steinmetz, 2011). Achieving this level of control has not been possible with inorganic or lipid nanoparticle materials (Pokorski and Steinmetz, 2011; Mateu, 2016).

Chemical and genetic modifications result in alteration of the surface characteristics of PVNs and lead to changes in their in vivo properties. Biodistribution, pharmacokinetics, and toxicology studies on several PVNs revealed that they are generally safe; well tolerated; and endowed with high bioavailability, broad biodistribution, and relatively short persistence time in animal models (Singh et al., 2007; Wen et al., 2012a; Shukla et al., 2014b; Bruckman et al., 2014a; Vishnu Vardhan et al., 2016). PVNs have already overcome many of the challenges of nanoparticle delivery, such as low stability in biological fluids, efficient delivery across the cellular membrane, avoidance of the exocytotic pathways, and target specificity (Young et al., 2008; Ren et al., 2010; Molino and Wang, 2014; Czapar and Steinmetz, 2017; Berardi et al., 2018). In the recent times, novel strategies have been established to engineer the capsid proteins of PVNs with new functionalities, which allows fine tuning of surface charges, drug encapsulation, ligand display, and particle stability (Wen et al., 2013, 2016; Steele et al., 2017; Hefferon, 2018). Therapies composed of functionalized PVNs shielded with stealth molecules (such as polyethylene glycol (PEG)) are increasingly being utilized in vivo, where PEGylation has been observed to reduce immunogenicity and nonspecific binding and increase circulation times (Steinmetz and Manchester, 2009; Wen et al., 2015a). Several research groups have produced a broad range of PVNs such as those based on *Brome mosaic virus* (BMV), *Cowpea chlorotic mottle virus* (CCMV), *Cowpea mosaic virus* (CPMV), *Tobacco mosaic virus* (TMV), and *Potato virus X* (PVX) for potential applications in nanobiosciences. Several other plant-virus based platforms are also being added to the list. Most of the well-developed plant virus-based examples have stable structures obtained in high yields, and the high-resolution crystal structure information is available for manipulations (reviewed in Steinmetz and Evans, 2007; Young et al., 2008; Lomonossoff and Evans, 2011; Culver et al., 2015; Wen et al., 2015a; Narayanan and Han, 2017a,b). In this review, the development of nanocontainers/nanodevices/nanocarriers based on icosahedral and rod-shaped plant viruses with the view of applications in the field of biomedicine is outlined, focusing primarily on imaging, drug delivery, and therapeutic applications.

6.2 ICOSAHEDRAL PVNs

6.2.1 Brome Mosaic Virus

Brome mosaic virus (BMV; *Bromovirus*, *Bromoviridae*) consists of a tripartite genome of positive-sense ssRNA encapsidated separately in $T=3$ (180 subunits) icosahedral particles of ~28 nm in diameter (Noueiry and Ahlquist, 2003). In vitro assembly of BMV empty capsids under certain conditions of temperature, ionic strength, and pH facilitated the encapsulation of metal nanoparticles inside BMV capsid cavity. Assembly of BMV CP subunits around preformed gold nanoparticles with a citrate intermediate layer between the Au and protein surface resulted in VLPs with properties similar to native virus (Dragnea et al., 2003). Encapsulation efficiency of gold nanoparticles could be improved by coating them with carboxyl-terminated PEG, where the functionalized gold particles provided a mimic for the electrostatic behavior of the nucleic acid component of the native BMV (Chen et al., 2006). Further, the core diameter of the nanogold was found to influence the assembly of the final capsids, specifically the size of the VLPs. Three-dimensional crystals of these VLPs represent a form of metallo-dielectric material that exhibits unique optical properties because of multipolar plasmonic coupling (Sun et al., 2007). Encapsulation of HS-PEG-COOH-coated CdSe/ZnS quantum dots within BMV particles by in vitro assembly resulted in stable and uniform functionalized BMV particles, which could be developed into high-performance luminescent probes (Dixit et al., 2006). Self-assembly of BMV capsids around iron oxide nanotemplates functionalized with PEGylated phospholipids, even though they were larger than the inner cavity of the native capsid, resulted in nanocomposite materials that displayed super paramagnetic properties for magnetic resonance imaging (MRI) and biosensing (Huang et al., 2011). Yildiz et al. (2012) demonstrated the application of maleimide-thiol conjugation and hydrazone ligation reactions to a cysteine-containing mutant of BMV (cBMV) chimera to display multiple copies of chemical modifiers, such as fluorescent dyes, PEG chains, tumor-homing proteins, chemotherapeutic moieties, and cell-penetrating peptides for various biomedical applications. Encapsulation of indocyanine green (ICG), a Food and Drug Administration (FDA)-approved near-infrared (NIR) chromophore by BMV protein shell as a contrast agent for NIR photoacoustic imaging (PAI), was demonstrated. These optical nanoconstructs are nontoxic and may serve as multifunctional nanoprobes for site-specific deep-tissue optical imaging and therapy of diseases (Jung et al., 2011; Gupta et al., 2013).

6.2.2 Cowpea Chlorotic Mottle Virus

Cowpea chlorotic mottle virus (CCMV; *Bromovirus, Bromoviridae*) is a tripartite virus that contains three unique, single-stranded RNA polymers encapsidated independently in separate capsid, composed of 180 identical CP subunits ($T=3$) that form an icosahedral shell with ~28 and 18 nm outer and inner diameters, respectively (Narayanan and Han, 2017a). CCMV particles are stabilized by RNA-protein interactions, and the highly negatively charged RNA is thought to interact strongly with the positively charged interior of the capsid surface (Culver et al., 2015). In vitro and in vivo self-assembly of recombinant CCMV CP resulted in VLPs (eCCMV) with high yields in *E. coli, Pseudomonas fluorescens*, and *Pichia pastoris* (Brumfield et al., 2004; Phelps et al., 2007). Biodistribution studies with intravenously administered CCMV demonstrated broad in vivo distribution with no toxicity and rapid clearance in a nonpathogenic model (Kaiser et al., 2007). Several conjugation approaches using naturally occurring external lysines, carboxylates, and introduced cysteines by genetic engineering on CCMV capsid outer surface were developed to attach a wide variety of ligands including small peptides, biotin, fluorescent dyes, organometallic photosensitizers, and intact IgG antibodies (Young et al., 2008; Lomonossoff and Evans, 2011; Wen et al., 2016; Narayanan and Han, 2017a).

A remarkable and most attractive feature of CCMV capsid is its structural dynamism. CCMV undergoes pH and metal ion-dependent disassembly (at higher pH 7.5), and self-assembly of the components is possible (at lower pH 5.0) in vitro. Further, it can undergo a reversible pH-dependent swelling at pH 6.5 to form a capsid with 60 pores of ~2 nm in size at the quasi-threefold axes (Speir et al., 1995; Uchida et al., 2007). This structural "gating" property of CCMV capsid was exploited to produce novel particle architectures that would not normally be possible. CCMV has 18 nm diameter inner cavity, and its inner surface possesses a high positive charge due to the presence of arginine and lysine residues. Thus, it provides a constrained reaction vessel for encapsulation of negatively charged materials and changes the charge of the interior surface (by replacing nine basic residues at the N-terminus of CCMV CP with negatively charged glutamic acids) by genetic engineering that resulted in encapsulation of positively charged materials. CCMV capsid has also been used as an artificial nanoreactor for the encapsulation of enzymes (such as horseradish peroxidase (HRP) and *Pseudozyma antarctica* lipase B) with improved activity. Anchoring of PEG on its outer surface and poly(styrene sulfonate) in its inner capsid leads to the synthesis of biohybrid nanostructured materials. Novel encapsulation methods were established to load CCMV capsids with various molecules such as enhanced green fluorescent protein (EGFP), hydrophobic drugs using DNA amphiphile micelle, 70 kDa negatively charged polystyrene sulfonate (PSS), Prussian blue (PB), digoxigenin, fluorophore (FAM, F5M, cadaverine, and SAMSA), ruthenium complexes (Ru(bpy2)phen-IA), and zinc phthalocyanine (ZnPc), and the resultant functionalized CCMV capsids were shown to have potential applications in pharmaceutics, nanoarrays, and electronics (reviewed in Culver et al., 2015; Lee et al., 2016b; Narayanan and Han, 2017a; Steele et al., 2017).

Suci et al. (2008) utilized CCMV capsids to assemble multilayer films and higher order structures using a layer-by-layer (LbL) approach to enhance signaling during imaging and to increase drug-loading capacity. Using gating and targeted genetic engineering strategies, encapsulation of the coiled form of polyanionic semiconducting polymer (poly-2-methoxy-5-propyloxy sulfonate phenylene vinlyene) (MPS-PPV) in CCMV capsid resulted in icosahedral particles resembling native virus, whereas changing the solution ionic strength and pH to get an extended conformation of the polymer resulted in rodlike CCMV structures (Ng et al., 2011). Both multifunctional MRI contrast and photodynamic therapy agents (chelated Gd3+ and Zn2+ phthalocyanine) were successfully encapsulated in CCMV as a first demonstration of theranostic protein cage nanoparticles (Millan et al., 2014; Koudelka et al., 2015).

To use CCMV capsid for therapeutic purposes, a 24-amino-acid antitumor peptide was attached to the surface-exposed lysine and cysteine residues (Gillitzer et al., 2002). In another study, eCCMV was used to deliver heterologous RNA originating from the mammalian *Sindbis virus* to cells leading to the expression of the reporter enhanced yellow fluorescent protein (Azizgolshani et al., 2013; Steele et al., 2017). eCCMV particles were used to encapsulate a soluble version of a bacterial cytochrome P450 CYP via electrostatic interactions. CCMV encapsulated CYP readily converted resveratrol and tamoxifen prodrug formulations to their active forms and at levels comparable or greater than that observed in human liver microsomal preparations (Sanchez-Sanchez et al., 2014; Rohovie et al., 2017). Genetically mutated CCMV capsids were self-assembled along with chemically conjugated moieties for visualization (fluorescent dye), targeting (folic acid, FA), and drug delivery (Doxorubicin, dox; anticancer drug) and the resulting chimeric nanoscaffold system showed highly specific receptor-mediated internalization (targeting) and ~300% more cytotoxicity to folate-receptor-positive Michigan Cancer Foundation-7 (MCF7) cell lines (Barwal et al., 2016).

6.2.3 Cowpea Mosaic Virus

Cowpea mosaic virus (CPMV; *Comovirus*, *Comoviridae*) is the most extensively studied platform for various conjugations and biotechnological applications. It is a 30 nm icosahedral virus and consists of bipartite positive-sense, single-stranded RNA genome with each RNA molecule encapsidated in a separate particle. CPMV capsid is composed of 60 copies each of two different CP subunits (termed as small, 24 kDa, and large, 42 kDa) with pseudo $T=3$ icosahedral symmetry. Due to the presence of highly symmetrical and heteromeric nature of the capsid, CPMV provides a remarkable opportunity to differentially introduce (at five-fold or threefold or both positions) multiple types of functionalities by specific genetic and chemical engineering with control over both the spatial distribution and the degree of multivalency (Uchida et al., 2007; Young et al., 2008; Steinmetz, 2010; Wen et al., 2016). CPMV can be purified in large quantities from infected leaves and is stable at wide range of temperatures (up to 60°C) and in the pH range of 3.0–9.0 and can withstand some organic solvents (Steinmetz et al., 2009). CPMV capsids are used for controlled immobilization of active enzymes. Utilizing CPMV capsid for templated fabrication of metallic nanoparticles by an electroless deposition (ELD) and mineralization by a variety of processes allowed the production of thin layers of metals and alloys with uniform thickness and composition. CPMV virions are used as nanobuilding blocks for the controlled construction of 2-D and 3-D structures using a bottom-up, layer-by-layer (LbL) approach for potential applications as nanoelectronic devices or electrochemical multiple biosensors (reviewed in Culver et al., 2015; Wen et al., 2015a, 2016; Narayanan and Han, 2017a).

Native CPMV and CPMV particles derivatized with lanthanide metal complexes were found to be safe and nontoxic, based on the biodistribution, toxicity, and pathology studies of these particles in vivo (Rae et al., 2005; Singh et al., 2007). CPMV was found to be associated with regions of inflammation and induced blood–brain barrier disruption during central nervous system (CNS) infection in mice (Shriver et al., 2009). CPMV with surface PEGylation has shown low immunogenicity, and internalization was prevented in several cell types due to reduction in nonspecific binding (Raja et al., 2003). Manchester and coworkers discovered that fluorescently labeled CPMV has an endogenous capacity to be taken up by vascular endothelial cells and can be used as an intravital imaging probe to visualize the vasculature and blood flow to a depth of up to 500 μm for 72 h in living mouse and chick embryos (Lewis et al., 2006). CPMV was found to have a natural ability to bind to surface vimentin, a cytoskeletal protein that modulates the architecture and dynamics of cells, and was found to be overexpressed in endothelial, cancer, and inflammatory cells (Koudelka et al., 2009). The biocompatible, natural interaction with vimentin and endothelium retention features of CPMV has been further exploited to image tumor angiogenesis and tumor homing in vivo (Leong et al., 2010; Steinmetz et al., 2011; Yildiz et al., 2011; Wen et al., 2015a).

The exterior surface of the CPMV capsid has been extensively modified using naturally occurring or genetically engineered exposed lysine, cysteine, tyrosines, and aspartic and glutamic acid residues by using standard chemical conjugation methods and conjugating cross-linkers to attach several reagents such as Au nanoparticles, redox reactive methyl viologen moieties, organometallic ferrocene carboxylate moieties, fluorophores, biotin, PEG chains, stilbene derivatives, carbohydrates, heterologous proteins (e.g., human holo-transferrin, T4 lysozyme, the LRR domian of internalin B and the Intron 8 gene product of the HER2 tyrosine kinase receptor), antibodies, oligonucleotides, semiconductor quantum dots and fullerenes. Incorporation of unnatural amino acids into CPMV capsid was carried through multiple orthogonal chemistries (such as chemoselective hydrazone ligation strategy and copper-catalyzed azide-alkyne cycloaddition (CuAAC) or "click" reaction). These efforts have facilitated the use of CPMV particles as optical probes, vaccine candidates, and memory and sensor nanodevices (reviewed in Young et al., 2008; Destito et al., 2009; Grasso and Santi,

2010; Steinmetz, 2010; Lomonossoff and Evans, 2011; Wen et al., 2015a; Zhang et al., 2016; Lee et al., 2016b; Narayanan and Han, 2017a). CPMV and TMV were loaded to carry contrast agents for dual-modality magnetic resonance (MR) and optical imaging, and both modalities demonstrated specificity of fibrin binding in vitro with the presence of targeting peptides. Preclinical studies in a carotid artery photochemical injury model of thrombosis confirmed thrombus homing of the nanoprobes, with the elongated TMV rods exhibiting significantly greater attachment to thrombi than icosahedral CPMV (Wen et al., 2015b,c). CPMV decorated with E7p72 peptides (high-affinity epidermal growth factor-like domain 7 (EGFL7) binding peptide that specifically targets human endothelial cells) targeted to tumor-associated neovasculature with high specificity as assessed by intravital imaging (Cho et al., 2017).

In vitro and in vivo assembly protocols for CPMV capsid subunits are not yet established well. This has limited its potential to use as a nanocontainer for encapsulation of cargo molecules. Native nucleic acid content of CPMV virions was used as an electrostatic sponge to attract imaging agents and therapeutic molecules via simple infusion technique (Yildiz et al., 2013). Transient coexpression of CPMV precursor (VP60) that consists of fused large and small proteins and 24K proteinase for proteolytic processing of VP60 in plants resulted in CPMV VLPs (eCPMV), which were completely devoid of RNA (either virus or host) (Saunders et al., 2009). Such empty capsids have been used to carry a wide range of cargo molecules, such as metals, fluorescent dyes, or drugs (Culver et al., 2015; Wen et al., 2015a; Narayanan and Han, 2017a). Fluorophores, biotin affinity tags, PEG, and various peptides have been selectively displayed on the interior surface of the empty CPMV by targeting reactive lysines (Wen et al., 2012b). The efficacy of eCPMV as an in situ vaccine was demonstrated in mouse models of metastatic melanoma, breast cancer, ovarian cancer, and colon cancer, where it induced antitumor immune response by triggering activation and infiltration of neutrophils, resulting in a chemo/cytokine profile that leads to activation of adaptive immunity. Most mice rechallenged with tumors on the opposite flank completely rejected the rechallenged tumor (Lizotte et al., 2016). Intratumoral delivery of magnetic nanoparticle (mNP) to induce hyperthermia and eCPMV in combination improved local and systemic tumor treatment efficacy (retardation of secondary tumor growth (abscopal effect) and resisting tumor rechallenge) in the C3H mouse/MTG-B mammary adenocarcinoma and C57-B6 mouse/B-16-F10 melanoma cancer cell models (Hoopes et al., 2017a). Using eight spontaneous canine cancers (two oral melanoma, three oral amelioblastomas, and one carcinoma), Hoopes et al. (2017b) demonstrated that hypofractionated radiation and mNP-induced hyperthermia and intratumoral eCPMV treatment in combination were shown to elicit enhanced immune responses.

To use CPMV for the targeted delivery of drugs to cancer cells and to overcome the natural interaction between CPMV and mammalian cells, a "click" reaction was used to conjugate PEG and folate ligand moiety (CPMV-PEG-FA) to CPMV. This conjugation enabled specific targeting of CPMV to folate receptors (FRs) expressing HeLa and KB cells (Destito et al., 2007). The targeting of CPMV to neuroblastoma tumor cells was achieved by modifying CPMV genetically to display the neuropeptide Y (NPY) analog as a targeting ligand. It was shown that the CPMV-NPY interacted specifically with SK-N-MC cells that overexpressed the Y1 receptor (Destito et al., 2009; Ma et al., 2012). Receptor-targeted imaging was achieved by anchoring short peptide sequences (F56 and bombesin specific to vascular endothelial growth factor receptor-1 (VEGFR-1)) and gastrin-releasing peptide receptors (GRPR), respectively, to CPMV. These CPMV-based sensors were found to be specific in binding to the cancer cells that are overexpressing these receptors on the cell surface (Brunel et al., 2010; Steinmetz et al., 2011). Wen and Steinmetz (2014) developed a bottom-up approach for the synthesis of CPMV nanoparticles, dimers and assemblies containing targeting ligands (cyclic RGD peptides), and fluorescent dyes and found that symmetrical presentation of RGD peptides and the dimerization of nanoparticles increased the efficacy of cancer cell targeting.

CPMV has been successfully used for display of several heterologous peptides to induce immune response. Immunization of animals with each of the chimera induced a strong humoral immune response that was protective against challenge by the respective pathogen. Multivalent display of antiviral molecules and carbohydrates on CPMV facilitated their usage for therapeutic applications (Lomonossoff and Evans, 2011; Koudelka et al., 2015; Hefferon, 2017). Aljabali et al. (2013) established covalent conjugation of dox to CPMV through either an amide bond or a disulfide bridge using lysine or cysteine residues on the external surface of the CPMV. The dox-loaded CPMV exhibited higher cell-killing effect than free dox even at low dosages. CPMV modified with a corona of negatively charged dendrons was used to load positively charged photosensitizers (for photodynamic therapy (PDT)). This hybrid PDT-CPMV carrier system was effective in killing both melanoma cells and macrophages in vitro (Wen et al., 2016). To overcome the dose-limiting nonspecific toxicities of trivalent chromium (a beneficial mineral nutrient), which can selectively inhibit the proliferation of high-glucose-induced human aortic smooth muscle cells (HASMC) in vitro, chromium chloride was loaded into the CPMV interior cavity via infusion, and the resultant $CrCl_3$-loaded CPMV (CPMV-Cr) particles significantly reduced glucose-induced HASMC proliferation and presented antiatherogenic effects under hyperglycemic conditions (Ganguly et al., 2016).

6.2.4 Other Icosahedral PVNs

Hibiscus chlorotic ringspot virus (HCRSV; *Carmovirus, Tombusviridae*) was shown to encapsulate polyprotic acid and dox inside the capsid, and folic acid was conjugated onto the capsids to impart cancer-targeting capability. The resultant conjugates (FA-HCRSV-Dox) improved the uptake and cytotoxicity of dox in the ovarian cancer cells, OVCAR-3 (Ren et al., 2007). Lu et al. (2012) demonstrated the encapsulation of fluorophore, oligonucleotide, and protein into *Cucumber mosaic virus* (CMV; *Cucumovirus, Bromoviridae*) VLPs. FA-CMV-Dox assemblies showed increased uptake and cytotoxicity in OVCAR-3 cells and antitumor effect in OVCAR-3 BALB/c nude mouse xenograft model (Zeng et al., 2013).

Red clover necrotic mosaic virus (RCNMV; *Dianthovirus, Tombusviridae*) capsid forms icosahedral particles with an outer diameter of 36 nm and an inner diameter of 17 nm. Similar to CCMV, RCNMV also undergoes reversible pore conversions in response to pH changes, providing a strategy for the loading and unloading of target molecules including gold nanoparticles, fluorescent dyes, and dox (Loo et al., 2008; Lockney et al., 2011; Honarbakhsh et al., 2013; Cao et al., 2014; Culver et al., 2015). The abamectin-loaded RCNMV was used to manage plant parasitic nematode infestation (Cao et al., 2015). To compare the pharmacokinetics and efficacy of dox containing PVNs with PEGylated liposomal dox (PLD) and dox alone in A375 melanoma and intraperitoneal SKOV3ip1 ovarian cancer xenografts, RCNMV was used as a PVN platform. It was found that PVNs were more efficacious than PLD and dox in the ovarian cancer model, but not in the melanoma cancer model. The pharmacokinetics profiles of the PVNs showed fast plasma clearance but with more efficient tumor delivery as compared with other carrier-mediated agents (Madden et al., 2017).

Chemoselective modification of *Turnip yellow mosaic virus* (TYMV; *Tymovirus, Tymoviridae*) was achieved using "click" chemistry to attach oligo(ethylene glycol) (OEG) short chain and RGD-containing peptide for cell targeting (Zeng et al., 2011). *Maize rayadofino virus* (MRFV; *Marafivirus, Tymoviridae*) VLPs expressed in tobacco plants were used as the platform for cysteine selective bioconjugation peptide display (Natilla and Hammond, 2011). *Artichoke mottled crinkle virus* (AMCV; *Tombusvirus, Tombusviridae*) capsid was used both as a carrier of immunogenic epitopes (HIV-1 2F5 neutralizing epitope) and for the delivery of dox molecules (Arcangeli et al., 2013).

In in vivo behavior of *Sesbania mosaic virus* (SeMV; *Sobemovirus*), a 30nm spherical plant *virus* was studied and found that it is a safe and nontoxic platform (Vishnu Vardhan et al., 2016). *E. coli*-expressed chimeric SeMV-derived VLPs displaying B domain of *Staphylococcus aureus* protein A (SpA) (termed as SLB) were shown to have higher affinity toward antibodies. Fluorescent wild-type SeMV VLPs and SLB generated using lysine-NHS chemistry were taken up by various types of mammalian cells. SLB could efficiently deliver three different monoclonal antibodies—D6F10 (targeting abrin), anti-α-tubulin (targeting intracellular tubulin), and Herclon (against HER2 receptor)—inside the cells (Abraham et al., 2016). Alemzadeh et al. (2017) explored the possibility of using a *Johnson grass chlorotic stripe mosaic virus* (JgCSMV; *Aureusvirus, Tombusviridae*)-derived VLPs as a nanocontainer for loading dox, taking advantage of the reversible swelling of VLPs in vitro. Outer and inner surfaces of *Physalis mottle virus* (PhMV; *Tymovirus, Tymoviridae*) VLPs (obtained from *E. coli*) were modified by chemical conjugation, infusion with fluorescent dyes and anticancer drugs, respectively, and the efficacy of these functionalized particles for imaging and drug delivery was demonstrated in vitro (Masarapu et al., 2017).

6.3 HELICAL PVNs

6.3.1 Tobacco Mosaic Virus

Tobacco mosaic virus (TMV; *Tobamovirus, Virgaviridae*) is a rodlike virus with a length of 300nm and diameter of 18nm. TMV capsids are composed of 2130 identical protein subunits, which assemble around the viral ssRNA to form a helical structure, with a hollow central cavity of 4nm diameter. The structural chirality and inherent asymmetry of the virus allowed chemical or genetic modifications spatially at one or the other end of the helical rod. TMV has a well-defined in vitro assembly system controlled by pH, ionic strength, and protein concentration. TMV particles can be produced in large quantities from infected tobacco plants (Young et al., 2008; Ma et al., 2012). TMV is both chemically and physically stable to solvents, pH (3.5–9.0), temperature (up to 90°C), and reducing agents. Due to this flexibility, tubelike TMV capsids were biochemically modified on both interior and exterior surfaces for the synthesis of metal nanoparticles, nanowires, and layer-by-layer assemblies that can be organized into higher order 2-D and 3-D structures through a variety of reactions (Young et al., 2008; Ma et al., 2012; Culver et al., 2015; Narayanan and Han, 2017b).

Blood clearance kinetics and biodistribution studies of radiolabeled TMV in both normal and tumor-bearing mice revealed their rapid clearance and nontoxicity in both models (Wu et al., 2013). TMV rods can also undergo thermal transition to form RNA-free densely packed spherical nanoparticles (SNPs) of size ~50 nm (Atabekov et al., 2011). Bruckman et al. (2014a) studied the biodistribution, pharmacokinetics, and blood compatibility of native and PEGylated TMV rods and spheres in mice. Both versions exhibited comparable in vivo profiles with few differences, where rods circulate longer than spheres illustrating the effect of shape on circulation. PEGylation increased circulation times. Spheres are more rapidly cleared from tissues compared with rods with no overt toxicity. The impact of aspect ratio on the biodistribution and tumor homing of PEGylated or receptor-targeted RGD-displaying TMV formulations was investigated, and it was found that the aspect ratio has a profound impact on the behavior of the nanoparticles in vivo and in vitro (Shukla et al., 2015a). In vivo implantation of RGD-TMV hosted bioadhesive peptide (RGD) hydrogels indicated that TMV is less immunogenic, nontoxic, and biodegradable in mice (Luckanagul et al., 2015).

TMV hosts well-defined amino acids (tyrosines and glutamates) at specific locations on its outer surface. Natural and genetically modified TMV particles with nonnative amino acids (cysteines and lysines) were used to attach new functionalities such as fluorophores, PEGs, and other nanoparticles using orthogonal synthetic methods to obtain biosensors, optoelectronic devices, catalysts, and imaging agents (reviewed in Culver et al., 2015; Wen et al., 2015a, 2016; Narayanan and Han, 2017b). The scope for genetic fusions to the termini of TMV capsid proteins is limited, as this can inhibit virus particle formation (Steele et al., 2017). TMV particles have been loaded with paramagnetic gadolinium ions selectively at the interior (iGd-TMV) or exterior (eGd-TMV) surface, and it was found that interior-labeled TMV rods can undergo thermal transition to form 170 nm-sized SNPs. These functionalized TMV rods and spheres showed high relaxivities to act as MRI contrast agents (Bruckman et al., 2013). Coupling of TMV particles to the two-photon dye BF3-NCS has been demonstrated to allow visualization of diseased brain vasculature in mice (Niehl et al., 2015). In a comparative study, fluorescently labeled CPMV and TMV were used to determine the impact of dye density, dye localization, conjugation chemistry, and microenvironment on the optical properties of the virus-based optical probes. The results indicated that both are extraordinarily robust under ultrashort, pulsed laser light conditions with a significant amount of excitation energy, maintaining structural and chemical stability (Wen et al., 2015b,c). In another study, CPMV and TMV were anchored with fibrin-binding peptides (CREKA and GPRPP) and contrast agents for dual-modality magnetic resonance (MR) and optical imaging. The resultant particles are found to be specific to fibrin binding in vitro. Preclinical studies in a carotid artery photochemical injury model of thrombosis confirmed greater attachment of elongated TMV rods to thrombi than CPMV (Wen et al., 2015b,c). Hu et al. (2017) prepared a bimodal contrast agent by loading the internal cavity of TMV nanoparticles with a dysprosium (Dy^{3+}) complex and NIR fluorescent dye Cy7.5, and the external surface of TMV was conjugated with an Asp-Gly-Glu-Ala (DGEA) peptide via a PEG linker to target integrin $\alpha_2\beta_1$. The resulting Dy-Cy7.5-TMV-DGEA was useful for ultra-high-field magnetic resonance and near-infrared fluorescence imaging of prostate cancer in vivo.

TMV has also been used as a vaccine vehicle to display various heterogeneous antigenic epitopes and immunization of animals with each of these chimeras induced a strong humoral protective immune response against the respective pathogen challenge (reviewed in Hefferon, 2017).

A tumor associated carbohydrate antigen (Tn; target for antitumor vaccine development) conjugated to TMV can elicit antigen-specific IgG and IgM responses (Yin et al., 2012). Engineering of TMV to target vascular cell adhesion molecule (VCAM)-1 (which is highly expressed on activated endothelial cells at atherosclerotic plaques) resulted in molecular targeting of atherosclerotic plaques by functionalized TMV particles in an atherosclerotic ApoE(−/−) mouse model (Bruckman et al., 2014b). Cysteine-inserted TMV mutants were cross-linked to methacrylated hyaluronic acid (MeHA) polymers by "click" chemistry and formed hydrogels under physiological condition, which could promote in vitro chondrogenesis of bone marrow mesenchymal stem cells (BMSCs) (Maturavongsadit et al., 2016). TMV was tailored with peroxidase-like inorganic nanoparticles (platinum nanoparticles) and cancer cell target groups (folic acid) to obtain TMV-FA-Pt nanoparticles for cancer cell detection using a cell-based ELISA (CELLISA) (Guo et al., 2018).

Self-assembling nanoscale disks derived from a double arginine mutant of recombinantly expressed TMV (RR-TMV) was functionalized with the dox and further modified with PEG for improved solubility. RR-TMV-DOX-PEG displayed cytotoxic properties like those of dox alone when incubated with U87MG glioblastoma cells (Finbloom et al., 2016). The cationic porphyrin-based photosensitizer was successfully and stably loaded into the interior channel of TMV via electrostatic interactions. The resulting TMV-photosensitizer exhibited improved cell uptake and efficacy when compared with free photosensitizer, making it a promising platform for improved PDT of melanoma (Lee et al., 2016a). Bruckman et al. (2016) demonstrated the loading of dox into rods and spheres by chemical conjugation and encapsulation approaches for targeted delivery to breast cancer. Tumor delivery and efficacy of TMV loaded with phenanthriplatin (a cationic monofunctional DNA-binding platinum (II) anticancer drug) was confirmed by using a triple-negative breast cancer mouse model (Czapar et al., 2016). The TMV-cisplatin conjugate (TMV-cisPt) was synthesized using a charge-driven reaction and it showed superior cytotoxicity and DNA double-strand breakage (DSB) in platinum-sensitive (PS) and platinum-resistant (PR) cancer cells when compared with free cisplatin (Franke et al., 2017). TMV conjugates loaded with streptokinase (STK) via intervening PEG linkers were synthesized, and the resultant TMV-STK formulations showed thrombolytic activity in vitro using static phantom clots (Pitek et al., 2017). TMV was used as a multivalent carrier for the delivery of the antimitotic drug valine-citrulline-monomethyl auristatin E (vcMMAE) targeting non-Hodgkin's lymphoma (Kernan et al., 2017).

Interaction studies between VNPs derived from TMV and plasma revealed that the VNP protein corona is significantly less abundant compared with the corona of synthetic particles. A library of functionalized TMV rods with PEG and peptide ligands targeting integrins or fibrin(ogen) showed different dispersion properties, cellular interactions, and in vivo fates depending on the properties of the protein corona, influencing target specificity, and nonspecific scavenging by macrophages (Pitek et al., 2016a). Serum albumin (SA) conjugation to TMV-based nanocarriers resulted in a "camouflage" effect more effective than PEG coatings. SA-"camouflaged" TMV particles exhibited decreased antibody recognition and enhanced pharmacokinetics in a BALB/c mouse model (Pitek et al., 2016b).

6.3.2 Potato Virus X

Potato virus X (PVX; *Potexvirus, Alphaflexiviridae*) is a monopartite positive-sense, ssRNA virus. It is a flexuous rod (13 nm diameter and 515 nm long) and composed of 1270 identical

CP (25 kDa) subunits (Lico et al., 2015). In vitro assembly of PVX CP subunits into VLPs has been established (Arkhipenko et al., 2011). Functional recombinant proteins can be fused to PVX CP through insertion of an internal ribosome entry site, which allows for initiation of translation in the middle of mRNA between the two genes, and this provides an opportunity for a wide range of biotechnological applications (Lico et al., 2015; Wen et al., 2016; Narayanan and Han, 2017b; Steele et al., 2017).

Biodistribution studies revealed that PVX particles are generally safe, well tolerated, with high bioavailability, broad biodistribution, and relatively short persistence in animal models (Wen et al., 2012a). Biodistribution and tumor-homing properties of fluorescently labeled PEGylated PVX in vivo are very different when compared with spherical CPMV in human tumor xenografts (fibrosarcoma, squamous sarcoma, and colon cancer). PVX exhibited higher tumor uptake and deeper penetration into tissues when compared with CPMV, particularly in the core of the tumor (Shukla et al., 2013). Genetic fusion of the green fluorescent protein (GFP) or the red fluorescent protein (mCherry) to the PVX CP resulted in fluorescent PVX filaments in plants. Application of mCherry-PVX filaments as optical probes for imaging in human cancer cells and a preclinical mouse model indicated the biocompatibility and rapid clearance of the particles (Shukla et al., 2014a). A detailed analysis of PVX biodistribution and clearance in healthy mice and mouse tumor xenograft models revealed retention of PVX signal up to 30% from the colon, mammary, and brain tumor tissues, and the remaining particles were cleared by the spleen, liver, and kidneys (Shukla et al., 2014b). Tailoring PEG chain length and conformation (brush vs mushroom) allows tuning of the pharmacokinetics, yielding long-circulating PVX stealth filaments with reduced immune recognition (Lee et al., 2015). Cyclic RGD peptides (which specifically bind to integrins upregulated on tumor cells, neovasculature, and metastatic sites) and PEG molecules were anchored to PVX filaments, and their behavior in vitro and in vivo was studied. Although the RGD-targeted filaments outperformed the PEGylated stealth filaments in vitro, enhanced tumor cell targeting did not translated into improved tumor homing in mouse tumor models. The RGD-PVX and PEG-PVX filaments showed contrast biodistribution profiles (Shukla et al., 2015a). In another study, repeated administration of PVX resulted in the generation of increased immune response, and PVX-immune complexes were observed in the mouse vasculature in vivo when imaged in real time using intravital two-photon laser scanning microscopy (2P-LSM) (Shukla et al., 2016). Lico et al. (2015) studied biodistribution of unlabeled, unpegylated, underivatized icosahedral *Tomato bushy stunt virus* (TBSV) and filamentous PVX by detecting antibodies and found that they are nontoxic and intrinsically endowed with peculiar properties.

PVX-based vectors are extensively used for the presentation of heterologous proteins and epitopes for vaccination purposes (reviewed in Lico et al., 2015, and Hefferon, 2017). Although the structure of PVX has not been determined at high resolution and chemical nature of the surface-exposed addressable amino acids is not known, amine-based conjugations and CuAAC "click" reaction strategy were applied to attach fluorescent dyes, biotins, and PEGs on the surface of PVX CP for imaging in cell cultures and to mark embryo vasculature and tumor tissue in chick and mouse systems (Steinmetz, 2010). In addition, extensive surface modifications have been made, and their effects have been investigated for the display of peptides and ligands on PVX through genetic and conjugation methods (Lee et al., 2014). A structural model of the PVX particle was built based on the available potexvirus structure, and multifunctional PVX nanoparticles were made by conjugating Gd-DOTA to lysines for MRI and fluorescent dyes to newly identified cysteines for optical imaging (Le et al., 2017b).

Display of a targeting 12-amino-acid peptide ligand (GE11) to the epidermal growth factor receptor (EGFR) on fluorescently labeled PVX via a complex two-step bioconjugation method resulted in targeted imaging in human skin epidermoid carcinoma, colorectal adenocarcinoma, and triple-negative breast cancer cell lines (A-431, HT-29, and MDA-MB-231), all of which upregulate EGFR to various degrees (Chariou et al., 2015). SpyTag/SpyCatcher technology is used to covalently attach the *Trichoderma reesei* endoglucanase Cel12A to PVX nanoparticles, and the conjugated PVX particles displaying Cel12A showed enhanced enzyme activity (Roder et al., 2017). Multivalent presentation of B domain of *S. aureus* protein A (SpAB) on PVX nanoparticles resulted in PVX chimeras that can function as an immunoabsorbent for antibody capture enabling biosensing (Uhde-Holzem et al., 2016). PVX was genetically engineered to present mineralization-inducing (MIP) and RGD peptides on its particle surface. Recombinant PVX-MIP/RGD particles showed enhanced focal adhesion-dependent cell adhesion and matrix mineralization (Lauria et al., 2017).

Bioconjugation was performed by EDC/sulfo-N-hydroxysuccinimide (sulfo-NHS) in a two-step protocol to conjugate the Herceptin (trastuzumab) monoclonal antibody to PVX. The efficiency of PVX-Herceptin to induce apoptosis was checked in HER2-positive SK-OV-3 and SK-BR-3 cells (Esfandiari et al., 2016). A simple in-solution protocol was followed to infuse dox into PVX, and it was found that PVX-DOX is an efficient drug delivery carrier both in vitro (ovarian cancer, breast cancer, and cervical cancer cells) and in vivo (athymic mice bearing human MDA-MB-231 breast cancer xenografts) (Le et al., 2017a). Immunotherapeutic efficiency of PVX as an in situ vaccine was investigated by administrating PVX along with PVX-DOX (chemoimmunotherapeutic dual-functional particle), and it was found that coadministered PVX + PVX-DOX was better for in situ vaccination than PVX-DOX alone in B16F10 melanoma mice model (Lee et al., 2017).

PVX-based vaccine carriers are found to be promising due to their tropism toward B cells and large surface area leading to high multivalency. This was demonstrated by administration of PVX-based vaccination platform with integrated human EGFR 2 (HER2) epitopes on the surface (via chemical conjugation). This led to higher antibody titers that were specific to HER2. PVX carrier acts as an adjuvant and improves stability and B-cell presentation of the epitopes (Shukla et al., 2014c). Steinmetz and coworkers evaluated in vivo capabilities of engineered CPMV and PVX particles delivering HER2 epitopes for therapy and prophylaxis of $HER2^+$ malignancies, and it was found that CPMV particles showed significantly enhanced lymph node transport and retention, greater uptake by/activation of APCs, and elevated HER2-specific antibody titers compared with filamentous PVX particles (Shukla et al., 2017).

Functionalization of PVX-based nanoparticles with protein A fragments was used to display whole antibody molecules by using protein A's property of binding to the Fc region of the heavy chain of IgGs (Uhde-Holzem et al., 2016). The PVX particles were used as a plug-and-play system for the display of a variety of antibodies, with a level of orientational control, not achievable with chemical conjugation. For the diagnosis of primary Sjögren's syndrome (pSjS), a chronic systemic autoimmune disease, lipopeptide derived from the human autoantigen lipocalin was displayed on PVX and found to be specific to pSjS patient sera and had greater reactivity than the peptide alone (Tinazzi et al., 2015). PVX nanoparticles carrying idiotypic (Id) tumor antigen-activated TLR7 induced high levels of protective antibody against murine B-cell lymphoma, where viral ssRNA genome acted as an adjuvant via TLR7 to stimulate IFN production (Jobsri et al., 2015). PVX particles displaying mineralization-inducing (MIP) and integrin binding motif (RGD) peptides were used to support hydroxyapatite crystal nucleation, indicating its potential for bone tissue substitutes (Lauria et al., 2017).

6.3.3 Other Helical PVNs

Papaya mosaic virus (PapMV; *Potexvirus, Alphaflexiviridae*) is a flexible rod-shaped (500 nm long and 14 nm diameter) virus with a neutrally charged capsid of 1400 subunits of CP (23 Kda) assembled around a positive-stranded ssRNA (Lecours et al., 2006). PapMV can be used as an immunotherapy agent due to its tremendous potential as (i) an immunostimulatory molecule, (ii) an adjuvant, and (iii) a vaccine platform through its intrinsic capacity to activate the innate immune response in an IFN-α-dependent manner (Savard et al., 2012; Mathieu et al., 2013; Therien et al., 2017). PapMV exhibited efficacy in treating a mouse model of melanoma when administered intratumorally, and the efficacy was improved by synergistic co-administration with dendritic cell vaccination and programmed cell death ligand 1 (Lebel et al., 2016; Czapar and Steinmetz, 2017; Hefferon, 2018). The potential of PapMV containing ssRNA for cancer immunotherapy was demonstrated by showing their ability to activate innate immunity responses (secretion of cytokines and chemokines), acting as TLR7/8 agonist in human peripheral blood mononuclear cells (PBMC) and mice (Carignan et al., 2017; Lebel et al., 2016). In another study, Chariou and Steinmetz (2017) demonstrated the potential of *Tobacco mild green mosaic virus* (TMGMV; *Tobamovirus*) as a carrier for anthelmintic drug to treat plants infected with parasitic nematodes.

The growing body of data on PVNs has indicated that each PVN is distinct with respect to its size, shape, and surface composition (surface charge and chemistry), which in turn seems to affect their in vivo behavior. Functionalization of PVNs can affect their physiochemical properties, biodistribution, pharmacokinetic behavior, immunogenicity, and toxicity. Steinmetz and coworkers demonstrated that the elongated rod-shaped and filamentous macromolecular nucleoprotein assemblies from plant viruses have longer circulation times and superior tumor homing and tissue penetration rates than spherical particles (Wen et al., 2015a, 2016). The aspect ratio (AR, defined as length divided by width) has a profound impact on the behavior of the PVNs in vivo and in vitro (Wen and Steinmetz, 2014; Shukla et al., 2015a). Wen et al. (2013, 2016) have discussed the design rules that need to be considered for the successful development and translation of virus-based nanomaterials from bench to bedside. As per the general rule, positively charged materials have longer circulation times. For example, negatively charged CCMV and CPMV tend to have short plasma half-lives (<15 min), while filamentous PVX with its positive surface charge has circulation time similar to CPMV. The other factors influencing the behavior of PVNs could be the complex biological barriers encountered and cellular interactions (Pitek et al., 2016a). In-depth in vivo investigations are required to shed more light on the influence of these factors on the fate of PVNs. PEGylation can be used to shield the PVNs, and receptor targeting is an effective strategy to confer cell-type specificity and to enhance cellular interactions (Lee et al., 2015). Czapar and Steinmetz (2016) have described in vivo protocols for preclinical evaluation of VNPs biological fate for their translational applications.

6.4 CONCLUSIONS

In summary, a range of different PVNs with distinct functionalities have been developed since a decade ago by following a wide variety of protocols and strategies for medical imaging, tissue targeting, and/or drug delivery. Most of the studies conducted so far have focused on the in vitro behavior of functionalized PVNs, either by biochemical assays or in cultured cells, and only limited data are available on the performance of specifically engineered PVNs in vivo (e.g., CPMV, PVX, and TMV). Although several new plant virus-derived nanoparticles are being added to the current array of virus-based nanotechnologies, they have not yet been investigated in vivo. Further studies are required to evaluate the behavioral properties of the functionalized PVNs in preclinical animal models including the assessment of their long-term fate to facilitate their translation from the research laboratory to the clinic. Development of multimodal PVN formulations for simultaneous disease diagnosis and therapy (theranostics) offers great promise for the development of next-generation nanodevices and therapeutics.

Acknowledgments

All the authors gratefully acknowledge the financial support received from the Department of Biotechnology (DBT) (Grant No. BT/PR6711/NNT/28/622/2012), New Delhi, India.

References

Abraham, A., Natraj, U., Karande, A.A., Gulati, A., Murthy, M.R.N., Murugesan, S., Mukunda, P., Savithri, H.S., 2016. Intracellular delivery of antibodies by chimeric Sesbania mosaic virus (SeMV) virus like particles. Sci. Rep. 6. 21803.

Alemzadeh, E., Izadpanah, K., Ahmadi, F., 2017. Generation of recombinant protein shells of Johnson grass chlorotic stripe mosaic virus in tobacco plants and their use as drug carrier. J. Virol. Methods 248, 148–153.

Aljabali, A.A., Shukla, S., Lomonossoff, G.P., Steinmetz, N.F., Evans, D.J., 2013. Mol. Pharm. 10, 3–10. CPMV-DOX delivers.

Arcangeli, C., Circelli, P., Donini, M., Aljabali, A.A., Benvenuto, E., Lomonossoff, G.P., Marusic, C., 2013. Structure-based design and experimental engineering of a plant virus nanoparticle for the presentation of immunogenic epitopes and a drug carrier. J. Biomol. Struct. Dyn. 32, 630–647.

Arkhipenko, M.V., Petrova, E.K., Nikitin, N.A., Protopopova, A.D., Dubrovin, E.V., Yaminskii, I.V., Rodionova, N.P., Atabekov, J.G., 2011. Characteristics of artificial virus-like particles assembled *in vitro* from potato virus X coat protein and foreign viral RNAs. Acta Nat. 3, 40–46.

Atabekov, J., Nikitin, N., Arkhipenko, M., Chirkov, S., Karpova, O., 2011. Thermal transition of native tobacco mosaic virus and RNA-free viral proteins into spherical nanoparticles. J. Gen. Virol. 92, 453–456.

Azizgolshani, O., Garmann, R.F., Cadena-Nava, R., Knobler, C.M., Gelbart, W.M., 2013. Reconstituted plant viral capsids can release genes to mammalian cells. Virology 441, 12–17.

Barwal, I., Kumar, R., Kateriya, S., Dinda, A.K., Yadav, S.C., 2016. Targeted delivery system for cancer cells consist of multiple ligands conjugated genetically modified CCMV capsid on doxorubicin GNPs complex. Sci. Rep. 6. 37096.

Berardi, A., Evans, D.J., BaldelliBombelli, F., Lomonossoff, G.P., 2018. Stability of plant virus-based nanocarriers in gastrointestinal fluids. Nanoscale 10, 1667–1679.

Bruckman, M.A., Yu, X., Steinmetz, N.F., 2013. Engineering Gd-loaded nanoparticles to enhance MRI via T_1 shortening. Nanotech. 24 (46), 462001. https://doi.org/10.1088/0957-4484/24/46/462001.

Bruckman, M.A., Randolph, L.N., VanMeter, A., Hern, S., Shoffstall, A.J., Turog, R.E., Steinmetz, N.F., 2014a. Biodistribution, pharmacokinetics and blood compatibility of native and PEGylated tobacco mosaic virus nano-rods and -spheres in mice. Virology 449, 163–173.

Bruckman, M.A., Jiang, K., Simpson, E.J., Randolph, L.N., Luyt, L.G., Yu, X., Steinmetz, N.F., 2014b. Dual-modal magnetic resonance and fluorescence imaging of atherosclerotic plaques *in vivo* using VCAM-1 targeted tobacco mosaic virus. Nano Lett. 14, 1551–1558.

Bruckman, M.A., Czapar, A.E., VanMeter, A., Randolph, L.N., Steinmetz, N.F., 2016. Tobacco mosaic virus-based protein nanoparticles and nanorods for chemotherapy delivery targeting breast cancer. J. Control. Release 231, 103–113.

Brumfield, S., Willits, D., Tang, L., Johnson, J.E., Douglas, T., Young, M., 2004. Heterologous expression of the modified coat protein of Cowpea chlorotic mottle bromovirus results in the assembly of protein cages with altered architectures and function. J. Gen. Virol. 85, 1049–1053.

Brunel, F.M., Lewis, J.D., Destito, G., Steinmetz, N.F., Manchester, M., Stuhlmann, H., Dawson, P.E., 2010. Hydrazone ligation strategy to assemble multifunctional viral nanoparticles for cell imaging and tumor targeting. Nano Lett. 10, 1093–1097.

Canizares, M.C., Nicholson, L., Lomonossof, G.P., 2005. Use of viral vectors for vaccine production in plants. Immunol. Cell Biol. 83, 263–270.

Cao, J., Guenther, R.H., Sit, T.L., Opperman, C.H., Lommel, S.A., Willoughby, J.A., 2014. Loading and release mechanism of red clover necrotic mosaic virus derived plant viral nanoparticles for drug delivery of doxorubicin. Small 10, 5126–5136.

Cao, J., Guenther, R.H., Sit, T.L., Lommel, S.A., Opperman, C.H., Willoughby, J.A., 2015. Development of Abamectin loaded plant virus nanoparticles for efficacious plant parasitic nematode control. ACS Appl. Mater. Interfaces 7, 9546–9553.

Carignan, D., Herblot, S., Laliberte-Gagne, M.E., Bolduc, M., Duval, M., Savard, P., Leclerc, D., 2017. Activation of innate immunity in primary human cells using a plant virus derived nanoparticle TLR7/8 agonist. Nanomedicine (17), 30195–30198. pii: S1549-9634.

Chariou, P.L., Steinmetz, N.F., 2017. Delivery of pesticides to plant parasitic nematodes using Tobacco mild green mosaic virus as a nanocarriers. ACS Nano 11, 4719–4730.

Chariou, P.L., Lee, K.L., Wen, A.M., Gulati, N.M., Stewart, P.L., Steinmetz, N.F., 2015. Detection and imaging of aggressive cancer cells using an epidermal growth factor receptor (EGFR)-targeted filamentous plant virus-based nanoparticle. Bioconjug. Chem. 26, 262–269.

Chen, C., Daniel, M.C., Quinkert, Z.T., De, M., Stein, B., Bowman, V.D., Chipman, P.R., Rotello, V.M., Kao, C.C., Dragnea, B., 2006. Nanoparticle-templated assembly of viral protein cages. Nano Lett. 6, 611–615.

Cho, C.F., Yu, L., Nsiama, T.K., Kadam, A.N., Raturi, A., Shukla, S., Amadei, G.A., Steinmetz, N.F., Luyt, L.G., Lewis, J.D., 2017. Viral nanoparticles decorated with novel EGFL7 ligands enable intravital imaging of tumor neovasculature. Nanoscale 9, 12096–12109.

Culver, J.N., Brown, A.D., Zang, F., Gnerlich, M., Gerasopoulos, K., Ghodssi, R., 2015. Plant virus directed fabrication of nanoscale materials and devices. Virology 480, 200–212.

Czapar, A.E., Steinmetz, N.F., 2016. Paving the way toward translational application of virus-based nanoparticles (VNPs): preclinical evaluation of their biological fates. In: Lu, Z.R., Sakuma, S. (Eds.), Nanomaterials in Pharmacology. Methods in Pharmacology and Toxicology. Humana Press, New York, NY.

Czapar, A.E., Steinmetz, N.F., 2017. Plant viruses and bacteriophages for drug delivery in medicine and biotechnology. Curr. Opin. Chem. Biol. 38, 108–116.

Czapar, A.E., Zheng, Y.R., Riddell, I.A., Shukla, S., Awuah, S.G., Lippard, S.J., Steinmetz, N.F., 2016. Tobacco mosaic virus delivery of phenanthriplatin for cancer therapy. ACS Nano 10, 4119–4126.

Destito, G., Yeh, R., Rae, C.S., Finn, M.G., Manchester, M., 2007. Folic acid-mediated targeting of cowpea mosaic virus particles to tumor cells. Chem. Biol. 14, 1152–1162.

Destito, G., Schneemann, A., Manchester, M., 2009. Biomedical nanotechnology using virus-based nanoparticles. Curr. Top. Microbiol. Immunol. 327, 95–122.

Dixit, S.K., Goicochea, N.L., Daniel, M.C., Murali, A., Bronstein, L., De, M., Stein, B., Rotello, V.M., Kao, C.C., Dragnea, B., 2006. Quantum dot encapsulation in viral capsids. Nano Lett. 6, 1993–1999.

Dragnea, B., Chen, C., Kwak, E.S., Stein, B., Kao, C.C., 2003. Gold nanoparticles as spectroscopic enhancers for *in vitro* studies on single viruses. J. Am. Chem. Soc. 125, 6374–6375.

Esfandiari, N., Arzanani, M.K., Soleimani, M., Kohi-Habibi, M., Svendsen, W.E., 2016. A new application of plant virus nanoparticles as drug delivery in breast cancer. Tumour Biol. 37, 1229–1236.

Finbloom, J.A., Han, K., Aanei, I.L., Hartman, E.C., Finley, D.T., Dedeo, M.T., Fishman, M., Downing, K.H., Francis, M.B., 2016. Stable disk assemblies of a tobacco mosaic virus mutant as nanoscale scaffolds for applications in drug delivery. Bioconjug. Chem. 27, 2480–2485.

Franke, C.E., Czapar, A.E., Patel, R.B., Steinmetz, N.F., 2017. Tobacco mosaic virus-delivered cisplatin restores efficacy in platinum-resistant ovarian cancer cells. Mol. Pharm. https://doi.org/10.1021/acs.molpharmaceut.7b00466.

Ganguly, R., Wen, A.M., Myer, A.B., Czech, T., Sahu, S., Steinmetz, N.F., Raman, P., 2016. Anti-atherogenic effect of trivalent chromium-loaded CPMV nanoparticles in human aortic smooth muscle cells under hyperglycemic conditions *in vitro*. Nanoscale 8, 6542–6554.

Gillitzer, E., Willits, D., Young, M., Douglas, T., 2002. Chemical modification of a viral cage for multivalent presentation. Chem. Commun. (Camb.) 21, 2390–2391.

Grasso, S., Santi, L., 2010. Viral nanoparticles as macromolecular devices for new therapeutic and pharmaceutical approaches. Int. J. Physiol. Pathophysiol. Pharmacol. 2, 161–178.

Guo, J., Zhao, X., Hu, J., Lin, Y., Wang, Q., 2018. Tobacco mosaic virus with peroxidase-like activity for cancer cell detection through colorimetric assay. Mol. Pharm. https://doi.org/10.1021/acs.molpharmaceut.7b00921.

Gupta, S., Chatni, M.R., Rao, A.L., Vullev, V.I., Wang, L.V., Anvari, B., 2013. Virus-mimicking nano-constructs as a contrast agent for near infrared photoacoustic imaging. Nanoscale 5, 1772–1776.

Hefferon, K., 2017. Plant virus expression vectors: a powerhouse for global health. Biomedicine 5, 44.

Hefferon, K.L., 2018. Repurposing plant virus nanoparticles. Vaccine 6, 11.

Honarbakhsh, S., Guenther, R.H., Willoughby, J.A., Lommel, S.A., Pourdeyhimi, B., 2013. Polymeric systems incorporating plant viral nanoparticles for tailored release of therapeutics. Adv. Healthc. Mater. 2, 1001–1007.

Hoopes, P.J., Mazur, C.M., Osterberg, B., Song, A., Gladstone, D.J., Steinmetz, N.F., Veliz, F.A., Bursey, A.A., Wagner, R.J., Fiering, S.N., 2017a. Effect of intra-tumoral magnetic nanoparticle hyperthermia and viral nanoparticle immunogenicity on primary and metastatic cancer. Proc. SPIE Int. Soc. Opt. Eng. . pii: 100660G.

Hoopes, P.J., Moodie, K.L., Petryk, A.A., Petryk, J.D., Sechrist, S., Gladstone, D.J., Steinmetz, N.F., Veliz, F.A., Bursey, A.A., Wagner, R.J., Rajan, A., Dugat, D., Crary-Burney, M., Fiering, S.N., 2017b. Hypo-fractionated radiation, magnetic nanoparticle hyperthermia and a viral immunotherapy treatment of spontaneous canine cancer. Proc. SPIE Int. Soc. Opt. Eng. . pii: 1006605.

Hu, H., Zhang, Y., Shukla, S., Gu, Y., Yu, X., Steinmetz, N.F., 2017. Dysprosium-modified tobacco mosaic virus nanoparticles for ultra-high-field magnetic resonance and near-infrared fluorescence imaging of prostate cancer. ACS Nano 11 (9), 9249–9258.

Huang, X., Stein, B.D., Cheng, H., Malyutin, A., Tsvetkova, I.B., Baxter, D.V., Remmes, N.B., Verchot, J., Kao, C., Bronstein, L.M., Dragnea, B., 2011. Magnetic virus-like nanoparticles in N. benthamiana plants: a new paradigm for environmental and agronomic biotechnological research. ACS Nano 5, 4037–4045.

Jeong, H., Seong, B.L., 2017. Exploiting virus-like particles as innovative vaccines against emerging viral infections. J. Microbiol. 55, 220–230.

Jobsri, J., Allen, A., Rajagopal, D., Shipton, M., Kanyuka, K., Lomonossoff, G.P., Ottensmeier, C., Diebold, S.S., Stevenson, F.K., Savelyeva, N., 2015. Plant virus particles carrying tumour antigen activate TLR7 and induce high levels of protective antibody. PLoS ONE 10, e0118096.

Jung, B., Rao, A.L., Anvari, B., 2011. Optical nano-constructs composed of genome-depleted brome mosaic virus doped with a near infrared chromophore for potential biomedical applications. ACS Nano 5, 1243–1252.

Kaiser, C.R., Flenniken, M.L., Gillitzer, E., Harmsen, A.L., Harmsen, A.G., Jutila, M.A., Douglas, T., Young, M.J., 2007. Biodistribution studies of protein cage nanoparticles demonstrate broad tissue distribution and rapid clearance *in vivo*. Int. J. Nanomedicine 2, 715–733.

Kan Davelaar, H.E., van Hest, J.C.M., Cornellisen, J.J.L.M., Koay, M.S.T., 2014. Using viruses as nanomedicines. Br. J. Pharmacol. 171, 4001–4009.

Kernan, D.L., Wen, A.M., Pitek, A.S., Steinmetz, N.F., 2017. Delivery of chemotherapeutic vcMMAE using tobacco mosaic virus nanoparticles. Exp. Biol. Med. (Maywood) 242, 1405–1411.

Koudelka, K.J., Manchester, M., 2010. Chemically modified viruses: principles and applications. Curr. Opin. Chem. Biol. 14, 810–817.

Koudelka, K.J., Destito, G., Plummer, E.M., Trauger, S.A., Siuzdak, G., Manchester, M., 2009. Endothelial targeting of cowpea mosaic virus (CPMV) via surface vimentin. PLoS Pathog. 5 (5), e1000417.

Koudelka, K.J., Pitek, A.S., Manchester, M., Steinmetz, N.F., 2015. Virus-based nanoparticles as versatile nanomachines. Annu. Rev. Virol. 2, 379–401.

Kushnir, N., Streatfield, S.J., Yusibov, V., 2012. Virus-like particles as a highly efficient vaccine platform: diversity of targets and production systems and advances in clinical development. Vaccine 31, 58–83.

Lauria, I., Dickmeis, C., Röder, J., Beckers, M., Rütten, S., Lin, Y.Y., Commandeur, U., Fischer, H., 2017. Engineered Potato virus X nanoparticles support hydroxyapatite nucleation for improved bone tissue replacement. Acta Biomater. 62, 317–327.

Le, D.H.T., Lee, K.L., Shukla, S., Commandeur, U., Steinmetz, N.F., 2017a. Potato virus X, a filamentous plant viral nanoparticle for doxorubicin delivery in cancer therapy. Nanoscale 9, 2348–2357.

Le, D.H.T., Hu, H., Commandeur, U., Steinmetz, N.F., 2017b. Chemical addressability of potato virus X for its applications in bio/nanotechnology. J. Struct. Biol. 200, 360–368.

Lebel, M.E., Chartrand, K., Tarrab, E., Savard, P., Leclerc, D., Lamarre, A., 2016. Potentiating cancer immunotherapy using papaya mosaic virus-derived nanoparticles. Nano Lett. 16, 1826–1832.

Lecours, K., Tremblay, M.H., Laliberte Gagne, M.E., Gagne SM Leclerc, D., 2006. Purification and biochemical characterization of a monomeric form of papaya mosaic potex virus coat protein. Protein Expr. Purif. 47, 273–280.

Lee, K.L., Uhde-Holzem, K., Fischer, R., Commandeur, U., Steinmetz, N.F., 2014. Genetic engineering and chemical conjugation of Potato virus X. Methods Mol. Biol. 1108, 3–21.

Lee, K.L., Shukla, S., Wu, M., Ayat, N.R., El Sanadi, C.E., Wen, A.M., Edelbrock, J.F., Pokorski, J.K., Commandeur, U., Dubyak, G.R., Steinmetz, N.F., 2015. Stealth filaments: polymer chain length and conformation affect the *in vivo* fate of PEGylated potato virus X. Acta Biomater. 9, 166–179.

Lee, K.L., Carpenter, B.L., Wen, A.M., Ghiladi, R.A., Steinmetz, N.F., 2016a. High aspect ratio of nanotubes formed by Tobacco mosaic virus for delivery of photodynamic agents targeting melanoma. ACS Biomater Sci. Eng. 2, 838–844.

Lee, K.L., Twyman, R.M., Fiering, S., Steinmetz, N.F., 2016b. Virus-based nanoparticles as platform technologies for modern vaccines. Wiley Interdiscip. Rev. Nanomed. Nanobiotechnol. 8, 554–578.

Lee, K.L., Murray, A.A., Le, D.H.T., Sheen, M.R., Shukla, S., Commandeur, U., Fiering, S., Steinmetz, N.F., 2017. Combination of plant virus nanoparticle-based in situ vaccination with chemotherapy potentiates antitumor response. Nano Lett. 17, 4019–4028.

Leong, H.S., Steinmetz, N.F., Ablack, A., Destito, G., Zijlstra, A., Stuhlmann, H., Manchester, M., Lewis, J.D., 2010. Intravital imaging of embryonic and tumor neovasculature using viral nanoparticles. Nat. Protoc. 5, 1406–1417.

Lewis, J.D., Destito, G., Zijlstra, A., Gonzalez, M.J., Quigley, J.P., Manchester, M., Stuhlmann, H., 2006. Viral nanoparticles as tools for intravital vascular imaging. Nat. Med. 12, 354–360.

Lico, C., Benvenuto, E., Baschieri, S., 2015. The two-faced Potato Virus X: from plant pathogen to smart nanoparticle. Front. Plant Sci. 6, 1009.

Liu, F., Ge, S., Li, L., Wu, X., Liu, Z., Wang, Z., 2012. Virus-like particles: potential veterinary vaccine immunogens. Res. Vet. Sci. 93, 553–559.

Lizotte, P.H., Wen, A.M., Sheen, M.R., Fields, J., Rojanasopondist, P., Steinmetz, N.F., Fiering, S., 2016. In situ vaccination with cowpea mosaic virus nanoparticles suppresses metastatic cancer. Nat. Nanotechnol. 11, 295–303.

Lockney, D.M., Guenther, R.N., Loo, L., Overton, W., Antonelli, R., Clark, J., Hu, M., Luft, C., Lommel, S.A., Franzen, S., 2011. The Red clover necrotic mosaic virus capsid as a multifunctional cell targeting plant viral nanoparticle. Bioconjug. Chem. 22, 67–73.

Lomonossoff, G.P., Evans, D.J., 2011. Applications of plant viruses in bionanotechnology. Curr. Top. Microbiol. Immunol. 375, 61–87.

Loo, L., Guenther, R.H., Lommel, S.A., Franzen, S., 2008. Infusion of dye molecules into Red clover necrotic mosaic virus. Chem. Commun. (1)88–90.

Lu, X., Thompson, J.R., Perry, K.L., 2012. Encapsidation of DNA, a protein and a fluorophore into virus-like particles by the capsid protein of cucumber mosaic virus. J. Gen. Virol. 93, 1120–1126.

Luckanagul, J.A., Lee, L.A., You, S., Yang, X., Wang, Q., 2015. Plant virus incorporated hydrogels as scaffolds for tissue engineering possess low immunogenicity *in vivo*. J. Biomed. Mater. Res. A 103, 887–895.

Ma, Y., Nolte, R.J., Cornelissen, J.J., 2012. Virus-based nanocarriers for drug delivery. Adv. Drug Deliv. Rev. 64 (9), 811–825.

Madden, A.J., Oberhardt, B., Lockney, D., Santos, C., Vennam, P., Arney, D., Franzen, S., Lommel, S.A., Miller, C.R., Gehrig, P., Zamboni, W.C., 2017. Pharmacokinetics and efficacy of doxorubicin loaded plant virus nanoparticles in preclinical models of cancer. Nanomedicine (London) 12, 2519–2532.

Masarapu, H., Patel, B.K., Chariou, P.L., Hu, H., Gulati, N.M., Carpenter, B.L., Ghiladi, R.A., Shukla, S., Steinmetz, N.F., 2017. Physalis mottle virus like particles as nanocarriers for imaging reagents and drugs. Biomacromolecules 18, 4141–4153.

Mateu, M.G., 2011. Virus engineering: functionalization and stabilization. Protein Eng. Des. Sel. 24, 53–63.

Mateu, M.G., 2016. Assembly, engineering and applications of virus-based protein nanoparticles. In: Cortajarena, A., Grove, T. (Eds.), Protein-Based Engineered Nanostructures. Advances in Experimental Medicine and Biology. vol. 940. Springer, Cham.

Mathieu, C., Rioux, G., Dumas, M.C., Leclerc, D., 2013. Induction of innate immunity in lungs with virus-like nanoparticles leads to protection against influenza and Streptococcus pneumoniae challenge. Nanomedicine 9, 839–848.

Maturavongsadit, P., Luckanagul, J.A., Metavarayuth, K., Zhao, X., Chen, L., Lin, Y., Wang, Q., 2016. Promotion of *in vitro* chondrogenesis of mesenchymal stem cells using in situ hyaluronic hydrogel functionalized with rod-like viral nanoparticles. Biomacromolecules 17, 1930–1938.

Millan, J.G., Brasch, M., Anaya-Plaza, E., de la Escosura, A., Velders, A.H., Reinhoudt, D.N., Torres, T., Koay, M.S., Cornellisen, J.J., 2014. Self-assembly triggered by self-assembly: optically active, paramagnetic micelles encapsulated in protein cage nanoparticles. J. Inorg. Biochem. 136, 140–146.

Molino, N.M., Wang, S.W., 2014. Caged protein nanoparticles for drug delivery. Curr. Opin. Biotechnol. 28, 75–82.

Narayanan, K.B., Han, S.S., 2017a. Icosahedral plant viral nanoparticles-bioinspired synthesis of nanomaterials/nanostructures. Adv. Colloid Interf. Sci. 248, 1–19.

Narayanan, K.B., Han, S.S., 2017b. Helical plant viral nanoparticles-bioinspired synthesis of nanomaterials and nanostructures. Bioinspir. Biomim. 12. 031001.

Narayanan, K.B., Han, S.S., 2017c. Genetic modifications of icosahedral plant virus-based nanoparticles for vaccine and immunotherapy applications. Curr. Protein Pept. Sci. 18, 1141–1151.

Natilla, A., Hammond, R.W., 2011. Maize rayado fino virus virus-like particles expressed in tobacco plants: a new platform for cysteine selective bioconjugation peptide display. J. Virol. Methods 178, 209–215.

Ng, B.C., Chan, S.T., Lin, J., Tolbert, S.H., 2011. Using polymer conformation to control architecture in semiconducting polymer/viral capsids assemblies. ACS Nano 5, 7730–7738.

Niehl, A., Appaix, F., Bosca, S., van der Sanden, B., Nicoud, J.F., Bolze, F., Heinlein, M., 2015. Fluorescent tobacco mosaic virus-derived bio-nanoparticles for Intravital two-photon imaging. Front. Plant Sci. 6, 1244.

Noueiry, A.O., Ahlquist, P., 2003. Brome mosaic virus RNA replication: revealing the role of the host in RNA virus replication. Annu. Rev. Phytopathol. 41, 77–98.

Palucha, A., Loniewska, A., Satheshkumar, S., Boguszewska-Chachulska, A.M., Umashankar, M., Milner, M., Haenni, A.L., Savithri, H.S., 2005. Virus-like particles: models for assembly studies and foreign epitope carriers. Prog. Nucelic Acid Res. Mol. Biol. 80, 135–168.

Phelps, J.P., Dao, P., Jin, H., Rasochova, L., 2007. Expression and self-assembly of cowpea chlorotic mottle virus-like particles in Pseudomonas fluorescens. J. Biotechnol. 128, 290–296.

Pichova, H., Cerovska, N., Vaculik, P., Moravec, T., 2017. Plant viruses as scaffolds for the presentation of vaccine epitops. Biol. Plant. 61, 1–12.

Pitek, A.S., Wen, A.M., Shukla, S., Steinmetz, N.F., 2016a. The protein corona of plant virus nanoparticles influences their dispersion properties, cellular interactions, and *in vivo* fates. Small 12, 1758–1769.

Pitek, A.S., Jameson, S.A., Veliz, F.A., Shukla, S., Steinmetz, N.F., 2016b. Serum albumin 'camouflage' of plant virus based nanoparticles prevents their antibody recognition and enhances pharmacokinetics. Biomaterials 89, 89–97.

Pitek, A.S., Wang, Y., Gulati, S., Gao, H., Stewart, P.L., Simon, D.I., Steinmetz, N.F., 2017. Elongated plant virus-based nanoparticles for enhanced delivery of thrombolytic therapies. Mol. Pharm. 14, 3815–3823.

Pokorski, J.K., Steinmetz, N.F., 2011. The art of engineering viral nanoparticles. Mol. Pharm. 8, 29–43.

Rae, C.R., Khor, I.W., Wang, Q., Destito, G., Gonzalez, M.J., Singh, P., Thomas, D., Estrada, M.N., Powell, E., Finn, M.G., Manchester, M., 2005. Systemic trafficking of plant virus nanoparticles in mice via the oral route. Virology 343, 224–235.

Raja, K.S., Wang, Q., Gonzalez, M.J., Manchester, M., Johnson, J.E., Finn, M.G., 2003. Hybrid virus-polymer materials. 1. Synthesis and properties of PEG-decorated cowpea mosaic virus. Biomacromolecules 4, 472–476.

Ren, Y., Wong, S.M., Lim, L.Y., 2007. Folic acid-conjugated protein cages of a plant virus: a novel delivery platform for doxorubicin. Bioconjug. Chem. 18, 836–843.

Ren, Y., Wong, S.M., Lim, L.Y., 2010. Application of plant viruses as nano drug delivery systems. J. Pharm. Res. 27, 2509–2513.

Roder, J., Fischer, R., Commandeur, U., 2017. Engineering Potato virus X particles for a covalent protein based attachment of enzymes. Small 13 (48). https://doi.org/10.1002/smll.201702151.

Rohovie, M.J., Nagasawa, M., Swartz, J.R., 2017. Virus-like particles: next-generation nanoparticles for targeted therapeutic delivery. Bioeng. Transl. Med. 2, 43–57.

Sanchez-Sanchez, L., Cadena-Nava, R.D., Palomares, L.A., Ruiz-Garcia, J., Koay, M.S., Cornelissen, J.J., Vazquez-Duhalt, R., 2014. Chemotherapy prodrug activation by biocatalytic virus-like nanoparticles containing cytochrome P450. Enzym. Microb. Technol. 60, 24–31.

Saunders, K., Sainsbury, F., Lomonossoff, G.P., 2009. Efficient generation of cowpea mosaic virus empty virus-like particles by the proteolytic processing of precursors in insect cells and plants. Virology 393, 329–337.

Savard, C., Laliberte-Gagne, M.E., Babin, C., Bolduc, M., Guerin, A., Drouin, K., Forget, M.A., Majeau, N., Lapointe, R., Leclerc, D., 2012. Improvement of the PapMV nanoparticle adjuvant property through an increased of its avidity for the antigen [influenza NP]. Vaccine 30, 2535–2542.

Scholthof, H.B., Scholthof, K.B.G., Jackson, A.O., 1996. Plant virus gene vectors for transient expression of foreign peptides in plants. Annu. Rev. Phytopathol. 34, 299–323.

Schwarz, B., Douglas, T., 2015. Development of virus-like particles for diagnostic and prophylactic biomedical applications. Nanomed. Nanobiochem. 7, 722–735.

Schwarz, B., Uchida, M., Douglas, T., 2017. Biomedical and catalytic opportunities of virus-like particles in nanotechnology. Adv. Virus Res. 97, 1–60. https://doi.org/10.1016/bs.aivir.2016.09.002.

Shirbaghaee, Z., Bolhassani, A., 2015. Different applications of virus-like particles in biology and medicine: vaccination and delivery systems. Biopolymers 105, 113–132.

Shriver, L.P., Koudelka, K.J., Manchester, M., 2009. Viral nanoparticles associate with regions of inflammation and blood brain barrier disruption during CNS infection. J. Neuroimmunol. 211, 66–72.

Shukla, S., Ablack, A.L., Am, W., Lee, K.L., Lewis, J.D., Steinmetz, N.F., 2013. Increased tumor homing and tissue penetration of the filamentous plant viral nanoparticles Potato virus X. Mol. Pharm. 10, 33–42.

Shukla, S., Dickmeis, C., Nagarajan, A.S., Fischer, R., Commandeur, U., Steinmetz, N.F., 2014a. Molecular farming of fluorescent virus-based nanoparticles for optical imaging in plants, human cells and mouse models. Biomater. Sci. 2, 784–797.

Shukla, S., Wen, A.M., Ayat, N.R., Commandeur, N., Gopalkrishnan, R., Broome, A.M., Lozada, K.W., Keri, R.A., Steinmetz, N.F., 2014b. Biodistribution and clearance of a filamentous plant virus in healthy and tumor bearing mice. Nanomedicine (London) 9, 221–235.

Shukla, S., Wen, A.M., Commandeur, U., Steinmetz, N.F., 2014c. Presentation of HER2 epitopes using a filamentous plant virus-based vaccination platform. J. Mater. Chem. 2, 6249–6258.

Shukla, S., Eber, F.J., Nagarajan, A.S., DiFranco, N.A., Schmidt, N., Wen, A.M., Eiben, S., Twyman, R.M., Wege, C., Steinmetz, N.F., 2015a. The impact of aspect ratio on the biodistribution and tumor homing of rigid soft-matter nanorods. Adv. Healthc. Mater. 4, 874–882.

Shukla, S., Dorand, R.D., Myers, J.T., Woods, S.E., Gulati, N.M., Stewart, P.L., Commandeur, U., Huang, A.Y., Steinmetz, N.F., 2016. Multiple administrations of viral nanoparticles alter *in vivo* behavior-insights from intravital microscopy. ACS Biomater. Sci. Eng. 2, 829–837.

Shukla, S., Myers, J.T., Woods, S.E., Gong, X., Czapar, A.E., Commandeur, U., Huang, A.Y., Levine, A.D., Steinmetz, N.F., 2017. Plant viral nanoparticles-based HER2 vaccine: immune response influenced by differential transport, localization and cellular interactions of particulate carriers. Biomaterials 121, 15–27.

Singh, P., Gonzalez, M.J., Manchester, M., 2006. Viruses and their uses in nanotechnology. Drug Dev. Res. 67, 23–41.

Singh, P., Prasuhn, D., Yeh, R.M., Destito, G., Rae, C.S., Osborn, K., Finn, M.G., Manchester, M., 2007. Biodistribution, toxicity and pathology of cowpea mosaic virus nanoparticles *in vivo*. J. Control. Release 120, 41–50.

Speir, J.A., Munshi, S., Wang, G., Baker, T.S., Johnson, J.E., 1995. Structures of the native and swollen forms of cowpea chlorotic mottle virus determined by X-ray crystallography and cryo-electron microscopy. Structure 3, 63–78.

Steele, J.F.C., Peyret, H., Saunders, K., Graells, R.C., Marsian, J., Meshcheriakova, Y., Lomonossof, G.P., 2017. Synthetic plant virology for nanotechnology and nanomedicine. Wiley Interdiscip. Rev. Nanomed. Nanobiotechnol. 9. e1447.

Steinmetz, N.F., 2010. Viral nanoparticles as platforms for next-generation therapeutics and imaging devices. Nanomedicine 6, 634–641.

Steinmetz, N.F., Evans, D.J., 2007. Utilization of plant viruses in bionanotechnology. Org. Biomol. Chem. 21, 2891–2902.

Steinmetz, N.F., Manchester, M., 2009. PEGylated viral nanoparticles for biomedicine: the impact of PEG chain length on VNP cell interactions *in vitro* and ex vivo. Biomacromolecules 13, 784–792.

Steinmetz, N.F., Lin, T., Lomonossoff, G.P., Johnson, J.E., 2009. Structure-based engineering of an icosahedral virus for nanomedicine and nanotechnology. In: Manchester, M., Steinmetz, N.F. (Eds.), Viruses and Nanotechnology. Current Topics in Microbiology and Immunology. vol. 327. Springer, Berlin, Heidelberg.

Steinmetz, N.F., Cho, C.F., Ablack, A., Lewis, J.D., Manchester, M., 2011. CPMV nanoparticles target surface vimentin on cancer cells. Nanomedicine 6, 351–364.

Strable, E., Finn, M.G., 2009. Chemical modification of viruses and virus-like particles. Curr. Top. Microbiol. Immunol. 327, 1–27.

Suci, P., Klem, M.T., Young, M., Douglas, T., 2008. Signal amplification using nanoplatform cluster formation. Soft Matter 4, 2519–2523.

Sun, J., DuFort, C., Daniel, M.C., Murali, A., Chen, C., Gopinath, K., Stein, B., De, M., Rotello, V.M., Holzenburg, A., Kao, C.C., Dragnea, B., 2007. Core-controlled polymorphism in virus-like particles. Proc. Natl. Acad. Sci. U. S. A. 104, 1354–1359.

Therien, A., Bedard, M., Carignan, D., Rioux, G., Gauthier-Landry, L., Laliberte-Gagne, M.E., Bolduc, M., Savard, P., Leclerc, D., 2017. A versatile papaya mosaic virus (PapMV) vaccine platform based on sortase-mediated antigen coupling. J. Nanobiotechnol. 15, 54.

Tinazzi, E., Merlin, M., Bason, C., Beri, R., Zampieri, R., Lico, C., Bartoloni, E., Puccetti, A., Lunardi, C., Pezzotti, M., Avesani, L., 2015. Plant-derived chimeric virus particles for the diagnosis of primary sjögren syndrome. Front. Plant Sci. 6, 1080.

Uchida, M., Klem, M.T., Allen, M., Suci, P., Flenniken, M.L., Gillitzer, E., Varpness, Z., Liepold, L.O., Young, M., Douglas, T., 2007. Biological containers: protein cages as multifunctional nanoplatforms. Adv. Mater. 19, 1025–1042.

Uhde-Holzem, K., McBurney, M., Tiu, B.D., Advincula, R.C., Fischer, R., Commandeur, U., Steinmetz, N.F., 2016. Production of immunoabsorbent nanoparticles by displaying single-domain protein A on potato virus X. Macromol. Biosci. 16, 231–241.

Vishnu Vardhan, G.P., Savithri, H.S., Murthy, M.R.N., Hema, M., 2016. Biodistribution and toxicity evaluation of sesbania mosaic virus in mice. Arch. Virol. 10, 2673–2681.

Wen, A.M., Steinmetz, N.F., 2014. The aspect ratio of nanoparticle assemblies and the spatial arrangement of ligands can be optimized to enhance the targeting of cancer cells. Adv. Healthc. Mater. 3, 1739–1744.

Wen, A.M., Steinmetz, N.F., 2016. Design of virus-based nanomaterials for medicine, biotechnology and energy. Chem. Soc. Rev. 45, 4074–4126.

Wen, A.M., Lee, K.L., Yildiz, I., Bruckman, M.A., Shukla, S., Steinmetz, N.F., 2012a. Viral nanoparticles for *in vivo* tumor imaging. J. Vis. Exp. 16, e4352.

Wen, A.M., Shukla, S., Saxena, P., Aljabali, A.A., Yildiz, I., Dey, S., Mealy, J.E., Yang, A.C., Evans, D.J., Lomonossoff, G.P., Steinmetz, N.F., 2012b. Interior engineering of a viral nanoparticle and its tumor homing properties. Biomacromolecules 13, 3990–4001.

Wen, A.M., Rambhia, P.H., French, R.H., Steinmetz, N.F., 2013. Design rules for nanomedical engineering: from physical virology to the applications of virus-based materials in medicine. J. Biol. Phys. 39, 301–325.

Wen, A.M., Cho, C.F., Lewis, J.D., Steinmetz, N.F., 2015a. The application of plant viral nanoparticles in tissue-specific imaging. In: Berezin, M.Y. (Ed.), Nanotechnology for Biomedical Imaging and Diagnostics: From Nanoparticle Design to Clinical Applications. John Wiley & Sons, Inc, Hoboken, NJ.

Wen, A.M., Podgornik, R., Strangi, G., Steinmetz, N.F., 2015b. Photonics and plasmonics go viral: self-assembly of hierarchical metamaterials. Rend. Fis. Acc. Lincei 26, 129–141.

Wen, A.M., Infusino, M., De Luca, A., Kernan, D.L., Czapar, A.E., Strangi, G., Steinmetz, N.F., 2015c. Interface of physics and biology: engineering virus-based nanoparticles for biophotonics. Bioconjug. Chem. 26, 51–62.

Wen, A.M., Lee, K.L., Cao, P., Pangilinan, K., Carpenter, B.L., Lam, P., Veliz, F.A., Ghiladi, R.A., Advincula, R.C., Steinmetz, N.F., 2016. Utilizing viral nanoparticle/Dendron hybrid conjugates in photodynamic therapy for dual delivery to macrophages and cancer cells. Bioconjug. Chem. 27, 1227–1235.

Wu, M., Shi, J., Fan, D., Zhou, Q., Wang, F., Niu, Z., Huang, Y., 2013. Biobehaviour in normal and tumor-bearing mice of tobacco mosaic virus. Biomacromolecules 14, 4032–4037.

Yildiz, I., Shukla, S., Steinmetz, N.F., 2011. Applications of viral nanoparticles in medicine. Curr. Opin. Biotechnol. 22, 901–908.

Yildiz, I., Tsvetkova, I., Wen, A.M., Shukla, S., Masarapu, M.H., Dragnea, B., Steinmetz, N.F., 2012. Engineering of Brome mosaic virus for biomedical applications. RSC Adv. 2, 3670–3677.

Yildiz, I., Lee, K.L., Chen, K., Shukla, S., Steinmetz, N.F., 2013. Infusion of imaging and therapeutic molecules into the plant virus-based carrier cowpea mosaic virus: cargo-loading and delivery. J. Control. Release 172, 568–578.

Yin, Z., Nguyen, H.G., Chowdhury, S., Bentley, P., Bruckman, M.A., Miermont, A., Gildersleeve, J.C., Wang, Q., Huang, X., 2012. Tobacco mosaic virus as a new carrier for tumor associated carbohydrate antigens. Bioconjug. Chem. 23, 1694–1703.

Young, M., Debbie, W., Uchida, M., Douglas, T., 2008. Plant viruses as biotemplates for materials and their use in nanotechnology. Annu. Rev. Phytopathol. 46, 361–384.

Yusibov, V., Rabindran, S., Commandeur, U., Twyman, R.M., Fischer, R., 2006. The potential of plant virus vectors for vaccine production. Drugs R. D. 7, 203–217.

Zdanowicz, M., Chroboczek, J., 2016. Virus-like particles as drug delivery vectors. Acta Biochim. Pol. 63, 469–473.

Zeltins, A., 2013. Construction and characterization of virus-like particles: a review. Mol. Biotechnol. 53, 92–107.

Zeng, Q., Saha, S., Lee, L.A., Bamhill, H., Oxsher, J., Dreher, T., Wang, Q., 2011. Chemoselective modification of turnip yellow mosaic virus by Cu(I) catalyzed azide-alkyne 1,3-dipolar cycloaddition reaction and its application in cell binding. Bioconjug. Chem. 22, 58–66.

Zeng, Q., Wen, H., Wen, Q., Chen, X., Wang, Y., Xuan, W., Liang, J., Wan, S., 2013. Cucumber mosaic virus as drug delivery vehicle for doxorubicin. Biomaterials 34, 4632–4642.

Zhang, Y., Ardejani, M.S., Orner, B.P., 2016. Design and applications of protein-cage-based nanomaterials. Chem. Asian. J. 11, 2814.

Further Reading

Shukla, S., DiFranco, N.A., Wen, A.M., Commandeur, U., Steinmetz, N.F., 2015b. To target or not to target: active vs. passive tumor homing of filamentous nanoparticles based on *Potato virus* X. Cell. Mol. Bioeng. 8, 433–444.

CHAPTER

7

Functional and Protective Role of Neutralizing Antibodies (NAbs) Against Viral Infections

Narayanaiah Cheedarla, Luke Elizabeth Hanna

Department of HIV/AIDS, National Institute for Research in Tuberculosis-ICMR, Chennai, India

OUTLINE

Recent Developments in Applied Microbiology and Biochemistry
https://doi.org/10.1016/B978-0-12-816328-3.00007-6

7.1 INTRODUCTION

Neutralizing antibodies (NAbs) play a major role and also strongly correlate with protection from viral infection after vaccination. At the same time, they are the markers of immunity against reinfection after clearance of an initial infection, and this kind of immunity can be lifelong (Klasse and Moore, 2012). Many vaccines against viral infections are good inducers of protective neutralizing antibody responses, but recalcitrant problems remain in the field of viral vaccination. One of the major problems is antigenic variability. The antigenic targets for neutralizing antibodies on influenza virus vary abundantly, and therefore, a new vaccine must be prepared every new season (Burton et al., 2012). Hepatitis C virus (HCV) and human immunodeficiency virus (HIV) are even more variable, and after years of research, researchers still cannot induce immune responses that protect against them (Klein et al., 2013). The sole response for the production of antibodies is the adaptive humoral immune response; the molecules that are recognized by antibodies are called antigens; the molecules that elicit the antibody response are immunogens. The specific portion on the surface of an antigen that is directly contacted by the antibody is called epitope, and the corresponding surface of the antibody in direct contact with the antigen is known as a paratope (Greenspan, 2010). Some successful vaccines against virus infections have been based on attenuated but replicating variants of the pathogenic virus, for example, the smallpox vaccine and one form of polio vaccine. Attenuated variants of the simian immunodeficiency virus (SIV), which is closely related to HIV, have provided stronger protection than nonreplicating experimental vaccines against the wild-type virus, but the mechanism of protection is not completely elucidated (Desrosiers, 1998). In other cases, recombinant proteins representing subunits of hepatitis B virus (HBV) and human papillomavirus (HPV) induce strong protection (Gerlich, 2013). The HPV vaccine consists of viruslike particles that may have the advantageous properties both antigenically and immunogenically: they may present native neutralization epitopes well and be seen by the innate immune system as pathogen-associated molecular patterns (Schiller and Chackerian, 2014). But subunit immunizations have failed to protect against HIV type 1 (HIV-1) (Burton et al., 2012). Only in the RV144 clinical trial, which combined with viral proteins expressed from a canary pox vector with recombinant subunit protein boosts, was some modest protection observed. But the vaccine had not induced NAbs (Meador et al., 2017). Therefore, the hunt is on for other antibodies and immune responses that might explain the limited protection. Many different antiviral effects of antibodies have been described that do not qualify as neutralization (Forthal et al., 2013). This brings us to some semantic clarifications.

7.2 VIRUS NEUTRALIZATION: DEFINITION

The reduction in viral infectivity by the binding of antibodies to the surface of viral particles (virions) is called neutralization, thereby blocking a step in the viral replication cycle that precedes virally encoded transcription or synthesis (Klasse and Sattentau, 2002). Classically, the term was applied only to antibodies and fragments of antibodies, Fab and F (ab′)2, but later, it has naturally been extended to single-domain antigen-binding recombinant fragments and natural nanobodies.

7.3 MECHANISMS OF NEUTRALIZATION

Neutralization has many mechanistic aspects: how the NAbs bind, whether they induce conformational changes, whether they irreversibly inactivate the viral proteins that mediate entry, and whether they are most effective against virions in suspension or after virion attachment to cells. Here, however, the mechanism of neutralization refers specifically to which early step in the viral replicative cycle is blocked (Fig. 7.1).

If the NAbs prevent the virions from attaching to the targets cells, then the process is called mechanism of neutralization. If the NAbs block essential receptor interactions after attachment, which is also referred to as mechanism of neutralization, so is interference with any other obligatory step in the entry of individual viruses, such as coreceptor engagement, endocytosis, fusion, or penetration. One way to analyze postattachment neutralization (PAN) is to let virus adsorb to cells at a low temperature that does not allow fusion or internalization and then to add the NAb. Further, HIV-1 infection is considered to explain more on antibody-mediated neutralization.

7.4 HOST IMMUNE RESPONSES TO HIV

The host humoral immune response to HIV infection has been the focus of intense interest since the virus was first identified in 1983 (Burton and Mascola, 2015). Detailed investigations carried out during the acute and chronic stages of HIV-1 infection have revealed that antibody responses to HIV-1 infection are complex and evolve over time. The latter characteristic is due to the changing nature of viral epitopes through mutation occurring during ongoing viral infection or possibly due to changes in $CD4^+$ T-cell help, because of either the

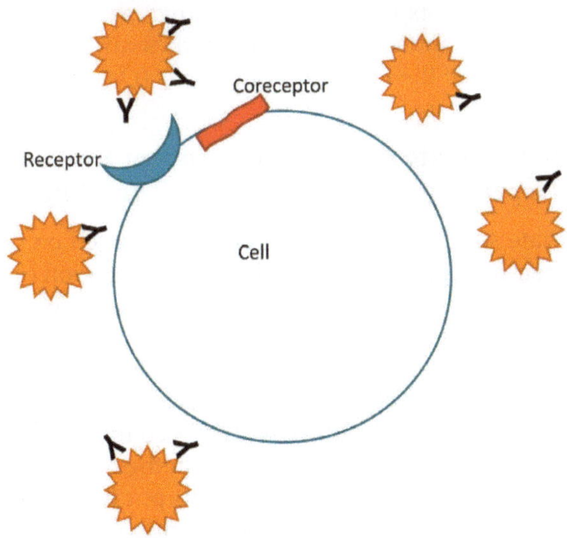

FIG. 7.1 The mechanism of neutralization: neutralization of enveloped viruses blocks viral attachment and entry into target cells. *Brown color* indicates the enveloped viral particles, which are bound by antibodies.

changes in T-cell epitopes as a result of mutation or the elimination of $CD4^+$ T cells through infection. Complexity and change are particularly apparent in antibody responses to the viral envelope glycoproteins, which are by far the most studied and most important functional responses and the only ones considered here. An abundance of HIV-specific antibodies develops throughout the course of HIV infection. These antibodies can be classified into three types, (i) binding antibodies, (ii) neutralizing antibodies, and (iii) opsonizing antibodies that work together with innate immune cells to kill virus-infected cells (Sarmay et al., 1992).

7.4.1 Binding Antibodies

Binding antibodies are produced at high levels throughout the life of an infected individual but are characterized by their inability to prevent viral infection. These antibodies identify highly immunogenic and variable regions of the HIV-1 virion. Even though they do not prevent viral entry, binding antibodies are useful as diagnostic indicators of whether an individual is infected or not. These kinds of antibodies typically bind to Env epitopes that are not presented on the native Env trimer spike (also known as the functional spike), which mediates entry into target cells. Because of their inability to bind to functional spikes, they are excluded from neutralizing activity. Hence, binding antibodies are also known as nonneutralizing antibodies (n-NAb). Neutralization is defined as the loss of infectivity that takes place when an antibody molecule binds to a virion and usually does not require the involvement of any other antibody activity. Because of the unstable nature of the functional Env spike, over a period of time, the functional spike structure may decay and reveal the epitopes that were recognized by these types of antibodies (Burton and Mascola, 2015). A common example of this type of antibody is directed against the immunodominant domain of glycoprotein 41 (gp41) that interacts with the HIV-1 envelope protein gp120 and is hidden from antibody recognition on the Env trimer (Blattner et al., 2014). If gp120 is shed from the trimer, gp41 is left in the viral membrane as six-helix bundles (known as "stumps"), which can also be recognized by antibodies. Such nonneutralizing antibodies can also act as antiviral agents by targeting nonfunctional forms of the Env on infected cells (Daar et al., 2001).

7.4.2 Antibodies That Mediate Antibody Dependent Cellular Cytotoxicity (ADCC)/Antibody Dependent Cell-Mediated Viral Inhibition (ADCVI)

Antibodies that mediate ADCC bind to viral epitopes exposed on the surface of HIV-1 infected cells and exert their biological activity through the Fc region of the antibody (Sarmay et al., 1992). Fc receptors are expressed on natural killer cells, monocytes, macrophages, dendritic cells, and neutrophils (Sarmay et al., 1992). ADCC occurs by a three-way interaction between an infected target cell that displays antigens on its surface, ADCC antibodies, and an effector cell expressing the Fc receptor. These interactions ultimately result in the killing of the infected target cell (Ackerman et al., 2016). ADCVI also involves the interaction between a target cell, an ADCVI antibody, and an effector cell. On the other hand, rather than causing cell death, ADCVI antibodies goal to reduce the viral output from infected target cells. It has been confirmed that HIV-specific ADCC and ADCVI antibodies are present in most HIV-1 infected individuals during early infection (Dugast et al., 2014). However, with disease progression, natural killer cells lose their function and become unable to mediate ADCC activity

(Jost et al., 2012). ADCC antibodies differ from neutralizing antibodies in that, while ADCC antibodies target infected cells and result in cell killing, neutralizing antibodies target cell-free virus and result in viral inhibition (Walker and Yu, 2013).

7.4.3 Neutralizing Antibodies

This group includes two types of antibodies known as autologous neutralizing antibodies and broadly cross clade neutralizing antibodies.

(i) *Autologous neutralizing antibodies*: Most HIV-1 infected individuals produce antibodies capable of inhibiting their own virus (autologous virus) but incapable of neutralizing heterologous viruses; these antibodies are also known as autologous neutralizing antibodies (ANAbs) (Gray et al., 2007). Some recent studies have reported that this kind of antibodies is usually produced early during HIV infection (~52 days) and can be identified within the first 3–12 months of infection (Wei et al., 2003). These antibodies target immunogenic, defined regions of the HIV-1 virion. Nevertheless, their neutralization ability is transient due to viral evolution (Burton and Mascola, 2015). The ANAbs recognize a particular viral variant (strain-specific) only, but not new viral variants that subsequently emerge due to de novo viral mutations induced by the error-prone RT enzyme (Moore et al., 2009) and host immune pressure. Interestingly, it has been noted that the potency of ANAbs is correlated with shorter Env variable region length and the presence of a smaller number of glycosylation sites in the V1-V2 and C3-V4 regions of the Env (Gray et al., 2007; Moore et al., 2009).

(ii) *Broadly neutralizing antibodies*: Neutralizing antibodies with cross-reactivity have been identified in 10%–30% of HIV-1 infected individuals. These antibodies have the unique ability to effectively neutralize multiple viral isolates across diverse strains and subtypes (Burton and Mascola, 2015). Target epitopes of neutralizing antibodies reside in the Env glycoprotein subunits, gp120 and gp41. Numerous studies have evaluated and characterized the neutralizing antibody responses elicited during natural infection among HIV-1 infected individuals worldwide (Fig. 7.2).

For example, a study by Doria-Rose et al. (2010) evaluated neutralizing antibody response in 103 participants in various stages of HIV-1 infection. The findings from this study suggest that neutralizing antibody titers increased with time since infection and that neutralization breadth correlated with viral load. The sole target of the bNAbs is the envelope of HIV-1. There are five major vulnerable sites on the HIV-1 envelope, namely, the CD4 binding site (CD4BS), V1V2 loop, V3 loop, membrane-proximal external region (MPER), and gp120/gp41 interface.

7.5 CLASSIFICATION OF BROADLY NEUTRALIZING ANTIBODIES

Broadly neutralizing antibodies have also been classified as first-generation bNAbs and second-generation bNAbs depending on when and how they were identified (Shcherbakov et al., 2015).

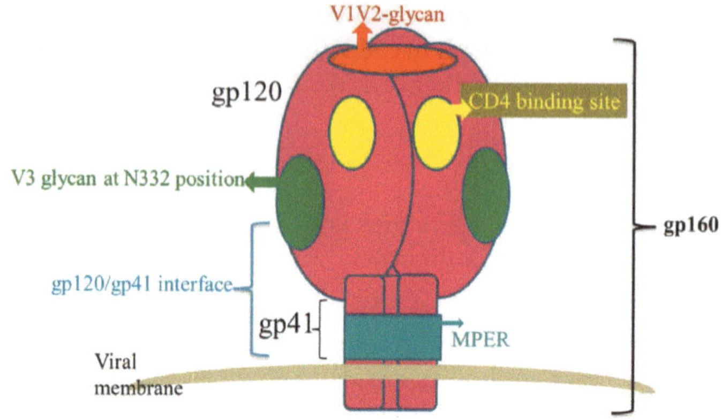

FIG. 7.2 Vulnerable sites for broadly neutralizing antibodies on the HIV-1 envelope glycoprotein. Epitopes in the V1/V2 loops *(brown)*, V3 loop *(green)*, CD4 binding site *(yellow)*, and gp120/gp41 bridging region *(blue)* are highlighted as defined by PG9, PGT128, VRC01, 8ANC195, and PGT151 binding positions, respectively. The MPER of gp41, for which only limited structural information is available in the context of the HIV-1 Env trimer, is also a site of vulnerability and is highlighted in *light blue*.

(i) First-generation broadly cross clade neutralizing antibodies: Before 2009, a finger-countable number of bNAbs were isolated using phage display technology, such as b12, 2G12, 2F5, 4E10, and Z13 (Pejchal et al., 2011). The b12 bNAb recognizes the conserved gp120 CD4BS; 2G12 bNAb mainly binds to the glycans; and 2F5, 4E10, and Z13 bNAbs recognize linear overlapping epitopes in the MPER of gp41.

(ii) Second-generation broadly cross clade neutralizing antibodies: In 2009–10, PG9/PG16 and VRC01 bNAbs were identified from chronically HIV-1 infected individuals by Walker et al. (2009) and Wu et al. (2010), respectively, and this came as a breakthrough in the field of discovery of bNAbs against HIV-1.

These bNAbs were isolated using new approaches such as B-cell selection, sorting, and high-throughput procedures for generating human monoclonal antibodies. These bNAbs showed superior efficacy and potency against a broad range of primary HIV-1 isolates as compared with the first-generation bNAbs. Since then, hundreds of bNAbs have been identified and isolated using the abovementioned methodologies; VRC02, PGV04, CH30–34, 3BNC117, 3BNC60, 3BNC55, 12A21, 12A12, 8ANC131, 8ANC134, NIH45–46, 1NC9, 1B2530, CH01–04, PGT141–145, 10E8, 8ANC195, and PGT151–158 are some of the second-generation bNAbs (Shcherbakov et al., 2015).

7.6 MAJOR TARGET SITES OF BROADLY NEUTRALIZING ANTIBODIES ON THE ENVELOPE OF HIV-1

7.6.1 CD4 Binding Site (CD4BS)

The CD4BS, one of the most prominent sites of vulnerability, harbors epitopes for a large number of bNAbs. It was a surprise to find that the VRC01 bNAb, isolated from a chronically infected individual, showed extraordinary neutralization ability (neutralizing over 90% of circulating HIV-1 pseudoviruses) by targeting the CD4BS on the HIV-1 gp120 trimeric molecule (Raja et al., 2003).

7.6.2 V1/V2 Loopglycansin the Quaternary Envelope of HIV-1

Glycans present in the V1/V2 loop of gp120 of HIV-1 are also targeted by bNAbs. For the first time, Walker et al. (2009) identified PG9 and PG16 bNAbs and isolated monoclonal antibodies (mAbs) from HIV-1 subtype A infected individual that showed 78% and 73% neutralization breadth, respectively. Glycosylation of N156 and N160 was subsequently shown to critically influence PG9 and PG16 interaction with the trimeric gp120. The third complementarity-determining region of the heavy-chain (CDR H3) loop of PG9 plays an essential role in stabilizing the PG9-gp120 complex. Its exceptional long loop of 30 amino acid residues penetrates the glycan shield on gp120 to allow access to the protein surface around the V2 and V3 loops. Soon after the identification of the PG9 and PG16 bNAbs, several bNAbs were isolated in quick succession, such as CH01–04, PGT141–145, VRC26.01–33, and CAP256-VRC26, which recognize the same epitope. The trademark of these antibodies is their extraordinarily long CDR H3, permitting penetration between the glycans and interaction with them, thus contributing to enhanced binding to gp120 (McLellan et al., 2011).

7.6.3 V3 Loop Glycanin the Quaternary Envelope of HIV-1

In 2011, two research groups identified new bNAbs that majorly target glycans surrounding the V3 loop (Pejchal et al., 2011). These bNAbs were found to be more potent than 2G12 that also targets the glycans around the V3 loop. Further studies showed that the newly identified bNAbs target the high-mannose glycans on residue N332 within the V3 loop (Kong et al., 2013). Walker et al. (2011) tested the potency of 2G12 and PGT series of antibodies against a panel of 162 pseudoviruses representing all major circulating HIV-1 subtypes. They found that 2G12 effectively neutralized 32% of the tested viruses (none of which were subtype C or E), while the PGT 121 series of monoclonal antibodies neutralized >50% of the tested viruses. PGT121, PGT128, and PGT145mAbs neutralized 70%, 72%, and 78% of the tested viruses, respectively. Interestingly, these mAbs potently neutralized over 30% of the clade C viruses in the panel. The evidence that 2G12 does not present cross neutralization against subtype C isolates may be due to the lack of a glycan at position 295 at the N-terminal base of the V3 loop in subtype C viruses (Binley et al., 2004).

7.6.4 Membrane Proximal External Region (MPER)

The fourth major site of vulnerability for bNAbs is the membrane-proximal external region (MPER) in the gp41 of the HIV-1 envelope. Similar to the CD4 binding site, this region is well conserved and has been well studied. The MPER is critical for viral fusion and cell entry, and hence, molecular conservation of MPER is highly important to maintain its functions. Five bNAbs, namely, 2F5, 4E10, Z13, Z13e1, and 10E8, recognizing overlapping epitopes in the conserved MPER at the base of gp41, have thus far been identified (Huang et al., 2012;

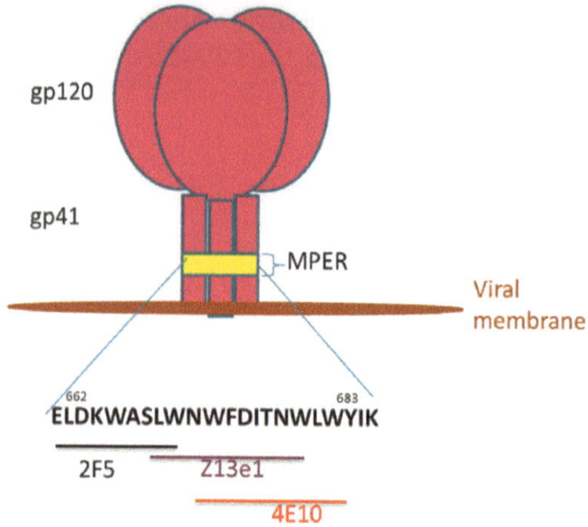

FIG. 7.3 The structure of the MPER and immersion depth changes induced by 2F5 and Z13e1 binding. Schematic diagram of HIV-1 gp160 trimer. The minimal peptide epitopes of bNAbs are indicated for 2F5 *(black)*, Z13e1 *(blue)*, and 4E10 *(red)*.

Shcherbakov et al., 2015). 2F5 recognizes the epitope ELDKWA and the DKW motif (positions 664–666 on gp41) that is the minimum core required for 2F5 neutralization (Fig. 7.3).

2F5 effectively neutralizes isolates belonging to subtypes A, B, D, and E but shows very poor neutralization of subtype C isolates, possibly due to the fact that most subtype C viruses have a DSW motif in place of DKW. 4E10 was previously described to recognize the NWFDIT epitope; however, Binley et al. (2004) showed that the minimum requirement for 4E10 neutralization is the WFXI motif. The authors also demonstrated that 4E10 has potent cross-reactive neutralization activity on 90 viruses belonging to subtypes A, B, C, D, F, G, and J and AC, AE, AG, BF, and BG recombinants. The extensive cross-reactive neutralization capacity of 2F5 was confirmed by its ability to recognize and bind to 31 variations of the MPER (Bryson et al., 2009). 2F5 also exhibits faster binding to the MPER and is more thermodynamically favorable as compared with 4E10. Dennison et al. (2009) showed that the neutralization breadth of antibodies targeting the MPER was dependent on the membrane immersion depth of their epitopes. The slow binding of 4E10 could be attributed to the membrane immersion depth of its epitope, as compared with the 2F5 epitope. Z13 was isolated in 2001 from an antibody phage display library and was shown to recognize an epitope similar to that of 4E10 but with an ability to neutralize a limited set of primary viruses (Zwick et al., 2001). Recently, a high-affinity variant of Z13, designated Z13e1, was isolated that recognizes an epitope overlapping with that of 2F5 and 4E10 (WASLWNWFDITN) and exhibits 100-fold better affinity for the MPER and enhanced neutralization capacity against sensitive strains of HIV-1 (Nelson et al., 2007). In 2012, Huang et al. isolated and characterized another highly potent broadly cross clade neutralizing antibody named 10E8, which neutralized ~98% of the HIV-1 isolates examined and did not show polyreactivity like the other four MPER-directed bNAbs. 10E8 recognized a novel epitope that overlaps with the known 4E10 and Z13e1 epitopes but differs in its critical dependence on Asn 671 and Arg 683, the last residues of the MPER, meaning 10E8 bound to the full 23 residues of MPER.

7.6.5 gp120/gp41 Interface Region

In the recent years, bNAbs have been identified that recognize both gp120 and gp41 proteins on the envelope of HIV-1 (Falkowska et al., 2014; Huang et al., 2014). Falkowska et al. (2014) identified a group of bNAbs (PGT-151 series) from a single donor that recognize a glycan-dependent epitope on the prefusion conformation of gp41. In this study, researchers isolated and characterized a monoclonal antibody called as 35022 that neutralized 62% of 181 tested pseudoviruses with a half-maximal inhibitory concentration (IC_{50}) < 50 μg mL^{-1}. Similar to the known major bNAbs (except for gp41 MPER-specific antibodies), these antibodies bind Env through glycan-mediated interactions, but in contrast, their critical glycan contacts are located on gp41. PGT151 has elongated CDRH3 and CDR L1 loops that are capable of recognizing Env in a conformation adapted prior to fusion of the viral membrane with the target cell membrane. Interestingly, PGT151–158-like antibodies can mediate antibody-dependent cellular cytotoxicity (Falkowska et al., 2014). The 35022 bNAb also possesses long CDR H3 and CDR L1 loops that are useful for recognizing both gp120 and gp41 on the HIV-1 Env.

7.7 CONCLUDING REMARKS

Intensive investigations in the last three decades provided a comprehensive knowledge about the role of broadly neutralizing antibodies against viral infections. Most of the viral infections are caused by three major viruses, influenza, hepatitis, and HIV. These are enveloped viruses that play a major role in the escape from the host immune system. However, recent studies are identified potent bNAbs that strongly inhibit viral entry by interacting major vulnerable sites on the envelope protein of the virus. For example, there are five vulnerable sites identified on the envelope protein of HIV-1, namely, CD4BS, PG9/PG16-like (N160-dependent), PGT121-like (N332-dependent), MPER, and gp120/gp41 interface. Further, few of these bNAbs were used as therapeutic vaccines in human trials and observed that combinational bNAb vaccines showed more protection than single bNAb vaccine. The quest to develop a successful HIV vaccine continues. There are many unsolved problems and challenges. Primarily, the extensive viral subtype and sequence diversity is generated by the "error-prone" HIV enzyme, reverse transcriptase, such that even within a particular subtype, the amino acid sequences of the *Env* gene could differ up to 20%, and the difference could reach up to 35% between subtypes. Secondly, extensive glycosylation of the HIV-1 envelope glycoprotein trimers effectively shields the conserved epitopes from antibody recognition. Therefore, further studies have to be done to improve strong antibody-mediated neutralization against globally circulating viral strains.

Acknowledgments

The authors are indebted to all the researchers whom we cited in this review for their significant and valuable research. No funding was received to perform this review.

Conflicts of interests

The authors declare no relevant competing financial interests to disclose.

References

Ackerman, M.E., Mikhailova, A., Brown, E.P., Dowell, K.G., Walker, B.D., Bailey-Kellogg, C., Suscovich, T.J., Alter, G., 2016. Polyfunctional HIV-specific antibody responses are associated with spontaneous HIV control. PLoS Pathog. 12, e1005315.

Binley, J.M., Wrin, T., Korber, B., Zwick, M.B., Wang, M., Chappey, C., Stiegler, G., Kunert, R., Zolla-Pazner, S., Katinger, H., Petropoulos, C.J., Burton, D.R., 2004. Comprehensive cross-clade neutralization analysis of a panel of anti-human immunodeficiency virus type 1 monoclonal antibodies. J. Virol. 78, 13232–13252.

Blattner, C., Lee, J.H., Sliepen, K., Derking, R., Falkowska, E., de la Pena, A.T., Cupo, A., Julien, J.P., van Gils, M., Lee, P.S., Peng, W., Paulson, J.C., Poignard, P., Burton, D.R., Moore, J.P., Sanders, R.W., Wilson, I.A., Ward, A.B., 2014. Structural delineation of a quaternary, cleavage-dependent epitope at the gp41-gp120 interface on intact HIV-1 Env trimers. Immunity 40, 669–680.

Bryson, S., Julien, J.P., Hynes, R.C., Pai, E.F., 2009. Crystallographic definition of the epitope promiscuity of the broadly neutralizing anti-human immunodeficiency virus type 1 antibody 2F5: vaccine design implications. J. Virol. 83, 11862–11875.

Burton, D.R., Mascola, J.R., 2015. Antibody responses to envelope glycoproteins in HIV-1 infection. Nat. Immunol. 16, 571–576.

Burton, D.R., Poignard, P.R., Stanfield, L., Wilson, I.A., 2012. Broadly neutralizing antibodies present new prospects to counter highly antigenically diverse viruses. Science 337 (6091), 183–186.

Daar, E.S., Little, S., Pitt, J., Santangelo, J., Ho, P., Harawa, N., Kerndt, P., Glorgi, J.V., Bai, J., Gaut, P., Richman, D.D., Mandel, S., Nichols, S., 2001. Diagnosis of primary HIV-1 infection. Los Angeles County primary HIV infection recruitment network. Ann. Intern. Med. 134, 25–29.

Dennison, S.M., Stewart, S.M., Stempel, K.C., Liao, H.X., Haynes, B.F., Alam, S.M., 2009. Stable docking of neutralizing human immunodeficiency virus type 1 gp41 membrane-proximal external region monoclonal antibodies 2F5 and 4E10 is dependent on the membrane immersion depth of their epitope regions. J. Virol. 83, 10211–10223.

Desrosiers, R.C., 1998. Prospects for live attenuated HIV. Nat. Med. 4 (9), 982.

Doria-Rose, N.A., Klein, R.M., Daniels, M.G., O'Dell, S., Nason, M., Lapedes, A., Bhattacharya, T., Migueles, S.A., Wyatt, R.T., Korber, B.T., Mascola, J.R., Connors, M., 2010. Breadth of human immunodeficiency virus-specific neutralizing activity in sera: clustering analysis and association with clinical variables. J. Virol. 84, 1631–1636.

Dugast, A.S., Stamatatos, L., Tonelli, A., Suscovich, T.J., Licht, A.F., Mikell, I., Ackerman, M.E., Streeck, H., Klasse, P.J., Moore, J.P., Alter, G., 2014. Independent evolution of Fc- and Fab-mediated HIV-1-specific antiviral antibody activity following acute infection. Eur. J. Immunol. 44, 2925–2937.

Falkowska, E., Le, K.M., Ramos, A., Doores, K.J., Lee, J.H., Blattner, C., Ramirez, A., Derking, R., et al., 2014. Broadly neutralizing HIV antibodies define a glycan-dependent epitope on the prefusion conformation of gp41 on cleaved envelope trimers. Immunity 40, 657–668.

Forthal, D., Hope, T.J., Alter, G., 2013. New paradigms for functional HIV-specific nonneutralizing antibodies. Curr. Opin. HIV AIDS 8 (5), 393–401.

Gerlich, W.H., 2013. Medical virology of hepatitis B: how it began and where we are now. Virol. J. 10, 239.

Gray, E.S., Moore, P.L., Choge, I.A., Decker, J.M., Bibollet-Ruche, F., Li, H., Leseka, N., Treurnicht, F., Mlisana, K., Shaw, G.M., Karim, S.S., Williamson, C., Morris, L., 2007. Neutralizing antibody responses in acute human immunodeficiency virus type 1 subtype C infection. J. Virol. 81, 6187–6196.

Greenspan, N.S., 2010. Cohen's conjecture, Howard's hypothesis, and Ptashnes ptruth: an exploration of the relationship between affinity and specificity. Trends Immunol. 31 (4), 138–143.

Huang, J., Ofek, G., Laub, L., Louder, M.K., Doria-Rose, N.A., Longo, N.S., Imamichi, H., et al., 2012. Broad and potent neutralization of HIV-1 by a gp41-specific human antibody. Nature 491, 406–412.

Huang, J., Kang, B.H., Pancera, M., Lee, J.H., Tong, T., Feng, Y., Imamichi, H., Georgiev, I.S., et al., 2014. Broad and potent HIV-1 neutralization by a human antibody that binds the gp41-gp120 interface. Nature 515, 138–142.

Jost, S., Altfeld, M., Chang, J.J., 2012. Harnessing innate and adaptive immunity for viral vaccine design. Expert Rev. Vaccines 11, 775–777.

Klasse, P.J., Moore, J.P., 2012. Good CoP, bad CoP? Interrogating the immune responses to primate lentiviral vaccines. Retrovirology 9, 80.

Klasse, P.J., Sattentau, Q.J., 2002. Occupancy and mechanism in antibody-mediated neutralization of animal viruses. J. Gen. Virol. 83, 2091–2108. Pt 9.

Klein, F., Mouquet, H., Dosenovic, P., Scheid, J.F., Scharf, L., Nussenzweig, M.C., 2013. Antibodies in HIV-1 vaccine development and therapy. Science 341, 1199–1204.

Kong, L., Lee, J.H., Doores, K.J., Murin, C.D., Julien, J.P., McBride, R., Liu, Y., Marozsan, A., et al., 2013. Supersite of immune vulnerability on the glycosylated face of HIV-1 envelope glycoprotein gp120. Nat. Struct. Mol. Biol. 20, 796–803.

McLellan, J.S., Pancera, M., Carrico, C., Gorman, J., Julien, J.P., Khayat, R., Louder, R., Pejchal, R., et al., 2011. Structure of HIV-1 gp120 V1/V2 domain with broadly neutralizing antibody PG9. Nature 480, 336–343.

Meador, L.R., Kessans, S.A., Kilbourne, J., Kibler, K.V., Pantaleo, G., Roderiguez, M.E., Blattman, J.N., Jacobs, B.L., Mor, T.S., 2017. A heterologous prime-boosting strategy with replicating Vaccinia virus vectors and plant-produced HIV-1 Gag/dgp41 virus-like particles. Virology 507, 242–256.

Moore, P.L., Ranchobe, N., Lambson, B.E., Gray, E.S., Cave, E., Abrahams, M.R., Bandawe, G., Mlisana, K., Abdool Karim, S.S., Williamson, C., Morris, L., 2009. Limited neutralizing antibody specificities drive neutralization escape in early HIV-1 subtype C infection. PLoS Pathog. 5, e1000598.

Nelson, J.D., Brunel, F.M., Jensen, R., Crooks, E.T., Cardoso, R.M., Wang, M., Hessell, A., Wilson, I.A., Binley, J.M., Dawson, P.E., Burton, D.R., Zwick, M.B., 2007. An affinity-enhanced neutralizing antibody against the membrane-proximal external region of human immunodeficiency virus type 1 gp41 recognizes an epitope between those of 2F5 and 4E10. J. Virol. 81, 4033–4043.

Pejchal, R., Doores, K.J., Walker, L.M., Khayat, R., Huang, P.S., Wang, S.K., Stanfield, R.L., et al., 2011. A potent and broad neutralizing antibody recognizes and penetrates the HIV glycan shield. Science 334, 1097–1103.

Raja, A., Venturi, M., Kwong, P., Sodroski, J., 2003. CD4 binding site antibodies inhibit human immunodeficiency virus gp120 envelope glycoprotein interaction with CCR5. J. Virol. 77 (1), 713–718.

Sarmay, G., Lund, J., Rozsnyay, Z., Gergely, J., Jefferis, R., 1992. Mapping and comparison of the interaction sites on the Fc region of IgG responsible for triggering antibody dependent cellular cytotoxicity (ADCC) through different types of human Fc gamma receptor. Mol. Immunol. 29, 633–639.

Schiller, J., Chackerian, B., 2014. Why HIV virions have low numbers of envelope spikes: implications for vaccine *development*. PLoS Pathog. 10 (8), e1004254.

Shcherbakov, D.N., Bakulina, A.Y., Karpenko, L.I., Ilyichev, A.A., 2015. Broadly neutralizing antibodies against HIV-1 as a novel aspect of the immune response. Acta Nat. 7, 11–21.

Walker, B.D., Yu, X.G., 2013. Unravelling the mechanisms of durable control of HIV-1. Nat. Rev. Immunol. 13, 487–498.

Walker, L.M., Phogat, S.K., Chan-Hui, P.Y., Wagner, D., Phung, P., Goss, J.L., Wrin, T., Simek, M.D., Fling, S., Mitcham, J.L., Lehrman, J.K., Priddy, F.H., Olsen, O.A., Frey, S.M., Hammond, P.W., Kaminsky, S., Zamb, T., Moyle, M., Koff, W.C., Poignard, P., Burton, D.R., 2009. Broad and potent neutralizing antibodies from an African donor reveal a new HIV-1 vaccine target. Science 326, 285–289.

Walker, L.M., Huber, M., Doores, K.J., Falkowska, E., Pejchal, R., Julien, J.P., Wang, S.K., et al., 2011. Broad neutralization coverage of HIV by multiple highly potent antibodies. Nature 477 (7365), 466–470.

Wei, X., Decker, J.M., Wang, S., Hui, H., Kappes, J.C., Wu, X., Salazar-Gonzalez, J.F., Salazar, M.G., Kilby, J.M., Saag, M.S., Komarova, N.L., Nowak, M.A., Hahn, B.H., Kwong, P.D., Shaw, G.M., 2003. Antibody neutralization and escape by HIV-1. Nature 422, 307–312.

Wu, X., Yang, Z.Y., Li, Y., Hogerkorp, C.M., Schief, W.R., Seaman, M.S., Zhou, T., Schmidt, S.D., Wu, L., Xu, L., Longo, N.S., McKee, K., O'Dell, S., Louder, M.K., Wycuff, D.L., Feng, Y., Nason, M., Doria-Rose, N., Connors, M., Kwong, P.D., et al., 2010. Rational design of envelope identifies broadly neutralizing human monoclonal antibodies to HIV-1. Science 329, 856–861.

Zwick, M.B., Labrijn, A.F., Wang, M., Spenlehauer, C., Saphire, E.O., Binley, J.M., Moore, J.P., Stiegler, G., Katinger, H., Burton, D.R., Parren, P.W., 2001. Broadly neutralizing antibodies targeted to the membrane-proximal external region of human immunodeficiency virus type 1 glycoprotein gp41. J. Virol. 75, 10892–10905.

SECTION 2

MICROBIAL BIOTECHNOLOGY

CHAPTER

8

Recent Advances in the Industrial Production of L-Lysine by Bacteria

Hamsa D. Tadepally

Scientific Curator (Consultant), Cleveland, OH, United States

OUTLINE

8.1 INTRODUCTION

Amino acids are the building blocks of proteins and are nutritionally important compounds for all living organisms. All the 20 common amino acids have different chemical properties with varied commercial interest. However, of the 20 amino acids, nine of them also called essential amino acids cannot be synthesized by animals and humans and need to be supplied through food or feed. There is a growing demand for amino acids in several markets, including animal feed, pharmaceutical industries, and cosmetics (Sanchez and Demain, 2014). The world market for L-lysine has been increasing exponentially from 550,000 tons in

Recent Developments in Applied Microbiology and Biochemistry
https://doi.org/10.1016/B978-0-12-816328-3.00008-8

2001 to above 1 million tons in 2009 (Wittmann and Becker, 2007). In general, there is low lysine content in the traditional animal feeds such as corn, wheat, and barley. The addition of L-lysine in its pure form increases the quality of the feed to a large extent (Brautaset et al., 2011). Lysine is also used as a fish feed additive (Ikeda, 2003, 2016). Interestingly, plants and bacteria can synthesize lysine. Lysine exists in two stereoisomeric forms—D-lysine and L-lysine. The D-lysine has industrial applications too. But for this review, only L-lysine will be discussed. L-lysine is the biologically active form found in cells and has the highest commercial value compared with all other lysine forms (Yoshida and Nagasawa, 2003). Interestingly, industrial production of L-lysine was the first amino acid to which the advantage of genetic tools was applied (Ikeda, 2016).

8.2 PRODUCTION STATISTICS OF L-LYSINE

Owing to the applications, demand, and growing market, the commercial production of L-lysine is increasing. Lysine takes second place with regard to worldwide annual production after glutamate (Wittmann and Becker, 2007). L-lysine in addition to the other amino acids in the aspartate pathway contributes to ~56% of the total amino acid market, which equates to ~4.5 billion US dollars (Razak and Viswanath, 2015a,b). The market is expected to be worth over 9.4 billion US dollars by 2018 (D'Este et al., 2018). About 4 million tons of amino acids were produced in 2009. L-lysine-HCl was second in terms of production after glutamate with ~1.3 million tons (Sanchez and Demain, 2008). Globally, there is an increase of about 10% annually in the demand for L-lysine with the market being 2.2 million metric tons per year (Eggeling and Bott, 2005).

8.3 L-LYSINE AND THE ASPARTATE FAMILY

To understand the impact of using genetic tools to increase industrial production, it is important to look at L-lysine in the context of its biosynthetic pathway (Fig. 8.1). L-lysine in addition to L-methionine, L-threonine, and L-isoleucine belongs to the aspartate amino acid family. Different bacterial species have different variants of this pathway; a common feature to all the organisms is feedback regulation. The aspartate biosynthetic pathway has particularly been studied extensively in the organism *Corynebacterium glutamicum* with the aim of identifying enzymes, rate-limiting steps that can be manipulated in the overproduction of L-lysine.

The first step in the aspartate biosynthetic pathway is the conversion of aspartate into aspartate-4-P using AK and then into aspartate semialdehyde using Asd. The step involving aspartate semialdehyde is the branching point in the pathway that can get converted into homoserine by HD that then gets converted into isoleucine, threonine, or methionine or into 2,3-dihydrodipicolinate by DapA to enter into the lysine pathway. The next step is the conversion of 2,3-dihydropicolinate by DapB into tetrahydropicolinate. There are several variants of the lysine biosynthetic pathway in bacteria, from the point of tetrahydropicolinate (Hermann, 2003; Wittmann and Becker, 2007) (Fig. 8.1).

In particular, the bacterium *C. glutamicum* has both the direct variant and the succinylase variants, which interestingly can operate together (Hermann, 2003; Li et al., 2017). In general,

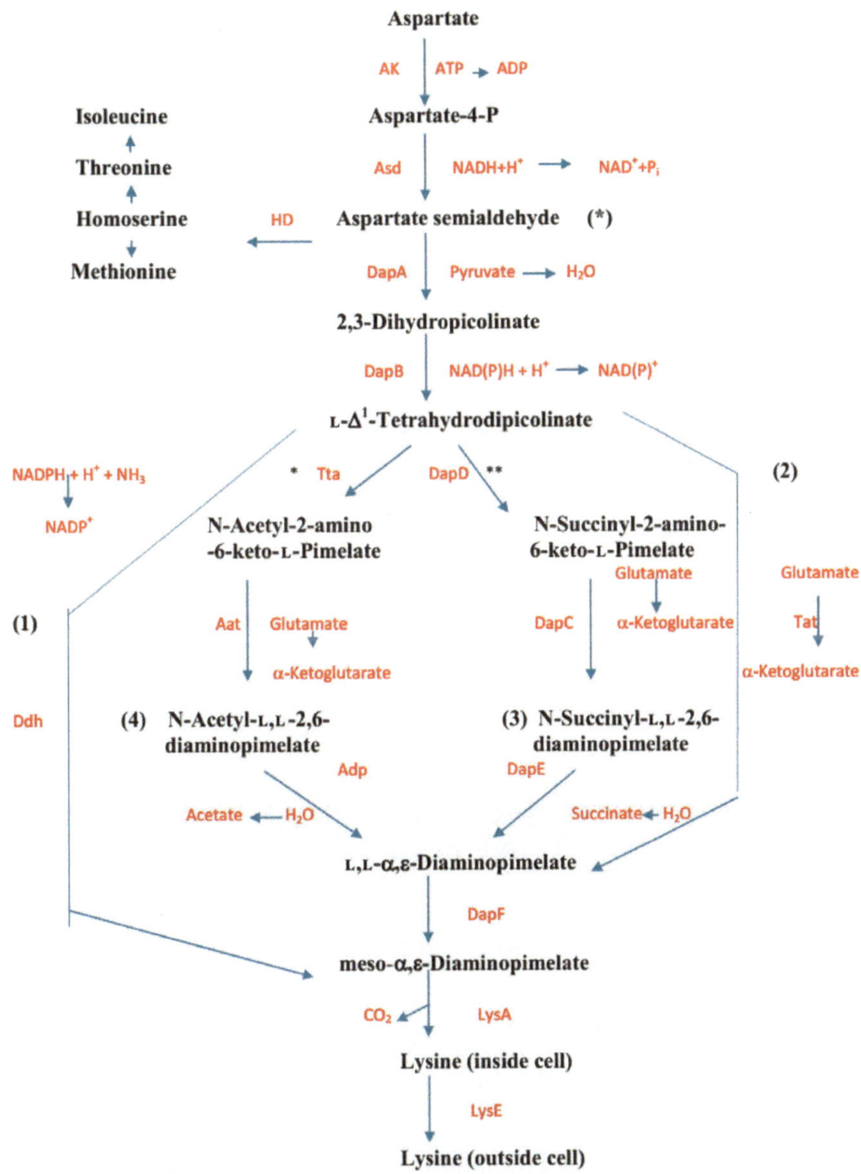

FIG. 8.1 The aspartate biosynthetic pathway and its components: steps leading to lysine production. (*)—Branching point of the pathway that leads to the production of lysine or other amino acids. *—H_2O + acetyl-CoA ≥ CoASH. **—H_2O + succinyl-CoA ≥ CoASH. (1) One variant is the conversion of tetrahydropicolinate directly into meso-DL-diaminopimelate that is the precursor of lysine using the enzyme diaminopimelate dehydrogenase (Ddh). (2) A second variant is where tetrahydropicolinate is converted into LL-diaminopimelate in a transamination step catalyzed by tetrahydropicolinate aminotransferase (Tat) that is then converted into the lysine precursor meso-DL-diaminopimelate using the enzyme diaminopimelate epimerase (DapF). (3) A third variant involves three enzymes, tetrahydropicolinate succinylase (DapD), succinyl-aminoketopimelate transaminase (DapC), and succinyl-diaminopimelate desuccinylase (DapE). This is a succinylase variant. (4) A fourth variant is an acetylase variant, catalyzed by three enzymes, tetradipicolinate acetylase (Tta), acetyl aminoketopimelate aminotransferase (Aat), and acetyl diaminopimelate deacetylase (Adp). *Modified from Brautaset, T., Ellingsen, T.E., Eggeling, L., 2011. Lysine: industrial uses and production. In: Brautaset, T. (Ed.), Reference Module in Life Sciences. Elsevier, Trondheim, pp. 542–554.*

all the enzymes involved in the aspartate pathway have to various extents been studied as important targets for developing bacterial strains in the overproduction of lysine.

8.4 FEEDBACK REGULATION IN DIFFERENT BACTERIA

The feedback inhibition step where the first enzyme of the pathway AK regulates the carbon flux entry is a common feature of the aspartate biosynthetic pathway that has been extensively used in the methods designed to overproduce lysine. This enzyme is in a concerted feedback inhibition loop with lysine and threonine. The aspartate biosynthetic pathway and the rate-limiting steps for the overproduction of L-lysine have been particularly well studied and documented in *C. glutamicum*. *C. glutamicum* has only one AK and is the only enzyme that is regulated (Eggeling and Bott, 2005). This is encoded by two genes *lysCα* and *lysCβ*. On the other hand, bacteria such as *Bacillus megaterium* and *Lactobacillus plantarum* have two AK enzymes, and *Escherichia coli*, *B. subtilis*, and *B. methanolicus* have three AK enzymes (Jakobsen et al., 2009).

In addition to the biosynthetic pathways steps, in *C. glutamicum*, a specific lysine exporter (secretion of L-lysine from the cells into the culture broth), LysE has been identified and has been proved to be important for efficient production of lysine (Brautaset et al., 2011). Overexpression of LysE, in a wild-type background led to an increased production of L-lysine (fivefold increase; 3.76 $nmol\,min^{-1}\,g^{-1}$) (Eggeling and Bott, 2005; Gunji and Yasueda, 2006).

8.5 HISTORICAL BACKGROUND AND METHODS OF LYSINE PRODUCTION

L-lysine was produced for the very first time by a company called Kyowa Hakko Kogyo, in Japan in 1978. Since then, L-lysine has been manufactured by bacterial fermentation of carbohydrates like beet and cane sugar or hydrolyzed starch. The most common bacterial strain used *C. glutamicum* (Ikeda, 2016; Anaya-Reza and Lopez-Arenas, 2017). There are three different ways in which amino acids are produced—using protein hydrolysates for extraction, chemical synthesis, and microbial processes (D'Este et al., 2018). Of the three processes, industrial production of L-lysine is almost entirely by microbial fermentation using *C. glutamicum*. In recent times, the main focus of research studies has been on improving the methods—production capacities, fermentation and downstream processes, and genomic refining of the strains used for fermentation (Brautaset et al., 2011).

- *Extraction:* Although not used for industrial production of lysine today, extraction was first used to produce L-lysine from casein in 1889. The major drawback of this method was the dependency on the availability of naturally available protein-rich raw materials such as keratin, blood meal, and soybean (Ikeda, 2003).
- *Chemical synthesis:* Chemical synthesis is a process that generates both the D- and L-forms of amino acids as a mixture. An additional optical resolution step to separate the D- and L-forms resulted in high-production costs. Hence, chemical synthesis is not a significant method used to produce L-lysine (Ikeda, 2003).

- *Biotechnological methods:* There are two types of biotechnological methods used in producing amino acids.
  - *Enzymatic synthesis:* In contrast to chemical synthesis, enzymatic synthesis provides the advantage of producing optically pure L-amino acids in higher concentrations. The downstream process is also simplified due to the presence of fewer by-products. However, the success of this method depends majorly on the substrate chosen and its cost. This method, therefore, is not very popular in the production of L-lysine (Brautaset et al., 2011).
  - *Fermentation:* In today's times, fermentation is the most popular and widely used method in the production of most L-amino acids.

8.6 INDUSTRIAL PRODUCTION OF L-LYSINE—TYPES OF FERMENTATION PROCESSES USED

Fermentation using overproducing mutant strains of corynebacteria (*C. glutamicum*) is almost exclusively the method used for industrial production of L-lysine.

There are several steps and processes involved in the entire fermentation process:

- *The fermentation medium:* For any successful industrial fermentation process, the medium, oxygen based on if it is aerobic or anaerobic, temperature, and the required pH of the medium during the fermentation process are of high importance. It is beneficial to use a well-defined and studied growth medium. Several growth conditions beneficial to the organism have to be taken into account. The source of vitamins, nitrogen, salts, and trace elements in the medium are all important factors to consider for a well-defined growth medium. In the case of nitrogen source, ammonia or ammonium sulfate is preferred. For *C. glutamicum*, biotin is an important source of vitamins (Brautaset et al., 2011). In addition to the individual nutrients, the main source of carbon is very crucial for the entire production of L-lysine. The source must be economical and add to the lowered production costs but more yield of the final product. In the case of L-lysine, some of the carbon sources used are also based on the local availability (Anastassiadis, 2007). Starch hydrolysates are popular in North America and Asia, while molasses are more in use in Europe and South America (Hermann, 2003). However, the most popular carbon sources used today owing to their efficient and economical usage in industrial production of L-lysine are glucose-rich starch hydrolysates from corn, wheat, and cassava. The economy of an industrial process also depends on the efficient use of the organism in the medium. Soybean meal and corn steep liquor are raw materials commonly added to the medium to meet the needs of the production strains for amino acids (Brautaset et al., 2011). Owing to the fact that L-lysine fermentation is an aerobic process, optimal maintenance of oxygen supply throughout the fermentation process is essential for a good yield. Generally, the oxygen transfer rate (OTR) is a good measure of the aerobic conditions in a medium that is increased by blowing oxygen-enriched air through the culture and then stirring vigorously. This breaks up the air bubbles to increase the gas-liquid interface area and OTR. One of the many challenges in the industrial production of lysine is to maintain continuous aerobic conditions in the fermentation medium, especially in the case of *C. glutamicum* that has been shown to need different levels of dissolved oxygen levels at different times (Hermann, 2003; D'Este et al., 2018). The pH of the cell culture is an important factor to consider in microbial fermentations. pH values in lysine fermentations in the case of *C. glutamicum* are maintained between 6 and 9. In lysine fermentations, alkaline compounds such as sodium hydroxide, potassium hydroxide, and ammonia are used to maintain the pH at desired levels. Optimal levels of temperature are also an important factor to consider in industrial production of amino acids. A patent literature study (Anastassiadis, 2007) reported that the operational temperatures for *C. glutamicum* strains in L-lysine production range from 20°C to 50°C.

- *Fermentation processes:* There are two processes involved in this step. The first step is the seed cultivation step that occurs in the seed reactor. Seed reactors are typically smaller, ranging from 1 to 50 m^3 as compared with the production reactor. This step helps obtain sufficient inoculation of the production reactor. Once the quality of the inoculum has been ascertained, there are different ways in which the lysine production can take place.
  - *Batch fermentation:* The batch fermentation is inoculated with the biomass from a seed reactor where all necessary nutrients for cell growth and L-lysine production are added from the start. This turns out to be a disadvantage as in the beginning only a limited amount of carbon source can be used in the growth medium that in turn will limit the maximum product concentration of L-lysine. This method is no longer a preferred method for the production of L-lysine, although it is still used in smaller fermentation plants worldwide (Brautaset et al., 2011).
  - *Fed-batch and repeated fed-batch fermentation:* This is the most preferred method for the production of L-lysine. This method basically involves feeding of more sugar with a predefined feeding profile into the production reactor, after the initial carbon source is completely consumed. In the case of L-lysine, when using dextrose as a carbon source, volumetric L-lysine-HCl concentrations above 160 $g L^{-1}$ have been observed (Eggeling and Bott, 2005). Another variant of the fed-batch method, called repeated fed-batch method, further improves the effectiveness of the fed-batch method. This method also called the "fill and draw" is a type of semicontinuous culture where, after the reactor is filled with the feeder, most of the cell culture is removed to downstream processing including product purification. But a fixed amount (~10%–20%) of the culture is left behind in the reactor. This serves as inoculum for the next round of the batch fermentation. Increased production period and reduced preparation times are the advantages of this method. However, the risk to this method is contamination and danger of changes in the production strain due to the high number of generations (Eggeling and Bott, 2005; Hermann, 2003; Brautaset et al., 2011).
  - *Continuous culture:* The continuous process is where the fresh medium is fed into the reactor and the same amount of culture liquid is taken out from the reactor. This keeps the entire process volume at a constant level. There are two methods in the continuous process—turbidostat and chemostat. The turbidostat keeps the biomass concentration constant, and the chemostat is where the limiting nutrient in the culture medium controls the cell concentration in the reactor. Owing to a major drawback that there is a possibility that the production strain will undergo spontaneous mutations because of multiple generations under substrate-limited growth conditions, there is also a higher contamination risk (Ikeda, 2003; Brautaset et al., 2011). The continuous

culture method was tested with *C. glutamicum* mutant B-6 and was reported to obtain a maximum product concentration of 105 $g L^{-1}$ L-lysine-HCl in 48 h and a productivity of 5.6 $g L^{-1} h^{-1}$.

- *Final production process—Downstream processes and recent developments:* The final step after the whole fermentation process is to obtain the final and main product—crystalline L-lysine-HCl. The overall cost efficiency of the industrial production of L-lysine also depends on the final downstream processes. The first step in the final process is to separate the product from the culture broth. Traditionally, the method used—centrifugation or filtration—is dependent on the desired final product—L-lysine, L-lysine-HCl, or Biolys (Brautaset et al., 2011). The final product could be in the form of liquid concentrates or as granulated solids typically having different concentrations of the product. Liquid lysine 60 is an example of liquid product that has both free lysine and lysine-HCl at 60% concentration. This is developed by Ajinomoto Co., Inc. Biolys another product is the granulated form of lysine. This is produced by evaporation and granulation of the final product without removing the cell mass. Although 54.6% pure, it does contain other amino acids and proteins (Ikeda, 2016). It also depends on what the final usage for the product is—pharmaceutical or used as feed and pertinent environmental regulations of disposal of the waste and raw material used. Interestingly, the disposed discharge liquid even though an industrial waste can be used as a fertilizer for plants (Brautaset et al., 2011). In recent times, however, more economical and simpler versions of downstream processing have been introduced. Once the cell biomass has been separated, the resulting filtrate is either spray—dried or evaporated. This reduces the production costs considerably (Ikeda, 2016).

8.7 ADVANCES AND DEVELOPMENTS IN THE RAW MATERIAL AND PRODUCTION STRAIN CHOICES THAT DETERMINE EFFECTIVE L-LYSINE PRODUCTION

Considering the importance and growing worldwide market of lysine, there is constant effort made to improve the entire production process—fermentation processes, optimizing the media conditions, and downstream processes to obtain the final product. Recent progress in metabolic engineering has improved mechanisms to produce lysine using more robust design and generating high-quality L-lysine-producing strains. In addition to genetically manipulating the bacterial strains and exploiting new bacterial strains, the possibility of using other carbohydrate substrates such a lignocellulose and methanol is being studied.

- *The carbon source (sugar):* For any industrial production process, the ultimate aim is to control and economize production costs and yet obtain a high yield. During fermentation processes, the primary important factor to consider that effects the entire production cost is the cost of the carbon source. L-lysine is a low-price bulk product, and the cost of the carbon source that also accounts for the raw materials is of utmost importance to be able to keep the production costs low. In general, for L-lysine, the main carbon sources used are glucose and sucrose (hexose sugars) that are derived from starch hydrolysates of corn, cassava, cane, and beet molasses. As mentioned earlier in the review, the choice of carbon source depends on the availability and the location of the fermentation plant. Owing to the rising cost of sugar sources, there have been recent efforts to look into alternative and inexpensive carbon sources such as methanol and lignocellulose (Jojima et al., 2010).

Methanol and methane are two compounds that are very important from a biotechnological and industrial standpoint. There are studies investigating its efficiency and usefulness in bacterial fermentation processes, more importantly in the production of L-lysine. Owing to its low cost, high availability, purity, and solubility in water, methanol is being considered as a useful alternative to sugar as a carbon source in microbial fermentations. Three methylotrophic bacteria have been studied for this purpose:

- Gram-negative, obligate *Methylophilus methylotrophus* and *Methylobacillus glycogenes*: In spite of there being several studies on manipulating the aspartate pathway in *M. glycogenes* for L-lysine overproduction, experimental knowledge on how the feedback process effects the pathway genes is not clear. A study by (Motoyama et al., 2001) where they looked at the overexpression of the DapA gene by recombinant strain AL119 yielded 8 g L^{-1} of L-lysine. In contrast, in another organism *M. methylotrophus*, the L-lysine biosynthetic pathway has been well characterized. About 11.3 g L^{-1} of L-lysine was produced from methanol by manipulating the feedback regulation in the DapA gene and the LysE24 gene in recombinant *M. methylotrophus* strain AS1 (pSEA10).
- Gram-positive and thermotolerant restricted methylotroph *B. methanolicus*: A study by using methanol as a raw material and AEC-resistant and auxotrophic *B. methanolicus* mutant NOA2#13A52-8A66 can secrete up to 47 g L^{-1} of L-lysine in optimized fed-batch fermentations (Brautaset et al., 2007). A major advantage of using *B. methanolicus* in L-lysine production is that it is a thermotolerant bacterium and can be cultivated at 50°C. This can significantly reduce the cooling costs caused otherwise by using methanol, which has high-oxygen requirements, as a carbon source (Brautaset et al., 2010).

Lignocellulose is another promising candidate as a feedstock for microbial fermentation of bulk products. There has been progress using this source; however, there is no industrial process in place taking advantage of this method (Brautaset et al., 2011).

8.8 APPLICATION OF GENETIC TOOLS AND METABOLIC ENGINEERING STRATEGIES TO INCREASE INDUSTRIAL PRODUCTION OF L-LYSINE

The majority of research conducted to manipulate and increase industrial L-lysine production has extensively been conducted with *C. glutamicum* as a model organism (Eggeling and Bott, 2005). The completely sequenced genome of the organism and application of several genetic tools have contributed to the increased progress toward being able to manipulate not only the L-lysine biosynthetic pathway but also the enzymes that play crucial roles and make efficient targets to increase the industrial production of L-lysine (Brautaset et al., 2011). In particular, enzymes active in the feedback regulation that when targeted can lead to overexpression have been studied (Anastassiadis, 2007; Eggeling and Bott, 2005; Wittmann and Becker, 2007). AK, the first enzyme in the pathway, is the main target for AEC resistance in

C. glutamicum. Also, it was shown that elevated gene copy numbers of DapA, DapB, DapC, and DapF lead to elevated enzyme levels and in turn increased production levels of L-lysine (Anastassiadis, 2007). A better understanding of the central metabolism of *C. glutamicum* includes studying the different mechanisms involved in sugar uptake. This also aids in being able to explore other alternatives for sugar sources. One important pathway in this process is the PP pathway (Li et al., 2017). The PP pathway is catalyzed by seven enzyme steps and represents a bypass of the glycolytic pathway at glucose-6-phosphate. A detailed explanation of this pathway can be found in recently published papers (Brautaset et al., 2011; D'Este et al., 2018). In general four molecules of NADPH are needed to produce one molecule of L-lysine. One example of manipulating the PP pathway is where a study identified a point mutation in one of the genes of the pathway, *gnd* gene, gnd361, which resulted in increased 6-phosphogluconate dehydrogenase activity and increased carbon flux by 8% into the PP pathway of the mutant strain (Ikeda et al., 2006). Another study engineered a new enzyme, glyceraldehyde 3-phosphate dehydrogenase (GAPDH), which produces two molecules of NADPH from one molecule of glucose, and this in turn results in an increased yield of the target amino acid (D'Este et al., 2018).

Once the wild-type organism has been manipulated using genetic tools to arrive at a beneficial mutation, it becomes more convenient to then look at different fermentation criteria to investigate more optimal conditions to be able to get higher yields of the product. A study by Razak and Viswanath (2015a) demonstrated that in the mutant *C. glutamicum* MH 20-22 B, it was seen that when these cells were immobilized and the following conditions were maintained—fermentation time 96h, pH 7.5, temperature 30°C, glucose concentration 80 g L^{-1}, airflow rate 1.25 vvm, and aeration rate 300 rpm—immobilized cells of *C. glutamicum* produced higher quantities of L-lysine (31.58 g L^{-1}) as compared with free cells (26.34 g L^{-1}) (Razak and Viswanath, 2015b).

8.9 CONCLUSIONS AND PERSPECTIVES

L-Lysine has a huge demand as a feed supplement in the pig and poultry industries. New applications for L-lysine in the pharmaceutical and cosmetic industries have been increasing too.

Comparative genomic analysis of wild-type and classical industrial mutants, including genome-based strain reconstruction, will be increasingly important to generate new production strains. This review is a brief summary of all the methods and latest developments in the industrial production of L-lysine.

Acknowledgments

The authors are indebted to all the researchers whom we cited in this review for their significant and valuable research. No funding was received to perform this review.

Conflicts of interests

The authors declare no relevant competing financial interests to disclose.

References

Anastassiadis, S., 2007. L-Lysine fermentation. Recent Pat. Biotechnol. 1, 11–24.

Anaya-Reza, O., Lopez-Arenas, T., 2017. Comprehensive assessment of the L-lysine production process from fermentation of sugarcane molasses. Bioprocess Biosyst. Eng. 40, 1033–1048.

Brautaset, T., Jakobsen, Ø.M., Josefsen, K.D., Flickinger, M.C., Ellingsen, T.E., 2007. *Bacillus methanolicus*: a candidate for industrial production of amino acids from methanol at 50 degrees centrigrade. Appl. Microbiol. Biotechnol. 74, 22–34.

Brautaset, T., Jakobsen, Ø.M., Degnes, K.F., Netzer, R., Naerdal, I., Krog, A., Dillingham, R., Flickinger, M.C., Ellingsen, T.E., 2010. *Bacillus methanolicus* pyruvate carboxylase and homoserine dehydrogenase I and II and their roles for L-lysine production from methanol at 50 degrees C. Appl. Microbiol. Biotechnol. 87, 951–964.

Brautaset, T., Ellingsen, T.E., Eggeling, L., 2011. Lysine: industrial uses and production. In: Brautaset, T. (Ed.), Reference Module in Life Sciences. Elsevier, Trondheim, pp. 542–554.

D'Este, M., Alvarado-Morales, M., Angelidaki, I., 2018. Amino acids production focusing on fermentation technologies—a review. Biotechnol. Adv. 36, 14–25.

Eggeling, L., Bott, M., 2005. Handbook of *Corynebacterium glutamicum*. CRC Press, Boca Raton, FL.

Gunji, Y., Yasueda, H., 2006. Enhancement of L-lysine production in methylotroph *Methylophilus methylotrophus* by introducing a mutant lysE exporter. J. Biotechnol. 127, 1–13.

Hermann, T., 2003. Industrial production of amino acids by coryneform bacteria. J. Biotechnol. 104, 155–172.

Ikeda, M., 2003. Amino acid production processes. Adv. Biochem. Eng. Biotechnol. 79, 1–35.

Ikeda, M., 2016. Lysine fermentation: history and genome breeding. Adv. Biochem. Eng. Biotechnol. 159, 73–102.

Ikeda, M., Ohnishi, J., Hayashi, M., Mitsuhashi, S., 2006. A genome-based approach to create a minimally mutated *Corynebacterium glutamicum* strain for efficient L-lysine production. J. Ind. Microbiol. Biotechnol. 33, 610–615.

Jakobsen, O.M., Brautaset, T., Degnes, K.F., Heggeset, T.M., Balzer, S., Flickinger, M.C., Valla, S., Ellingsen, T.E., 2009. Overexpression of wild-type aspartokinase increases L-lysine production on the thermotolerant methylotrophic bacterium *Bacillus methanolicus*. Appl. Environ. Microbiol. 75, 652–661.

Jojima, T., Omumasaba, C.A., Inui, M., Yukawa, H., 2010. Sugar transporters in efficient utilization of mixed sugar substrates: current knowledge and outlook. Appl. Microbiol. Biotechnol. 85, 471–480.

Li, Y., Wei, H., Wang, T., Xu, Q., Zhang, C., Fan, X., Ma, Q., Chen, N., Xie, X., 2017. Current status on metabolic engineering for the production of L-aspartate family amino acids and derivatives. Bioresour. Technol. 245, 1588–1602.

Motoyama, H., Yano, H., Terasaki, Y., Anazawa, H., 2001. Overproduction of L-lysine from methanol by *Methylobacillus glycogenes* derivatives carrying a plasmid with a mutated *dapA* gene. Appl. Environ. Microbiol. 67, 3064–3070.

Razak, M.A., Viswanath, B., 2015a. Comparative studies for the biotechnological production of L-lysine by immobilized cells of wild-type *Corynebacterium glutamicum* ATCC 13032 and mutant MH 20-22 B. 3 Biotech 5, 765–774.

Razak, M.A., Viswanath, B., 2015b. Optimization of fermentation upstream parameters and immobilization of *Corynebacterium glutamicum* MH 20-22 B cells to enhance the production of L-lysine. 3 Biotech 5, 531–540.

Sanchez, S., Demain, A.L., 2008. Metabolic regulation and overproduction of primary metabolites. Microb. Biotechnol. 1, 283–319.

Sanchez, S., Demain, A.L., 2014. Fermentation (industrial): production of amino acids. In: Sanchez, S. (Ed.), Encyclopedia of Food Microbiology. Elsevier, Mexico, pp. 778–784.

Wittmann, C., Becker, J., 2007. The L-lysine story: from metabolic pathways to industrial production. In: Wittmann, C., Becker, J. (Eds.), Amino Acid Biosynthesis – Pathways, Regulation and Metabolic Engineering. Springer, Berlin, Heidelberg, pp. 39–70.

Yoshida, Y., Nagasawa, Y., 2003. Epsilon-poly-L-lysine: microbial production, biodegradation and application potential. Appl. Microbiol. Biotechnol. 62, 21–26.

CHAPTER

9

Emerging Trends in the Industrial Production of Chemical Products by Microorganisms

Pandeeti Emmanuel Vijay Paul, Veeraiah Sangeetha, Routhu Gyana Deepika

Aurobindo Pharma Ltd, APL Research Center-2, Sanga Reddy Mandal, India

OUTLINE

https://doi.org/10.1016/B978-0-12-816328-3.00009-X

9.1 INTRODUCTION

Application of microbes in the form of microbiological processes, systems, and substances to produce process materials and chemicals results in new and improved methods to make industrial raw materials and intermediate and consumer goods or to manage waste when compared with the conventional chemical manufacturing process. The process of producing chemicals, materials, and energy through the application of microbes and other life systems is known as industrial microbiology or industrial biotechnology or white biotechnology. In 2009, the global chemical market is estimated to be over $100 billion, and only 3% of it is derived from industrial biotechnology. It is estimated that the bio-based global chemical sales would grow to about 15% by 2025 (C&News, 2009).

The global market was valued at US$75.81 billion in 2010, and the major market segments are distributed as shown below:

- Biofuels—US$39.69 billion
- Pharmaceutical ingredients—US$22.06 billion
- Food and feed—US$9.27 billion
- Enzymes—US$4.32 billion
- Bioplastics—US$0.24 billion
- Personal care ingredients—US$0.23 billion

The use of whole microorganisms or their enzymes or enzymes from other sources expressed in microbes to make products allows easy access to building blocks and materials that are used as feed stocks or raw materials in sectors such as chemistry, food and feed, paper and pulp, textiles, and energy.

Countries like the United States have invested in massive research programs to develop industrial biotechnology especially in biofuels and bio-based chemicals, in order to achieve energy security and energy independence and to regenerate their agriculture industry to produce feed stocks for bio-based industries (Lorenz and Zinke, 2005). For the European Union, industrial biotechnology is their key strategic area that is necessary for them to have a competitive advantage in their chemical industry. Developing countries have very good prospects in industrial biotechnology in fields of biofuels and also in the production of fine and bulk chemicals (Zhang et al., 2011a,b; Mahanta et al., 2008; Balamurugan et al., 2010; Sanda et al., 2011; Hassan and Yaacob, 2009; Lin et al., 2010; Hniman et al., 2011; Thanh et al., 2008). Traditionally chemicals were synthesized by well-established routes of synthesis that often are multistep reactions and face difficulties such as unstable reactants or intermediates, complex process controls, and formation of undesired by-products. These factors led researchers to look for alternate means of synthesis of these chemicals through novel strategies. The microbial production of these chemicals is a very promising technology for biosynthesis and production of bulk and fine chemicals. This technology is not adopted for the manufacture of chemicals and biofuels (alcohols, fatty acids, and alkanes), bulk chemicals (diols, organic acids, etc.), pharmaceuticals, and nutraceuticals (amino acids, hydroxycinnamic acids, flavonoids, stilbenoids, coumarins, isoprenoids, etc.)

9.2 BREAKTHROUGH IN ENABLING TECHNOLOGIES

The microbial production of chemicals is possible now because of the advancements in technologies such as genomics, metagenomics, quantitative proteomics, metabolic engineering, and synthetic biology that fill the gaps in our knowledge and also provide tools to manipulate the microbial systems (Tyo et al., 2007). Novel fermentation, downstream process technologies, and process engineering have made the industrial production of chemicals more efficient and commercially viable and competitive (Villadsen, 2008).

(1) DNA sequencing: The next-generation high-throughput sequencing technology provides vast genomic data in an efficient and inexpensive way providing knowledge of the DNA sequence and their regulatory elements. Advancements in bioinformatics such as sequence assembly and analysis tools also greatly aid in understanding the genetic architecture of the microbial systems that allows efficient manipulation of the genome.

(2) Recombinant DNA technology

Recombinant DNA technology provides the procedures to manipulate the DNA through cloning procedures either to bring together different genes and their regulatory elements for expression of various proteins or to knockout undesirable genes for the purpose of eliminating their activities. The various technologies for gene assembly paved the way for construction of pathways for production of desired chemicals. Some of the tools for gene assembly and manipulation of DNA are listed in Table 9.1.

TABLE 9.1 Gene Assembly Types and Methods That Are Useful for Modification of the Pathways

Type	Assembly Method	Reference
Restriction enzyme-based methods	BioBrick standard (*Eco*RI, *Xba*I, *Spe*I, and *Pst*I)	Shetty et al. (2008)
	BglBrick standard (*Eco*RI, *Bgl*II, *Bam*HI, and *Xho*I)	Anderson et al. (2010)
	ePathBrick standard (*Spe*I, *Xba*I, *Nhe*I, and *Avr*II)	Xu et al. (2012)
	Standard European Vector Architecture (SEVA)	Martínez-García et al. (2015)
	HomeRun Vector Assembly System (HVAS)	Li et al. (2014a,b)
Restriction enzyme-based scarless methods	Golden Gate method	Engler et al. (2008)
	GoldenBraid 2.0 standard	Sarrion-Perdigones et al. (2013)
	Modular cloning system (MoClo)	Weber et al. (2011)
	Methylation-assisted tailorable ends rational (MASTER)	Chen et al. (2013)
Sequence homology-based methods	Overlap extension polymerase chain reaction (OE-PCR)	Horton et al. (1989)
	Circular polymerase extension cloning (CPEC)	Quan and Tian (2009)
	Sequence and ligation-independent cloning (SLIC)	Li and Elledge (2007)

Continued

TABLE 9.1 Gene Assembly Types and Methods That Are Useful for Modification of the Pathways—cont'd

Type	Assembly Method	Reference
	Nicking endonucleases for ligation-independent cloning (NE-LIC)	Wang et al. (2013)
	Gibson assembly method	Gibson et al. (2009)
	Seamless ligation cloning extract (SLiCE)	Zhang et al. (2012)
	Uracil-specific excision reagent cloning (USER)	Nour-Eldin et al. (2010)
	Serine integrase recombinational assembly (SIRA)	Colloms et al. (2014)
	Isothermal assembly	Chao et al. (2014)
	In-Fusion kit, Gateway kit	Casini et al. (2015)
Homology recombination-based methods	Yeast DNA assembler, 196, 197	de Kok et al. (2014) and Shao et al. (2008)
	Bacillus subtilis DNA assembly, 198, 199	Yonemura et al. (2007)
	Escherichia coli RecET system, 200	Itaya et al. (2007)
	Escherichia coli Redαβ system, 201	Sharan et al. (2009)

9.2.1 Microarray Technology

DNA microarray technology is currently used as a lab-based method for profiling gene expression for a number of different genes and a given tissue type. In the near future, improved microarrays will be used by researchers to better understand gene regulation, cell proliferation, and disease progression.

9.3 PRODUCTS OF MICROBIAL CONVERSIONS

9.3.1 Biofuels

Biofuels has a great potential as an alternate to fossil fuels, and investment in the commercial production of biofuels began in the 1970s to reduce carbon dioxide emissions. Biofuel has been rapidly growing, and in 2010, the total quantity of biofuels produced was over 100 billion liters. Biofuel technology can be classified into a number of generations of technological development as summarized in Fig. 9.1.

- First generation—conventional biofuels are manufactured from sugar, starch, and vegetable oil, derived primarily from food crops. The majority of biofuels currently produced in commercial quantities fall into this category.
- Second generation—advanced biofuels are those manufactured from nonfood crops and lignocellulose wastes. The manufacturing process requires enzymatic digestion and fermentation.
- Third generation—algae biofuels are sometimes referred to as third-generation biofuel and are manufactured from photosynthetic algae.

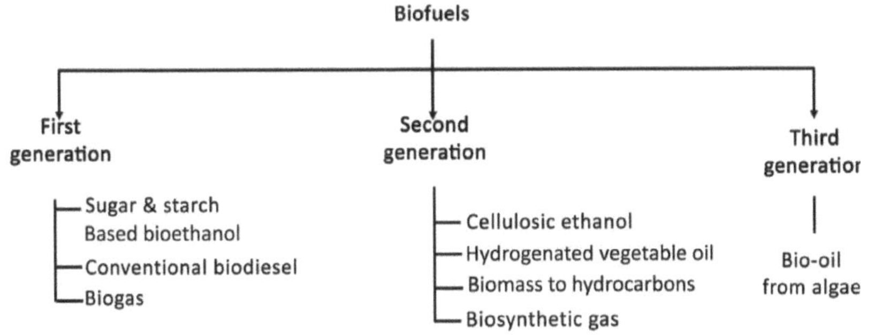

FIG. 9.1 Technological generations of biofuel technology.

Microbes play a major role in the pretreatment of the lignocellulosic feedstock where the raw material is converted by the microbial enzymes into a hydrolysate that is a mixture of sugars and other products such as furans, organic acids, and phenolic compounds (Olsson and Hagerdal, 1996; Kumar et al., 2009). *Escherichia coli*, *Saccharomyces cerevisiae*, and *Zymomonas mobilis* are recognized as the most promising microorganisms for industrial biofuel production (Dien et al., 2003). However, these organisms have limitations in ethanol tolerance and also from inhibitory products present in the lignocellulosic hydrolysate (Zaldivar et al., 1999, 2000). Tolerance engineering was performed on *E. coli* to generate a strain fit for ethanol production by employing different strain engineering methods (Mills et al., 2009). Global transcription machinery engineering (gTME) was successfully employed by mutating RpoD gene to develop a high-ethanol-producing strain of *Z. mobilis* (Tan et al., 2016). The 2,3-BDO pathways from *B. licheniformis* and *Enterobacter cloacae* have been successfully expressed in *E. coli* and in *Z. mobilis* (Xu et al., 2012; He et al., 2012). The 2,3-BDO biosynthesis pathway in these microorganisms utilizes three enzymes—acetolactate synthase (Als), acetolactate decarboxylase (AldC), and butanediol dehydrogenase (Bdh)—which convert, sequentially, pyruvate to α-acetolactate, acetoin, and then 2,3-butanediol. The expression of the these three enzymes in *Z. mobilis* has resulted in high titer of 2,3-BDO under fed-batch fermentation conditions (Yang et al., 2016). Table 9.2 summarizes the list of biofuel-producing microorganisms.

9.3.2 Biorefineries

Biorefineries are the facilities and plants that are used to convert biomass resources into energy (biofuels/power/heat) and value-added products. Biorefineries will allow for a more efficient use of resources than current biofuel production units and will reduce competition among different sources of biomass. The production of high-value coproducts (such as propylene glycol, a platform chemical with many applications including in resins, paints, cosmetics, and flavoring) is an essential part of the biorefineries business model. There are two main categories of biorefineries: those that focus on creating energy (including biofuel plants) and those that focus on producing food, feed, chemicals, and other materials and might create power or heat as a coproduct. Drivers for the development and implementation of biorefineries are the growth in demand for energy, fuels, and chemicals and the ongoing price increases

TABLE 9.2 List of Biofuel-Producing Microorganisms

Product	Strain	Titer (g/L)	Reference
1,4-Butanediol	*Escherichia coli* ECKh-422	18	Yim et al. (2011)
1,3-Propanediol	*Escherichia coli*	135.3	Nakamura and Whited (2003)
2,3-Butanediol	*Escherichia coli* BL21(DE3)	1.04	Park et al. (2014a,b)
1,2-Propanediol	*Escherichia coli* MG1655	5.6	Clomburg and Gonzalez (2011)
1,2-Propanediol	*Escherichia coli* W3110	0.59	Conrado et al. (2012)
Isobutanol	*Escherichia coli* JCL260	50.8	Baez et al. (2011)
Isobutanol	*Saccharomyces cerevisiae*	0.638	Avalos et al. (2013)
Ethanol	*Escherichia coli* KO12	54	Ohta et al. (1991)
Ethanol	*Saccharomyces cerevisiae* BY4741		Ohta et al. (1991)
Myo-inositol	*Escherichia coli* L19S	>1.7	Gupta et al. (2017)
Free fatty acids	*Escherichia coli* RB03	7	Dellomonaco et al. (2011)
Free fatty acids	*Escherichia coli*	8.6	Xu et al. (2013)
Triacylglycerol	*Phaeodactylum tricornutum*	1.5 mg/109 cells	Daboussi et al. (2014)

of fossil resources. Future biorefineries will play a major role in the replacement of some chemicals and materials traditionally produced from petroleum.

9.3.3 Bioplastics

Bioplastics are derived from renewable biological materials rather than petrochemicals. Emerging technological developments in bioplastics will improve product quality, reduce production cost, and create a more viable substitute to traditional plastic materials. Future technological development will produce bioplastics with novel properties and improved functionality, such as flame retardant bioplastics and biodegradable plastics suitable for use in food packaging.

9.3.4 Biocatalysts

Enzyme-catalyzed processes are generally more efficient than conventional chemical processes, as fewer steps are involved and input yields are higher. Over the coming decade the discovery of novel biocatalysts will drive improvements in a variety of industrial processes. Biocatalysts are typically manufactured using transgenic organisms that secrete the catalyst, requiring only renewable inputs, namely, sugars and amino acids.

In the near term, the increasing range of bioenzymes available will enable safer and more efficient industrial processes. Biocatalysts are highly selective and can carry out difficult chemical reactions with fewer synthesis steps, eliminating waste in multistep reactions. Biocatalysts also usually use water as a solvent, eliminating the production of organic solvent waste and enabling safer and less hazardous chemical and product synthesis. In addition to this, biocatalysts are biodegradable, some are even edible, and their production requires only renewable feedstock. Broadly, biocatalysts are driving a trend toward more efficient and environmentally friendly manufacturing processes.

9.3.5 Biocatalysis

Biocatalysis is defined as the use of natural substances that include enzymes from biological sources or whole cells to speed up chemical reactions. Enzymes have pivotal role in the catalysis of hundreds of reactions that include production of alcohols from fermentation and cheese by breakdown of milk proteins. Recent advances in the field of scientific research has helped to understand the structure and functional activities of enzymes, which has in turn led to an increase in their stability, activity, sustainability, and substrate specificity. Currently, there are hundred different biocatalytic processes that have been implemented in various pharma, chemical, food, and agro-based industries (biocatalysis Tyler Johannes). Biocatalytic processes are similar to conventional processes in many ways and the major factors to be accounted for are reaction kinetics and stability for both single and multistep reactions.

The process of biocatalysis starts with identifying target reaction, discovery of biocatalyst, characterization, engineering, and process modeling. Biocatalyst engineering requires rational design and directed evolution. The impact of biocatalysis in the future will be precisely this: the increasing ability to use enzymes to catalyze chemical reactions in industrial processes, including the production of drug substances, flavors, fragrances, electronic chemicals, and polymers—chemicals that literally impact almost every facet of your life. In adopting biocatalysis as a mainstream technology for chemical production, we will be introducing a technology that is greener, reduces pollution and cost, and creates greater sustainability.

9.4 ADVANTAGES

In industrial production, the major limiting factors are a number of reactions toward end product, separation of the final product from other contaminants, and finally the chemical effluents that are released after the process completion. Biocatalysts have unique features over chemical catalysts: They speed up the reaction without altering the thermodynamics of the reaction. They are highly selective in terms of substrate, binding sites, and functional groups that are essential in chemical synthesis. Biocatalysts also reduce the side reactions, thus reducing the time of synthesis. They do not require additional proactive groups, thus making separation process simpler. For commercial application, mild operating conditions, that is, pH range between 5 and 8 and temperature range of 20–40°C, and high-catalytic efficiency so that low concentration of enzyme might be sufficient for the entire process are required, which is successfully achieved through biocatalysis. Biocatalysis constitutes a significant portion of process economics by constructing immobilized biocatalysts.

Compared with whole cells, isolated enzymes have several benefits, which include simpler reaction apparatus, higher productivity through higher concentration of biocatalyst, and simpler product purification. Recently, via recombinant techniques, large amounts of enzymes are produced by identifying the DNA sequence that code for the enzyme and subsequently expressing them in *E. coli* or *S. cerevisiae*. The overexpressed enzymes are purified based on the physical and chemical properties. Whole cells are used when the specific target enzyme is difficult to isolate or when specific biotransformation requires multiple enzymes. Further, whole cells do not require cofactor regeneration, and the process can be carried out with cheap and abundant raw materials. The major disadvantage is that due to uncontrolled metabolic reaction, there might be other side reactions during cell growth. Due to accumulation of both desirable and undesirable products, the separation process becomes difficult.

Biocatalysts resuspended in organic solvents have low activity compared with water, but due to its high-boiling point and high heat of vaporization, it is not desirable to use water. In recent years through protein engineering and directed evolution, it has been easier to resuspend biocatalysts in organic solvents without altering the activity. Lyophilization of biocatalysts in the presence of both organic and inorganic material is called excipients, which also include nonbuffer salts, crown ethers, cyclodextrins, and solid-state buffers, thus, providing an advantage over maintaining biocatalysts in nonaqueous phase. Immobilized biocatalysts have been employed in the industrial process due to easier separation of biocatalysts from the final product, immobilized biocatalysts are more stable, and reactor configuration is more flexible. Apart from biocatalysts, whole cells can also be immobilized. Though immobilization of biocatalysts is well established, recently nanotube biotechnology has created another possible means of immobilizing biocatalyst.

9.5 EXAMPLES FOR BIOCATALYSIS REACTIONS

9.5.1 Oxidoreductases

Oxidoreductases catalyze oxidation and reduction reactions that occur within the cell. However, it often needs expensive cofactors such as nicotinamide adenine dinucleotides (e.g., NAD^+/NADH) and flavines (e.g., FAD/$FADH_2$) in the reactions. In fact, nicotinamide adenine dinucleotides are required by about 80% of oxidoreductases. Several NAD (H) regeneration systems have been developed; the most widely used is formate-formate dehydrogenase (FDH) system. An example of a pharmaceutical synthesis reaction involving an oxidoreductase is the synthesis of 3,4-dihydroxylphenyl alanine (DOPA). 3,4-Dihydroxylphenyl alanine is a chemical used in the treatment of Parkinson's disease. The industrial process that synthesizes DOPA utilizes the oxidoreductase polyphenol oxidase. Another example is the use of leucine dehydrogenase coupled with FDH for the reductive amination of trimethylpyruvate to L-tert-leucine. The whole process is carried out in a membrane reactor in which the NAD^+ is regenerated by FDH.

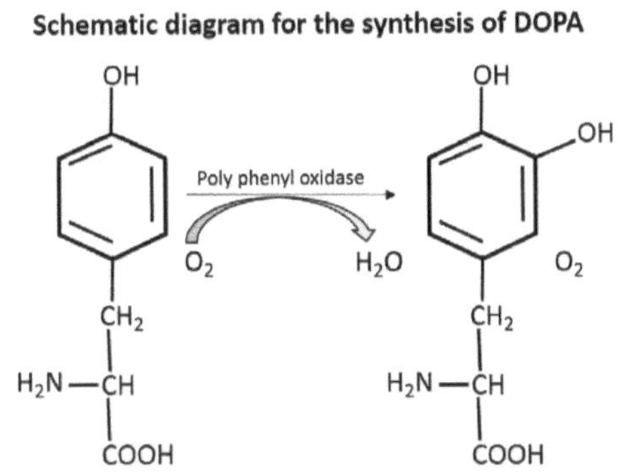

Synthesis of DOPA.

9.5.2 Hydrolases

Hydrolases catalyze the addition of water to a substrate by means of a nucleophilic substitution reaction. Hydrolases are the biocatalysts most commonly used in organic synthesis. They have been used to produce intermediates for pharmaceuticals and pesticides and chiral synthons for asymmetrical synthesis. Among hydrolases, amidases, proteases, esterases, and lipases are of particular interest. These enzymes catalyze the hydrolysis and formation of ester and amide bonds. Lipases can hydrolyze triglycerides into fatty acids and glycerol. They have been used extensively to produce optically active alcohols, acids, esters, and lactones by kinetic resolution. A classic example is the use of a lipase for the production of (*S*,*R*)-2,3-*p*-methoxyphenylglycyclic acid, an intermediate for diltiazem.

Proteases such as α-chymotrypsin, papain, and subtilisin are also useful biocatalysts for regioselective or stereoselective hydrolytic biotransformations. For example, dibenzyl esters of aspartic and glutamic acid can be selectively deprotected at the 1-position by subtilisin-catalyzed hydrolysis. In addition, α-chymotrypsin is used in the kinetic resolution of α-nitro-α-methyl carboxylates, which results in L-configured enantiomers of the unhydrolyzed esters with high-optical purity.

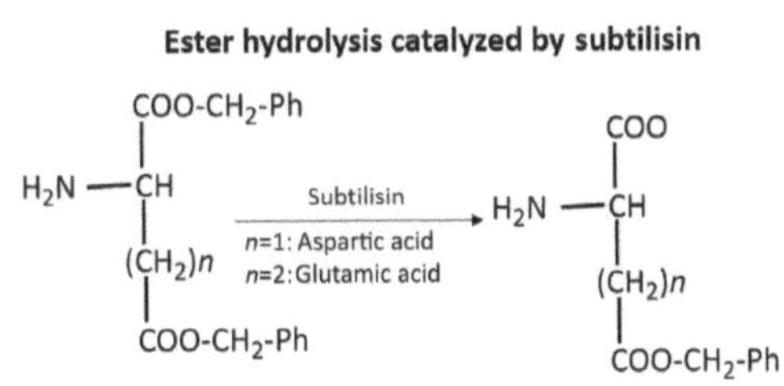

Ester hydrolysis by subtilisin.

9.5.3 Lyases

Lyases are the enzymes responsible for catalyzing addition and elimination reactions. Lyase-catalyzed reactions break the bond between a carbon atom and another atom such as oxygen, sulfur, or another carbon atom. They are found in cellular processes, such as the citric acid cycle, and in organic synthesis, such as in the production of cyanohydrins. Another example is the use of a fumarase for the production of (*S*)-malic acid from fumaric acid. A third example is biocatalytic production of a cyanohydrin from a ketone that is catalyzed by oxynitrilase.

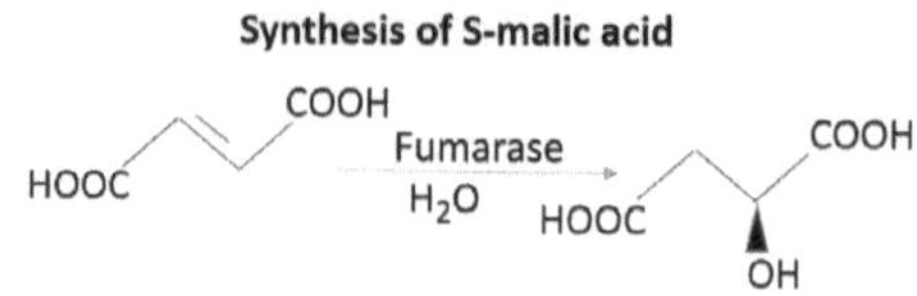

Synthesis of S-malic acid by fumarase.

9.5.4 Transferase

Transferases catalyze the transfer of functional groups such as methyl, hydroxymethyl, formal, glycosyl, acyl, alkyl, phosphate, and sulfate groups by means of a nucleophilic substitution reaction. They are not widely used in industrial processes. Industrial application of transferases is the use of various glycosyltransferases for the synthesis of oligosaccharides. Oligosaccharides and polysaccharides are important classes of naturally occurring compounds, which play important roles in cellular recognition and communication processes. Another example is the use of a glucokinase (a transferase) in combination with an acetate kinase for the production of glucose-6-phosphate.

9.5.5 Isomerases

Isomerases catalyze isomerization reactions such as racemization and epimerization. They have not been used in many industrial applications. However, one of the most successful enzyme-based biocatalytic processes involves an isomerase: the use of glucose isomerase for the production of high-fructose corn syrup (HFCS). HFCS is used as an alternative sweetener to sucrose in the food and beverage industry. The isomerization of glucose to HFCS on an industrial scale is carried out in continuous fixed-bed reactors using immobilized glucose isomerases. The total amount of HFCS produced by glucose isomerase exceeds a million tons per year.

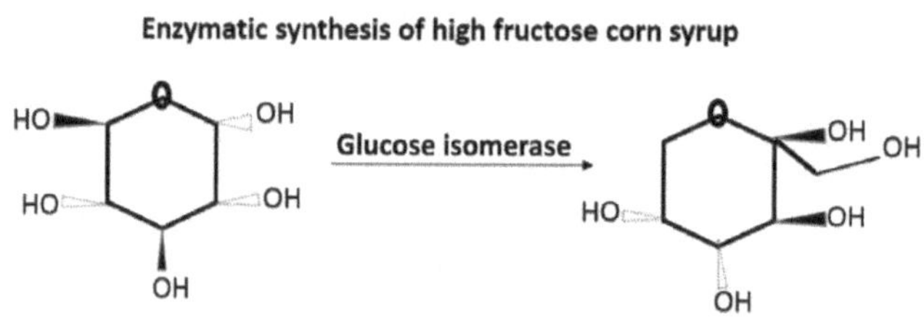

9.5.6 Ligases

Ligases catalyze reactions that involve the creation of chemical bonds with nucleotide triphosphates. They are important in certain cellular processes, such as connecting nucleotides in DNA replication. However, similar to isomerases, ligases have very few industrial applications. It is important to note that DNA ligases are essential tools in recombinant DNA technology and are used almost in every biology-related laboratory.

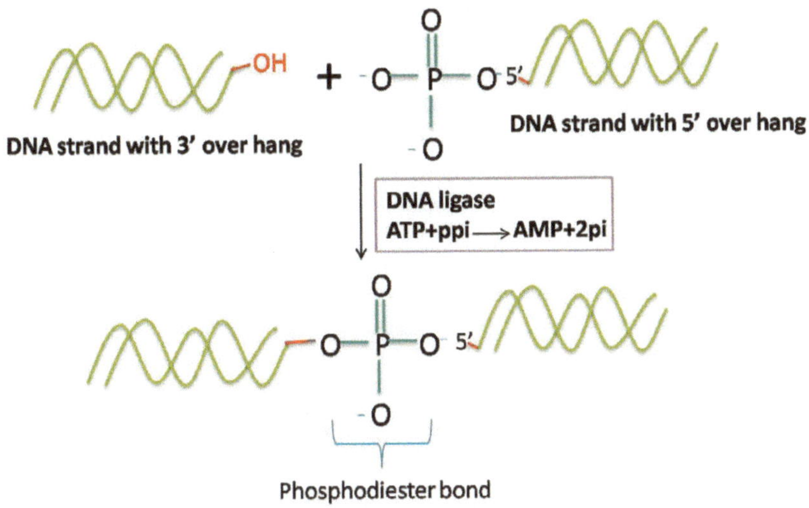

9.5.7 Amination

Aminations have been performed by the two different routes of transamination and reductive amination. Transaminases have obtained increased interest for the asymmetrical synthesis of amines from prochiral ketones, and amination is becoming a key reaction in industrial biotransformations. The best engineered enzyme could convert 200 g/L prositagliptin ketone to sitagliptin with an excellent ee of >99.95%. A 53% productivity increase, 19% waste reduction, elimination of heavy metals, cost reductions, and avoiding specialized high-pressure hydrogenation equipment have been found as specific advantages of the biocatalytic process.

9.6 MICROBIAL PRODUCTION OF ORGANIC ACIDS

A wide range of organic acids have been produced with the use of microorganisms. Though these processes seem to be feasible, large-scale production of these organic acids is not successful. The market share of biosynthetic production of chemical products using microorganisms is expected to substantially increase. Among them, organic acids constitute a vital role since their functional groups are useful as starting material for chemical industry. Few examples for organic acids are listed below.

9.6.1 Polylactic Acid

Polylactic acid (PLA) produced through microorganisms has an added advantage over other polyesters due to its ample availability in the market and low price. It has high-environmental characteristics and established global market with 20 applications over 70,000 stores. Ingeo brand of PLA is the world's leading biopolymer.

9.6.2 Citric Acid

Citric acid is traditionally produced from *Aspergillus niger.* Citric acid production per annum is 1.6 tons. *Aspergillus* accumulate this acid under high-substrate concentration, low and finite content of nitrogen, and few trace elements under maintenance of high DO and low pH. Citric acid production is still under improvisation where the traditional modification techniques like mutagenesis and culture condition optimization no longer give good yield.

Apart from *Aspergillus,* filamentous fungal species like *Yarrowia lipolytica* has been developed for citric acid production. *Y. lipolytica* can utilize *n*-paraffins and fatty acids as carbon sources and can result in increased yield of citric acid. The yield achieved by this microorganism is 140 g/L.

9.6.3 Lactic Acid

Lactic acid bacteria were traditionally used for the production of lactic acid industrially. The production of biodegradable plastic polylactide has led to increased production of lactic acid. About 150,000 tons of lactic acid was produced in 2002, and 90% of it was from the fermentation of lactic acid bacteria. Apart from lactic acid bacteria, *Rhizopus oryzae* is also widely used for lactic acid production, since it can grow in mineral medium and carbon sources such as starch and xylose.

9.6.4 Succinic Acid

Succinic acid is also known as amber acids, which is derived from petroleum feedstocks. Succinic acid has a pivotal role as starting material for secondary chemical production that is used in pharmaceutical, food, and agriculture. Succinic acid produced biochemically from glucose using a genetically modified organism is cost-effective and renders high functionality. Myriant and BioAmber are two good examples for succinic acids that are produced commercially using genetically engineered organisms.

9.7 MICROBIAL PRODUCTION OF AMINO ACIDS

9.7.1 L-Threonine

L-threonine belongs to the aspartate family of amino acids that has nutritional values and is used in pharmaceutical and chemical reagents. Its biosynthetic pathway is composed of five enzymatic reactions that are catalyzed by aspartokinase I, II, and III, that is, ThrA, met, LysC; aspartate semialdehyde dehydrogenase; homoserine dehydrogenase I and II; homoserine kinase; and threonine synthase. Strains involved in the production of L-threonine have been modified by eliminating feedback inhibition by making mutations in S345F in thrA and T342I in lysC. Enzyme deregulation is another approach where the native promoter is replaced by tac promoter. Competitive pathway elimination was employed to delete genes lysA, metA, and tdh. Overexpression of L-threonine exporter (rhtABC) has also been one of the processes for the production of L-threonine. With the help of these strategies, the finally modified *E. coli* strain MT201 produced 52.0 g/L of L-threonine. In silico analysis also led to identifying the target genes that are essential for growth and increase of pyruvate excretion. Thus, modified final *E. coli* strain was able to produce about 82.4 g/L of L-threonine.

9.7.2 L-Tyrosine

The aromatic amino acid L-tyrosine is a precursor for synthesis of drugs, polymers, melanin, and phenylpropanoids. The biosynthetic pathway of L-tyrosine is highly regulated in *E. coli* and is a branch of the shikimate pathway that mainly consists of phosphoenolpyruvate (PEP) and mainly contains nine enzymes, 3-deoxy-D-arabino-heptulosonate (DAHP) synthase (aroG), dehydroquinate (DHQ) synthase (aroB), DHQ dehydratase (aroD), shikimate dehydrogenase (aroE/ydiB), shikimate kinase I/II (aroK/L), 5-enolpyruvoylshikimate-3-phosphate (EPSP) synthase (aroA), chorismate synthase (aroC), chorismate mutase/prephenate dehydrogenase (CM/PDH), and tyrosine aminotransferase (TyrB). To achieve high-yielding L-tyrosine producers, many approaches were employed such as removal of feedback inhibition and overexpression of key enzymes *aroK/L*, ydiB, tkt, etc. However, these approaches did not provide much success in obtaining the L-tyrosine-overproducing strain. Cellular transcriptome engineering was performed using global transcription machinery engineering by using mutated forms of RNA polymerase components, rpoA and rpoD, which led to isolation of strains that produce yields and titers of L-tyrosine.

9.7.3 L-Valine

L-Valine is an essential amino acid that has pharmaceutical, cosmetics, and nutritional values. L-valine from pyruvate has four reactions that are catalyzed by acetohydroxy acid synthase (*ilvBN)*, acetohydroxy acid isomeroreductase (*ilv* C), dihydroxy acid dehydratase (*ilvD)*, and transaminase B (*ilvE)*. Production of valine from *Corynebacterium glutamicum* includes removing feedback inhibition in *ilvN*, deleting *panB*, *ilvA*, and *ilvBNC*, and *ilvGMDA* operon. Rational metabolic engineering based on transcriptome analysis has led to increased production of L-valine. Comparative transcriptome analysis during batch fermentation has shown that expression levels of *ilvCED* genes were increased but lower than *ilvBN* genes. Also, as an activator of the *ilvIH* operon, the expression of *lrp* was downregulated. L-valine exporter *ygaZ and ygaH* expression levels were decreased. The final strain of *E. coli* produced 7.61 g/L of L-valine, which was 113% more than control strain.

9.7.4 L-Phenylalanine

L-Phenylalanine has many applications in food and pharmaceutical industries. Biosynthetic pathways include ten enzymatic reactions out of which five enzymes are feedback-regulated: aroE, aroK/L, tyrB, aroG, and CM-PDT. Feedback inhibition can be

TABLE 9.3 List of Chemical Products Produced by Microorganisms

Product	Strain	Titer (g/L)	Reference
Malate	*Escherichia coli* XZ658	34	Zhang et al. (2011a,b)
Malate	*Saccharomyces cerevisiae*	59	Zelle et al. (2008)
Fumarate	*Escherichia coli* EF02	41.5	Li et al. (2014c)
Fumarate	*Candida glabrata*	15.76	Chen et al. (2015)
D-Lactate	*Escherichia coli* B0013–070B	122.8	Zhou et al. (2012)
D-Lactate	*Escherichia coli* ALS974	13.8	Zhu et al. (2007)
L-Lactate	*Escherichia coli* 090B3	142.2	Niu et al. (2014)
Butyrate	*Escherichia coli* DSM03	7.2	Baek et al. (2013)
Butyrate	*Escherichia coli*	10	Saini et al. (2015)
Butyrate	*Escherichia coli*	12.3	Kataoka et al. (2017)
D-Glucaric acid	*Escherichia coli* JT9	2.5	Moon et al. (2010)
Adipic acid	*Escherichia coli* AA7	0.639 mg/L	Yu et al. (2014)
3-Dehydroshikimic acid	*Escherichia coli* NR7	19	Ran and Frost (2007)
Mevalonate	*Escherichia coli* DH10B	0.596 mM/OD	Pfleger et al. (2007)
Itaconic acid	*Aspergillus terreus*	40.2	Gao et al. (2014)
3-Hydroxypropionate	*Escherichia coli* SH501	71.9	Chu et al. (2015)
Glycolic acid	*Escherichia coli* EYX-2	56.44	Deng et al. (2015)
L-Valine	*Escherichia coli* VAMF	7.61	Park et al. (2007)
L-Valine	*Escherichia coli* WLA	60.7	Park et al. (2011)
L-Threonine	*Escherichia coli* TH28C	82.4	Lee et al. (2006)
L-Ornithine	*Saccharomyces cerevisiae*	5.1	Qin et al. (2015)
L-Arginine	*Corynebacterium glutamicum* AR6	92.5	Park et al. (2014a,b)
L-Homoalanine	*Escherichia coli* ATCC 98082	5.4	Zhang et al. (2010a,b)
L-Phenylalanine	*Escherichia coli* W14	47	Liu et al. (2014)
L-Phenylalanine	*Escherichia coli* YP1617	56.2	Yuan et al. (2015)
L-Tyrosine	*Escherichia coli* P2	13.8	Santos et al. (2012)

overcome by overexpression of these genes under a strong promoter. In order to further increase the production, protein directed evolution was adopted. By this, there was a large decrease in feedback inhibition by L-Phe, and the yield was increased by 0.21 g/g of glucose. Further precursor supply was increased, and PTS system was inactivated to produce a final yield of L-Phe of 47.0 g/L. List of various chemical products produced by microorganisms are summarized in Table 9.3.

9.8 CONCLUSION

Microbial processes for industrial production of chemicals and a variety of low-molecular-weight compounds are quickly gaining practical importance for synthesis of high-purity products, in an environmentally acceptable manner while realizing energy savings. To improve the productivity and yield of useful chemicals, novel approaches must be developed. Progresses in metabolic engineering and synthetic biology have facilitated us to produce a range of industrially useful chemicals and various amino acids, which have typically been studied to expand productivity.

Acknowledgments

The authors are indebted to all the researchers whom we cited in this review for their significant and valuable research. No funding was received to perform this review.

Conflicts of interests

The authors declare no relevant competing financial interests to disclose.

References

Anderson, J.C., Dueber, E.J., Leguia, M., Wu, C.G., Goler, A.J., Arkin, P.A., Keasling, D.J., 2010. BglBricks: a flexible standard for biological part assembly. J. Biol. Eng. 4 (1), 1.
Avalos, J.L., Fink, G.R., Stephanopoulos, G., 2013. Compartmentalization of metabolic pathways in yeast mitochondria improves the production of branched-chain alcohols. Nat. Biotechnol. 31 (4), 335–341.
Baek, J.M., Mazumdar, S., Lee, S.W., Jung, M.Y., Lim, J.H., Seo, S.W., Jung, G.Y., Oh, M.K., 2013. Butyrate production in engineered *Escherichia coli* with synthetic scaffolds. Biotechnol. Bioeng. 110 (10), 2790–2794.
Baez, A., Cho, K.M., Liao, J.C., 2011. High-flux isobutanol production using engineered *Escherichia coli*: a bioreactor study with in situ product removal. Appl. Microbiol. Biotechnol. 90 (5), 1681–1690.
Balamurugan, A.N., Breite, A.G., Anazawa, T., Loganathan, G., Wilhelm, J.J., Papas, K.K., Dwulet, F.E., McCarthy, R.C., Hering, B.J., 2010. Successful human islet isolation and transplantation indicating the importance of class 1 collagenase and collagen degradation activity assay. Transplantation 89 (8), 954–961.
C&News, 2009. Bio products from bio refineries—trends, challenges and opportunities. Bio Refineries 28–29.
Casini, A., Storch, M., Baldwin, G.S., Ellis, T., 2015. Bricks and blueprints: methods and standards for DNA assembly. Nat. Rev. Mol. Cell Biol. 16 (9), 568–576.
Chao, R., Yuan, Y., Zhao, H., 2014. Recent advances in DNA assembly technologies. FEMS Yeast Res. 15 (1), 1–9.
Chen, W.H., Qin, Z.J., Wang, J., Zhao, G.P., 2013. The MASTER (methylation-assisted tailorable ends rational) ligation method for seamless DNA assembly. Nucleic Acids Res. 41 (8), e93.
Chen, X., Wu, J., Song, W., Zhang, L., Wang, H., Liu, L., 2015. Fumaric acid production by Torulopsis glabrata: engineering the urea cycle and the purine nucleotide cycle. Biotechnol. Bioeng. 112 (1), 156–167.
Chu, H.S., Kim, Y.S., Lee, C.M., Lee, J.H., Jung, W.S., Ahn, J.H., Song, S.H., Choi, I.S., Cho, K.M., 2015. Metabolic engineering of 3-hydroxypropionic acid biosynthesis in *Escherichia coli*. Biotechnol. Bioeng. 112 (2), 356–364.
Clomburg, J.M., Gonzalez, R., 2011. Metabolic engineering of *Escherichia coli* for the production of 1,2-propanediol from glycerol. Biotechnol. Bioeng. 108 (4), 867–879.
Colloms, S.D., Merrick, C.A., Olorunniji, F.J., Stark, W.M., Smith, M.C., Osbourn, A., Keasling, J.D., Rosser, S.J., 2014. Rapid metabolic pathway assembly and modification using serine integrase site-specific recombination. Nucleic Acids Res. 42 (4), e23.
Conrado, R.J., Wu, G.C., Boock, J.T., Xu, H., Chen, S.Y., Lebar, T., Turnsek, J., Tomsic, N., Avbelj, M., Gaber, R., 2012. DNA-guided assembly of biosynthetic pathways promotes improved catalytic efficiency. Nucleic Acids Res. 40 (4), 1879–1889.

Daboussi, F., Leduc, S., Maréchal, A., Dubois, G., Guyot, V., Perez-Michaut, C., Amato, A., Falciatore, A., Juillerat, A., Beurdeley, M., Voytas, D.F., Cavarec, L., Duchateau, P., 2014. Genome engineering empowers the diatom *Phaeodactylum tricornutum* for biotechnology. Nat. Commun. 5, 3831.

Dellomonaco, C., Clomburg, J.M., Miller, E.N., Gonzalez, R., 2011. Engineered reversal of the β-oxidation cycle for the synthesis of fuels and chemicals. Nature. 476 (7360), 355–359.

de Kok, S., Stanton, L.H., Slaby, T., Durot, M., Holmes, V.F., Patel, K.G., Platt, D., Shapland, E.B., Serber, Z., Dean, J., 2014. Rapid and reliable DNA assembly via ligase cycling reaction. ACS Synth. Biol. 3 (2), 97–106.

Deng, Y., Mao, Y., Zhang, X.J., 2015. Metabolic engineering of *E. coli* for efficient production of glycolic acid from glucose. Biochem. Eng. J. 103, 256–262.

Dien, B.S., Cotta, M.A., Jeffries, T.W., 2003. Bacteria engineered for fuel ethanol production: current status. Appl. Microbiol. Biotechnol. 63 (3), 258–266.

Engler, C., Kandzia, R., Marillonnet, S., 2008. A one pot, one step, precision cloning method with high throughput capability. PLoS ONE 3 (11), e3647.

Gao, Q., Liu, J., Liu, L.M., 2014. Relationship between morphology and itaconic acid production by *Aspergillus terreus*. J. Microbiol. Biotechnol. 24 (2), 168–176.

Gibson, D.G., Young, L., Chuang, R.Y., Venter, J.C., Hutchison, C.A., Smith, H.O., 2009. Enzymatic assembly of DNA molecules up to several hundred kilobases. Nat. Methods 6 (5), 343–345.

Gupta, M., Prasad, N.G., Dey, S., Joshi, A., Vidya, T.N.C., 2017. Niche construction in evolutionary theory: the construction of an academic niche? J Genet. 96 (3), 491–504.

Hassan, M.A., Yaacob, W.A.R.W., 2009. Current status of industrial biotechnology in Malaysia. J. Biosci. Bioeng. 108, S1–S3.

He, Z., Zhong, C., Su, S., Xu, M., Wu, H., Cao, Y., 2012. Enhanced power-conversion efficiency in polymer solar cells using an inverted device structure. Nat. Photonics 6, 591–595.

Hniman, A., O-Thong, S., Prasertsan, P., 2011. Developing a thermophilic hydrogen-producing microbial consortia from geothermal spring for efficient utilization of xylose and glucose mixed substrates and oil palm trunk hydrolysate. Int. J. Hydrog. Energy 36, 8785–8793.

Horton, R.M., Hunt, H.D., Ho, S.N., Pullen, J.K., Pease, L.R., 1989. Engineering hybrid genes without the use of restriction enzymes: gene splicing by overlap extension. Gene 77 (1), 61–68.

Itaya, M., Fujita, K., Kuroki, A., Tsuge, K., 2007. Bottom-up genome assembly using the *Bacillus subtilis* genome vector. Nat. Methods 5 (1), 41–43.

Kataoka, N., Vangnai, A.S., Pongtharangkul, T., Yakushi, T., Matsushita, K., 2017. Production of 1,3-diols in *Escherichia coli*. Bioresour. Technol. 245, 1538–1541.

Kumar, P., Barrett, D.M., Delwiche, M.J., Stroeve, P., 2009. Methods for pretreatment of lignocellulosic biomass for efficient hydrolysis and biofuel production. Ind. Eng. Chem. Res. 48 (8), 3713–3729.

Lee, M.H., Lee, H.W., Park, J.H., Ahn, J.O., Jung, J.K., Hwang, Y.I., 2006. Improved L-threonine production of *Escherichia coli* mutant by optimization of culture conditions. J. Biosci. Bioeng. 101 (2), 127–130.

Li, M.Z., Elledge, S.J., 2007. Harnessing homologous recombination in vitro to generate recombinant DNA via SLIC. Nat. Methods 4 (3), 251–256.

Li, M.V., Shukla, D., Rhodes, B.H., Lall, A., Shu, J., Moriarity, B.S., Largaespada, D.A., 2014a. HomeRun vector assembly system: a flexible and standardized cloning system for assembly of multi-modular DNA constructs. PLoS ONE 9 (6), e100948.

Li, G.W., Burkhardt, D., Gross, C., Weissman, J.S., 2014b. Quantifying absolute protein synthesis rates reveals principles underlying allocation of cellular resources. Cell 157 (3), 624–635.

Li, N., Zhang, B., Wang, Z., Tang, Y.J., Chen, T., Zhao, X., 2014c. Engineering *Escherichia coli* for fumaric acid production from glycerol. Bioresour. Technol. 174, 81–87.

Lin, L., Cunshan, Z., Vittayapadung, S., Xianggian, S., Mingdong, D., 2010. Opportunities and challenges for biodiesel fuel. Appl. Energy 88, 1020–1031.

Liu, S.P., Liu, R.X., Xiao, M.R., Zhang, L., Ding, Z.Y., Gu, Z.H., Shi, G.Y., 2014. A systems level engineered *E. coli* capable of efficiently producing L-phenylalanine. Process Biochem. 49 (5), 751–757.

Lorenz, P., Zinke, H., 2005. White biotechnology: differences in US and EU approaches? Trend. Biotechnol. 23 (12), 570–574.

Mahanta, N., Gupta, A., Khare, S.K., 2008. Production of protease and lipase by solvent tolerant Pseudomonas aeruginosa PseA in solid-state fermentation using Jatropha curcas seed cake as substrate. Bioresour. Technol. 99 (6), 1729–1735.

Martínez-García, E., Aparicio, T., Goñi-Moreno, A., Fraile, S., de Lorenzo, V., 2015. SEVA 2.0 an update of the standard European vector architecture for de-/re-construction of bacterial functionalities. Nucleic Acids Res. 43, 1183–1189.

Mills, T.Y., Sandoval, N.R., Gill, R.T., 2009. Cellulosic hydrolysate toxicity and tolerance mechanisms in *Escherichia coli*. Biotechnol. Biofuels 2, 26.

Moon, T.S., Dueber, J.E., Shiue, E., Prather, K.L., 2010. Use of modular, synthetic scaffolds for improved production of glucaric acid in engineered *E. coli*. Metab. Eng. 12 (3), 298–305.

Nakamura, C.E., Whited, G.M., 2003. Metabolic engineering for the microbial production of 1,3-propanediol. Curr. Opin. Biotechnol. 14 (5), 454–459.

Niu, D., Tian, K., Prior, B.A., Wang, M., Wang, Z., Lu, F., Singh, S., 2014. Highly efficient L-lactate production using engineered *Escherichia coli* with dissimilar temperature optima for L-lactate formation and cell growth. Microb. Cell Fact. 13, 78.

Nour-Eldin, H.H., Geu-Flores, F., Halkier, B.A., 2010. USER cloning and USER fusion: the ideal cloning techniques for small and big laboratories. Methods Mol. Biol. 643, 185–200.

Ohta, M., Ina, K., Kusuzaki, K., Kido, N., Arakawa, Y., Kato, N., 1991. Cloning and expression of the rfe-rff gene cluster of *Escherichia coli*. Mol. Microbiol. 5, 1853–1862.

Olsson, L., Hagerdal, B.H., 1996. Fermentation of lignocellulosic hydrolysates for ethanol production. Enzym. Microb. Technol. 18 (5), 312–331.

Park, J.H., Lee, K.H., Kim, T.Y., Lee, S.Y., 2007. Metabolic engineering of *Escherichia coli* for the production of L-valine based on transcriptome analysis and in silico gene knockout simulation. Proc. Natl. Acad. Sci. U. S. A. 104 (19), 7797–7802.

Park, J.H., Lee, K.H., Kim, T.Y., Lee, S.Y., 2011. *Escherichia coli* W as a new platform strain for the enhanced production of L-valine by systems metabolic engineering. Biotechnol. Bioeng. 108 (5), 1140–1147.

Park, S.Y., Kim, B., Lee, S., Oh, M., Won, J.I., Lee, J., 2014a. Increased 2,3-butanediol production by changing codon usages in *Escherichia coli*. Biotechnol. Appl. Biochem. 61 (5), 535–540.

Park, S.H., Kim, H.U., Kim, T.Y., Park, J.S., Kim, S.S., Lee, S.Y., 2014b. Metabolic engineering of *Corynebacterium glutamicum* for L-arginine production. Nat. Commun. 5, 4618.

Pfleger, B.F., Pitera, D.J., Newman, J.D., Martin, V.J., Keasling, J.D., 2007. Microbial sensors for small molecules: development of a mevalonate biosensor. Metab. Eng. 9 (1), 30–38.

Qin, J., Zhou, Y.J., Krivoruchko, A., Huang, M., Liu, L., Khoomrung, S., Siewers, V., Jiang, B., Nielsen, J., 2015. Modular pathway rewiring of *Saccharomyces cerevisiae* enables high-level production of L-ornithine. Nat. Commun. 6, 8224.

Quan, J., Tian, J., 2009. Circular polymerase extension cloning of complex gene libraries and pathways. PLoS ONE 4 (7), e6441.

Ran, N., Frost, J.W., 2007. Directed evolution of 2-keto-3-deoxy-6-phosphogalactonate aldolase to replace 3-deoxy-D-arabino-heptulosonic acid 7-phosphate synthase. J. Am. Chem. Soc. 129 (19), 6130–6139.

Saini, M., Hong, M., Chung-Jen, C., Chao, C.Y.-P., 2015. Potential production platform of n-butanol in *Escherichia coli*. Metab. Eng. 2, 76–82.

Sanda, T., Hasunuma, T., Matsuda, F., Kondo, A., 2011. Repeated-batch fermentation of lignocellulosic hydrolysate to ethanol using a hybrid *Saccharomyces cerevisiae* strain metabolically engineered for tolerance to acetic and formic acids. Bioresour. Technol. 102, 7917–7924.

Santos, C.N., Xiao, W., Stephanopoulos, G., 2012. Rational, combinatorial, and genomic approaches for engineering L-tyrosine production in *Escherichia coli*. Proc. Natl. Acad. Sci. U. S. A. 109 (34), 13538–13543.

Sarrion-Perdigones, A., Vazquez-Vilar, M., Palaci, J., Castelijns, B., Forment, J., Ziarsolo, P., Blanca, J., Granell, A., Orzaez, D., 2013. GoldenBraid 2.0: a comprehensive DNA assembly framework for plant synthetic biology. Plant Physiol. 162 (3), 1618–1631.

Shao, Z., Zhao, H., Zhao, H., 2008. DNA assembler an in vivo genetic method for rapid construction of biochemical pathways. Nucleic Acids Res. 37 (2), e16.

Sharan, S.K., Thomason, L.C., Kuznetsov, S.G., Court, D.L., 2009. Recombineering: a homologous recombination-based method of genetic engineering. Nat. Protoc. 4 (2), 206–223.

Shetty, R.P., Endy, D., Knight Jr., T.F., 2008. Engineering BioBrick vectors from BioBrick parts. J. Biol. Eng. 14, 2–5.

Tan, F., Wu, B., Dai, L., Qin, H., Shui, Z., Wang, J., Zhu, Q., Hu, G., He, M., 2016. Using global transcription machinery engineering (gTME) to improve ethanol tolerance of *Zymomonas mobilis*. Microb. Cell Factories 15, 4.

Thanh, L.T., Okitsu, K., Sadanaga, Y., Takenaka, N., Bandow, H., 2008. Biodiesel production from virgin and waste oils using ultrasonic reactor in pilot scale. Proc. Symp. Ultrason Electron. 29, 395–396.

Tyo, K.E., Alper, H.S., Stephanopoulos, G.N., 2007. Expanding the metabolic engineering toolbox: more options to engineer cells. Trends Biotechnol. 25, 132–137.

Villadsen, K., 2008. 'Polyphonic' welfare: Luhmann's systems theory applied to modern social work. Int. J. Soc. Welf. 17, 65–73.

Wang, R.Y., Shi, Z.Y., Guo, Y.Y., Chen, J.C., Chen, G.Q., 2013. DNA fragments assembly based on nicking enzyme system. PLoS ONE 8 (3), e57943.

Weber, E., Engler, C., Gruetzner, R., Werner, S., Marillonnet, S., 2011. A modular cloning system for standardized assembly of multigene constructs. PLoS ONE 6 (2), e16765.

Xu, P., Vansiri, A., Bhan, N., Koffas, M.A., 2012. ePathBrick: a synthetic biology platform for engineering metabolic pathways in *E. coli*. ACS Synth. Biol. 1 (7), 256–266.

Xu, Z., Graham, K., Foote, M., Liang, F., Rizkallah, R., Hurt, M., Wang, Y., Wu, Y., Zhou, Y., 2013. 14-3-3 protein targets misfolded chaperone-associated proteins to aggresomes. J. Cell Sci. 126, 4173–4186.

Yang, S., Mohagheghi, A., Franden, M.A., Chou, Y.C., Chen, X., Dowe, N., Himmel, M.E., Zhang, M., 2016. Metabolic engineering of *Zymomonas mobilis* for 2,3-butanediol production from lignocellulosic biomass sugars. Biotechnol. Biofuels 9, 189.

Yim, H., Haselbeck, R., Niu, W., Pujol-Baxley, C., Burgard, A., Boldt, J., Khandurina, J., Trawick, J.D., Osterhout, R.E., Stephen, R., 2011. Metabolic engineering of *Escherichia coli* for direct production of 1,4-butanediol. Nat. Chem. Biol. 7 (7), 445–452.

Yonemura, I., Nakada, K., Sato, A., Hayashi, J., Fujita, K., Kaneko, S., Itaya, M., 2007. Direct cloning of full-length mouse mitochondrial DNA using a *Bacillus subtilis* genome vector. Gene 391 (1–2), 171–177.

Yu, J.L., Xia, X.X., Zhong, J.J., Qian, Z.G., 2014. Direct biosynthesis of adipic acid from a synthetic pathway in recombinant *Escherichia coli*. Biotechnol. Bioeng. 111 (12), 2580–2586.

Yuan, P., Cao, W., Wang, Z., Chen, K., Li, Y., Ouyang, P., 2015. Enhancement of L-phenylalanine production by engineered *Escherichia coli* using phased exponential L-tyrosine feeding combined with nitrogen source optimization. J. Biosci. Bioeng. 120 (1), 36–40.

Zaldivar, J., Martinez, A., Ingram, L.O., 1999. Effect of selected aldehydes on the growth and fermentation of ethanologenic *Escherichia coli*. Biotechnol. Bioeng. 65 (1), 24–33.

Zaldivar, J., Martinez, A., Ingram, L.O., 2000. Effect of alcohol compounds found in hemicellulose hydrolysate on the growth and fermentation of ethanologenic *Escherichia coli*. Biotechnol. Bioeng. 68 (5), 524–530.

Zelle, R.M., de Hulster, E., van Winden, W.A., de Waard, P., Dijkema, C., Winkler, A.A., Geertman, J.-M.A., van Dijken, J.P., Pronk, J.T., van Maris, A.J.A., 2008. Malic acid production by *Saccharomyces cerevisiae*: engineering of pyruvate carboxylation, oxaloacetate reduction and malate export. Appl. Environ. Microbiol. 74 (9), 2766–2777.

Zhang, K., Li, H., Myung, K.C., Liao, J.C., 2010a. Expanding metabolism for total biosynthesis of the nonnatural amino acid L-homoalanine. PNAS 107 (14), 6234–6239.

Zhang, K., Li, H., Cho, K.M., Liao, J.C., 2010b. Expanding metabolism for total biosynthesis of the nonnatural amino acid L-homoalanine. Proc. Natl. Acad. Sci. U. S. A. 107 (14), 6234–6239.

Zhang, F., Rodriguez, S., Keasling, J.D., 2011a. Metabolic engineering of microbial pathways for advanced biofuels production. Curr. Opin. Biotechnol. 22, 775–783.

Zhang, L., Zhao, H., Gan, M., Jin, Y., Gao, X., Chen, Q., Guan, J., Wang, Z., 2011b. Application of simultaneous saccharification and fermentation (SSF) from viscosity reducing of raw sweet potato for bioethanol production at laboratory, pilot and industrial scales. Bioresour. Technol. 102 (6), 4573–4579.

Zhang, Y., Werling, U., Edelmann, W., 2012. SLiCE: a novel bacterial cell extract-based DNA cloning method. Nucleic Acids Res. 40 (8), e55.

Zhou, L., Niu, D., Tian, K., Chen, X., Prior, B.A., Shen W Shi, G.Y., Singh, S., Wang, Z.X., 2012. Genetically switched D-lactate production in *Escherichia coli*. Metab. Eng. 14, 560–568.

Zhu, Y., Eiteman, M.A., DeWitt, K., Altman, E., 2007. Homolactate fermentation by metabolically engineered *Escherichia coli* strains. Appl. Environ. Microbiol. 73 (2), 456–464.

Further Reading

Beloqui, A., María, P.D., Golyshin, P.N., Ferrer, M., 2008. Recent trends in industrial microbiology. Curr. Opin. Microbiol. 11 (3), 240–248.

Choi, J.M., Han, S.S., Kim, H.S., 2015. Industrial applications of enzyme biocatalysis: current status and future aspects. Biotechnol. Adv. 33 (7), 1443–1454.

Erickson, B., Nelson, J.E., Winters, P., 2012. Perspective on opportunities in industrial biotechnology in renewable chemicals. Biotechnol. J. 7, 176–185.

Johannes, T., Simurdiak, M.R., Zhao, H., 2006. Bio catalysis. In: Encyclopedia of Chemical Processing. Taylor & Francis, USA.

Luengo, J.M., García, B., Sandoval, A., Naharroy, G., Olivera, E.R., 2003. Bioplastics from microorganisms. Curr. Opin. Microbiol. 6 (3), 251–260.

Mahanta, N., Gupta, A., Khare, S.K., 2011. Production of protease and lipase by solvent tolerant *Pseudomonas aeruginosa* PseA in solid-state fermentation using *Jatropha curcas* seed cake as substrate. Bioresour. Technol. 99, 1729–1735.

Shen, C.R., Lan, E.I., Dekishima, Y., Baez, A., Cho, K.M., Liao, J.C., 2011. Driving forces enable high-titer anaerobic 1-butanol synthesis in *Escherichia coli*. Appl. Environ. Microbiol. 77 (9), 2905–2915.

Wohlgemuth, R., 2010. Biocatalysis—key to sustainable industrial chemistry. Curr. Opin. Biotechnol. 21, 713–724.

Youngquist, J.T., Schumacher, B.M.H., Rose, J.P., Raines, T.C., Politz, M.C., Copeland, M.F., Pflegera, B.F., 2013. Production of medium chain length fatty alcohols from glucose in *Escherichia coli*. Metab. Eng. 20, 177–186.

Zhang, Y., Muyrers, J.P., Testa, G., Stewart, A.F., 2000. DNA cloning by homologous recombination in *Escherichia coli*. Nat. Biotechnol. 18 (12), 1314–1317.

Zhang, L., Zhao, H., Gan, M., Jin, Y., Gao, X., Chen, Q., Guan, J., Wang, Z., 2015. Benchmark data for identifying N(6)-methyladenosine sites in the *Saccharomyces cerevisiae* genome. Data Brief 5, 376–378.

CHAPTER

10

Current Advances in the Protection of Viral Diseases in Aquaculture With Special Reference to Vaccination

*Bestha Lakshmi**, *Shameer Syed*†, *Viswanath Buddolla*‡

*Department of Zoology, Sri Krishnadevaraya University, Anantapur, India †Nanotechnology Laboratory, Institute of Frontier Technology, Regional Agricultural Research Station, Acharya N.G. Ranga Agricultural University, Tirupati, India ‡Department of Bionanotechnology, Gachon University, South Korea

OUTLINE

Recent Developments in Applied Microbiology and Biochemistry
https://doi.org/10.1016/B978-0-12-816328-3.00010-6

10.1 INTRODUCTION

Fish farming is becoming more important to the world's food sector as the global fish production has grown steadily in the last five decades and food fish supply has been increasing at an average annual rate of 3.2%, outpacing world population growth of 1.6%. Apparently, the world per capita fish consumption increased from an average of 9.9kg in the 1960s to 19.2kg in 2012 (FAO, 2014). This remarkable development has been made possible by the expansion of fish production and more efficient distribution and marketing approaches. As intensive fish farming develops, so do the hurdles, like pathogens causing serious diseases. The most common causative agents of infectious diseases in aquaculture are bacteria (54.9%), followed by viruses (22.6%), parasites (19.4%), and fungi (3.1%) (Kibenge et al., 2013). Among these, viral diseases are more difficult to control due to their rapid spread and high morbidity and severe mortality rates attributed to the lack of specific therapeutics, high susceptibility of aquatic animals, and limited knowledge regarding antiviral mechanisms in fish. In attempts to improve profitability of fish farming, several strategies that were successfully implemented for terrestrial farm animals were employed in the fish farms including selective breeding for increasing resistance and vaccination (Gjedrem, 2013). In spite of these measures, viruses set in as sudden outbreaks creating enormous economic losses in fish farming.

Viral hemorrhagic septicemia (VHS), infectious hematopoietic necrosis (IHN), infectious pancreatic necrosis (IPN), spring viremia of carp (SVC), infectious salmon anemia (ISA), and pancreatic disease (PD) are some of the commercially important farmed fish diseases listed by the World Organization for Animal Health (OIE, 2012). Infectious viral diseases are responsible for shattering economic losses in fish farming; the best instances include outbreaks of ISA in Norway, Canada, Scotland, the Faroe Islands, the United States, and Chile and the outbreak of PD in Norway (Pettersen et al., 2012). In the past few years, multiple infections are emerging simultaneously, a major reason of concern in fish farming as emergence of viral disease occurs often in association with other pathogens. *Cyprinid herpesvirus 2* infection in goldfish (*Carassius auratus*) occurred in association with multidrug-resistant *Aeromonas hydrophila*, from commercial farms (Sahoo et al., 2016). Similarly, disease outbreaks in cultured Nile tilapia (*Oreochromis niloticus*) farms showed natural contemporaneous infections of bacterial pathogens, and *Iridovirus* caused total mortality of fish. The pathogen identification studies found that the prevalence of virus was up to 40% in the infected tilapia farms and *Iridovirus* was proved to be the sole perpetrator for the mortality in tilapia (Dong et al., 2015). This paper gives an updated view of the literature available on the recent outbreaks of viral diseases of farmed fish and their control strategies, particularly with reference to vaccination.

10.2 RECENT VIRAL DISEASE OUTBREAKS IN FARMED FISH

Increased aquaculture practices and global movement of the aquatic animals stand as the major influencing factors for viral disease outbreaks in fish farming (Kibenge et al., 2012; Lakshmi et al., 2013). The spread of virus is the important factor to be controlled to prevent the disease reoccurrence and emergence of new serotypes. Different factors leading to virus spreading and their transmission are summarized below.

10.2.1 Spread of Virus

Contagiousness of a virus is considered as the foremost threatening factor in the disease outbreaks of aquaculture. Viruses can spread through different ways, but translocation of live aquatic animals with low prevalence of disease or asymptomatic carrier hosts to new destinations is a major route leading to the entry of pathogen to new uninfected farms. Transport of koi herpesvirus-infected carp (*Cyprinus carpio*) for ornamental pet trade from Israel to the world leads to mortality and morbidity in carp farms and fisheries, cumulating to economic losses (Rodgers et al., 2011). The common origin of the virus revealed from molecular studies of IPNV isolates of Chile and Europe is an outcome of the introduction of infected/contaminated fish eggs from former to latter (Mutoloki and Evensen, 2011). Feeding farmed fish with wild fish has also been believed to be responsible for the origin of VHS in Europe (Snow, 2011). The spreading of Baltic strains of viral hemorrhagic septicemia virus by herring off the UK coast establishes migratory fish and birds as vectors of viral pathogens (Matejusova et al., 2010). Wild or escaped fish are considered as potential reservoirs for *salmonid alphavirus* found in wild fish in Scotland, irrespective to the aquaculture activities (Snow et al., 2010). Human handling of the aquatic animals also leads to the transport of viruses and exposure of the animals to the native viruses. Different ways contributing to the spread of virus in fish farming are showed in Fig. 10.1.

10.2.2 Transmission of Virus

As water currents move viral particles over many kilometers, water is considered as an active energetic environment for the transmission of viruses in aquaculture, and the best instances for this assumptions are the spread of pancreatic disease in salmon farms, outbreak of infectious salmon anemia (ISA) in Atlantic salmon (*Salmo salar* L.) farms, and detection of IPNV isolates in a river located downstream to an infected hatchery (Murray, 2013). Translocation of a large number of aquatic animals, vessels, and equipment between farms also results in the transmission of pathogens if the source site is infected (Munro and Gregory, 2009). Trading a large number of infected ova over long distances leads to vertical transmission

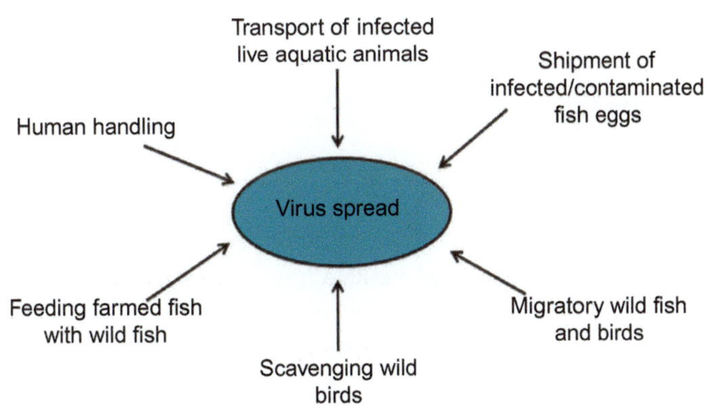

FIG. 10.1 Different modes of viral pathogen spreading in fish farming.

of virus, and the best instance is the presence of IPNV in ovarian and seminal fluids of Atlantic salmon (Smail and Munro, 2008).

Virus transmission under aquaculture is studied by SIR model (Murray, 2013), where S represents the susceptible group, I stands for the infected cohort, and R stands for removed cohort. The model generates a relationship $R_0=\beta S_0/\gamma$ where R_0 is the basic reproductive rate of the pathogen, β is the transmission coefficient, S_0 is S before infection is introduced, and γ is the rate of recovery or death from infection. If R_0 for a pathogen is low, it will be unable to infect an adequate number of hosts to establish infection and thus will be eradicated; if R_0 is near 1, the pathogen will infect a proportion of the susceptible group but not enough to deplete the susceptible population to the point of elimination; if and only if $R_0 \geq 1$, then the disease becomes progressive.

10.2.3 Recent Viral Disease Outbreaks

Most common and commercially important viral diseases of farmed fish are discussed below. Major fish viral diseases along with their causative agents are presented in Table 10.1 and their respective target species and clinical symptoms in Table 10.2.

10.2.4 Viral Haemorrhagic Septicaemia (VHS)

Viral hemorrhagic septicemia is one of the most fatal viral diseases of cultured rainbow trout (*O. mykiss*) and is caused by VHS virus (VHSV). VHSV is a member of the genus *Novirhabdovirus* of the family *Rhabdoviridae* and contains a single linear, negative-sense ss-RNA as genetic material of approximately 11.1 kb. Molecular epidemiology of freshwater and marine VHSV isolate gene sequences has revealed the existence of four genotypes.

Recent outbreak of VHS virus occurred in rainbow trout farms in Iran. Mortality rate up to 70% was recorded from October 2014 to July 2015, majorly affecting trout production. Tissue

TABLE 10.1 List of Major Viral Diseases of Farmed Fish

Disease	Virus	Genome	Family	Genus
Viral hemorrhagic septicemia	*Viral hemorrhagic septicemia virus (VHSV)*	−ssRNA	*Rhabdoviridae*	*Novirhabdovirus*
Infectious hematopoietic necrosis	*Infectious hematopoietic necrosis virus (IHNV)*	−ssRNA	*Rhabdoviridae*	*Novirhabdovirus*
Infectious pancreatic necrosis	*Infectious pancreatic necrosis virus (IPNV)*	dsRNA	*Birnaviridae*	*Aquabirnavirus*
Infectious salmon anemia	*Infectious salmon anemia virus (ISAV)*	−ssRNA	*Orthomyxoviridae*	*Isavirus*
Spring viremia of carp	*Spring viremia of carp virus (SVCV)*	−ssRNA	*Rhabdoviridae*	*Vesiculovirus*
Pancreas disease	*Salmonid alphavirus (SAV)*	+ssRNA	*Togaviridae*	*Alphavirus*

TABLE 10.2 Clinical Signs and Target Species of Viral Diseases

Disease	Species Affected	Clinical Signs
Viral hemorrhagic septicemia	*Oncorhynchus mykiss, Scophthalmus maximus, Paralichthys olivaceus*	Rapid onset of mortality; lethargy; darkening of the skin; exophthalmia; anemia; hemorrhages at the base of the fins, gills, eyes, and skin; abnormal swimming such as flashing and spiraling
Infectious hematopoietic necrosis	*Oncorhynchus tshawytscha, O. nerka, O. keta, Clupea harengus*	Lethargy, darkening of the skin, pale gills, ascites, distended abdomen, exophthalmia, petechial hemorrhages internally and externally
Infectious pancreatic necrosis	All farmed salmonid fish	Abnormal swimming behavior, exophthalmos, distended abdomen, darkened pigmentation, focal necrotic lesions in the exocrine pancreatic tissue, and severe liver necrosis
Infectious salmon anemia	Marine-farmed Atlantic salmon	Exophthalmos, petechial hemorrhages on the abdomen, congestion and enlargement of the liver
Spring viremia of carp	*Cyprinus carpio carpio* and other wild common carp	Distended abdomen, exophthalmia, and petechial hemorrhages of the skin
Pancreas disease	All farmed salmonids	Sleeping behavior, exophthalmos, and swollen abdomens

homogenates of diseased fish induced cytopathic effects (CPE) in Chinook salmon embryo (CHSE-214) cells, further confirmed by RT-PCR (Ahmadivand et al., 2016). A wide sampling of 39 different species in a research study conducted during 2009–11 in Norwegian coastal waters and fjord systems detected VHS virus in all fish samples (Sandlund et al., 2014). Virus spread has been studied in VHSV-targeted surveillances in the state of Michigan during 2005–10. Of the 1823 cases tested, 30 cases from 19 fish species tested positive for VHSV by tissue culture and RT-PCR (Faisal et al., 2012). Among three European genotypes of VHSV, genotype III (GIII) has been recently studied for its virulence factors. VHSV isolate named NO-2007-50-385 was isolated from an outbreak in sea-farmed rainbow trout in Norway, and it was comparatively analyzed for its virulence with a nonvirulent GIII isolate, 4p168. Rainbow trout were treated with VHSV NO-2007-50-385 for 1 and 6h, resulting in mortalities of 5% and 35%, respectively, and no mortality was observed in the similarly treated rainbow trout with the genotype III VHSV isolate 4p168. The viral titers in the in vitro studies on the rainbow trout cell line RTG-2 showed lower titer of 4p168 isolate compared with the virulent NO-2007-50-385 VHSV isolate. Comparative analysis of the entire genome of the genotype III isolates suggests that substitutions of amino acids in positions 118–123 of the nucleoprotein are essential to the virulence of VHSV GIII in rainbow trout (Ito et al., 2016).

10.2.5 Infectious Hematopoietic Necrosis (IHN)

Infectious hematopoietic necrosis virus (IHNV), a member of the genus *Novirhabdovirus* and family *Rhabdoviridae*, causes infectious hematopoietic necrosis in fish. The virus has the typical bullet shape of members of the *Rhabdoviridae*, and the genome consists of a single molecule of negative-sense ssRNA with varying strain pathogenicity. IHNV isolates can be grouped into three genetic types, which are correlated mainly with geographic regions. The U genotype group includes isolates from Alaska, British Columbia, and watersheds of coastal Washington and the Columbia River basin and isolates from Oregon, California, and Japan. The L genotype contains most of the viruses from California and the Oregon coast. The M group contains isolates from Idaho, the Columbia River basin, and Europe and a virus from the Washington coast. The genetic diversity of M group is significantly higher than the L or U groups. An IHNV database is available containing records for >1000 individual field isolates, which is updated annually (Kurath, 2012).

The first IHN outbreak occurred in sockeye salmon at fish hatcheries in Oregon and Washington in 1953. In an experimentally induced outbreak of IHNV, the virus was detected in the fish even after 9 months of IHNV exposure but with no symptoms, and mortality was observed (Müller et al., 2015). The results from finite-volume ocean circulation and particle tracking models of waterborne transmission of IHNV among Atlantic salmon (*S. salar*) reveal that neighboring naive farms are at higher risk of contamination in early summer and IHNV was the major perpetrator in the outbreaks of 2001–02 (Foreman et al., 2015). Similar outbreak of IHNV occurred in cultured yearling rainbow trout along Southwest China causing severe mortality (>40%). The analysis of the moribund and dead fish showed analogous symptoms with the naturally IHNV-infected fish (Yu et al., 2014). In May 2012, during a routine health screening program of a company in Atlantic salmon farms, IHNV was diagnosed in three farm sites on the west coast of Vancouver Island. Virus detection triggered a quarantine of the farm and its entire stock of 560,000 one-kilo-sized salmon (Anonymous, 2012).

10.2.6 Infectious Pancreatic Necrosis (IPN)

Infectious pancreatic necrosis (IPN) is a highly transmittable viral contagion principally of trout and salmon, but the virus was also been isolated from a wide variety of fish species. The infection is characteristically seen in trout as an acute disease with high mortality in fry and fingerlings. The causative agent is IPN virus, a member of the genus *Aquabirnavirus* and family *Birnaviridae* with nonenveloped, bisegmented, double-stranded RNA genome; serologically, the IPNV strains are divided into groups A and B and further classified on the basis of deduced amino acid similarities of VP2.

IPN is considered as a highly contagious viral disease due to its severe mortalities in adult fish and 100% mortalities in fry and fingerlings. The virus is transmitted through different modes as mentioned above; in addition to that, practices like cocultivation also lead to the transmission of virus from one species to the other susceptible species. In an experimental study of the transmission of IPNV in integrated multitrophic aquaculture system, IPNV was found to be transmitted from mussels to Atlantic salmon smolts. The preexposed blue mussel (*Mytilus edulis*) serves as a reservoir host for IPNV and infects its cohabiting species Atlantic salmon smolts (Molloy et al., 2013). IPNV is a very tough virus, is able to withstand desiccation, survives in both fresh water and seawater for considerable periods, is resistant to high temperature, and is 100 times more resistant to ultraviolet radiation than many other viruses. Thermal inactivation assay of IPNV in a medium mimicking the water-soluble phase of hydrolyzed fish by-products at different temperatures ranging from 60°C to 90°C revealed that 2h of treatment of fish silage at 60°C is sufficient for virus inactivation, allowing the silage to be used for feedstock (Nygaard et al., 2012). Farmed fish show lifelong persistent infection and considered to be the most important reservoir of IPNV in the aquatic environment (Johansen et al., 2011).

10.2.7 Infectious Salmon Anemia (ISA)

Infectious salmon anemia is a disease of sea-farmed Atlantic salmon (*S. salar*), and the perpetrator is ISA virus, a member of family *Orthomyxoviridae* and genus *Isavirus*, and virions are pleomorphic and enveloped, containing eight negative-sense ssRNA molecules that encode at least 10 proteins. RNA segments 5 and 6 code for a fusion protein and hemagglutinin esterase (HE), respectively, while two polypeptides coded by segments 7 and 8 inhibit interferon induction. Genomic analysis has shown the existence of a high polymorphism region (HPR) in segment 6 (Cottet et al., 2010). Based on the sequence differences in the HE gene, ISAV occurs in two basic genotypes, North American and European. The latter genotype shows more genetic variation than the former genotype. A small, highly polymorphic region (HPR) of the viral HE is considered to be an important virulence factor. HPR0 group is known to be nonpathogenic variants of ISAV indicating that they have a "full-length" highly polymorphic region (HPR) of their surface envelope glycoprotein or HE, but their evolutionary status is not clearly understood (Kibenge et al., 2012).

Most of the ISAV isolates belong to European genotype. Kibenge et al. (2016) gave the first published report of the detection of ISAV sequences in fish from British Columbia (BC), Canada. This was found to be a new ISAV variant of European genotype from classical ISAV isolates. In a study on recent outbreaks in Chile, two strains ISAV-HPR7a and ISAV-HPR7b were isolated from infected salmon tissues, and these were considered as the causative agents of the 2013 and 2014 outbreaks in Chile, respectively (Godoy et al., 2014). Segments 5 and 6 sequence analysis showed that ISAV-HPR7a and ISAV-HPR7b belong to the European genotype I strains only found in Europe and Chile and show high similarity in segments 5 and 6 with 95%–96% identity. In comparison with the 2007 ISAV-HPR7b isolate, the 2014 ISAV-HPR7b does not have an insertion in segment 5 and was associated with low mortality, which suggests that ISAV virulence was attenuated by the absence of the insertion in segment 5; contrastingly, the highly virulent ISAV-HPR14 from April 2013 outbreak did not have the insertion in segment 5 either. As mentioned earlier, the ISAV-HPR0 is considered as nonpathogenic variant, but the possibility that it could mutate to virulent ISAV-HPRa was also reported in Chilean strains. A low-pathogenic ISAV-HPR0 variant and virulent ISAV-HPRa that caused two new ISA outbreaks in April 2013 were comparatively studied. The reports provide strong evidence supporting the contention that ISAV-HPR0 shows a strong relationship to virulent ISAV-HPRa viruses and the possibility that it could mutate to virulent ISAV-HPRa (Godoy et al., 2013). A possible explanation for the emergence of ISAV-HPRa from ISAV-HPR0 was given by the European Food Safety Authority in the year 2012; virulent ISAV has deletions in the HPR region of the HE gene, and they have either an insertion or the Q266L mutation in the F gene. This is further supported by the close association between the genetic relatedness and spatiotemporal distances of virus strains in solitary outbreaks. Area contact networks and spatiotemporal relation were known to be major reasons for the spread and outbreak of ISAV. Spatiotemporal analyses identified hydrographic linkages, shared ports, and fish transfers from areas with recent occurrence of ISAV as the strongest predictors of virus spread between areas (Gustafson et al., 2016). A panel of fish health experts identified barrio processing plant biosecurity, fallowing strategies, adult live fish transfers, fish and site density, smolt quality, hydrographic connection with other neighborhoods, and the presence of sea lice (*Caligus rogercresseyi*) as factors with the greatest predictive strength for ISAV virulent genotype occurrence, whereas fish and site density, adult live fish transfers, and smolt quality facilitate the HPR0 prevalence (Gustafson et al., 2014). The nonpathogenic HPR0 also causes ISAV outbreaks, but the infection was only subclinical and transient one. In an Atlantic salmon farming industry in the Faroe Islands, analysis of healthy Atlantic salmon tested positive for ISAV by RT-PCR and sequence analysis revealed the alleged low-pathogenic ISAV-HPR0 subtype in all cases (Christiansen et al., 2011).

10.2.8 Spring Viraemia of Carp (SVC)

Spring viremia of carp (SVC) is an infectious viral disease common to carp (*C. carpio*) and other cyprinid fish species, caused by SVC virus (SVCV), a member of the family *Rhabdoviridae*, with negative-sense ssRNA encoding five structural proteins—nucleoprotein (N), phosphoprotein (P), matrix protein (M), glycoprotein (G), and RNA-dependent polymerase (L) approximately about 11 kb—but is distinct from VHSV and IHNV and the genus *Vesiculovirus*. Common carp is considered as the natural host and vector of SVC, but experimental susceptibility was observed in Caspian white fish (*Rutilus frisii kutum*) (Zamani et al., 2014). Related studies were done on related fish like fathead minnow (*Pimephales promelas* Rafinesque), emerald shiner (*Notropis atherinoides* Rafinesque), and white sucker (*Catostomus commersonii* (Lacepede)), with lower mortality rates (Misk et al., 2016). Among the five structural proteins, matrix protein (M) and glycoprotein (G) were studied widely for their significant role in viral pathogenesis. M protein causes apoptosis and inhibits host cell transcription, whereas G protein was concerned with viral endocytosis. Monoclonal antibodies were generated against M and G proteins that are useful in the development of SVCV diagnostic methods. On the basis of G protein gene sequences, SVCV has been divided into four subgroups, Ia, Ib, Ic, and Id. An improved RT-PCR has been developed for specific detection of these SVCV subtypes where a new primer set was designed from a G gene sequence of SVCV (Shimahara et al., 2016). SVCV was known to induce autophagy in the infected cells. Epithelioma papulosum cyprini (EPC) cells when infected with SVCV showed the activation of autophagy pathway. Autophagy promoted the survival of SVCV-infected cells by eliminating damaged mitochondrial DNA generated during viral infection (Liu et al., 2015). Synthetic siRNA molecules were intended to target SVCV-N and SVCV-P transcripts to hamper SVCV reproduction, and the inhibition of gene transcription was measured by RT-qPCR (Gotesman et al., 2015).

10.2.9 Pancreas Disease (PD)

One of the serious illnesses of farmed salmonids is pancreas disease caused by an enveloped, single-stranded, positive-sense RNA virus *salmonid alphavirus*, a new member of the genus *Alphavirus* in the family *Togaviridae*. This distinctive group of viruses has been classified into six subtypes (SAVs 1–6) based on sequence analysis of the most variable genes, nsP3 and E2, SAVs notorious for causing pancreas disease (PD) in cultured Atlantic salmon (*S. salar* L.) and sleeping disease (SD) in rainbow trout (*O. mykiss* Walbaum) (Fringuelli et al., 2008).

SAV subtypes were reported for clinical outbreaks of marine PD leading to severe financial losses to commercial aquaculture. Subtypes 1, 2, 4, and 5 caused an epidemic in the United Kingdom; SAVs 1, 2, 4, and 6 in Ireland; and SAVs 2 and 3 in Norway. Clinical outbreaks of SD in rainbow trout in fresh water, caused by SAV 2, have been reported in several continental European countries including France, Germany, Italy, Poland, and the United Kingdom (Herath et al., 2016). These six subtypes in experimental challenge against Atlantic salmon in a freshwater cohabitation trial showed that SAV subtypes 1 and 3 cause synchronous infection in the cohabitant fish and showed the highest virus load and the most extensive histopathologic changes. In difference, subtype 2 and 6 strains were consistent with asynchronous infection in the cohabitant fish that is slow to spread and shows low virus titer and only mild histopathologic changes, but the subtypes 4 and 5 occupied an intermediate position in this regard (Graham et al., 2011). SAV infection causes severe histopathologic changes in the heart, pancreas, and skeletal muscles of diseased fish as evident from in vivo studies by Herath et al. (2016), and Bruno et al. (2014) identified common dab (*Limanda limanda*) as a wild reservoir of *salmonid alphavirus*. Random sampling of common dab by quantitative real-time RT-qPCR showed the presence of SAV RNA from common dab in regions standing apart from salmon aquaculture. SAV is cultured in mosquito cell lines as a substitute fabrication system for vaccine applications that still need further research (Hikke et al., 2014). For the spread and outbreak of SAV, many factors contribute, but infected farmed salmonids are the main source, but transmission between seawater sites occurs passively through water currents or actively through human activity together with inadequate biosecurity measures (Jansen et al., 2017).

10.3 ONTOGENY OF THE FISH IMMUNE SYSTEM

Excluding certain differences physiologically, the immune system of fish resembles that of higher vertebrates in the development of the immune organs, but onset of immune response is earlier than in vertebrates because of their free-living habitat, which makes the fish susceptible to different pathogens starting from their larval stages (Uribe et al., 2011). The thymus, kidney, and spleen are the largest lymphoid organs in teleosts, and their developmental sequence is as follows—the kidney, spleen, and thymus—the kidney in teleost fish is similar to that of bone marrow in vertebrates (Zapata et al., 2006). The hematopoiesis at the primitive level occurs in the intermediate cell mass, and the myelopoiesis initiates in the anterior lateral mesoderm that is located intraembryonically (Chen and Zon, 2009). After 24h of postfertilization (hpf), hematopoiesis shifts to the posterior blood island for a brief instance, while cells of myeloid lineage were found around the yolk sac. After 48 hpf, young myeloblasts and neutrophilic myelocytes were observed perceptibly and found to migrate through blood vessels, whereas erythrocytes were found circulating in the blood vessels and mature granulocytes in the blood system of connective tissue surrounding the yolk sac. Mature neutrophils were observed in various tissues after 72 hpf (Chen and Zon, 2009).

Humoral immune response in fish includes the production of autologous and maternal antibodies, complement systems, lysozymes, and other lytic enzymes. The appearance of autologous humoral IgM and maternal IgM was found to be varied in different fish like teleost fish that produce IgM, IgD, and IgT antibodies. In the case of salmonid fish, maternal IgM was found in the eggs and embryos at very low concentrations (0.3–6.3 mg/g wet weight), about one thousandth of maternal plasma; higher levels of maternal IgM were noticed in the ovary during vitellogenesis, reached maximum concentration during ovulation, and found to be quickly reduced after hatching just prior to the onset of feeding and the autologous IgM production (Olsen and Press, 1994). IgD antibody was known to be an enigmatic molecule that is found to be ancient Ig isotype found in teleost fish and other vertebrates and shows some species-specific adaptations; research findings till date suggest that IgD plays a direct role in immune defense in teleost fish and some other vertebrates (Edholm et al., 2011). IgT is another isotype of gut-associated Ig found in teleost fish when challenged with *viral hemorrhagic septicemia virus* (VHSV). It is produced by subsets of B cells and found to play a key role in systemic immunity (Castro et al., 2013). Complement system was similar as that in higher vertebrates expect for which the teleost fish show multiple forms of C3. Other immune parameters like protection by lysozyme and other enzymes were also seen in the eggs of teleosts offering resistance against certain bacterial infections preventing vertical transfer of pathogens.

Cellular immune response of fish also demonstrates strong similarities with mammals. Several in vitro studies like allograft rejection, graft-versus-host reaction, and delayed hypersensitivity reaction showed specific cell-mediated cytotoxic reactions suggesting the presence of cytotoxic T cells in fish. T cells were known to effectively combat against virus infections in mammals, and it was found to show similar mechanism in fish too. The CD8+ MHC class I presentation was found to be similar to that that occurs in higher vertebrates. In the T-cell lineage, in salmonids and some other teleosts, T-helper and cytotoxic T cells are not well described, while the available markers for some species suggest the presence of T-cell subsets like Th0, Th1, Th2, Th17, and Tregs. (Fast et al., 2009). Transcripts of CD8+ T cells and MHC class I were detected in young rainbow trout after 1 week of insemination, and the 14-day larvae were capable of destroying skin allografts, a process accompanied by infiltration of cytotoxic T lymphocytes (Tatner and Manning, 1993). Further, the onset of antibody production was quicker to T-independent antigens than to T-dependent antigens.

10.3.1 Immunity of Farmed Fish to Viral Diseases

A well-adapted mechanism is found in the fish against viral infections. Studies showed that salmon display high resistance against IPN virus (IPNV) and PD virus (PDV), and recent research revealed that they have a strong interferon system, which is the most significant constituent in the first line of defense against viruses and found to be comparable with that of mammals.

10.3.2 Innate Antiviral Responses

As mentioned above, fish interferon system is analogous to mammals. In teleost fish based on the receptors in signaling pathway, two groups of IFNs were classified, group I (IFNφ1 and IFNφ4) and group II (IFNφ2 and IFNφ3), but the crystal structure studies showed a similar architecture of type I IFN to both groups of IFNφs; based on these considerations, fish IFNs were named as type I IFNs, virus-induced IFNs, and IFNλ, among which virus-induced IFNs are mostly the type I. Upon viral infection, the virus transfers its genome into the host cells where it undergoes replication and transcription to fabricate viral nucleic acids such as double-stranded (ds) RNA, single-stranded (ss)RNA, and dsDNA by using RNA polymerases; these nucleic acids coalesce with viral glycoproteins and form "virus-associated molecular patterns" (VAMPs). These VAMPs are detected by specific pattern recognition receptors (PRRs) such as RIG-I-like receptors (RLRs) and toll-like receptors (TLRs) and are highly conserved in mammals and fish. Activation of the signaling pathway leads to the expression of some IFN-stimulated genes like Mx and PKR, leading to the synthesis of antiviral proteins such as Mx, ISG-15, ISG-56, PKR, and OAS. Pepscan mapping of VHSV glycoprotein G (gpG) revealed two regions of amino acid residues 280–310 and 340–370 have the ability to increase the basal expression level of the IFN-stimulated Mx gene and to restrict the spread of virus among responder cells signifying the vital role of type I IFN in protective-antigen-specific immune responses. Investigative studies on fish IFN induction in zebra fish (*Danio rerio* (Dr)) revealed that IFN regulatory factor (IRF), DrIRF1, acts as positive regulator of the IFN antiviral response, can be found abundantly in zebra fish immune tissues upon virus infection, and mediates the activation of zebra fish IFNϕ1 and IFNϕ3 promoters (Feng et al., 2015). Once the IFN-signaling pathway is activated, it leads to the production of inflammatory cytokines and chemokines and increased surface expression of costimulatory molecules (Kibenge et al., 2012). Contemporary research has identified the role of mitochondria in innate immune response of fish as an adaptor protein, mitochondrial antiviral signaling (MAVS) protein, participates in RLR signaling (Su et al., 2011). MAVS acts in alliance with tumor necrosis factor (TNF) receptor-associated factor (TRAF) 3 and activates the pathway by TANK-binding kinase 1 (TBK1) via phosphorylation of IFN regulatory factor (IRF) (Xiang et al., 2011). A transmembrane protein, STING, acts as adaptor and mediates the signaling between MAVS and downstream kinases and activates the RLR signaling.

Toll-like receptors (TLRs) are cellular sensors directly involved in the activation of IFN system and are more diversified in fish. >17 different TLRs have been identified in different fish species among which specific TLRs share functional properties with their mammalian counterparts. TLR3, TLR7, and TLR8 in mammals have good orthologues in fish. Mammalian TLR3 is involved in the recognition of double-stranded RNA (dsRNA), whereas piscine TLR3 besides detecting viral genome also detects bacterial molecular patterns. Similarly, conserved counterparts of TLR7 and TLR8 were observed between mammals and fish. Piscine TLR7 and TLR8 act as sensors of viral ligands and induce typical IFN response in salmonid leukocytes. TLR22 is a fish-specific TLR that is functionally active responding to virus infections and dsRNA, while it is a nonfunctional pseudogene in humans (Rebl et al., 2010).

10.3.3 Immune Enhancement Through Feed Additives

Though fish possess active defense system against viral attacks, they are more prone to infections. The concept of immune enhancement through feed additives is a much intense area of research that is commercially significant. The negative effects associated with the use of antibiotics and the high-priced commercial vaccines have raised the necessity of using immune-stimulatory feed additives. Feed additives are primarily intended for prophylactic use in order to avert the disease occurrence by supporting and enhancing the immune system and making it strong enough to fight the infection. Feed additives like phytobiotics are well reported in aquaculture for preventing and controlling the viral and bacterial diseases.

Phytobiotics, a group of plant-based, natural compounds, are added into diets resulting in improved animal productivity and immunostimulation against viruses, bacteria, and parasitic infections in carp and reduce losses. Harikrishnan et al. (2010) reported that intraperitoneal administration of the leaf extracts of *Punica granatum* at 50 or 100 mg kg^{-1} dose enhanced the innate immune responses and disease resistance after 8 weeks in *P. olivaceus* against natural LDV infection. Antiviral activity was reported for *Psidium guajava* extracts against the fish pathogenic viruses, IHNV, IPNV, and *O. masou virus* (OMV) using plaque reduction in CHSE-214 cell lines. Plaque reduction by 65%–100% and 21%–100% was observed for IHNV and OMV, respectively (Sivasankar et al., 2015). Direkbusarakom (2004) reported that *Phyllanthus amarus* is very effective against the fish viruses such as INHV and OMV and shrimp virus YHV. Behera et al. (2014) reported that the addition of nano-Fe as feed additive to *Labeo rohita* increased muscle iron and hemoglobin contents leading to enhanced antioxidant enzymatic activities. Some of the reported plant products that are capable of enhancing immunity in fish include *Eclipta alba, Aloe vera, Ocimum sanctum, Viscum album, Urtica dioica, Zingiber officinale, Solanum trilobatum, Astragalus radix, Scutellaria radix*, and *Achyranthes aspera*.

10.4 VACCINATION OF FISH

As aquaculture is a rapidly growing food-producing sector, it has been continuously exposed to many constraints among which disease is the most important limiting factor (Gudding and Van Muiswinkel, 2013; Munang'andu and Evensen, 2018). Though prophylactic treatments and good management practices can usually prevent or reduce the vulnerability to diseases, nonetheless, established viral diseases cannot be treated (Lafferty et al., 2015). Therefore, vaccination is the best alternative to combat bacterial and viral diseases and plays an imperative role in large-scale commercial fish farming (Mutoloki et al., 2015). Different types of fish vaccines and their route of administration are shown in Fig. 10.2.

The most efficient practice will depend upon the pathogen and its natural mode of infection, age of the fish, production techniques, and other logistic considerations (Muktar et al., 2016). A precise route of administering or even manifold applications using diverse methods may be necessary for adequate protection. The three major routes of vaccine delivery include injection, immersion, and oral; the advantages and limitations are shown in Table 10.3. The

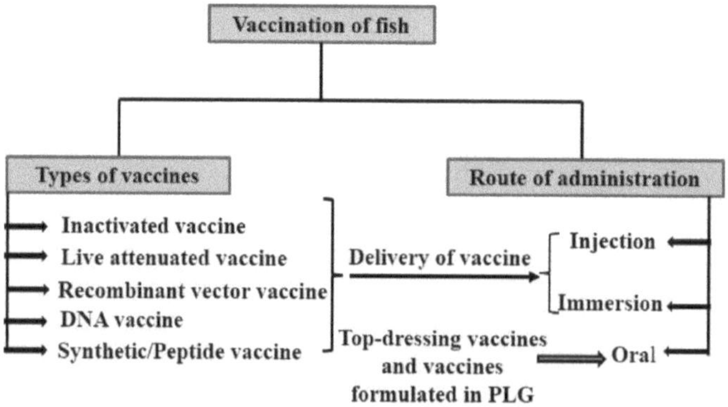

FIG. 10.2 Types of fish vaccines and their route of administration.

TABLE 10.3 Routes of Administration of Vaccines With Their Advantages and Limitations

Route of Administration	Advantages	Limitations
Injection 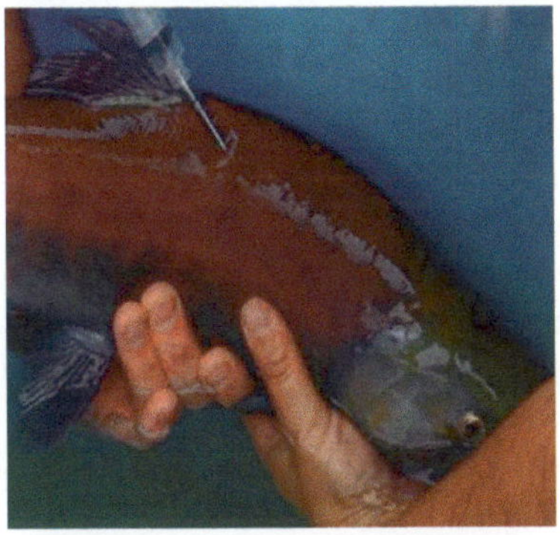	• Most potent route of vaccination • Cost-effective method for large fish • More direct stimulation of systemic response • Multiple antigens can be delivered at the same time • Long-term protection	• Stressful method (anesthesia, handling) • Require more time, labor, and skilled personnel • Incite a more severe reaction if it is injected into the wrong portion of the fish • Smaller-sized fish (under 10 g) may not respond well
Immersion 	• Cost-effective method • Widely used for vaccinating fry • Immune cells get directly exposed to antigen • More systemic response is developed	• Immunity may not last long where booster doses are required • Impractical for large-size fish due to cost-effectiveness
Oral 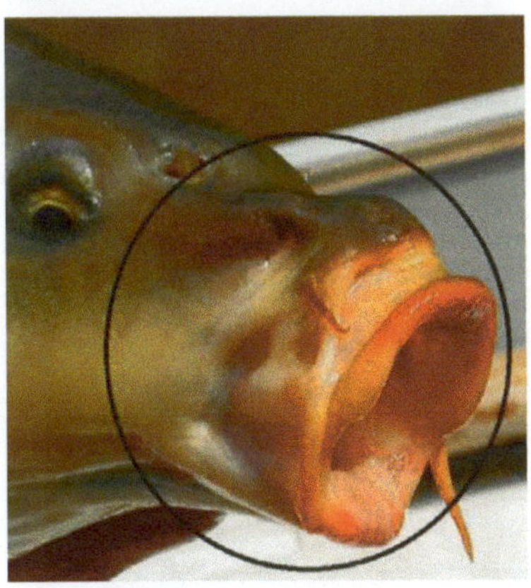	• Direct delivery of antigen through the digestive system • Easiest method • No stress on the fish • Allows mass vaccination • Low labor costs	• Short immunity • Large amounts of vaccine are required • Intact passage of vaccine through the intestine is not assured

most successful way to deliver fish vaccine relies upon the pathogen and its route of infection, life stage of the fish, and other logical consideration.

Major obstacle in the development of efficacious vaccines for aquaculture is the cost, which include requirements for specialists (e.g., labor-intensive, time-consuming handling of fish in injecting vaccination) and the costs of equipment (Muktar et al., 2016). The adverse effects of injection vaccination include stress, injection site necrosis, adhesions of tissues, reduced growth rates, and even death. Substituting methods of administration have to be developed for small fish, for which IM injection is not practical and/or cost-effective due to their high dosage requirements. For the application of DNA technology in the development of fish vaccine, important issues like safety and production cost will be judgmental, and the knowledge regarding piscine immune system is also limited. List of commercially available fish vaccines is summarized in Table 10.4.

TABLE 10.4 List of Commercially Available Fish Vaccines

S. No.	Vaccine
1.	*Aeromonas hydrophila* vaccine
2.	Carp erythrodermatitis/ulcer disease vaccine
3.	Combined *Pasteurella*/vibriosis vaccine
4.	Combined vibriosis/furunculosis vaccine
5.	Combined vibriosis/furunculosis/cold-water vibriosis/*Moritella viscosa* vaccine
6.	Combined vibriosis/furunculosis/cold-water vibriosis/*M. viscosa*/IPNV vaccine
7.	*Edwardsiella ictaluri* vaccine
8.	Enteric redmouth (ERM) vaccine
9.	*Flavobacterium psychrophilum* vaccine
10.	Furunculosis vaccine
11.	Gaffkaemia vaccine
12.	IPN virus vaccine
13.	ISA virus vaccine
14.	KHV vaccine
15.	*Lactococcus garvieae*/*Streptococcus iniae* vaccine
16.	Nodavirus vaccine
17.	Pancreas disease virus vaccine
18.	*Pasteurella* vaccine
19.	PD vaccine
20.	*Piscirickettsia salmonis* vaccine
21.	Shrimp vibriosis vaccine
22.	*Streptococcus agalactiae* vaccine

TABLE 10.4 List of Commercially Available Fish Vaccines—cont'd

S. No.	Vaccine
23.	SVC virus vaccine
24.	*Vibrio anguillarum* vaccine
25.	*Vibrio salmonicida* vaccine
26.	Vibriosis vaccine for cod
27.	Warmwater *Vibrio* spp. vaccine

10.4.1 Therapeutics for Viral Diseases of Farmed Fish

Chemotherapy has been effectively utilized in aquaculture for bacterial, fungal, and parasitic infections but not in the case of viral diseases. Though FDA has approved many drug products for use in aquaculture, their application as antiviral drugs is endorsed only to laboratory experiments, and their commercial use is yet to be permitted. As the virus replication and pathology are usually advanced by the time clinical signs appear, antiviral compounds against infections in fish should be given as prophylactic use or incorporated into the feed (recommendable dosage). This process of drug administration may work out, but the major issue concerned is the residues remaining in the tissues, as the aquatic animals farmed are for human consumption. Some of the drugs that are experimentally used are presented in Table 10.5.

10.5 IMPORTANCE OF PROBIOTICS TO TREAT VIRAL DISEASES

Utilization of probiotics in fish farming has been successfully reported, where probiotics are given in the feed that showed improvement of attributes in the fish (Gatesoupe, 2008). The conformist definition of probiotics is "live microorganisms when added to food help to reconstruct a balanced indigenous microflora in the GIT of host," but considering certain

TABLE 10.5 List of Experimental Antiviral Drugs

Antiviral Drug	Target Virus	Experimental Setup
Acyclovir	*Oncorhynchus masou virus* (OMV), *channel catfish virus* (CCV)	Rainbow trout gonad cells
Ribavirin, amantadine, methisazone, bis-benzimidazole	*Infectious hematopoietic necrosis virus* (IHNV)	Rainbow trout cell cultures
Ribavirin	*Infectious pancreatic necrosis virus* (IPNV), *infectious salmon anemia virus* (ISAV)	Rainbow trout cell cultures
46NW-04A (compound extracted from *Pseudomonas fluorescens*)	IHNV, OMV	Rainbow trout gonad cells
Methisoprinol	*Spring viremia of carp virus* (SVCV), *pike fry rhabdovirus* (PFRV)	Endothelial progenitor cell culture
Purified recombinant IFN	SVCV, IHNV	Endothelial progenitor cell culture

particular factors in aquaculture, the original definition can be tailored to probiotic as "a live microbial addition that has an advantageous outcome on the host by altering the host ambient microbial commune, by enhancing the feed nutritional value and/or immune responsiveness, toward diseases" (Lakshmi et al., 2013). Use of antimicrobials in aquaculture to control the pathogen attacks is associated with many backdrops; hence, probiotics are preferred as superlative alternatives. Making the being resistant is the best way of defense from pathogen attacks rather than cure; in this regard, probiotics are well reported for their immune augmentation. Selected bacterial probiotics when supplemented with feed were reported to cause a boost in some cellular and humoral parameters (Villamil et al., 2006). *Lactococcus lactis* was known to enhance immune functions of turbot, *S. maximus*. The whole-cell products and fractions of the whole-cell and the extracellular products of *L. lactis* such as nisin act as immunomodulator in turbot and augment the nitric oxide production (Liu et al., 2012). Endurance of probiotic bacteria in the host is an utmost significant factor to be considered in order to flaunt its effects. *Bacillus subtilis* was able to survive in grouper (*Epinephelus coioides*) posterior intestines during the feeding period and improved the survival proportion of the fish challenged with *Streptococcus* spp. and *Iridovirus* in a time- and dose-reliant manner. The survived fish showed better respiratory bursts, phagocytic activity, superoxide dismutase (SOD) level of leukocytes, and serum alternative complement activity when compared with controls (Balcazar, 2003). Probiotics are capable of stimulating diverse immunohematologic parameters in teleosts, including mononuclear phagocytic cells (monocytes and macrophages) and polymorphonuclear leukocytes (neutrophils) and NK cells (Balcazar, 2003). Existing literature quotes numerous reports on immune improvement by probiotic administration where probiotics stimulate increased respiratory burst activity and bactericidal enzyme production (lysozyme and peroxidase) and can effectively amend the production of pro-inflammatory cytokines such as interleukin-1 (IL-1), IL-6, IL-12, tumor necrosis factor-α (TNF-α), and gamma interferon (IFN-γ) and anti-inflammatory cytokines such as IL-10 and transforming growth factor-β (TGF-β) in many animals (Ibrahem, 2015). Some of the immunologic effects caused by the administration of probiotics in fish are mentioned in Table 10.6.

TABLE 10.6 Influence of Probiotic Bacteria on Immune Parameters of Fish

Probiotics	Experimental Animal	Influenced Immune Parameters
Bacillus subtilis, *L. lactis*, and *Saccharomyces cerevisiae*	*Labeo rohita*	Significantly higher rates of respiratory burst assay, blood glucose, erythrocyte count, total serum protein, albumin, alkaline phosphatase, and acid phosphatase
Lactococcus lactis BFE920	*Paralichthys olivaceus*	Increased lysosomal activities and production of IL-12 and IFN-γ
Bacillus amyloliquefaciens FPTB16	*Catla catla*	Enhanced production of superoxide anion, nitric oxide, myeloperoxidase content, and lysozyme activity
Lactococcus lactis ssp. *lactis* CLFP 100, *Leuconostoc mesenteroides* CLFP 196, and *Lactobacillus sakei* CLFP 202	*Oncorhynchus mykiss*	Increased superoxide anion production and lysozyme activity
Lactobacillus rhamnosus (ATCC 53103)	*Oncorhynchus mykiss*	Increased serum immunoglobulins and respiratory burst activity

10.6 CONCLUSIONS

Fish farming is a swiftly rising sector with potential for economic success for farmers. Having said that, aquaculture is faced with multiple restrains that have to be conquered to get to the full potential of aquaculture. Most burdening among them are diseases, measures to thwart diseases, and economic burden it has on the farmer, reducing the overall profitability of fish farming. Due to many reasons like immunoboosters, healthy growth enhancers, and least side effects, probiotics are being used to increase the productivity of the farmed fish on large scale. Yet, there is a limited usage of probiotics in fish farming despite their tremendous potential, owing to the lack of wide stretch awareness and lapse in commercializing and marketing of the probiotics from lab to farms. Vaccination is one of the best methods to increase survival rate and profitability in aquaculture when used in combination with several factors that are necessary to guarantee the highest possible survival rate. Even probiotics are not the cure for all, but they are the best shot at many hurdles in fish farming predominantly in underdeveloped and developing countries where economic feasibility is the most important priority before anything else. Still, probiotics need to be developed and evaluated for multiple applications, longer survivability, stability, and more economical means for bulk production.

Acknowledgments

The authors are indebted to all the researchers whom we cited in this review for their significant and valuable research. No funding was received to perform this review.

Conflicts of interests

The authors declare no relevant competing financial interests to disclose.

References

Ahmadivand, S., Soltani, M., Mardani, K., Shokrpoor, S., Rahmati-Holasoo, H., et al., 2016. Isolation and identification of viral hemorrhagic septicemia virus (VHSV) from farmed rainbow trout (*Oncorhynchus mykiss*) in Iran. Acta Trop. 156 (2016), 30–36.

Anonymous, 2012. IHN Outbreak May Be Spreading. http://fis.com/fis/worldnews.

Balcazar, J.L., 2003. Evaluation of probiotic bacterial strains in *Litopenaeus vannamei* (Final report). National Center for Marine and Aquaculture Research, Guayaquil, Ecuador.

Behera, T., Swain, P., Rangacharulu, P.V., Samanta, M., 2014. Nano-Fe as feed additive improves the hematological and immunological parameters of fish, *Labeo rohita* H. Appl. Nanosci. 4, 687–694.

Bruno, D.W., Noguera, P.A., Black, J., Murray, W., Macqueen, D.J., Matejusova, I., 2014. Identification of a wild reservoir of salmonid alphavirus in common dab *Limanda limanda*, with emphasis on virus culture and sequencing. Aquac. Environ. Interact. 5, 89–98.

Castro, R., Jouneau, L., Pham, H.P., Bouchez, O., Giudicelli, V., et al., 2013. Teleost fish mount complex clonal IgM and IgT responses in spleen upon systemic viral infection. Plos Pathol 9, e1003098.

Chen, A.T., Zon, L.I., 2009. Zebrafish blood stem cells. J. Cell. Biochem. 1, 35–42.

Christiansen, D.H., Østergaard, P.S., Snow, M., Dale, O.B., Falk, K., 2011. A low-pathogenic variant of infectious salmon anemia virus (ISAV-HPR0) is highly prevalent and causes a non-clinical transient infection in farmed Atlantic salmon (*Salmo salar* L.) in the Faroe Islands. J. Gen. Virol. 92, 909–918.

Cottet, L., Cortez-San Martin, M., Tello, M., Olivares, E., Rivas-Aravena, A., et al., 2010. Bioinformatic analysis of the genome of infectious salmon anemia virus associated with outbreaks with high mortality in Chile. J. Virol. 84, 11916–11928.

Direkbusarakom, S., 2004. Application of medicinal herbs to aquaculture in Asia. Walailak J. Sci. Tech. 1, 7–14.

Dong, H.T., Nguyen, V.V., Le, H.D., Sangsuriya, P., Jitrakorn, S., Saksmerprome, V., Senapin, S., Rodkhuma, C., 2015. Naturally concurrent infections of bacterial and viral pathogens in disease outbreaks in cultured Nile tilapia (*Oreochromis niloticus*) farms. Aquaculture 448, 427–435.

Edholm, E.S., Bengten, E., Wilson, M., 2011. Wilson, insights into the function of IgD. Dev. Comp. Immunol. 35, 1309–1316.

Faisal, M., Shavalier, M., Kim, R.K., Millard, E.V., Gunn, M.R., Winters, A.D., et al., 2012. Winton, spread of the emerging viral hemorrhagic septicemia virus strain, genotype IVb, in Michigan, USA. Viruses 4, 734–760.

FAO, 2014. Aquaculture Department, in Brief the State of World Fisheries and Aquaculture, World Review of Fisheries and Aquaculture, p. 8.

Fast, M.D., Tse, B., Boyd, J.M., Johnson, S.C., 2009. Mutations in the *Aeromonas salmonicida* subsp. Salmonicida type III secretion system affect Atlantic salmon leucocyte activation and downstream immune responses. Fish Shellfish Immunol. 27, 721–728.

Foreman, M.G., Guo, M., Garver, K.A., Stucchi, D., Chandler, P., et al., 2015. Modelling infectious hematopoietic necrosis virus dispersion from marine salmon farms in the discovery islands, British Columbia, Canada. PLoS One 10, e0130951.

Fringuelli, E., Rowley, H.M., Wilson, J.C., Hunter, R., Rodger, H., Graham, D.A., 2008. Phylogenetic analyses and molecular epidemiology of European salmonid alphaviruses (SAV) based on partial E2 and nsP3 gene nucleotide sequences. J. Fish Dis. 31, 811–823.

Gatesoupe, F.J., 2008. Updating the importance of lactic acid bacteria in fish farming: natural occurrence and probiotic treatments. J. Mol. Microbiol. Biotechnol. 14, 107–114.

Gjedrem, T., 2013. The first family-based breeding program in aquaculture. Rev. Aquac. 2, 2–15.

Godoy, M.G., Kibenge, M.J., Suarez, R., Lazo, E., Heisinger, A., et al., 2013. Infectious salmon anaemia virus (ISAV) in Chilean Atlantic salmon (*Salmo salar*) aquaculture: emergence of low pathogenic ISAV-HPR0 and re-emergence of virulent ISAV-HPRΔ: HPR3 and HPR14. Virol. J. 10, 344.

Godoy, M.G., Suarez, R., Lazo, E.S., Llegues, K.O., Kibenge, M.J., et al., 2014. Genetic analysis and comparative virulence of infectious salmon anemia virus (ISAV) types HPR7a and HPR7b from recent field outbreaks in Chile. Virol. J. 11, 204.

Graham, D.A., Frost, P., McLaughlin, K., Rowley, H.M., Gabestad, I., 2011. A comparative study of marine salmonid alphavirus subtypes 1–6 using an experimental cohabitation challenge model. J. Fish Dis. 34, 273–286.

Gudding, R., Van Muiswinkel, W.B., 2013. A history of fish vaccination: science-based disease prevention in aquaculture. Fish Shellfish Immunol. 35, 1683–1688.

Gustafson, L., Antognoli, M., Lara Fica, M., Ibarra, R., Mancilla, J., et al., 2014. Risk factors perceived predictive of ISA spread in Chile: applications to decision support. Prev Vet Med 117, 276–285.

Gustafson, L., Remmenga, M., Sandoval Del Valle, O., Ibarra, R., Antognoli, M., et al., 2016. Area contact networks and the spatio-temporal spread of infectious salmon anemia virus (ISAV) in Chile. Prev Vete Med 125, 135–146.

Harikrishnan, R., Balasundaram, C., Heo, M.S., 2010. Effect of probiotics enriched diet on *Paralichthys olivaceus* infected with lymphocystis disease virus (LCDV). Fish Shellfish Immunol. 29, 868–874.

Herath, T.K., Ferguson, H.W., Weidmann, M.W., Bron, J.E., Thompson, K.D., et al., 2016. Pathogenesis of experimental salmonid alphavirus infection in vivo: an ultrastructural insight. Vet. Res. 47, 7.

Hikke, M.C., Verest, M., Vlak, J.M., Pijlman, G.P., 2014. Salmonid alphavirus replication in mosquito cells: towards a novel vaccine production system. Microb. Biotechnol. 7, 480–484.

Ibrahem, M.D., 2015. Evolution of probiotics in aquatic world: potential effects, the current status in Egypt and recent prospective. J. Adv. Res. 6, 765–791.

Ito, T., Kurita, J., Mori, K., Olesen, N.J., 2016. Virulence of viral haemorrhagic septicaemia virus (VHSV) genotype III in rainbow trout. Vet. Res. 47, 4.

Jansen, M.D., Bang Jensen, B., McLoughlin, M.F., Rodger, H.D., Taksdal, T., et al., 2017. The epidemiology of pancreas disease in salmonid aquaculture: a summary of the current state of knowledge. J. Fish Dis. 40, 141–155.

Johansen, L.H., Jensen, I., Mikkelsen, H., Bjørn, P.A., Jansen, P.A., Bergh, O., 2011. Disease interaction and pathogens exchange between wild and farmed fish populations with special reference to Norway. Aquaculture 315, 167–186.

Kibenge, F.S., Godoy, M.G., Fast, M., Workenhe, S., Kibenge, M.J., 2012. Countermeasures against viral diseases of farmed fish. Antivir. Res. 95, 257–281.

Kibenge, M.J., Iwamoto, T., Wang, Y., Morton, A., Routledge, R., Kibenge, F.S., 2016. Discovery of variant infectious salmon anaemia virus (ISAV) of European genotype in British Columbia, Canada. Virol. J. 13, 3.

Kibenge, M.J.T., Iwamoto, T., Wang, Y., Morton, A., Routledge, R., Kibenge, F.S.B., 2013. Discovery of variant infectious salmon anaemia virus (ISAV) of European genotype in British Columbia, Canada. Virol. J. 13, 3–13.

Kurath, G., 2012. An Online Database for IHN Virus in Pacific Salmonid Fish: MEAPIHNV. http://pubs.usgs.gov/fs/2012/3027/pdf/fs20123027.pdf.

Lafferty, K.D., Harvell, C.D., Conrad, J.M., Friedman, C.S., Kent, M.L., et al., 2015. Infectious diseases affect marine fisheries and aquaculture economics. Annu. Rev. Mar. Sci. 7, 471–496.

Lakshmi, B., Viswanath, B., Sai Gopal, D.V.R., 2013. Probiotics as antiviral agents in shrimp aquaculture. J. Pathog. 2013, 424123. https://doi.org/10.1155/2013/424123.

Liu, C.H., Chiu, C.H., Wang, S.W., Cheng, W., 2012. Dietary administration of the probiotic, Bacillus subtilis E20, enhances the growth, innate immune responses, and disease resistance of the grouper, Epinephelus coioides. Fish Shellfish Immunol. 33, 699–706.

Liu, L., Zhu, B., Wu, S., Lin, L., Liu, G., et al., 2015. Spring viraemia of carp virus induces autophagy for necessary viral replication. Cell. Microbiol. 17, 595–605.

Gotesman, M., Soliman, H., Besch, R., El-Matbouli, M., 2015. Inhibition of spring viraemia of carp virus replication in an *Epithelioma papulosum* cyprini cell line by RNAi. J. Fish Dis. 38, 197–207.

Matejusova, I., McKay, P., Bland, F., Snow, M., 2010. Application of a sensitive, specific and controlled real-time PCR assay to surveillance indicates a low prevalence of viral haemorrhagic septicaemia virus (VHSV) in wild herring, *Clupea harengus* L., in Scottish waters. J. Fish Dis. 33, 841–847.

Misk, E., Garver, K., Nagy, E., Isaac, S., Tubbs, L., et al., 2016. Pathogenesis of spring viremia of carp virus in emerald shiner *Notropis atherinoides* Rafinesque, fathead minnow *Pimephales promelas* Rafinesque and white sucker *Catostomus commersonii* (Lacepede). J. Fish Dis. 39, 729–739.

Molloy, S.D., Pietrak, M.R., Bricknell, I., Bouchard, D.A., 2013. Experimental transmission of infectious pancreatic necrosis virus from the blue mussel, *Mytilus edulis*, to cohabitating Atlantic Salmon (*Salmo salar*) smolts. Appl. Environ. Microbiol. 79, 5882–5890.

Muktar, Y., Tesfaye, S., Tesfaye, B., 2016. Present status and future prospects of fish vaccination: a review. Vet. Sci. Technol. 7 (2). https://doi.org/10.4172/2157-7579.1000299.

Müller, A., Sutherland, B.J., Koop, B.F., Johnson, S.C., Garver, K.A., 2015. Infectious hematopoietic necrosis virus (IHNV) persistence in sockeye Salmon: influence on brain transcriptome and subsequent response to the viral mimic poly (I:C). BMC Genomics 16, 634.

Munang'andu, H.M., Evensen, Ø., 2018. Correlates of protective immunity for fish vaccines. Fish Shellfish Immunol. S1050-4648 (18), 30184–30190.

Munro, L.A., Gregory, A., 2009. Application of network analysis to farmed salmonid movement data from Scotland. J. Fish Dis. 32, 641–644.

Murray, A.G., 2013. Epidemiology of the spread of viral diseases under aquaculture. Curr. Opin. Virol. 3, 74–78.

Mutoloki, S., Evensen, Ø., 2011. Sequence similarities of the capsid gene of Chilean and European isolates of infectious pancreatic necrosis virus point towards a common origin. J. Gen. Virol. 92 (2011), 1721–1726.

Mutoloki, S., Munang'andu, H.M., Evensen, Ø., 2015. Oral vaccination of fish-antigen preparations, uptake, and immune induction. Front. Immunol. 6, 519.

Nygaard, H., Modahl, I., Myrmel, M., 2012. Thermal inactivation of infectious pancreatic necrosis virus in a peptone-salt medium mimicking the water-soluble phase of hydrolyzed fish by-products. Appl. Environ. Microbiol. 78, 2446–2448.

OIE, 2012. OIE Listed Diseases 2012. www.oie.int/animal-health-in-the-world/oie-listed-diseases-2012.

Olsen, Y.A., Press, C., 1994. Degradation kinetics of immunoglobulin in the egg, alevin and fry of Atlantic salmon, *Salmo salar* L., and the localisation of immunoglobulin in the egg. Fish Shellfish Immunol. 7, 81e.

Pettersen, J.M., Osmundsen, T., Aunsmo, A., Mardones, F.O., Rich, K.M., 2012. Controlling emerging infectious diseases in salmon aquaculture. Rev. Sci. Tech. 34, 923–938.

Rebl, A., Goldammer, T., Seyfert, H.M., 2010. Toll-like receptor signalling in bony fish. Vet. Immunol. Immunopathol. 134, 139–150.

Rodgers, C.J., Mohan, C.V., Peeler, E.J., 2011. The spread of pathogens through trade in aquatic animals and their products. Rev. Sci. Tech. 30, 241–256.

Sahoo, P.K., Swaminathan, T.R., Abraham, T.J., Kumar, R., Pattanayak, S., et al., 2016. Detection of goldfish haematopoietic necrosis herpes virus (*Cyprinid herpesvirus*-2) with multi-drug resistant *Aeromonas hydrophila* infection in goldfish: first evidence of any viral disease outbreak in ornamental freshwater aquaculture farms in India. Acta Trop. 161, 8–17.

Sandlund, N., Gjerset, B., Bergh, Ø., Modahl, I., Olesen, N.J., Johansen, R., 2014. Screening for viral hemorrhagic septicemia virus in marine fish along the Norwegian coastal line. PLoS One 9, e108529.

Shimahara, Y., Kurita, J., Nishioka, T., Kiryu, I., Yuasa, K., et al., 2016. Development of an improved RT-PCR for specific detection of spring viraemia of carp virus. J. Fish Dis. 39, 269–275.

Sivasankar, P., Santhiya, A.A.V., Kanaga, V., 2015. A review on plants and herbal extracts against viral diseases in aquaculture. J. Med. Plants Stud. 3, 75–79.

Smail, D.A., Munro, E.S., 2008. Isolation and quantification of infectious pancreatic necrosis virus from ovarian and seminal fluids of Atlantic salmon, *Salmo salar* L. J. Fish Dis. 31, 49–58.

Snow, M., 2011. The contribution of molecular epidemiology to the understanding and control of viral diseases of salmonid aquaculture. Vet. Res. 42, 56.

Snow, M., Black, J., Matejusova, I., McIntosh, R., Baretto, E., Wallace, I.S., Bruno, D.W., 2010. Detection of salmonid alphavirus RNA in wild marine fish: implications for the origins of salmon pancreas disease in aquaculture. Dis. Aquat. Org. 91, 177–188.

Su, J., Huang, T., Yang, C., Zhang, R., 2011. Molecular cloning, characterization and expression analysis of interferon-β promoter stimulator 1 (IPS-1) gene from grass carp *Ctenopharyngodon idella*. Fish Shellfish Immunol. 30, 317–323.

Tatner, M.F., Manning, M.J., 1993. The ontogeny of cellular immunity in the rainbow trout, Salmo gairdneri Richardson, in relation to the stage of development of the lymphoid organs. Dev. Comp. Immunol. 7, 69–75.

Uribe, C., Folch, H., Enriquez, R., Moran, G., 2011. Innate and adaptive immunity in teleost fish: a review. Vet. Med. 56, 486–503.

Villamil, L., Tafalla, C., Figueras, A., Novoa, B., 2006. Evaluation of immunomodulatory effects of some lactic acid bacteria in turbot (*Scophthalmus maximus*). Clin. Diagn. Lab. Immunol. 9, 1318–1323.

Yu, Z., Deng, M., Geng, Y., Zhou, Y., Wang, K., et al., 2014. An outbreak of infectious haematopoietic necrosis virus (IHNV) infection in cultured rainbow trout (*Oncorhynchus mykiss*) in Southwest China. Aquac. Res. 47, 2355–2362.

Zamani, H., Ghasemi, M., Hosseini, S.M., Haghighi Karsidani, S., 2014. Experimental susceptibility of Caspian white fish, *Rutilus frisii* kutum to Spring viraemia of carp virus. Virus Dis. 25, 57–62.

Zapata, A., Diez, B., Cejalvo, T., Gutiérrez-de Frías, C., Cortés, A., 2006. Ontogeny of the immune system of fish. Fish Shellfish Immunol. 20, 126–136.

Further Reading

Luo, P., Ruan, X., Zhang, Q., Li, Z., Wang, M., Liu, X., 2014. Monoclonal antibodies against g protein of spring viremia of carp virus. Monoclon. Antib. Immunodiagn. Immunother 33, 340–343.

CHAPTER

11

Comparative Biochemistry and Kinetics of Microbial Lignocellulolytic Enzymes

*Muni Ramanna Gari Subhosh Chandra**, *Mekapogu Madakka*[†]

[*]Department of Microbiology, Yogi Vemana University, Kadapa, India [†]Department of Biotechnology & Bioinformatics, Yogi Vemana University, Kadapa, India

OUTLINE

Recent Developments in Applied Microbiology and Biochemistry
https://doi.org/10.1016/B978-0-12-816328-3.00011-8

11.1 INTRODUCTION

As a result of industrialization and rapid population growth, utilization of natural resources has been increased tremendously during the past few decades. Increasing concern regarding the environmental pollutions and the depletion of fossil fuels compelled us to exploit alternative renewable energy resources so as to meet the ever-increasing energy requirements (Dashtban et al., 2009). Degradation of lignocellulosic biomass is carried out primarily by microbial involvement—that is, utilize it as carbon and nutrient/energy source for their growth. They include species of bacteria, fungi, and actinomycetes. Among these, fungi have received much attention in recent years for their potent extracellular oxidative and hydrolytic enzymes that degrade lignocellulosic biomass. These extracellular enzymes include ligninolytic enzymes (laccase, manganese peroxidase, lignin peroxidase, and versatile peroxidase) and cellulases. Lignocellulosic biomass, either from agricultural or forestry wastes, is abundant, low-cost feedstock alternatives in nature but requires hydrolysis into simple sugars for biofuel production. As biochemistry and enzyme kinetics of microbial lignocellulose-degrading enzyme knowledge is appropriate for understanding metabolic regulations that occur under distinctive conditions, this chapter focuses on the biochemical aspects of the lignocellulose-degrading enzyme systems from microbial sources.

11.2 LIGNOCELLULOSE DEGRADATION BY FUNGI

The microorganisms predominantly responsible for lignocellulose degradation are fungi in which white-rot fungi break down the lignin in wood, leaving the lighter-colored cellulose behind; some of these fungi break down both lignin and cellulose by producing important extracellular oxidative and hydrolytic enzymes, whereas, brown-rot fungi helps in hydrolysis of wood polysaccharides, while partially modifying lignin. Whereas brown-rot fungi help in hydrolysis of wood polysaccharides, while partially modifying lignin. As a result of this type of decay, the wood shrinks, shows a brown discoloration due to oxidized lignin, and cracks into rough cubical pieces. Soft-rot fungi secrete cellulase, an enzyme that breaks down cellulose, but not lignin in wood, from their hyphae, which leads to the formation of microscopic cavities inside the wood and sometimes to a discoloration and cracking pattern similar to that of brown-rot fungi. Soft-rot fungi can usually be found in dry environments but are mostly known to occur where brown-rot and white-rot fungi are inhibited by factors such as high moisture, low aeration, the presence of preservatives, or high temperatures (Abdel-Hamid et al., 2013).

11.3 LIGNOCELLULOLYTIC ENZYME-PRODUCING FUNGI

Lignocellulolytic enzyme-producing fungi are widespread and include species from Ascomycetes (e.g., *Trichoderma reesei*), Basidiomycetes including white-rot fungi (e.g., *Phanerochaete chrysosporium*), brown-rot fungi (e.g., *Fomitopsis palustris*), and finally a few anaerobic species (e.g., *Orpinomyces* sp.) that degrade cellulose in gastrointestinal tracts of ruminant animals (Ljungdahl, 2008). Degradation by these fungi is performed by a complex of cellulases (Chandra and Reddy, 2013), hemicellulases (Ljungdahl, 2008), and ligninases (Weng et al., 2008), reflecting the complexity of the materials. Cellulases and most hemicellulases belong to a group of enzymes known as glycoside hydrolases (GH). Currently, more than 2500 GH have been identified and classified into 115 families (Cantarel et al., 2009). Interestingly, the same enzyme family may contain members from bacteria, fungi, and plants with several different activities and substrate specifications.

11.3.1 Cellulases Production

In the recent years, solid-state fermentation (SSF) is gaining interest as a suitable strategy for the recycling of nutrient-rich wastes such as lignocelluloses. Not only SSF facilitates the possibilities for the bioconversion of agroresidues to value-added products, but also it enables the efficient recycling of lignocellulosic materials with the expenditure of less energy. Due to the distinct nature of SSF, the physicochemical characteristics of substrate such as crystallinity, bed porosity, and enormous surface area can influence the production of the cellulolytic enzyme system in fungal cultures. In SSF, the operating conditions like pH, temperature, and moisture content are fundamental factors influencing the microbial growth and production of cellulase. Various agricultural substrates/by-products and microbial cultures have been used in SSF for the production of cellulase (Reddy et al., 2015). Among the four lignocellulosic substrates like cassava bagasse, sugarcane bagasse, rice straw, and wheat bran tested in SSF by *T. reesei* NRRL 11460, sugarcane bagasse yielded maximum titers of FPase after 96h of incubation (Singhania et al., 2006). According to Reddy et al. (2015), maximum cellulase activity was obtained on rice bran as against on corncobs and sawdust. Production of cellulase was made of lignocellulosic materials such as sugar beet pulp, alkaline extracted beet pulp, and cellulose (Olsson et al., 2003). It is complicated to compare cellulolytic enzymes reported in literature, as different authors established the activity of these enzymes in different units. According to Yadav et al. (2016) and Shruthi et al. (2018) the maximum cellulase activity recorded on rice husk and groundnut fodder by *Aspergillus protuberus* and *A. unguis* in SSF, respectively.

11.3.2 Laccases

Laccases (benzenediol: oxygen oxidoreductases, EC 1.10.3.2) are classified as multicopper oxidases. These glycoproteins have the redox ability of copper ions to catalyze the oxidation of a broad range of aromatic substrates where water is obtained as a by-product from the reduction of molecular oxygen. Laccases has been observed in plants, insects, and bacteria, but the most studied are from the fungi classified as white-rot fungi, which are considered as ligninolytic enzymes because lignin sources are the best substrate for the growth of these fungi. Laccase production was enhanced by phenolic compounds contained in rice bran, leading to increasing of laccase production. Rice bran was also used as a sole carbon source for xylanase production by *Streptomyces actuosus* A-151 (Wang et al., 2003). It also interferes with the oxygen diffusion in the substrate, especially at the basement part of the flask where the substrate was not fully fermented or utilized. A thinner bed height allows for better heat removal than a thicker bed height. Ergun and Urek (2017) reported that the maximum laccase activity in SSF by *Pleurotus ostreatus* was observed as 6708.3 ± 75 (17th day) in dry potato peel waste, pretreated with distilled water. Highest laccase activity (14,189 U/g of dry substrate)

TABLE 11.1 Production of Laccase by Fungi on Various Lignocellulosic Substrates Using Solid-State Fermentation

Fungi	Substrate	Laccase Activity (U/mL)	References
Trametes pubescens MB89	Wheat bran flakes	2.14	Osma et al. (2011)
Pleurotus ostreatus DSM 11191	Wheat bran flakes	2.78	Osma et al. (2011)
Pleurotus flabellatus	Coffee pulp	4.08	Parani and Eyini (2012)
Ganoderma lucidum	Pineapple leaf	317.8	Hariharan and Nambisan (2012)
Irpex lacteus F17	Sawdust, rice straw, and soybean powder	0.017	Zhao et al. (2015)
Pleurotus ostreatus	Potato	6.71	Ergun and Urek (2017)
Coriolus versicolor	Rice bran	0.98	Nasreen et al. (2015)
Coriolus versicolor	Peanut shell	0.79	Nasreen et al. (2015)
Trametes versicolor and *Funalia trogii*	Wheat bran	4.97 and 4.10	Boran and Yesilada (2011)
Marasmius sp.	Rice straw	1.12	Hendro et al. (2012)
Trametes hirsuta	Wheat bran	9.3	Bakkiyaraj et al. (2013)
Oudemansiella radicata	Rice bran	1.48	Balaraju et al. (2010)
Trametes versicolor	Corncob	0.39	Emre and Ozfer (2013)
Trichoderma harzianum	Wheat straw powder	5.24	Huiju et al. (2013)
Trichoderma muroiana	Rubber wood dust	5.8	Jaber et al. (2017)

was obtained using 0.28 mM $CuSO_4$ under optimized conditions. In a study by Ramirez et al. (2003), it was found that *P. ostreatus* grown in SSF on wheat bran and vinasse produced laccase activity (20 U/mL). Table 11.1 summarizes the production of laccase activity on different lignocellulosic substrates by various species of fungi, and the activities are expressed in units per milliliter of fermented substrates.

11.3.3 Manganese Peroxidase

Manganese peroxidase (MnP) (EC 1.11.1.13, Mn^{2+}: H_2O_2 oxidoreductases) belongs to the family of oxidoreductases. MnP is a specific enzyme that can oxidize Mn^{2+} to Mn^{3+}, which diffuses from the enzyme surface and in turn oxidizes the phenolic substrate, including lignin model compounds and some organic pollutants (Zhou et al., 2013). The highest MnP activity in SSF by *P. ostreatus* was found as 2503 ± 50 U/L (17th day) in dry potato peel waste, pretreated with distilled water report noticed by Ergun and Urek (2017). Maximum enzyme production was obtained on the 13th day of incubation for both organisms, and the activity of enzyme produced by *P. chrysosporium* was 42.08 IU/g of husk, and that

TABLE 11.2 Production of Manganese Peroxidase by Fungi on Various Lignocellulosic Substrates

Fungi	Substrate	MnP Activity	References
Phanerochaete chrysosporium	Coffee pulp	3.1 IU/mL	Parani and Eyini (2012)
Pleurotus flabellatus	Coffee pulp	2.83 IU/mL	Parani and Eyini (2012)
Ganoderma lucidum	Pineapple leaf	658.3 IU/mL	Hariharan and Nambisan (2012)
Trametes trogii MYA 28–11	Poplar wood	333.4 nkat/g	Levin et al. (2008)
Phanerochaete chrysosporium ATCC 24725	Rice straw	46 nkat/g	Liang et al. (2010)
Phanerochaete chrysosporium ATCC 24725	Wheat straw	133 nkat/g	Pinto et al. (2012)
Phanerochaete chrysosporium	Arecanut husk	42.08 IU/g	Rajan et al. (2010)
Irpex lacteus F17	Sawdust, rice straw, and soybean powder	950 U/L	Zhao et al. (2015)
Phanerochaete chrysosporium	Wheat bran	793 U/g	Siva Ranjanee and Banu (2011)
Pleurotus ostreatus	Potato peel waste	2503.6 ± 50 U/mL	Ergun and Urek (2017)

produced by *Phanerochaete* sp. had an activity of 47.20 IU/g reported by Rajan et al. (2010). The activity of MnP was approximately 950 U/L, with the maximum activity observed at the bottom layer of the bed at 84 h (Zhao et al., 2015). In addition to this, Table 11.2 shows the production of laccase activity on different lignocellulosic substrates by various species of fungi.

11.3.4 Lignin Peroxidase

The highest lignin peroxidase activity in SSF by *P. ostreatus* was found as 231.2 ± 9 U/L (17th day) in dry potato peel waste, pretreated with distilled water reported by Ergun and Urek (2017). Zhao et al. (2015) noticed the maximum activity (31 U/L) of LiP at 84 h of incubation. The higher LiP activity (1126 IU/mL) was obtained by using pineapple leaves as a substrate by *Ganoderma lucidum* (Hariharan and Nambisan, 2012).

11.4 PURIFICATION AND CHARACTERIZATION OF LIGNOCELLULOLYTIC ENZYMES

Enzyme purification and characterization is of huge significance in obtaining information about structural and functional properties and in predicting its applications. The eventual quantity of purity of a particular enzyme depends upon its end use. The purpose behind deciding the approach for purification is to attain the maximum possible yield of the enzyme with the highest catalytic activity and the maximum possible purity.

11.4.1 Cellulase

Different types of columns are used for the purification of cellulase, among which Sephadex with different sieve sizes is the most popular matrix used for gel-exclusion chromatography (Tao et al., 2010). The efficiency of purification is generally analyzed in terms of purification folds and yield. Table 11.3 represents the various columns used, purification folds, and yield of cellulase produced by different species of fungi. The molecular weight (MW) of cellulase produced by different fungal species may differ from 12 to 126 kDa (Bai et al., 2013). Fungal cellulase may be of monomeric (Naika et al., 2007) or dimeric (Chaabouni et al., 2005) in nature. Cellulase produced by *T. viride* was purified to homogeneity using DEAE-Sepharose column, and the MW was estimated as 87 kDa by SDS-PAGE (Yasmin et al., 2013). *Penicillium pinophilum* MS 20 produced a monomeric cellulase with MW of 42 kDa, which appeared as a single band on SDS-PAGE (Pol et al., 2012). In all these studies, the purified cellulase appeared as a single band upon SDS-PAGE, indicating that the cellulase produced by these fungi is active in solution as monomers or homodimers and consequently migrates through the SDS-PAGE according to their MW so as to segregate as a single band. In contrast, some studies reported the identification of heterodimeric cellulases or isoforms that appeared as separate bands upon SDS-PAGE. For instance, *A. niger* Z10 produced two cellulase bands on SDS-PAGE gel with MWs of 50 and 83 kDa (Coral et al., 2002, 2002). Similarly, another strain of *A. niger* was also reported to have produced dimeric cellulase with MWs of 23 and 36 kDa, whereas *A. fumigatus* produced dimeric cellulase with MWs of 21 and 32 kDa (Immanuel et al., 2007). The molecular mass of the cellulase *T. longibrachiatum* was found to be 67 ± 1 kDa (Pachauri et al., 2017). The molecular mass of the cellulase by *A. niger* subsp. *awamori* was determined as 66 ± 1 kDa (Pachauri et al., 2018).

TABLE 11.3 Purification Fold and Yield of Fungal Cellulase Based on Various Column Packing Materials

Fungi	Column	Yield (%)	Purification Fold	Reference
Aspergillus aculeatus	DEAE-Sephadex	25	4	Naika et al. (2007)
Aspergillus glaucus XC9	Sephadex G-00	22.3	21.5	Tao et al. (2010)
Aspergillus terreus AN1	DEAE-Sepharose	1.3	40	Nazir et al. (2009)
Aspergillus terreus DSM	Sepharose-4B column F11	16.6	15.4	Elshafei et al. (2009)
Trichoderma harzianum	Sephadex G-50	10.3	21.9	Ahmed et al. (2009)
Trichoderma viride	Sephadex G-100	2.1	2.3	Nasir et al. (2011)
Aspergillus flavus	DEAE-Cellulose	75	2	Gudi et al. (2016)
Trichoderma viride	Sephadex G-100	8.1	5.1	Irshad et al. (2013)
Trichoderma reesei	Sephadex G-100	23.85	8.52	Abdu-Hadi et al. (2016)
Trichoderma longibrachiatum	Sephadex G-100	25.8	14.82	Pachauri et al. (2017)
Aspergillus niger subsp. *awamori*	Sephadex G-100	26.5	42.5	Pachauri et al. (2018)

11.4.2 Laccase

In a study, laccase produced from *Marasmius scorodonius* was purified, and the molecular mass was found to be 67 kDa by SDS-PAGE (Jeon and Lim, 2017). Laccase produced by *Marasmius* sp. was purified to homogeneity using DEAE-Sepharose column, and the MW was estimated by SDS-PAGE and found as 75 kDa (Vantamuri and Kaliwal, 2016). The MW was estimated as 58.6 kDa by SDS-PAGE by *Cerrena* sp. (Yang et al., 2014). Laccase produced by *Pleurotus* sp. was purified to homogeneity using DEAE-cellulose column, and the MW was estimated as 40 ± 1 kDa by SDS-PAGE (More et al., 2011). The MW was estimated as 68.42 kDa by SDS-PAGE by *P. ostreatus* HP-1 (Patel et al., 2014). The recombinant laccase was purified and visualized on SDS-PAGE as a single band with an apparent molecular weight of 71.5 kDa (Ma et al., 2018). Table 11.4 shows the columns, purification fold, and yield of laccase produced by different species of fungi.

11.4.3 Manganese Peroxidase

Molecular masses of fungal MnPs usually range from 32.0 to 75.0 kDa (Shin et al., 2005; Champagne and Ramsay, 2005; De Oliveira et al., 2009). According to Asgher et al. (2013), a homogenous single band of 40 kDa for purified *Schizophyllum commune* IBL-06 MnP was obtained after gel documentation of Native-PAGE that was further confirmed by SDS-PAGE, signifying that the enzyme was a single-polypeptide protein. The molar masses for MnPs from *Bacillus pumilus* and *Paenibacillus* sp. on SDS-PAGE were 25 and 40 kDa, respectively (De Oliveira et al., 2009). Cheng et al. (2007) reported 48.7 kDa molar mass for MnP from *Schizophyllum* sp. The purified enzyme by *Trametes versicolor* IBL-04 elucidated a single band in the 43 kDa region on SDS-PAGE (Asgher et al., 2016). The MW was estimated as 36 kDa by SDS-PAGE by *Penicillium* sp. CHY-2 (Govarthanan et al., 2017). Table 11.5 represents the columns, purification fold, and yield of manganese peroxidase produced by different species of fungi.

TABLE 11.4 Purification Fold and Yield of Fungal Laccase Based on Various Column Packing Materials

Fungi	Column	Yield (%)	Purification Fold	Reference
Pleurotus ostreatus HP-1	DEAE-Sepharose column	77.63	13.13	Patel et al. (2014)
Pleurotus sp.	DEAE-Cellulose	50	12.4	More et al. (2011)
	Sephadex G-100	22.4	72.2	
Perenniporia tephropora	Superdex 200	8	4.17	Younes et al. (2007)
Cerrena sp. HYB07	DEAE FF	45.2	2.8	Yang et al. (2014)
Marasmius sp. BBKAV79	Sephadex G-100	23.65	318.6	Vantamuri and Kaliwal (2016)
	DEAE-Cellulose	13.5	376.66	
Stereum ostrea	Sephadex G-100	2.85	70.0	Viswanath et al. (2008)

TABLE 11.5 Purification Fold and Yield of Fungal Manganese Peroxidase Based on Various Column Packing Materials

Fungi	Column	Yield (%)	Purification Fold	Reference
Irpex lacteus	HiPrep Q	41.3	6.1	Shin et al. (2005)
	HiPrep S-200	24.3	11.0	
Schizophyllum commune IBL-06	Sephadex G-100	22	1.8	Asgher et al. (2013)
Bacillus pumilus	Q-Sepharose	88.37	5.44	De Oliveira et al. (2009)
Paenibacillus sp.	Q-Sepharose	80.00	0.96	
Trametes versicolor IBL-04	DEAE	71.26	5.60	Asgher et al. (2016)
EB-60 strain	DEAE	71.51	4.35	
EMS-90 strain	DEAE	62.19	6.34	
Phanerochaete chrysosporium	DEAE-Sephadex	5.56%	2.68	Rajan et al. (2010)

11.4.4 Lignin Peroxidase

SDS-PAGE of purified lignin peroxidase revealed a single protein of 38,000 and 40,000 kDa reported by Singh et al. (2016). Roushdy et al. (2011) purified LiP from *Cunninghamella elegans*, which had a subunit molecular mass of 50 kDa. Regarding the chemical nature, it has been proved that LiPs are glycoproteins with an average molecular weight of 38–46 kDa (Sahadevan et al., 2016). The purified LiP from *T. versicolor* IBL-04 by SDS-PAGE was found to be a homogenous monomeric protein of 30 kDa (Asgher et al., 2012).

11.5 KINETICS OF LIGNOCELLULOLYTIC ENZYMES

11.5.1 Cellulase

The maximum velocity (V_{max}) and Michaelis-Menten constant (K_m) are the two constants describing the kinetic characteristics of the enzyme action. The V_{max} describes the specific point in an enzymatic reaction at which the rate of the reaction is catalyzed to the maximum, if other factors are optimum; the V_{max} is attained only when the substrate concentration is sufficiently available to fill the enzyme's active site. K_m is the substrate concentration required to fill half of the enzyme's active site. High K_m describes that the enzyme has less affinity toward substrate, whereas low K_m indicates high affinity toward substrate, thereby higher activity. The K_m and V_{max} of two cellulases produced by *A. fumigatus* Z5 as 37.8 mg/mL and 437.3 μmol/min/mg; 51.8 mg/mL and 652.7 μmol/min/mg, respectively (Liu et al., 2011). The K_m and V_{max} of cellulase produced by *A. niger* BCRC31494 were found to be 134 mg/mL and 4.6 U/min/mg (Li et al., 2012), while the K_m and V_{max} of cellulase produced by *P. pinophilum* were 4.8 mg/mL and 72.5 U/mg, respectively (Pol et al., 2012).

11.5.2 Laccase

The K_m and V_{max} values of laccase from *Pleurotus* sp. were found to be 250 mM and 0.33 μmole/min of protein, respectively (More et al., 2011). Catechol had the maximum K_m value (528.8 μmole), followed by guaiacol (299.8 μmole) and 2,6-DMP (200.4 μmole) with purified laccases from *Cerrena* sp. (Yang et al., 2014). DMP had the higher K_m value (400 mmol), followed by guaiacol (100 mmol), ABTS (46.51 mmole), and (*o*-dianisidine) (23.52 mmol) with of purified laccases from *Pleurotus ostreatus* HP-1 (Patel et al., 2014). The K_m and V_{max} values for the purified laccase of *Marasmius* sp. were 3.03 mM and 5 μmole/min (Vantamuri and Kaliwal, 2016). The K_m and V_{max} values for guaiacol were found to be 13.25 mM and 255 η kat/mg of protein, respectively (Viswanath et al., 2008). In case of native *Melanocarpus albomyces*, the K_m value was 910 ± 80 μM, while, with recombinant *M. albomyces*, the K_m value was 890 ± 80 μM (Kiiskinen et al., 2004).

11.5.3 Manganese Peroxidase

The K_m and V_{max} values of the MnP were 0.4 mM and 410 mM/min, indicating a very high affinity and catalytic efficiency of *S. commune* IBL-06 MnP (Asgher et al., 2013). Boer et al. (2006) noted the K_m value of MnP_2 for $MnSO_4$ was 22.2×10^{-3} mM, while the MnP from *Bjerkandera* sp. strain BOS55 showed the K_m value of 51 μM and turnover number of 59/s (Mester and Field, 1998). The K_m values of MnP for H_2O_2 and 2.6-dimetoxyphenol were 71.4 and 28.57 μM, respectively, at pH of 4.5 (Urek and Pazarlioglu, 2004). The K_m values for *P. chrysosporium* obtained using the H_2O_2 showed a variation of between 5 and 71 μM (Urek and Pazarlioglu, 2004) and that of *Anthracophyllum discolor* Sp4 (Diez et al., 2011) that could be considered higher as compared with the manganese peroxidase from *Rigidoporus lignosus*.

11.5.4 Lignin Peroxidase

The different LiP isoenzymes showed different K_m and V_{max} values representing the varied affinities of the four isoenzymes for veratryl alcohol as a substrate. The K_m values for H1, H2, H3, and H4 were 5.56, 1.42, 1.25, and 1.4 μM, respectively, while the V_{max} values for the same isoenzymes were 0.53, 0.60, 0.45, and 0.58 U/mg, respectively. The values of K_m were lesser than the values obtained with *Bjerkandera* sp. and *P. chrysosporium* indicating the high affinity of *Humicola grisea* isoenzymes toward the substrate veratryl alcohol (Have and Teunissen, 2001). The K_m and V_{max} values for LiP by *T. versicolor* IBL-04 were 70 μM and 417 U/mg (Asgher et al., 2012).

11.6 CONCLUSION

Overwhelming demand for lignocellulose has elevated the significance of industrial enzymes, among which lignocellulolytic enzymes occupy a key position. However, the utilization of lignocellulosic wastes has proved as a main competitor to overcome the problem to a great extent. Still, the exploration of sustainable substrates, microorganisms, and fermentation strategies is to be evolved so as to achieve higher productivity and economic feasibility.

In addition, further studies should be accelerated for the expansion of lignocellulose research by manipulating the ability of microorganism via gene/protein engineering for the effective utilization of biomass, facilitating better methods for bioconversion and solid waste management. In fact, the upper hand of SSF in alleviating the environmental burden due to the heaping up of lignocellulosic biomass has to be exploited with an economic and industrial perspective.

References

Abdel-Hamid, A.M., Solbiati, J.O., Cann, I.K.O., 2013. Insights into lignin degradation and its potential industrial applications. Adv. Appl. Microbiol. 82, 1–28.

Abdu-Hadi, S.Y., Al-Saffar, F.A., Al-Bayyar, A.H., 2016. Purification and characterization of cellulase from *Trichoderma reesei*. J. Genet. Environ. Resour. Conserv. 4 (3), 230–236.

Ahmed, S., Bashir, A., Saleem, H., Saadia, M., Jamil, A., 2009. Production and purification of cellulose-degrading enzymes from a filamentous fungus *Trichoderma harzianum*. Pak. J. Bot. 41, 1411–1419.

Asgher, M., Iqbal, H.M.N., Irshad, M., 2012. Characterization of purified and xerogel immobilized novel lignin peroxidase produced from *Trametes versicolor* IBL-04 using solid state medium of corncobs. BMC Biotechnol. 12 (1), 46.

Asgher, M., Irshad, M., Nasir Iqbal, H.M., 2013. Purification and characterization of novel manganese peroxidase from *Schizophyllum commune* IBL-06. Int. J. Agric. Biol. 15 (4), 749–754.

Asgher, M., Ramzan, M., Bilal, M., 2016. Purification and characterization of manganese peroxidases from native and mutant *Trametes versicolor* IBL-04. Chin. J. Catal. 37 (4), 561–570.

Bai, H., Wang, H., Sun, J., Irfan, M., Han, M., 2013. Production, purification and characterization of novel beta glucosidase from newly isolated *Penicillium simplicissimum* H-11 in submerged fermentation. EXCLI J. 12, 528–540.

Bakkiyaraj, S., Aravindan, R., Arrivukkarasan, S., Viruthagiri, T., 2013. Enhanced laccase production by *Trametes hirusta* using wheat bran under submerged fermentation. Int. J. ChemTech Res. 5 (3), 1224–1238.

Balaraju, K., Kyungseok, P., Shamarao, J., Kaviyarasan, V., 2010. Production of cellulase and laccase enzymes by *Oudemansiella radicata* using agro wastes under solid-state and submerged conditions. Res. Biotechnol. 1, 21–28.

Boer, C.G., Obici, L., Marques de Souza, C.G., Peralta, R.M., 2006. Purification and some properties of Mn peroxidase from *Lentinula edodes*. Process Biochem. 41, 1203–1207.

Boran, F., Yesilada, O., 2011. Enhanced production of laccase by fungi under solid substrate fermentation condition. BioResources 6 (4), 4404–4416.

Cantarel, B.L., Coutinho, P.M., Rancurel, C., Bernard, T., Lombard, V., Henrissat, B., 2009. The carbohydrate-active enzymes database (CAZy): an expert resource for glycogenomics. Nucleic Acids Res. 37, D233–D238.

Chaabouni, S.E., Mechichi, T., Limam, F., Marzouki, N., 2005. Purification and characterization of two low molecular weight endoglucanases produced by *Penicillium occitanis* mutant Pol 6. Appl. Biochem. Biotechnol. 125, 99–112.

Champagne, P.P., Ramsay, J.A., 2005. Contribution of manganese peroxidase and laccase to dye decoloration by *Trametes versicolor*. Appl. Microbiol. Biotechnol. 69, 276–285.

Chandra, M.S., Reddy, B.R., 2013. Exoglucanase production by *Aspergillus niger* grown on wheat bran. Ann. Microbiol. 63, 871–877.

Cheng, X.B., Jia, R., Li, P.S., Tu, S.Q., Zhu, Q., Tang, W.Z., 2007. Purification of a new manganese peroxidase of the white-rot fungus *Schizophyllum* sp F17, and decolorization of azo dyes by the enzyme. Enzym. Microb. Technol. 41, 258–264.

Coral, G., Arikan, B., Unaldi, M.N., Guvenmez, H., 2002. Some properties of crude carboxymethyl cellulase of *Aspergillus niger* Z10 wild-type strain. Turk. J. Biol. 26, 209–213.

Dashtban, M., Schraft, H., Qin, W., 2009. Fungal bioconversion of lignocellulosic residues; opportunities and perspectives. Int. J. Biol. Sci. 5, 578.

De Oliveira, P.L., Duarte, M.C.T., Ponezi, A.N., Durrant, L.R., 2009. Purification and partial characterization of manganese peroxidase from *Bacillus pumilus* and *Paenibacillus* sp. Braz. J. Microbiol. 40 (4), 818–826.

Diez, M., AcevedoI, F., Pizzul, L., Castillo, M., Rubilar, O., Lienqueo, O., Tortella, G., 2011. A practical culture technique for enhanced production of manganese peroxidase by *Anthracophyllum discolor* Sp4. Braz. Arch. Biol. Technol. 54 (6), 1175–1186.

Elshafei, A.M., Hassan, M.M., Haroun, B.M., Abdel-Fatah, O.M., Atta, H.M., 2009. Purification and properties of an endoglucanase of *Aspergillus terreus* DSM 826. J. Basic Microbiol. 49, 426–432.

Emre, B., Ozfer, Y., 2013. The utilization of lignocellulosic wastes for laccase production under semisolid-state and submerged fermentation conditions. Turk. J. Biol. 37, 450–456.

Ergun, S.O., Urek, R.O., 2017. Production of ligninolytic enzymes by solid state fermentation using *Pleurotus ostreatus*. Ann. Agric. Sci. 15 (2), 273–277.

Govarthanan, M., Fuzisawa, S., Hosogaia, T., Chang, Y.C., 2017. Biodegradation of aliphatic and aromatic hydrocarbons using the filamentous fungus *Penicillium* sp. CHY-2 and characterization of its manganese peroxidase activity. RSC Adv. 7, 20716–20723.

Gudi, S.K., Gurramkonda, C., Chandra, S., Prasad, S., 2016. Thermostable β-D-glucosidase from *Aspergillus flavus*: production, purification and characterization. Int. J. Clin. Biol. Sci. 1 (2), 1–15.

Hariharan, S., Nambisan, P., 2012. Optimization of lignin peroxidase, manganese peroxidase, and lac production from *Ganoderma lucidum* under solid state fermentation of pineapple leaf. BioResources 8 (1), 250–271.

Have, R., Teunissen, P.J., 2001. Oxidative mechanisms involved in lignin degradation by white-rot fungi. Chem. Rev. 101, 3397–3413.

Hendro, R., Elis, S., Sri, H.S., Tjandra, S., 2012. Optimisation of laccase production using white rot fungi and agriculture wastes in solid state fermentation. ITB J Eng Sci 44 (2), 93–105.

Huiju, G., Xiang, C., Yanwen, W., Fei, Z., Kai, Z., Zhimei, M., Qingxin, L., 2013. Media optimization for laccase production by *Trichoderma harzianum* ZF-2 using response surface methodology. J. Microbiol. Biotechnol. 23 (12), 1757–1764.

Immanuel, G., Bhagavath, C., Raj, P.I., Esakkiraj, P., Palavesam, A., 2007. Production and partial purification of cellulase by *Aspergillus niger* and *A. fumigatus* fermented in coir waste and sawdust. Internet J Microbiol 3, 1–20.

Irshad, M., Anwar, Z., Ramzan, M., Mahmood, Z., Nawaz, H., 2013. Characterization of purified β-glucosidase produced from *Trichoderma viride* through bio-processing of orange peel waste. Adv. Biosci. Biotechnol. 4 (10), 941.

Jaber, S.M., Shah, U.K.M., Asaari, A.Z.M., Ariff, A.B., 2017. Optimization of laccase production by locally isolated *Trichoderma muroiana* IS1037 using rubber wood dust as substrate. BioResources 12 (2), 3834–3849.

Jeon, S.J., Lim, S.J., 2017. Purification and characterization of the laccase involved in dye decolorization by the white-rot fungus *Marasmius scorodonius*. J. Microbiol. Biotechnol. 27 (6), 1120–1127.

Kiiskinen, L.L., Ratto, M., Kruus, K., 2004. Screening for novel laccase-producing microbes. J. Appl. Microbiol. 97, 640–646.

Levin, L., Herrmann, C., Papinutti, V.L., 2008. Optimization of lignocellulolytic enzyme production by the white-rot fungus *Trametes trogii* in solid-state fermentation using response surface methodology. Biochem. Eng. J. 39 (1), 207–214.

Li, C.H., Wang, H.R., Yan, T.R., 2012. Cloning, purification, and characterization of a heat and alkaline-stable endoglucanase B from *Aspergillus niger* BCRC31494. Molecules 17, 9774–9789.

Liang, Y.S., Yuan, X.Z., Zeng, G.M., 2010. Biodelignification of rice straw by *Phanerochaete chrysosporium* in the presence of dirhamnolipid. Biodegradation 21, 615–624.

Liu, D., Zhang, R., Yang, X., Xu, Y., Tang, Z., 2011. Expression, purification and characterization of two thermostable endoglucanases cloned from a lignocellulosic decomposing fungi *Aspergillus fumigatus* Z5 isolated from compost. Protein Expr. Purif. 79, 176–186.

Ljungdahl, L.G., 2008. The cellulase/hemicellulase system of the anaerobic fungus *Orpinomyces* PC-2 and aspects of its applied use. Ann. N. Y. Acad. Sci. 1125, 308–321.

Ma, S., Liu, N., Jia, H., Dai, D., Zang, J., Cao, Z., Dong, J., 2018. Expression, purification, and characterization of a novel laccase from *Setosphaeria turcica* in *Escherícha coli*. J. Basic. Microbiol. 58, 68–75.

Mester, T., Field, J.A., 1998. Characterization of a novel manganese peroxidase lignin peroxidase hybrid isozyme produced by *Bjerkandera* species strain BOS55 in the absence of manganese. J. Biol. Chem. 273, 15412–15417.

More, S.S., Renuka, P.S., Pruthvi, K., Swetha, M., Malini, S., Veena, S.M., 2011. Isolation, purification, and characterization of fungal laccase from *Pleurotus* sp. https://doi.org/10.4061/2011/248735.

Naika, G.S., Kaul, P., Prakash, V., 2007. Purification and characterization of a new endoglucanase from *Aspergillus aculeatus*. J. Agric. Food Chem. 55, 7566–7572.

Nasir, I.H.M., Ishtiaq, A., Muhamma Anjum, Z., Muhammad, I., 2011. Purification and characterization of the kinetic parameters of cellulase produced from wheat straw by *Trichoderma viride* under SSF and its detergent compatibility. Advan. BioSci. Biotechnol. 2, 149–156.

Nasreen, Z., Usman, S., Yasmeen, A., Nazir, S., Yaseen, T., Ahmad, S., 2015. Production of laccase enzyme by *Basidiomycetes Coriolus versicolor* through solid state fermentation. Int. J. Curr. Microbiol. App. Sci. 4 (8), 1069–1078.

Nazir, A., Soni, R., Saini, H., Manhas, R., Chadha, B., 2009. Purification and characterization of an endoglucanase from *Aspergillus terreus* highly active against barley β-glucan and xyloglucan. World J. Microbiol. Biotechnol. 25, 1189–1197.

Olsson, L., Christensen, T.M., Hansen, K.P., Palmqvist, E.A., 2003. Influence of the carbon source on production of cellulases, hemicellulases and pectinases by *Trichoderma reesei* Rut C-30. Enzym. Microb. Technol. 33 (5), 612–619.

Osma, J.F., Moilanen, U., Toca-Herrera, J.L., Rodriguez-Couto, S., 2011. Morphology and laccase production of white-rot fungi grown on wheat bran flakes under semi-solid-state fermentation conditions. FEMS Microbiol. Lett. 318 (1), 27–34.

Pachauri, P., Aranganathan, V., More, S., Sullia, S.B., Deshmukh, S., 2017. Purification and characterization of cellulase from a novel isolate of *Trichoderma longibrachiatum*. Biofuels. https://doi.org/10.1080/17597269.2017.1345357.

Pachauri, P., More, S., Aranganathan, V., Sullia, S.B., Deshmukh, S., 2018. Kinetic study and characterization of cellulase enzyme from isolated *Aspergillus niger* subsp. *awamori* for cellulosic biofuels. J. Sci. Ind. Res. 77, 55–60.

Parani, K., Eyini, M., 2012. Production of ligninolytic enzymes during solid state fermentation of coffee pulp by selected fungi. Sci. Res. Rep. 2 (3), 202–209.

Patel, H., Gupte, S., Gahlout, M., Gupte, A., 2014. Purification and characterization of an extracellular laccase from solid-state culture of *Pleurotus ostreatus* HP-1. 3 Biotech 4 (1), 77–84.

Pinto, P.A., Dias, A.A., Fraga, I., Marques, G., Rodrigues, M.A., Colaco, J., Sampaio, A., Bezerra, R.M., 2012. Influence of ligninolytic enzymes on straw saccharification during fungal pretreatment. Bioresour. Technol. 111, 261–267.

Pol, D., Laxman, R.S., Rao, M., 2012. Purification and biochemical characterization of endoglucanase from *Penicillium pinophilum* MS 20. Indian J. Biochem. Biophys. 49, 189–194.

Rajan, A., Kurup, J.G., Abraham, T.E., 2010. Solid state production of manganese peroxidases using arecanut husk as substrate. Braz. Arch. Biol. Technol. 53 (3), 555–562.

Ramirez, N.E., Vargas, M.C., Ariza, J.C., Martínez, C., 2003. Characterization of laccase obtained by two production methods with *Pleurotus ostreatus*. Rev. Colomb. Biotecnol. 2, 64–72.

Reddy, G.P.K., Narasimha, G., Kumar, K.D., Ramanjaneyulu, G., Ramya, A., Kumari, B.S., Reddy, B.R., 2015. Cellulase production by *Aspergillus niger* on different natural lignocellulosic substrates. Int. J. Curr. Microbiol. App. Sci. 4 (4), 835–845.

Roushdy, M.M., Abdel-Shakour, E.H., El-Agamy, E.I., 2011. Biotechnological approach for lignin peroxidase (LiP) production from agricultural wastes (Rice husk) by *Cunninghamella elegans*. J. Am. Sci. 7, 6–13.

Sahadevan, L.D.M., Misra, C.S., Thankamani, V., 2016. Characterization of lignin degrading enzymes (LDEs) from a dimorphic novel fungus and identification of products of enzymatic break down of lignin. 3 Biotech 6 (1), 1–16.

Shin, K.S., Kim, Y.H., Lim, J.S., 2005. Purification and characterization of manganese peroxidase of the white-rot fungus *Irpex lacteus*. J. Microbiol. 43 (6), 503–509.

Shruthi, K., Suresh Yadav, P., Siva Prasad, B.V., Subhosh Chandra, M., 2018. Production of cellulase by a local isolate of *Aspergillus unguis* on different lignocellulosic substrates in solid state fermentation. J. Forest Res. https://doi.org/10.1007/s11676-018-0619-4.

Singh, P., Jain, P., Verma, R., Jagadish, R.S., 2016. Characterization of lignin peroxidase from *Paecilomyces* species for decolorisation of pulp and paper mill effluent. J. Sci. Ind. Res. 75, 500–505.

Singhania, R.R., Sukumaran, R.K., Pillai, A., Prema, P., Szakacs, G., 2006. Solid-state fermentation of lignocellulosic substrates for cellulase production by *Trichoderma reesei* NRRL 11460. Indian J. Biotechnol. 5, 332–336.

Siva Ranjanee, S., Banu, N., 2011. Screening of manganese peroxidase production through solid state fermentation and textile dyes decolourization by *Phenarochaete chrysosporium*. Res. J. Sci. Technol. 3 (1), 33–43.

Tao, Y.M., Zhu, X.Z., Huang, J.Z., Ma, S.J., Wu, X.B., 2010. Purification and properties of endoglucanase from a sugar cane bagasse hydrolyzing strain *Aspergillus glaucus* XC9. J. Agric. Food Chem. 58, 6126–6130.

Urek, R.O., Pazarlioglu, N.K., 2004. Purification and partial characterization of manganese peroxidase from immobilized *Phanerochaete chrysosporium*. Process Biochem. 39, 2061–2068.

Vantamuri, A.B., Kaliwal, B.B., 2016. Purification and characterization of laccase from *Marasmius* species BBKAV79 and effective decolorization of selected textile dyes. 3 Biotech. 6 (189), 1–10.

Viswanath, B., Chandra, M.S., Kumar, K.P., Reddy, B.R., 2008. Production and purification of laccase from *Stereum ostrea* and its ability to decolorize textile dyes. Dyn. Biochem. Process. Biotechnol. Mol. Biol. 2 (1), 19–25.

Wang, S.L., Yen, Y.H., Shih, L., Chang, A.C., Chang, W.T., Wu, W.C., Chai, Y.D., 2003. Production of xylanases from rice bran by *Streptomyces actuosus* A-151. Enzym. Microb. Technol. 33 (7), 917–925.

Weng, J.K., Li, X., Bonawitz, N.D., Chapple, C., 2008. Emerging strategies of lignin engineering and degradation for cellulosic biofuel production. Curr. Opin. Biotechnol. 19, 166–172.

Yadav, P.S., Shruthi, K., Prasad, B.S., Chandra, M.S., 2016. Enhanced production of β-glucosidase by new strain *Aspergillus protuberus* on solid state fermentation in rice husk. Int. J. Curr. Microbiol. App. Sci. 5 (12), 551–564.

Yang, J., Lin, Q., Ng, T.B., Ye, X., Lin, J., 2014. Purification and characterization of a novel laccase from *Cerrena* sp. HYB07 with dye decolorizing ability. PLoS One 9 (10), 1–13.

Yasmin, S., Mattoo, R., Nehvi, F., 2013. Isolation, characterization and molecular weight determination of cellulase from *Trichoderma viride*. Afr. J. Biotechnol. 12, 4503–4511.

Younes, B.S., Mechichi, T., Sayadi, S., 2007. Purification and characterization of the laccase secreted by the white rot fungus *Perenniporia tephropora* and its role in the decolourization of synthetic dyes. J. Appl. Microbiol. 102 (4), 1033–1042.

Zhao, X.H., Yao, J., Zhou, Y., Jia, R., 2015. Fungal growth and manganese peroxidase production in a deep tray solid-state bioreactor, and *in vitro* decolorization of Poly R-478 by MnP. J. Microbiol. Biotechnol. 25 (6), 803–813.

Zhou, X.W., Cong, W.R., Su, K.Q., Zhang, Y.M., 2013. Ligninolytic enzymes from *Ganoderma* spp: current status and potential applications. Crit. Rev. Microbiol. 39, 416–426.

Further Reading

Bocchini, D.A., Oliveira, O.M.M.F., Gomes, E., Da Silva, R., 2005. Use of sugarcane bagasse and grass hydrolysates as carbon sources for xylanase production by *Bacillus circulans* D1 in submerged fermentation. Process Biochem. 40 (12), 3653–3659.

Duenas, R., Tengerdy, R.P., Gutierrez-Correa, M., 1995. Cellulase production by mixed fungi in solid-substrate fermentation of bagasse. World J. Microbiol. Biotechnol. 11, 333–337.

Hafiz Muhammad Nasir, I., Ishtiaq, A., Muhammad Anjum, Z., Muhammad, I., 2011. Purification and characterization of the kinetic parameters of cellulase produced from wheat straw by *Trichoderma viride* under SSF and its detergent compatibility. Arch. Biochem. Biophys. 2, 149–156.

Hakala, T.K., Hilden, K., Maijala, P., Olsson, C., Hatakka, A., 2006. Differential regulation of manganese peroxidases and characterization of two variable MnP encoding genes in the white-rot fungus *Physisporinus rivulosus*. Appl. Microbiol. Biotechnol. 73 (4), 839–849.

Irshad, M., Asgher, M., Scheikh, M.A., Nawaz, H., 2011. Purification and characterization of laccase produced by *Schyzophylum commune* IBL-06 in solid state culture of banana stalks. BioResources 6 (3), 2861–2873.

Kang, S.W., Park, Y.S., Lee, J.S., Hong, S.I., Kim, S.W., 2004. Production of cellulases and hemicellulases by *Aspergillus niger* KK2 from lignocellulosic biomass. Bioresour. Technol. 91, 153–156.

Knezevic, A., Milovanovic, I., Stajic, M., Vukojevic, J., 2013. *Trametes suaveolens* as ligninolytic enzyme producer. Zbornik Matice srpske za prirodne nauke (124), 437–444.

Moubasher, H., Mostafa, F.A., Wahsh, S., Haroun, O., 2017. Purification and characterization of lignin peroxidase isozymes from *Humicola grisea* (Traaen) and its application in bioremediation of textile dyes. Egypt. J. Bot. 57, 335–343.

Silva, M.L.C., De Souza, V.B., Da Silva Santos, V., Kamida, H.M., De Vasconcellos-Neto, J.R.T., Goes-Neto, A., Koblitz, M.G.B., 2014. Production of manganese peroxidase by *Trametes villosa* on un expensive substrate and its application in the removal of lignin from agricultural wastes. Adv. Biosci. Biotechnol. 5 (14), 1067.

CHAPTER

12

Future Prospects of Biodiesel Production by Microalgae: A Short Review

Myla Sunil Kumar[*,†], *Viswanath Buddolla*[‡]

[*]Department of Virology, Sri Venkateswara University, Tirupati, India [†]NextBio Research Pvt., Ltd., Bengaluru, India [‡]Department of Bionanotechnology, Gachon University, South Korea

OUTLINE

12.1 INTRODUCTION

Cultivation of microalgae is a promising long-term, sustainable candidate for lipid production, and the lipids from microalgae can be easily converted into biodiesel, a green and sustainable fuel, via transesterification process (Chen et al., 2018). These lipids are soluble in organic solvents, and the main constituents of lipids are neutral and polar lipid molecules of saturated and unsaturated types (Mubarak et al., 2015). Because of their relatively high lipid content, rapid growth properties, and high biomass production, microalgae are recognized as potentially good sources for biofuel production (Tan et al., 2018). The important advantages of microalgae over agricultural crops are the quick growth rate and no arable land necessity.

https://doi.org/10.1016/B978-0-12-816328-3.00012-X

Additionally, carbon sequestration and burning clean (of microalgae biodiesel) are also the magnetizing aspects of exploitation microalgae for producing biodiesel (Chen et al., 2018). There is a plenty of literature on microalgae as feedstock of biodiesel production and the approaches of lipid extraction from microalgae, various transesterification processes, and advantages and limitations (Chisti, 2007; Rawat et al., 2013; Chen et al., 2018; Mata et al., 2010; Rawat et al., 2013; Sibi et al., 2016). This article presents the recent trends into the development of biodiesel production from microalgae along with future prospects.

12.2 SUMMARY OF PATHWAYS FOR MICROALGAL BIODIESEL/FUEL PRODUCTION

Microalgae have been used for centuries to offer nourishment to humans and animals, but recently, they have become much more widely cultured and harvested at a large industrial scale (Garcia et al., 2017). The biology of microalgae has been studied for more than 50 years now and more intensely since the 1970s, when genetics and molecular biology approaches were integrated into the research programs. Recently, energy and resource creation using microalgae has established an immense deal of consideration. Microalgae grow quickly and contain high oil content compared with terrestrial crops, which take a season to grow and only contain a maximum of about 5% dry weight of oil (Chisti, 2007). Commonly, doubling time for microalgae is 24 h. During the peak growth phase, some microalgae can double every 3.5 h (Chisti, 2007). The oil content of microalgae is usually between 20% and 50% (dry weight, Table 12.1), while some strains can reach as high as 80%.

The prospect of microalgal biotechnology depends on the progress of large-scale photobioreactors capable of operating under distinct optimal conditions with negligible contamination risk. Compared with open-air systems, closed systems can avoid most of their problems; however, development of more economical and efficient closed culturing systems is required (Garcia et al., 2017). The set of approaches that range from microalgae cultivation to collection and harvest and to fuel conversion is termed the pathway for microalgal biodiesel/biofuel

TABLE 12.1 Percentage of Lipids in Dry Biomass of Various Easily Existing Microalgae

Microalgae	% of Lipid (by Dry Mass)
Botryococcus braunii	25–75
Chlorella vulgaris	28–32
Crypthecodinium cohnii	20
Cylindrotheca sp.	16–37
Nitzschia sp.	45–47
Phaeodactylum tricornutum	20–30
Schizochytrium sp.	50–77
Tetraselmis suecia	15–23

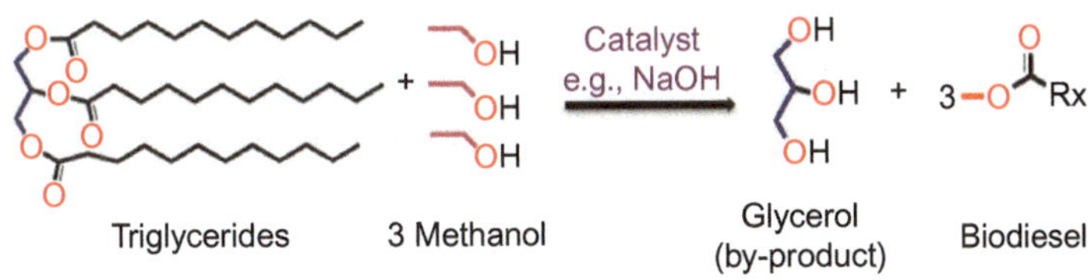

FIG. 12.1 Base-catalyzed transesterification of a triglyceride with alcohol (Buddolla et al., 2010).

production. Chaudry et al. (2015) summarized the most promising conversion pathways of microalgae and compared based on significant assessment parameters of product quality and yield, nutrient recovery, GHG emissions, energy, and the cost connected with the production of fuel from microalgae, in order to understand the advantages and disadvantages of each method. In addition, an overview of the potential environmental impacts of large-scale microalgae cultivation for various applications was summarized by Usher et al. (2014) and Sutherland et al. (2017). As summarized by Hadrich et al. (2018), three stages are particularly considered in order to produce microalgal lipids for a variety of applications, that is, controlling the microalgae cultivation via experimental and modeling investigations, optimizing culture conditions to maximize lipid production and to determine the fatty acid profile the most suitable for biodiesel synthesis, and optimizing the extraction of the lipids accumulated in the microalgal cells. However, environmental factors and process conditions can vary the quality and the quantity of lipid produced by microalgae, which can be dangerous for the overall production of biodiesel and required proper attention (Shin et al., 2018). Biodiesel is produced through the transesterification of these microalgal lipids, which are subjected to a reaction with alcohol in the presence of a catalyst. The process removes roughly 10% of the original weight of the oil as glycerin, which is the backbone of the triglyceride molecule as summarized in Fig. 12.1.

12.3 RECENT INSIGHTS INTO THE DEVELOPMENT OF BIODIESEL PRODUCTION FROM MICROALGAE AND FUTURE PROSPECTS

Recently, a research demonstrated that the oleaginous microalgae have the ability to purify biogas along with enhanced production of lipids. Here, the researchers have followed the approach of the stepwise increasing of significant growth factors, including, nitrogen source, light intensity, and CO_2 supply to enhance the microalgae performance in the successful manner. This strategy helped not only in biomass and lipid production but also in the removal of CO_2 (Srinuanpan et al., 2018). The combined processes and approaches applied in this research are very much helpful for the production of biogas and biodiesel along with the removal of CO_2. In another study, researchers have focused on the biorefinery strategy for simultaneous production of biodiesel and bioethanol from microalgae feedstock (Sivaramakrishnan and Incharoensakdi, 2018a,b). In this study, different pretreatment techniques were employed to confirm the maximum recovery of sugars from *Scenedesmus* sp., and the total sugar yield of more than 90% was attained when the biomass was pretreated by acid hydrolysis. In addition, the hydrolysate produced 86% of ethanol after the fermentation using yeast. Moreover, this approach of bioethanol and biodiesel production was optimally attained when direct transesterification was completed first, followed by ethanol fermentation, yielding 92% and 93% of methyl ester and ethanol, correspondingly.

A novel direct saponification-esterification conversion strategy was developed for biodiesel production from wet microalgae with a water content higher than 90 wt% by Fang et al. (2018). This approach is very simple and can replace conventional feedstock drying and lipid extraction methods for enhanced biodiesel production at low cost. Moreover, the proposed kinetic model with the trial results specified that the system is mainly by means of saponification/esterification conversion, which principally converted lipid released from microalgae cell into soap in the saponification step followed by the esterification into biodiesel in the next step. Similarly, Sandmann et al. (2018) applied a novel single-cell analytic technology to study a well-established model experiment and to understand the single-cell dynamics during lipid production in microalgae, and these studies are helpful for deeper investigation of the microalgal lipid accumulation process. In another study by Jiang et al. (2018), struvite precipitation was selected as a digestate pretreatment technique to verify the optimal combination of chemical additives and reaction conditions and to examine appropriate microalgal species. This provides a novel approach for combining liquid digestate treatment and microalgal cultivation for enhanced biodiesel production. Mureed et al. (2018) conducted to optimize the growth of microalgal strains and to evaluate biodiesel production potential of algae using untreated food industry wastewater as a source of nutrients. From the experimental results, it was confirmed that microalgae have an abundant potential to produce biodiesel from spent wash of the food industry. Additionally, this approach is cheaper and presents an environmentally friendly way to handle food industry wastewater along with biodiesel. In a study, a two-step processing approach, that is, hydrothermal liquefaction followed by catalytic upgrading was employed to produce upgraded bio-oil in a successful manner from microalgae. In addition, the extracted upgraded bio-oils have similar properties to those of naphtha and jet fuel (Xu et al., 2018). The application of ultrasound in biodiesel production from microalgae has freshly appeared as an innovative technology. In general, ultrasound treatment augments the mass-transfer characteristics leading to the increased reaction rate with short reaction time and potentially reduces the production cost. However, it requires further examination particularly in the area of technoeconomic feasibility (Sivaramakrishnan and Incharoensakdi, 2018b).

Zhou et al. (2017) conducted a study for continuous production of biodiesel from two kinds of microalgae Chrysophyta and *Chlorella* sp. In this study, coupling with the supercritical carbon dioxide extraction, the lipids of microalgae were extracted initially and then transferred to the downstream production of biodiesel. The residue after decompression can be reused as the material for pharmaceuticals and nutraceuticals. The observations of these experiments proved that the particle size of microalgae, temperature, pressure, molar ratio of methanol to oil, and flow of CO_2 and *n*-hexane all have an influence on the biodiesel production. In another study, a combinatorial strategy integrating toxic and carcinogenic heavy-metal arsenic alleviation coupled to biodiesel production using oleaginous microalgae grown in synthetic soft water (SSW) was undertaken, and this integrative technology has a strong potential path for verdurous environment and renewable fuels resulting in socioeconomic welfare (Arora et al., 2017). In a study, Wang et al. (2016) designed an approach based on using formic acid aided with minute amounts of hydrochloric acid, to treat water-containing microalgae and extract lipid consequently. The results of this study confirmed that the dosage of formic acid and hydrochloric acid, liquid/solid (l/s) ratio, and temperature had a considerable influence on lipid extraction from water-containing microalgae. Gupta et al. (2016) developed a superstructure-based optimization model to optimize the microalgae for biodiesel production. This model contains the key processing steps of converting microalgae into biodiesel, especially, their growth, harvesting, lipid extraction, and transesterification of lipid. Here, various options to execute these steps are considered, and the mass and volumetric balance for each procedure and equipment and the equipment capacity limitations comprise the significant model constraints. Moreover, the decision variables include growth duration, medium, and techniques and specifications to be followed in each of the downstream steps. Shin et al. (2018) summarized the strategies for increasing overall economic viability of lipid production and confirmed that the microalgal bioprocess of microalgal strain with desirable characteristics is essential for decreasing the overall cost of biodiesel production. In general, biodiesel from microalgae is the present research role that addresses the concerns of energy and environmental sustainability. Developing a large-scale microalgae cultivation system is necessary without delay to explore the economic feasibility of this new feedstock for alternative energy and valuable by-product production (Tan et al., 2018).

12.4 CONCLUSIONS

The lipid, an energy reserve in the microalgal cell, is a promising feedstock for biodiesel production; therefore, these organisms have been rising as a feedstock for biofuel in response to the energy crisis. But it was observed that there were still several obstacles to make biodiesel production from commercialized microalgae, and constructive research is essential to overcome these obstacles, particularly, to get biodiesel in low cost with environmental sustainability.

Acknowledgments

The authors are indebted to all the researchers whom we cited in this review for their significant and valuable research. No funding was received to perform this review.

Conflict of Interest

The authors declare no relevant competing financial interests to disclose.

References

Arora, N., Gulati, K., Patel, A., Pruthi, P.A., Poluri, K.M., Pruthi, V., 2017. A hybrid approach integrating arsenic detoxification with biodiesel production using oleaginous microalgae. Algal Res. 24, 29–39.

Buddolla, V., Mutanda, T., White, S., Bux, F., 2010. The microalgae—a future source of biodiesel. Dyn. Biochem. Process. Biotechnol. Mol. Biol. 4, 37–47.

Chaudry, S., Bahri, P.A., Moheimani, N.R., 2015. Pathways of processing of wet microalgae for liquid fuel production: a critical review. Renew. Sust. Energ. Rev. 52, 1240–1250.

Chen, J., Li, J., Dong, W., Zhang, X., Tyagi, R.D., Drogui, P., Surampalli, R.Y., 2018. The potential of microalgae in biodiesel production. Renew. Sust. Energ. Rev. 90, 336–346.

Chisti, Y., 2007. Biodiesel from microalgae. Biotechnol. Adv. 25, 294–306.

Fang, Y.R., Yeh, Y., Liu, H.S., 2018. A novel strategy of biodiesel production from wet microalgae by direct saponification–esterification conversion (DSEC). J. Taiwan Inst. Chem. Eng. 83, 23–31.

Garcia, J.L., de Vicente, M., Galán, B., 2017. Microalgae, old sustainable food and fashion nutraceuticals. Microb. Biotechnol. 10, 1017–1024.

Gupta, S.S., Shastri, Y., Bhartiya, S., 2016. Model-based optimisation of biodiesel production from microalgae. Comput. Chem. Eng. 89, 222–249.

Hadrich, B., Akremi, I., Dammak, M., Barkallah, M., Fendri, I., Abdelkafi, S., 2018. Optimization of lipids' ultrasonic extraction and production from Chlorella sp. using response-surface methodology. Lipids Health Dis. 17, 87.

Jiang, Y., Pu, X., Zheng, D., Zhu, T., Wang, S., Deng, L., Wang, W., 2018. Cultivation of lipid-producing microalgae in struvite-precipitated liquid digestate for biodiesel production. Biotechnol. Biofuels 11, 101.

Mata, T.M., Martins, A.A., Caetano, N.S., 2010. Microalgae for biodiesel production and other applications: a review. Renew. Sust. Energy Rev. 14, 217–232.

Mubarak, M., Shaija, A., Suchithra, T.V., 2015. A review on the extraction of lipid from microalgae for biodiesel production. Algal Res. 7, 117–123.

Mureed, K., Kanwal, S., Hussain, A., Noureen, S., Hussain, S., Ahmad, S., Ahmad, M., Waqas, R., 2018. Biodiesel production from algae grown on food industry wastewater. Environ. Monit. Assess. 190, 271.

Rawat, I., Ranjith Kumar, R., Mutanda, T., Bux, F., 2013. Biodiesel from microalgae: a critical evaluation from laboratory to large scale production. Appl. Energy 103, 444–467.

Sandmann, M., Schafberg, M., Lippold, M., Rohn, S., 2018. Analysis of population structures of the microalga *Acutodesmus obliquus* during lipid production using multi-dimensional single-cell analysis. Sci. Rep. 8 (1), 6242.

Shin, Y.S., Choi, H.I., Choi, J.W., Lee, J.S., Sung, Y.J., Sim, S.J., 2018. Multilateral approach on enhancing economic viability of lipid production from microalgae: a review. Bioresour. Technol. 258, 335–344.

Sibi, G., Shetty, V., Mokashi, K., 2016. Enhanced lipid productivity approaches in microalgae as an alternate for fossil fuels—a review. J. Energy Inst. 89, 330–334.

Sivaramakrishnan, R., Incharoensakdi, A., 2018a. Utilization of microalgae feedstock for concomitant production of bioethanol and biodiesel. Fuel 217, 458–466.

Sivaramakrishnan, R., Incharoensakdi, A., 2018b. Microalgae as feedstock for biodiesel production under ultrasound treatment—a review. Bioresour. Technol. 250, 877–887.

Srinuanpan, S., Cheirsilp, B., Prasertsan, P., 2018. Effective biogas upgrading and production of biodiesel feedstocks by strategic cultivation of oleaginous microalgae. Energy 148, 766–774.

Sutherland, D.L., Turnbull, M.H., Craggs, R.J., 2017. Environmental drivers that influence microalgal species in fullscale wastewater treatment high rate algal ponds. Water Res. 124, 504–512.

Tan, X., Lam, M.K., Uemura, Y., Lim, J.W., Wong, C.Y., Lee, K.T., 2018. Cultivation of microalgae for biodiesel production: a review on upstream and downstream processing. Chin. J. Chem. Eng. 26, 17–30.

Usher, P.K., Ross, A.B., Camargo-Valero, M.A., Tomlin, A.S., Gale, W.F., 2014. An overview of the potential environmental impacts of large-scale microalgae cultivation. Biofuels 5, 331–349.

Wang, S., Zhu, J., Dai, L., Zhao, X., Liu, D., Du, W., 2016. A novel process on lipid extraction from microalgae for biodiesel production. Energy 115, 963–968.

Xu, Y.P., Duan, P.G., Wang, F., Guan, Q.Q., 2018. Liquid fuel generation from algal biomass via a two-step process: effect of feedstocks. Biotechnol. Biofuels 11, 83.

Zhou, D., Qiao, B., Li, G., Xue, S., Yin, J., 2017. Continuous production of biodiesel from microalgae by extraction coupling with transesterification under supercritical conditions. Bioresour. Technol. 238, 609–615.

CHAPTER

13

Genetic and Metabolic Engineering of Microorganisms for the Production of Various Food Products

Nimgampalle Mallikarjuna, Kuna Yellamma

Department of Zoology, Sri Venkateswara University, Tirupati, India

OUTLINE

https://doi.org/10.1016/B978-0-12-816328-3.00013-1

13.1 INTRODUCTION

Genetic modification is a biological practice that alters the genetic material of all types of living organisms. According to the World Health Organization (WHO), genetically modified organism (GMO) can be defined as organisms in which the genetic material (DNA) is altered artificially by using rDNA technology methods. Similarly, the Food and Agriculture Organization (FAO) and the European Commission define GMO as a product "which does not occur naturally by mating and/or natural recombination." In the food industry, GM foods refer to foods produced from genetically modified microbes, which are responsible for the enhancement of nutritional values, change the taste and texture, increase yields and shelf life, and reduce maturation time. GM foods that are available in the market include soy, milk, cheeses, yogurt, some other dairy products, food additives, amino acids, vitamins, wine, and valuable nutrition foods. Currently, the production of microbial food products is increased, and its consumption by the common people is also tremendously enhanced for the improvement of health conditions with nutritional values. Here, the microorganisms not only enrich the quality of foods but also further increase the bioavailability of nutrients, surface (texture), aroma to the nourishments, and also the delivery of certain inhibitory compounds that assist in causing food deterioration thus in this way expanding the capacity and security of food product (Tamang et al., 2016). In addition, microorganisms have the ability to grow on a wide variety of substrates and ferment them into desired products with biological activity and improve the biofunctionality of food products. Moreover, various biological activities of microbes used in the fermented foods include probiotic properties, antimicrobial properties, antioxidant, peptide production, polyglutamic acid, degradation of antinutritive compounds, and fibrinolytic activity, which might be critical criteria for the choice of starter culture(s) to be utilized in the production of functional foods. The functional properties of microbes in the various fermented food products have been promoting potential health benefits to consumers. In view of the above multiple beneficial qualities of GMM, the present chapter is focused mainly to explore the details of recent developments and the production of genetic and metabolically engineered microbes for the improvement of diverse food products in the food industry.

13.2 ROLE OF MICROORGANISMS IN FOOD INDUSTRY

Different kinds of fermented food products are produced by altering the various metabolic activities in microorganisms. Microbes are the modest assets that consist of abundant enzymes, which can convert large chemical substrates into simple edible foods with a high nutritive quality. For example, lactic acid bacteria (LAB), commonly found in milk, fruit juice, plant products, the intestine, and mucosa, are essential in the food processing industry that require lactic acid fermentation, prominently dairy products (yogurt and cheese), fermented vegetables (olives, pickles, and sauerkraut), fermented meats (salami), and sourdough bread. The genera imperative individuals from this group include *Lactobacillus*, *Leuconostoc*, *Pediococcus*, and *Streptococcus*. These bacteria are heterotrophic and have complex healthy prerequisites because of the lack of numerous biosynthetic abilities (Kumar, 2016). Therefore, most species have different prerequisites for amino acids and vitamins. Another type of microbe, yeast, also has been used worldwide in the baking and beverage industries for the production of different fermented food items. Baker's yeast is available in two forms, either as a dry powder or as compressed cakes. On the other hand, there is also a profitable intermediate of the process called "cream yeast." Actually, yeasts are eukaryotes and belong to ascomycetes, possessing rich source of vitamin B and proteins. The genera of yeast consist of *Saccharomyces*, *Candida*, *Torulopsis*, and *Hansenula*. Yeasts and their metabolic items have been utilized as a part of various types of food preparations particularly for baking and brewer industries. Currently, yeast biotechnology is a section of monetarily vital divisions, including foods, drinks, pharmaceuticals, and modern compounds. Among others, *Saccharomyces cerevisiae, S. boulardii, S. cerevisiae* var. *boulardii, and S. carlsbergensis* etc. are the principal organisms of yeast used in baking and beverage industries (Ali et al., 2012; Lara-Hidalgo et al., 2017). In the same way, molds are important microbes for several food products, such as types of cheese, sausages, and soy sauce. Nowadays, determination and utilization of molds are a vast expanding field of innovative work in the food industry. From the earlier days to present days, there is a continuous demand of molds for the production of several food products, food additives, enzymes, and organic acids. Some of the molds such as *Aspergillus niger, A. nidulans, Rhizomucor miehei, Endothia parasitica,* and *Mucor* have been tremendously utilized in the food processing industries (Pai, 2003). The present research evidences described that the genetically modified molds were tremendously used in the commercial production of various foods and ingredients.

13.3 IMPORTANCE OF GENETIC AND METABOLICALLY ENGINEERED MICROBES IN FOOD INDUSTRY

13.3.1 Genetically Engineering of Microorganisms

In general, modern biotechnology made the utilization of microorganisms, for example, yeast in bread making and lactobacilli in the production of dairy items. Biotechnologists have confronted a few difficulties in using the wild microorganisms. Since these wild microbial floras were typically unstable and not predictable in their capacity to create the desired food product, it was exceptionally troublesome for biotechnologists to create homogenous foods within the time bond and subsequently unable to achieve the quality of their foods. In some cases, the microorganisms are not capable to stay alive in the fermenters. The conditions inside the fermenters controlled or altered the standard work of the microorganisms and their survival; as a result, the metabolites or products produced by microbes are lethal to themselves (feedback suppression). Mostly, the food processing industries had been getting low productivity because of the low growth rate leading to increase in the cost of production. All these complications have seriously affected the productivity of food industries financially and economically. Hence, there is a need for stable microorganisms that can have resistance to fermentation conditions and also can enhance production.

With the innovation of genetic engineering (GE), the scientists have observed that genetically modified microbes have the ability to resist the fermentation conditions. For instance, cloning a gene that encodes for thermoresistance gene to the microbes can be made to survive at high temperature in fermenters. As a result, genetic engineering could likewise create clones of a specific microorganism that is steady and could grow quickly. Based on this concept, GMM had been preferred over wild microflora and were quickly implemented in several biotechnological purposes. Hence, microorganisms are responsible for the production of desired food or any compound, cultured with a suitable nutritive medium under defined conditions in the fermenter. After the fermentation process, the desired products are purified by various downstream procedures.

13.3.1.1 *Applications of Genetically Modified Microorganism (GMM)*

In modern days, GMM has been widely used in various industries like dairy, chemical, pharmaceutical, biotech, and agrochemical. In the western countries, most of the foodstuffs such as bread, wine, cheese, butter, yogurts, kefir, fermented meats (dry-cured sausage or salami), and fermented vegetables (sauerkraut and olives) are produced by the action of GMM. Some dairy products can be produced by complex or minimal microbial vegetation considered as wild microbial population found in crude materials and nature. These products include unpasteurized cheeses, beers, and sourdough bread. Others were produced by using genetically modified industrial starter cultures and may incorporate numerous cheese produced using pasteurized milk. The organoleptic constituents of genetically modified foods fill in as an extra additional advantage of genetically modified microbes over the wild microorganism. In the chemical industries, GMMs have been utilized to deliver numerous molecules, for example, enzymes, organic acids, and biofuels. A larger part of these products fill in as reagents for different companies. Certain enzymes are produced and utilized as a part of agroalimentary enterprises to process nourishment items and catalyze the production of different items, for example, alcohol or organic acids, in distillery businesses. Some other products produced by the pharmaceuticals act as end-product for human consumption or its environment. These incorporate beer, wine, alcohol, amino acids, vitamins or fuel utilized by vehicles.

In the pharmaceutical and biotech organizations, microorganisms normally delivered a few biologically active substances (antibiotics) and proteins (enzymes and bacteriocins) that release them ordinarily. There are also different proteins or enzymes whose genes have been cloned in microorganisms (e.g., human insulin, development hormone, and hepatitis B immunization). These products have been promoted for quite a long period and are responsible for development of various countries with therapeutically active substances (recombinant insulin has been conveyed since 1983). Owing to their capacity to survive or go through human and animal mucosa, microbes can be utilized to treat or prevent certain sicknesses. For instance, *L. jensenii* has been engineered to produce CD4 protein utilized by the HIV infection in the vaginal mucosa to infiltrate lymphocytes.

13.3.2 Metabolic Engineering of Microorganisms

The term metabolic engineering (ME) is itself tells about the strategy. Here, the metabolism is a set of enzyme-catalyzed chemical process that occurs in the cell of an organism, and engineering is to control things to make it more productive. Metabolic engineering is usually stated to as the targeted and determined change of metabolic pathways found in an organism in order to better understanding and utilization of cellular pathways for chemical transformation, energy transduction, and supramolecular assembly (Kumar and Prasad, 2011). It is appropriate for the coordinated change of cell properties through the adjustment of particular biochemical responses or the introduction of new genes with the utilization of different systems (rDNA technology). The objective of metabolic engineering is precise investigation of metabolic and different pathways with biotechnological methods to enhance cell properties for the desired product development by designing and implementing rational genetic modifications. Microorganisms are utilized for the creation of numerous profitable and desirable products in pharmaceutical, agriculture, food industries etc. due to widespread use of microbes. There is a need of metabolic engineering to accomplish numerous objectives, (i) increases product formation, (ii) speed up the process, (iii) saves energy, (iv) stops production of by-product, and (v) develop the strains resistance to environmental stress. One way of grouping the different targets for metabolic engineering is as follows: extension of substrate range; improvements in productivity and yield; elimination of by-products; improvement of process performance; improvements of cellular properties; and extension of the product range including heterologous protein production (Anesiadis et al., 2008).

13.3.2.1 *Applications of Metabolically Modified Microorganisms (MMM)*

Metabolically engineered microbes such as lactic acid bacteria and yeast have also been applied in the various chemical, pharmaceutical, biotech, dairy, food and brewery industries for the production of biofuels, organic acids, antibiotics, chemicals, drugs, enzymes, amino acids, vitamins, sweeteners, flavoring agents and valuable nutritional foods, various types of wines, etc. Further, the details of different food products produced by various metabolic changes of microbes were described in the section below, that is, recent developments in the production of engineered microbes in the food industry.

13.4 STRATEGIES INVOLVED IN GENETIC AND METABOLIC ENGINEERING OF MICROORGANISMS

Worldwide, the population has been increasing day-to-day and also in the present modern days, most of the people prefer ready-made/instant foods enriched with high nutritional values. Technologically, scientists have also developed various methods for the improvement of genetically modified microorganisms (GMOs) and production of various desirable food products to meet the needs of normal peoples. Genetically modified methods include gene transformation, electroporation, transduction, protoplast fusion, conjugation, and mutagenesis (Adrio and Demain, 2006) (Fig. 13.1).

Genetically modified foods are synthesized by using recombinant DNA technology, also called as "genetic engineering," genetic modification, and transgenic technology. In this technology, when the gene of interest that encodes desired trait is inserted into nuclear DNA, it leads to the genetic modification known as recombinant DNA, which translates the desired food product. In food processing industries, genetic engineering is helpful in the production of desired products with enriched taste, nutritive value, shell-life and yield. Another approach, namely, conjugation, depends on normal techniques for hereditary trade whereby DNA is exchanged starting with one strain then onto the next. Conjugation can occur between LAB strains and in addition among LAB and other microorganisms, despite the fact that the subsequent strains could possibly be marked as recombinant. The way that this procedure can happen normally goes around the utilization of the genetic engineered microbes. Rarely,

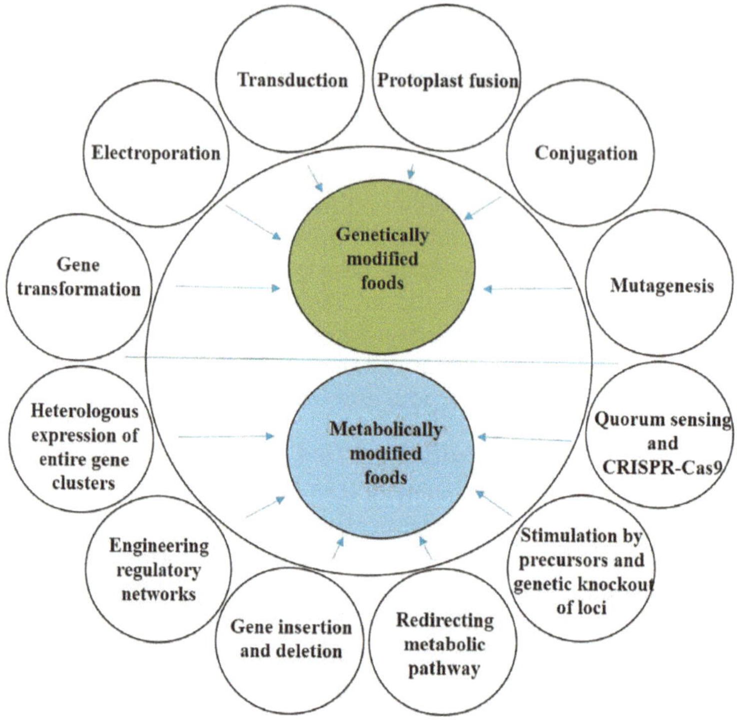

FIG. 13.1 Various strategies applied for the improvement of genetically and metabolically modified microorganisms for the development of GMFs.

this technique has been used to utilize protoplast fusion to encourage recombination between two strains with unraveled however unique characteristics, delivering a strain that has the coveted qualities of both parents. Protoplast fusion was traditionally utilized as a mapping technique in microorganisms and of late, has been utilized effectively to create strains of LAB with desired traits (Patnaik et al., 2002).

The most surely understood method used to bring recombinant DNA into microorganisms is transformation, whereby gene of interest is brought particularly into recipient cells by making them penetrable using chemical agents, proteins, or electroporation. The primary strategy produced for LAB was plasmid protoplast combination, in which beneficiary cells are stripped of walls and in this way fused with polyethylene glycol, catching the recently presented DNA between the cells. Electroporation was created for LAB in the midst of the late 1980s and uses electric streams to make pores in the cell envelope, empowering DNA from various sources to enter. This strategy is presumably the most broadly utilized to explore because of its effortless. Recombinant DNA additionally can be brought into LAB utilizing a strategy called transduction, in which a bacteriophage is utilized to move DNA from one strain into another. Dissimilar to transformation, transduction can be laden with issues that cause deletion inside the plasmid. Further, mutations are also practiced to produce the desired microbial food product. In this approach, microbes are treated with mutagenic agents (nitrosoguanidine (NTG), 4-nitroquinoline-1-oxide, methylmethane sulfonate (MMS), ethylmethane sulfonate (EMS), hydroxylamine (HA), and ultraviolet light) to get a desired product.

Diverse strategies have been utilized for the improvement of secondary metabolites created by specific microorganisms in metabolic engineering (Fig. 13.1). Some of these strategies include the following:

(a) *Heterologous expression of the entire gene clusters*: Recent advances have been developed for the production of various metabolites in the biosynthetic pathway of different microbial strains. This was achieved by the transfer of complete gene cluster of the biosynthetic pathway into new hosts for the heterologous production of bioactive compounds. For example, both tetrangulol and tetrangomycin have been produced in *Streptomyces rimosus* NRRL 3016. The cluster gene encoding for the polyhydroxybutyrate (PHB) biosynthetic pathway of *Ralstonia eutropha* was transferred into *S. cerevisiae*. The *phaABCRe* operon containing genes *phb*A (b-ketothiolase), *phb*B (NADPH-linked acetoacetyl-CoA), and *phb*C (PHA synthase) of the PHA biosynthetic pathway of *R. eutropha* has been metabolically engineered in the cytoplasm of the yeast strains.

(b) *Engineering regulatory networks*: Genes for various metabolites in the single biosynthetic pathway are commonly gathered together in clusters on the chromosome along with their pathway-specific regulatory genes, either positive (activators) or negative (repressors) effects on the expression of gene cluster elements. Generally, in microbes, all genes are not expressed at a time. The microbes produce valuable products only for their own beneficial needs; hence, they don't overproduce their metabolites. However, in the industries, we need continuous production of metabolites; for this purpose, microbes are engineered by up- and downregulation strategies. For instance, in downregulation of engineered *St. griseus*, we can observe the continuous production of chromomycin due to the lack of the specific inhibitor gene in the biosynthetic pathway. Similarly, in upregulation, continuous expression of streptomyces antibiotic regulatory protein positive (SARP) regulators in *St. coelicolor* has been reported to increase continuous production of actinorhodin.

(c) *Gene insertion and deletion*: The genes responsible for intermediate metabolites could be deleted and inserted into another organism to get the desired product.

(d) *Redirecting metabolic pathway*: Most of the microbes have the ability to produce different products along with the desired product. For instance, *Escherichia coli* produced minor product of succinic acid along with a much more quantity of other organic acids like acetic acid, formic acid, lactic acid, and ethanol during anaerobic fermentation. So, it is essential to redirect metabolic pathways for increasing the production of succinic acid besides decreasing the formation of other metabolites.

(e) *Stimulation by precursors*: Various stimulators have been used to induce the biosynthetic pathway for the production of a particular product. Stimulators such as inducers include various amino acids, namely, leucine for bacitracin synthetase and tryptophan for dimethylallyl tryptophan synthase in ergot alkaloid biosynthesis.

(f) *Genetic knockout of loci*: These approaches to "cell condition" dynamically change intracellular architecture in response to environmental cues and hence affect heterologous product stability.

(g) *Quorum sensing (cell-to-cell communication)*: It enables bacterial population density-based control of gene transcription via the production, release, and sensing of low-molecular-weight compounds. Extracellular signaling provides a new basis for control over molecular and cellular processes, as well as population behavior.

(h) *Clustered regularly interspaced short palindromic repeats-Cas9 (CRISPxR-Cas9)*: This tool has been developed to engineer *Clostridium beijerinckii* in order to optimize the production of ester compounds by removing competing pathways and by increasing the flux toward the substrates of ethanol acetyltransferase 1.

13.5 RECENT DEVELOPMENTS IN THE PRODUCTION OF ENGINEERED MICROBES IN FOOD INDUSTRY

Lactic acid bacteria (LAB) are considered as genetically safe. The genetically modified LAB strains include *Lactobacillus, Leuconostoc, Pediococcus, Bifidobacterium*, and *Lactococcus* and have been used to enrich the flavor, text, nutritive values, and preservation of dairy and fermented products that include yogurt, butter, milk, cheese, and pickled vegetables (Table 13.1) (Carminati et al., 2015).

13.5.1 Cheese

Traditionally, cheese is prepared by using a protease called calf rennet. Several microorganisms, namely, *R. miehei, E. parasitica*, and *R. pusillus*, are mutated for calf rennet production for the development of cheese production. Several microbes (*E. coli, K. lactis, A. niger, and yeast*) have been genetically modified with calf rennet gene (chymosin) for the improvement of cheese products (Pai, 2003). For the acceleration of cheese ripening, *L. lactis* DN209 has

TABLE 13.1 Genetically Modified Microorganisms and Their Foods

S. No.	Name of the Genetically Modified Microorganisms	Genetically Modified Food
1.	*Rhizomucor miehei, Endothia parasitica, Rhizomucor pusillus, E. coli, K. lactis, A. niger, yeast*, and *L. lactis* DN209	Cheese
2.	*Kluyveromyces lactis, Kluyveromyces marxianus, Kluyveromyces (Kazachstania) lodderae, S. cerevisiae, Saccharomyces boulardii, I. orientalis, Yarrowia lipolytica, Debaryomyces hansenii, Candida inconspicua*, and *Candida maris*	Yogurt and kefir
3.	*Saccharomyces cerevisiae, Saccharomyces exiguus, Candida humilis, Candida krusei, Candida milleri, Issatchenkia orientalis, Pichia anomala, Pichia subpelliculosa*, and *Torulaspora delbrueckii*	Baking products
4.	*Leuconostoc mesenteroides, Lactobacillus brevis, Pediococcus pentosaceus*, and *Lactobacillus plantarum*	Sauerkraut and pickles
5.	*Lactobacillus sake, Lactobacillus curvatus, Lactobacillus plantarum, Pediococcus pentosaceus*, and *Pediococcus acidilactici*, which are used together with *Micrococcus* and *Staphylococcus*. *L. sake* LTH673 and *L. curvatus* LTH 1432	Sausages

been genetically engineered with various genes *PepN*, *PepC*, *PepX*, and *PepI* peptidases from a highly proteolytic *L. helveticus* strain (Joutsjoki et al., 2002). Further, transformation of *gdh* gene (glutamate dehydrogenase) from *Peptoniphilus asaccharolyticus* into *L. lactis* is responsible for the production of α-ketoglutarate, which can enhance the cheese ripening process (Rijnen et al., 2000).

13.5.2 Yogurt and Kefir

Different strains of yeast species have been identified in dairy products that are inadequate. Their occurrence differs with the source of isolation: *K. lactis* and *K. marxianus* are seldom found in yogurt and kefir; *K. (Kazachstania) lodderae* is found in kefir; *S. cerevisiae* is found in raw cheese, yogurt, and kefir; *S. boulardii* is found in milk, yogurt, and ice cream; *I. orientalis* is found in cheese, kefir, and yogurt; *Y. lipolytica* and *D. hansenii* are found in kefir and milk; and *C. inconspicua* and *C. maris* are found in kefir.

13.5.3 Baking

Saccharomyces cerevisiae is the yeast species used in bread and baking products. Nevertheless, other strains of yeast were also isolated from sourdough including *S. exiguus*, *C. humilis*, *C. krusei*, *C. milleri*, *I. orientalis*, *P. anomala*, *P. subpelliculosa*, and *T. delbrueckii* (Pulvirenti et al., 2004).

13.5.4 Fermented Vegetables

Fermentation of various vegetables like cabbage (sauerkraut), pickles, and olives is carried out by various LAB species, namely, *L. mesenteroides*, *L. brevis*, *P. pentosaceus*, and *L. plantarum* (Boyacioglu et al., 2010). Additionally, different types of LAB have been found in sauerkraut, which include *L. lactis* subsp. *lactis*, *L. sakei*, *L. curvatus*, and *L. fallax* (Plengvidhya et al., 2007).

13.5.5 Meat Products

Different types of sausages are fermented by using various LAB as starter cultures that include *L. sake*, *L. curvatus*, *L. plantarum*, *P. pentosaceus*, and *P. acidilactici*, which are used together with *Micrococcus* and *Staphylococcus*. The main function of starter cultures is to release flavor compounds during fermentation and provide meat products with specific tastes and aroma (Boyacioglu et al., 2010). It has been described that two LAB stains *L. sake* LTH673 and *L. curvatus* LTH 1432, which were cloned with the genes, encode for catalase (katA) and lysostaphin (lys) for their potential antibacterial activity against *Staphylococcus aureus* in the fermentation of sausages (Csutak and Sarbu, 2018). Mostly, the metabolically engineered microbes have been extensively used for the production of various drugs, antibiotics, hormones, chemicals, and biofuels than that of food products and nutrients. However, some of the fermented dairy food additives, vitamins, amino acids, organic acids, and exopolysaccharides are developed by the metabolically engineered microorganisms as summarized in Table 13.2.

TABLE 13.2 Metabolically Modified Microorganisms and Their Useful Food Products

S. No.	Name of the Microorganism	Gene or Metabolism of Interest Engineered	Desired Product
1.	*L. lactis biovar. Diacetylactis*	α-Acetolactate decarboxylase	Diacetyl
2.	*Lactobacillus lactis*	NADH-oxidase and α-acetolactate decarboxylase	Diacetyl
3.	*Streptococcus thermophilus*	*GlyA* gene	Acetaldehyde
4.	*L. lactis*	*Pdc* gene—pyruvate dehydrogenase	Acetaldehyde
5.	*E. coli*	Biotin operon (*bioABFCD*) cluster gene	Biotin
6.	*Corynebacterium ammoniagenes*	Overexpression of own riboflavin biosynthesis genes	Riboflavin
7.	*B. subtilis* mutant	*rib* operon	Riboflavin
8.	*Erwinia citreus*	2,5-Diketo-D-gluconate	Vitamin C
9.	*S. thermophilus*	Genes of PGM and GalU	Exopolysaccharides
10.	*Corynebacterium glutamicum*	Vitreoscilla hemoglobin gene *vgb*	Glutamate
11.	*C. glutamicum*	*xfp* gene	Glutamate
12.	*C. glutamicum*	Alterations in various genes in lysine metabolism	Lysine
13.	*Serratia marcescens*	Control of feedback mechanisms of threonine	L-Threonine
14.	*E. coli*	Inactivation of threonine dehydrogenase enzyme	L-Threonine
15.	*C. glutamicum* ssp. *lactofermentum*	*hom* (homoserine dehydrogenase), *thrB* (homoserine kinase), and *thrC* (L-threonine synthase)	L-Threonine
16.	*E. coli*	*thrABC* operon	L-Threonine
17.	*C. glutamicum* VLA1	*ilvA* gene	L-Valine
18.	*C. glutamicum* and *C. lactofermentum*	*ilvA* gene	Isoleucine
19.	*E. coli*	Deletion of genes *aceF* and *ldhA*, replaced with gene of *alaD*	L-Alanine
20.	*S. marcescens*	Gene of dehydroproline resistance	L-Proline
21.	*E. coli*	Cloning of *aroF* and feedback-resistant *pheA* genes	Phenylalanine
22.	*C. glutamicum*	Amplification of *aro* II and *csm* genes	L-Tyrosine

13.5.6 Dairy Flavoring Agents

Diacetyl is an important flavoring agent, which gives aroma to the dairy fermented food products. It is naturally produced by lactic acid bacteria (LAB), especially *L. lactis biovar. Diacetylactis*. This flavoring component production was enhanced by metabolic modification in microbes. For instance, lactic acid bacteria, cloned metabolically with the gene α-acetolactate decarboxylase, are responsible for the release of α-acetolactate and increase the accumulation of this product that leads to higher production of diacetyl in dairy products. Further, this approach was also employed for the overproduction of diacetyl in *L. lactis* by modifying the metabolism of NADH-oxidase and α-acetolactate decarboxylase. Acetaldehyde is a major component of the dairy products, produced by LAB, especially *S. thermophilus* and *L. bulgaricus*. In the LAB, acetaldehyde is released from the intermediates such as pyruvate and Acetyl-CoA of glycolysis metabolism, and it is also produced through the conversion of threonine into glycine and acetaldehyde. This reaction is catalyzed by threonine aldolase. *Streptococcus thermophilus* was metabolically modified by the insertion of *GlyA* gene encoding threonine aldolase that results in the overproduction of acetaldehyde (Chaves et al., 2002). Another challenge was also made for the improvement of acetaldehyde production in *L. lactis* by introducing the gene (*pdc*) responsible for pyruvate dehydrogenase metabolism from *Zygomonas mobilis* (Bongers et al., 2005).

13.5.7 Vitamins

Currently, the production of vitamins by recombinant and metabolic microbes has made competitive economic approach than that of traditional chemical synthesis. For example, more biotin production was achieved by cloning of biotin operon (*bioABFCD*) cluster gene into *E. coli*. Similarly, the mutant strains of *S. marcescens*, resistant to the biotin antimetabolite acidomycin, ethionine, and ET2 to S-2-aminoethylcysteine, have enhanced biotin yield. Another vitamin, namely, riboflavin, produced by *C. ammoniagenes*, was improved by cloning and overexpression of own riboflavin biosynthesis genes. *Bacillus subtilis* mutant, having *rib* operon at two different locations in the chromosome, made resistant against purine analogue leads to the production of GTP and riboflavin analogue (roseoflavin) resistance mutation in *ribC* that deregulated the entire pathway; finally, the riboflavin production is over 25 g/L. Vitamin C was also produced by the metabolic engineering of 2,5-diketo-D-gluconate reductase from *Corynebacterium* sp. into *E. citreus*. Further, folic acid is an essential vitamin found in yogurt released by the metabolically modified *L. lactis*, *L. gasseri*, and *L. reuteri*.

13.5.8 Exopolysaccharides

Exopolysaccharides (EPS) were also produced by metabolically modified LAB such as *Lactococcus* and *Streptococcus* and have widespread uses in the structural characteristics of dairy products. In situ exopolysaccharide synthesis permits the variation of rheology and enriches sensory properties, for example, mouth feel and texture of food products, and the acquisition of some health-promoting properties (prebiotics). Frequently, food grade EPS such as xanthan and gurdlan were produced more by nondairy bacteria than the natural production of EPS by LAB. The metabolic strategies were mainly focused on precursors for the development of EPS by LAB. Consequently, the research has focused on enhancing the production of various nucleotide sugars involved in the biosynthesis of EPS, namely, UDP-glucose, UDP-galactose, and dTDP rhamnose. Phosphoglucomutase (PGM) and UDP-glucose phosphorylase (GalU) are controlled enzymes in the biosynthesis of EPS. As a results, metabolic engineering of *S. thermophilus* with the genes of PGM and GalU leads to the increased production of EPS.

13.5.9 Amino Acids

Gene cloning of microbes has been making a huge production of amino acids in the field of fermentation industries. The microbial strains, namely *Corynebacterium*, *Brevibacterium*, and *Serratia*, have been widely used in the production of different types of amino acids. Glutamate was the first amino acid produced by fermentation due to the widespread usage as a flavoring agent in food industries. Insertion of *Vitreoscilla* hemoglobin gene *vgb* into *C. glutamicum* strains enhanced the cell growth and glutamic acid, and glutamine production is improved through oxygen uptake. It has been described that cloning of the *xfp* gene (from *Bifidobacterium animalis*) into *C. glutamicum*, encoding phosphoketolase, leads to the improved production of glutamate.

L-Lysine is one of the frequently producing amino acid through different approaches of metabolically engineered microbes, especially the mutants of *C. glutamicum* that have been routinely utilized for the production of L-lysine as follows: deletion of PEP carboxykinase and glucose-6-phosphate dehydrogenase, distraction of malate quinone oxidoreductase, decreasing the activity of citrate synthase by mutation, and increasing the availability of the precursor pyruvate by removing pyruvate dehydrogenase activity. Overexpression of pyruvate carboxylase or DAP dehydrogenase genes.

L-Threonine has also been produced by the *Serratia marcescens*, in which mutations were created with transductional crosses responsible for the control of the various feedback mechanisms of threonine. Mutant *E. coli*, in which the enzyme threonine dehydrogenase was inactivated, yielded more threonine in the fermentation. Further, L-threonine is also overproduced by the *C. glutamicum* ssp. *lactofermentum*; this strain was metabolically engineered with a recombinant plasmid carrying its own hom (encoding homoserine dehydrogenase), *thrB* (encoding homoserine kinase), and *thrC* (encoding L-threonine synthase) genes. Cloning of extra copy of *thrABC* operon along with the isoleucine auxotrophy in *E. coli* leads to the inactivation of threonine-degrading gene *tdh* and results in the increased production of threonine. *Corynebacterium glutamicum* VLA1 mutant strain has been developed through the insertion of plasmid with *ilvA* gene encoding threonine dehydratase by eliminating two genes of pantothenate biosynthesis. This strain has the ability to enhance the production of L-valine. In addition, L-isoleucine feedback regulation in *C. glutamicum* was replaced by native threonine dehydratase gene *ilvA* from *E. coli* that results in the enhanced production of respective amino acid (L-isoleucine). Amplification of the wild-type threonine dehydratase gene *ilvA* in a threonine-producing strain of *C. lactofermentum* led to isoleucine production.

Further, L-alanine was also produced by the mutant strain *E. coli*, constructed by the deletion of genes encoding a protein of the pyruvate dehydrogenase complex (*aceF*) and lactate dehydrogenase (*ldhA*), and replaced with the plasmid vector containing the *B. sphaericus* alanine dehydrogenase gene (*alaD*). And also L-proline was produced by mutant strain *S. marcescens* and cloned with the gene of dehydroproline-resistance locus on a plasmid vector. Aromatic amino acids, namely, phenylalanine and tyrosine, have been also produced by the metabolically engineered microbes. Phenylalanine production was achieved by genetic engineering-based cloning of *aroF* and feedback-resistant *pheA* genes into *E. coli* strain. L-Tyrosine overproduction was gained by the amplification of *aro* II (encoding desensitized type II DS) and *csm* in a tryptophan producer of *C. glutamicum*.

13.5.10 Organic Acids

These include acetate, pyruvate, succinate, and lactate that have been extensively produced via microbial production to meet synthesis of building blocks of chemicals from renewable carbon sources. In metabolic engineering of *Lactobacillus* species, *E. coli* and *C. glutamicum* have produced high levels of lactate, acetate, pyruvate, succinate, etc. (Wendisch et al., 2006; John et al., 2008).

13.6 HEALTH BENEFITS OF VARIOUS MICROBIAL FOOD PRODUCTS

The process of food fermentation is developed from ancient times, and in the present generation, fermentation has explored from the household to commercialized and industrial level by capturing a huge marketplace due to its role in the maintenance of health and prevention of various diseases in the modern era. Fermented foods have been enzymatically altered by microorganisms in a manner that it maintains taste and smell liked by humans. Some of the potential health benefits of fermented foods that have been explored in the recent past were based on an extensive body of anecdotal information and included such benefits as antihypertensive activity (Koyama et al., 2014; Ahren et al., 2014), blood glucose-lowering benefits (Kamiya et al., 2013; Oh et al., 2014), and antidiarrheal and antithrombotic properties (Kamiya et al., 2013). Fermented foods in animal and human health are still exploring for achieving food and nutritional security globally (Quave and Pieroni, 2014). Fermented foods help in health promotion and disease prevention by alternating metabolism, oxidative stress, and inflammation (Camps and Garcia-Heredia, 2014). Concentrations of fermented food intake will be useful to coexamine its influences on the regulation of lipid profiles, cellular energetics and metabolism, and oxidative stress, as well as immune system and cognitive support. Fermented legumes and soybeans consumed in China, Japan, Korea, and Indonesia, as a major staple food, exhibit antidiabetic effects. Fermented foods may exert health benefits through bioactive compounds, which may synthesize during the fermentation process (Martins et al., 2011). Fermented rice bran fractions, with *S. boulardii*, showed significant differences in metabolite composition compared with nonfermented (Ryan et al., 2011) by reducing the growth of human lymphomas. Tea leaves are fermented with the fungus, *Eurotium cristatum*, and showed increased amounts of dodecanamide, linoleamide, stearamide, and epicatechin gallate when compared with a nonfermented green tea (Keller et al., 2013).

Blueberries fermented by probiotic strains of *L. plantarum* significantly lower blood pressure (Ahren et al., 2014). A fermented papaya preparation shows strong antioxidant and immune-modulating potentials. LAB strains in fermented foods showed successful modulation of the immune response against the influenza virus (Kawashima et al., 2011). *Lactobacillus plantarum* produces trace portions of phenyl acids that have strong antifungal elements. A fungal fermentation of separated polysaccharides from rice bran is a therapeutic approach intended for activating the natural killer (NK) cells that mediate anticancer activity through a nonspecific immune response (Choi et al., 2014). Lactic acid bacteria participate in the predigestion of food, which makes the digestion get easier; hence, fermentation food products will help consumers to overcome the digestion problem. The fermented foods consist millions of bacterial load that produces more vitamins and nutrients by digesting sugars and starch. The lactic acid, produced during fermentation, helps healthy bacteria already present in the gut to proliferate and is associated with better gut health. Some experts say that each small half cup serving of fermented foods can contain up to 10 trillion probiotic organisms. Literally, one serving of vegetables was equal to an entire bottle of a high potency probiotic. Eighty percent of our immune system resides in the gut. Therefore, healthy gut is equal to a healthy immune system. Adding fermented foods to the consumer's diet will ensure a healthy gut.

13.7 FUTURE PROSPECTIVE

Genetic and metabolically engineered microbes will play a major role in the production of several functional metabolites in the foods, which have large structures and difficulty to synthesize chemically. Globally, GMFs are solving many problems of the people with the intention of providing sufficient food supplementation and health benefit aspects. Most of the microorganisms used in the food and chemical industries are nonpathogenic. However, gene manipulation of these microbes might have associated with the certain human health and environment complications. Hence, GMM should be assessed for risk monitoring by either "step-by-step" or "case-by-case" approaches (Engwa, 2014). The main intention of the risk assessment is to detect potential adverse effects of GMM on human health issues and the environment. To overcome this type of complications, after risk assessment, only accepted GMFs are introduced into the market for safety use. Additionally, the future strategies of gene manipulation are predicted to be rapidly productive, inexpensive, adequately specific, and quantifiable and must allow multiplexing and quantitation (Holst-Jensen, 2009; Pandey et al., 2010) in order to offer an extensive and efficient perspective on the consumption of novel GMFs.

13.8 CONCLUSION

Genetic and metabolically engineered microbes have made progress in various food industries that leads to the production of innovative genetically modified foods. Further, their role in the enrichment of sensory qualities, bioavailability of nutrients, texture, and aroma to the GMFs has made the people fascinated to more consumption for potential health benefits.

Acknowledgment

N. Mallikarjuna is thankful to the University Grants Commission, India, for the financial support through BSR Fellowship.

References

Adrio, J.L., Demain, A.L., 2006. Genetic improvement of processes yielding microbial products. FEMS Microbiol. Rev. 30, 187–214.

Ahren, I.L., Xu, J., Önning, G., Olsson, C., Ahrne, S., Molin, G., 2014. Antihypertensive activity of blueberries fermented by *Lactobacillus plantarum* DSM 15313 and effects on the gut microbiota in healthy rats. Clin. Nutr. 34, 719–726.

Ali, A., Shehzad, A., Khan, M.R., Shabbir, M.A., Amjid, M.R., 2012. Yeast, its types and role in fermentation during bread making process. Pak. J. Biol. Sci. 22 (3), 171–179.

Anesiadis, N., Cluerr, W.R., Mahadevan, R., 2008. Dynamic metabolic engineering for increasing bioprocess productivity. Metab. Eng. 10, 255–266.

Bongers, R.S., Hoefnagel, M.H.N., Kleerebezem, M., 2005. High-level acetaldehyde production in *Lactococcus lactis* by metabolic engineering. Appl. Environ. Microbiol. 71, 109–1113.

Boyacioglu, D., Nilufer, D., Capanoglu, E., 2010. Flavor compounds in foods. In: Yildiz, F. (Ed.), Advances in Food Biochemistry. CRC Press, Boca Raton, pp. 301–302 (Chapter 9).

Camps, J., Garcia-Heredia, A., 2014. Introduction: oxidation and inflammation, a molecular link between non-communicable diseases. Adv. Exp. Med. Biol. 824, 1–4.

Carminati, D., Giraffa, G., Zago, M., Marco, M.B., Guglielmotti, D., Binetti, A., et al., 2015. Lactic acid bacteria for dairy fermentations. In: Mozzi, F., Raya, R., Vignolo, G.M. (Eds.), Biotechnology of Lactic Acid Bacteria: Novel Applications. Wiley-Blackwell, London, UK, pp. 199–200 (Chapter 12).

Chaves, A.C.S.D., Fernandez, M., Lerayer, A.L.S., Mierau, I., Kleerebezem, M., Hugenholtz, J., 2002. Metabolic engineering of acetaldehyde production by *Streptococcus thermophilus*. Appl. Environ. Microbiol. 68, 5656–5662.

Choi, J.Y., Paik, D.J., Kwon, D.Y., Park, Y., 2014. Dietary supplementation with rice bran fermented with Lentinus edodes increases interferon-gamma activity without causing adverse effects: a randomized, double-blind, placebo-controlled, parallel-group study. Nutr. J. 13, 35.

Csutak, O., Sarbu, I., 2018. Genetically modified microorganisms: harmful or helpful? In: Genetically Engineered Foods, A Volume in Handbook of Food Bioengineering. Academic Press Publishers, pp. 143–175. https://doi.org/10.1016/B978-0-12-811519-0.00006-6.

Engwa, A.G., 2014. Genetic engineering on microorganism: the ecological and bioethical implications. Eur. J. Biotechnol. 1 (3), 27–33.

Holst-Jensen, A., 2009. Testing for genetically modified organisms (GMOs): past, present, and future perspectives. Biotechnol. Adv. 27 (6), 1071–1082.

John, R.P., Gangadharan, D., Madhavan Nampoothiri, K., 2008. Genome shuffling of *Lactobacillus delbrueckii* mutant and *Bacillus amyloliquuefaciens* through protoplastic fusion for L-lactic acid production from starchy wastes. Bioresour. Technol. 99, 8008–8015.

Joutsjoki, V., Luoma, S., Tamminen, M., Kilpi, M., Johansen, E., Palva, A., 2002. Recombinant *Lactococcus* starters as a potential source of additional peptidolytic activity in cheese ripening. J. Appl. Microbiol. 92, 1159–1166.

Kamiya, S., Owasawara, M., Arakawa, M., Hagimori, M., 2013. The effect of lactic acid bacteria-fermented soybean milk products on carrageenan-induced tail thrombosis in rats. Biosci. Microbiota Food Health 32, 101–105.

Kawashima, T., Hayashi, K., Kosaka, A., Kawashima, M., Igarashi, T., Tsutsui, H., Tsuji, N.M., Nishimura, I., Hayashi, T., Obata, A., 2011. Lactobacillus plantarum strain YU from fermented foods activates Th1 and protective immune responses. Int. Immunopharmacol. 11, 2017–2024.

Keller, A.C., Weir, T.L., Broeckling, C.D., Ryan, E.P., 2013. Antibacterial activity and phytochemical profile of fermented *Camellia sinensis* (fuzhuan tea). Food Res. Int. 53, 945–949.

Koyama, M., Hattori, S., Amano, Y., Watanabe, M., Nakamura, K., 2014. Blood pressure-lowering peptides from neo-fermented buckwheat sprouts: a new approach to estimating ACE-inhibitory activity. PLoS One 9. e105802.

Kumar, A., 2016. Role of microbes in food and industrial microbiology. J. Food Ind. Microbiol. 2, e101.

Kumar, R.R., Prasad, S., 2011. Metabolic engineering of bacteria. Indian J. Microbiol. 51 (3), 403–409.

Lara-Hidalgo, C.E., Hernández-Sánchez, H., Hernández-Rodríguez, C., Dorantes-Álvarez, L., 2017. Yeasts in fermented foods and their probiotic potential. Austin J. Nutr. Metab. 4 (1), 1045.

Martins, S., Mussatto, S.I., Martinez-Avila, G., Montanez-Saenz, J., Aguilar, C.N., Teixeira, J.A., 2011. Bioactive phenolic compounds: production and extraction by solid-state fermentation. A review. Biotechnol. Adv. 29, 365–373.

Oh, M.R., Park, S.H., Kim, S.Y., Back, H.I., Kim, M.G., Jeon, J.Y., Ha, K.C., Na, W.T., Cha, Y.S., Park, B.H., Park, T.S., Chae, S.W., 2014. Postprandial glucose-lowering effects of fermented red ginseng in subjects with impaired fasting glucose or type 2 diabetes: a randomized, doubleblind, placebo-controlled clinical trial. BMC Complement. Altern. Med. 14, 237.

Pai, J.S., 2003. Applications of microorganisms in food biotechnology. Indian J. Biotechnol. 2, 382–386.

Pandey, A., et al., 2010. Genetically modified food: its uses, future prospects, and safety assessments. Biotechnology 9, 444–458.

Patnaik, R.S., Louie, S., Gavrilovic, V., Perry, K., Stemmer, W.P.C., Ryan, C.M., del Cardayre, S., 2002. Genome shuffling of *Lactobacillus* for improved acid tolerance. Nat. Biotechnol. 20, 707–712.

Plengvidhya, V., Breidt Jr., F., Zhongjing, L., Fleming, H.P., 2007. DNA fingerprinting of lactic acid bacteria in sauerkraut fermentations. Appl. Environ. Microbiol. 73, 7697–7702.

Pulvirenti, A., Solieri, L., Gullo, M., De Vero, L., Giudici, P., 2004. Occurrence and dominance of yeast species in sourdough. Lett. Appl. Microbiol. 38, 113–117.

Quave, C.L., Pieroni, A., 2014. Fermented foods for food security and food sovereignty in the Balkans: a case study of the Gorani people of Northeastern Albania. J. Ethnobiol. 34, 28–43.

Rijnen, L., Courtin, P., Gripon, J.-C., Yvon, M., 2000. Expression of a heterologous glutamate dehydrogenase gene in *Lactococcus lactis* highly improves the conversion of amino acids to aroma compounds. Appl. Environ. Microbiol. 66 (4), 1354–1359.

Ryan, E.P., Heuberger, A.L., Weir, T.L., Barnett, B., Broeckling, C.D., Prenni, J.E., 2011. Rice bran fermented with *Saccharomyces boulardii* generates novel metabolite profiles with bioactivity. J. Agric. Food Chem. 59, 1862–1870.

Tamang, J.P., Shin, D.-H., Jung, S.-J., Chae, S.-W., 2016. Functional properties of microorganisms in fermented foods. Front. Microbiol. 7, 578.

Wendisch, V.F., Bott, M., Eikmanns, B.J., 2006. Metabolic engineering of *Escherichia coli* and *Corynebacterium glutamicum* for biotechnological production of organic acids and amino acids. Curr. Opin. Microbiol. 9, 268–274.

SECTION 3

SOIL AND AGRICULTURE MICROBIOLOGY

CHAPTER

14

Effective Role of Microorganisms in Livestock Development

Dilip Reddy Gunturu, Muralidhar Yegireddy*, Sravanthi Mannem*, Alpha Raj Mekapogu*, N.V.K.V. Prasad Tollamadugu*[†]

*Department of Veterinary Pharmacology and Toxicology, College of Veterinary Science, Tirupati, India [†]Nanotechnology Laboratory, Institute of Frontier Technology, Regional Agricultural Research Station, Acharya N.G. Ranga Agricultural University, Tirupati, India

OUTLINE

14.1 INTRODUCTION

Globally, livestock is an important source of food like meat and butter and power, and it plays a symbolic appearance in rural livelihoods and the economies of developing countries (Herrero et al., 2013). Moreover, livestock often form the most important capital reserve of

Recent Developments in Applied Microbiology and Biochemistry
https://doi.org/10.1016/B978-0-12-816328-3.00014-3

husbandry households, and, in broad spectrum, it improves the economic viability and sustainability of a farming system (Windsor et al., 2018). Most researchers trust that research will support the advancement of livestock production. But regrettably, this acceptance is not regularly shared by the decision-makers who assign the resources (Thornton, 2010). Further, livestock development programs up to now were hardly based on the understanding of the livestock production systems, multipurpose production, and its complex relationships between the biological, technical, and social components of these systems (Janssen et al., 2017). Therefore, livestock development projects have not always led to substantial or sustainable long-term increases in productivity. It is therefore of importance to demonstrate that research is justified by the more effective development. An essential component of a successful research strategy is to ensure favorable social, economic, and environmental impact. Currently, research on livestock developing microorganisms is in progress as these organisms are playing an important role in livestock development. In this chapter, we focus more on the effective role of microorganisms in livestock development.

14.2 IMPORTANCE OF PROBIOTICS IN LIVESTOCK DEVELOPMENT

As per the definition of FAO/WHO Working Group (2002) and Bajagai et al. (2016), probiotics can be defined as "live microorganisms which, when administered in adequate amounts, confer a health benefit on the host." This definition is also accepted by the International Scientific Association for Probiotics and Prebiotics (Hill et al., 2014). The probiotics can be classified as, first, bacterial and nonbacterial probiotics and, second, spore- and nonspore-forming probiotics. Majority of the probiotics used in livestock are bacterial species like *Bacillus* (Mingmongkolchai and Panbangred, 2018), *Bifidobacterium* (Khaksar et al., 2012; Pedroso et al., 2013), *Enterococcus* (Mountzouris et al., 2010), and *Lactobacillus* (Mookiah et al., 2014; Bouchard et al., 2016). Nonbacterial probiotics primarily consist of yeast or fungi like *Aspergillus oryzae* (Daskiran et al., 2012; Shim et al., 2012), *Candida pintolopesii* (Daskiran et al., 2012), *Saccharomyces boulardii* (Rahman et al., 2013), and *Saccharomyces cerevisiae* (Bai et al., 2013). The probiotics further can be classified as spore- or nonspore-forming probiotics, multispecies (or multistrain) or single-species (or single-strain) probiotics, and allochthonous (the probiotic strains not native to GIT) or autochthonous probiotics (the probiotic strains native to GIT) (Bajagai et al., 2016).

14.3 ROLE OF MICROORGANISMS IN RUMINANT ANIMALS

At birth, the rumen is sterile but is quickly colonized by environmental bacteria like facultative organisms in the sequential manner, where anaerobic bacteria colonize within few hours after birth, cellulolytic bacteria and methanogenic organisms by 2–4 days of age, anaerobic fungi by the second week, and ciliate protozoa during the third week (Fonty et al., 1987; Jami et al., 2013). The characteristic reductive environment of the rumen is created by bacterial action, which subsequently results in the establishment of the strict anaerobes and later by protozoa (Cunningham and Klein, 2013). Ruminants are able to degrade and use the fibrous feed offered, which will be converted as source of energy by the presence of colonized anaerobic microbiota in the rumen like bacteria, fungi, and protozoa (Chaucheyras-Durand and Ossa, 2014). Rumen bacteria comprise around 50% of the total microbial biomass. Bacteria species themselves are an important source of protein, which accounts for 75%–80% of animal's metabolizable protein. Bacteria are also involved in producing enzymes that digest fiber (cellulose and hemicellulose), starch, and sugar. Ciliate protozoa account for up to 50% of the total microbial biomass. They have various activities like digestion of plant particles and utilization of lactic acid with subsequent reduction or control of acidosis. Rumen fungi make around 8%–10% of microbial biomass and are anaerobic, which play a pivotal role in fiber digestion (McSweeney and Mackie, 2012). At many instances, the transition from milk to solids will be done before the complete microbial colonization of rumen, which results in digestive disorders in young ruminants (Jami et al., 2013). In dairy animals, progressive development of microbiota associated with GIT is important for survival and milk production, as this microbiota is responsible for synthesizing nutrients from plant-based feeds. In recent study, calves exhibited similar microbial families and genera between weaning and 1 year of age, which is clearly suggesting that modulation of the microbiota immediately following the weaning transition is helpful for improving milk production in the later stages (Dill-McFarland et al., 2017). Direct-fed microbials (DFM) are very well-established practice in ruminant nutrition, where nonbacterial cultures like *Aspergillus oryzae* fermentation extract or *Saccharomyces cerevisiae* cultures were reported to increase the nutrients like lactate to ruminal bacteria like *Selenomonas ruminantium* and increase the cell yield; production of acetate, propionate, and succinate; and total VFA (Martin and Nisbet, 1992). Feeding of probiotic species (DFM) like *Lactobacillus, Bifidobacterium, Enterococcus, Streptococcus, Bacillus, Propionibacterium, Megasphaera elsdenii,* and *Prevotella bryantii* and yeast such as *Saccharomyces* and *Aspergillus* resulted in immune system modulation; microbial balance in the GIT; and increase in dry matter intake, milk yield, fat-corrected milk yield, and milk fat in ruminants (Elghandour et al., 2015). There are wide range of applications of probiotics in animal production (Fig. 14.1). The uses of probiotics for growth and production have been well documented and discussed previously (Wallace and Newbold, 1992). Probiotics, *Bacillus toyonensis,* and *Saccharomyces boulardii* are reported to improve the immune response to vaccine in sheep (Roos et al., 2018). The fungal probiotics like *Aspergillus* and *Saccharomyces* have modulatory effect on the rumen fermentation products in sheep, where they increased the concentrations of acetate, butyrate, and isoacids like isobutyrate and isovalerate (Mathieu et al., 1996). The recent reports also suggest the use of lactic acid bacilli on reducing the methane production in ruminants (Ellis et al., 2016).

14.4 APPLICATIONS OF MICROBES/PROBIOTICS IN OTHER ANIMALS

In pigs, the use of probiotic *Lactobacillus reuteri* has been extensively reviewed by Hou et al. (2015), and they reported its positive effects on performance, prevention of diarrhea, stress relief, altered gut microbiota, and immunomodulation. In another study, the

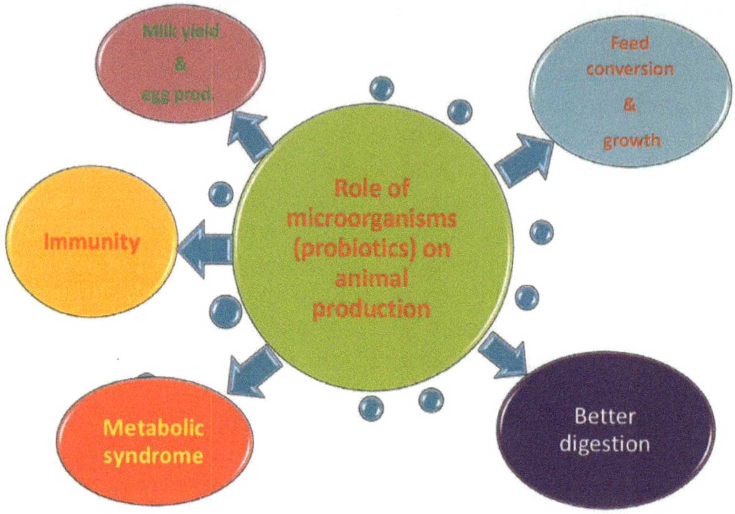

FIG. 14.1 Important uses of probiotics in animal production.

incidence of postweaning diarrhea was significantly reduced by feeding *Enterococcus faecium* and *B. cereus* to piglets (Simon, 2010). The piglets received nontoxigenic *Clostridium difficile* as probiotic supplement and exhibited lower prevalence, signs, and symptoms of a common bacterial infection caused by *Clostridium difficile* (Arruda et al., 2016). Administration of *Bacillus* sp. probiotics significantly improved body weights, daily weight gains, and feed conversion of the experimental pigs with protective effect against diarrhea, indicating a possible replacement of antibiotic growth promotors with probiotics in pigs (Link et al., 2005).

The use of probiotics in poultry production was well reviewed (Jernigan et al., 1985; Khan and Naz, 2013). Probiotic supplemented diet reduced the mortality at high altitudes and improved the economic performance of broiler chicken (Kalia et al., 2017). The addition of probiotics in poultry is reported to improve meat quality with lesser *Salmonella* contamination (Park et al., 2016) and *Campylobacter* contamination (Oyarzabal and Backert, 2016). The use of subtherapeutic doses of antibiotics in poultry as growth promotors has raised the concerns about antibiotic resistance in humans. A better and productive alternative approach to this subtherapeutic antibiotic use in livestock and poultry is the use of probiotic microorganisms. Lactobacilli and bifidobacteria have been associated with improved health in poultry (Patterson and Burkholder, 2003). Direct-fed microbials (probiotics) in poultry are reported to improve growth and feed conversion in broilers, egg mass, egg weight, and egg size in layers by various modes of action like maintaining normal intestinal microflora, increasing digestive enzyme activity and decreasing bacterial enzyme activity and ammonia production, improving feed intake, and neutralizing enterotoxins and immune stimulation (Jin et al., 1997). The lactic acid bacteria are reported to replace various enteric pathogens like *Salmonella enterica*, *E. coli*, and *Campylobacter jejuni* by competitive exclusion in the intestinal tract and improve the intestinal health of poultry (Harimurti and Hadisaputro, 2015).

14.5 OTHER APPLICATIONS OF MICROORGANISMS IN ANIMAL PRODUCTION

14.5.1 Meat Quality

The addition of probiotics in poultry is reported to improve meat quality with lesser *Salmonella* (Park et al., 2016) and *Campylobacter* contamination (Oyarzabal and Backert, 2016). Better poultry meat quality was reported in birds fed with probiotics in water and feed when compared with control birds (Pelicano et al., 2003). Bacteriocins produced by probiotic *Lactobacillus* in packaging can be used to improve meat safety. The use of several bacteriocins for application in meat, poultry, ready-to-eat meat products, and sausage casings has been reported (Woraprayote et al., 2016). The effect of bacteriocins in reducing *Listeria monocytogenes* levels in some meat products is reported by Lücke (2000). *B. subtilis* B2A supplementation in broilers improved meat quality parameters. The probiotic significantly improved water-holding capacity and reduced 1-day drip loss (Park and Kim, 2014).

14.5.2 Feed

Animal feed and fodder consists of a range of microbiota as natural constituents or as contaminants. The beneficial effects are due to the presence of lactic acid bacteria during the fermentation of forages for silage. Lactic acid bacteria and yeast cultures are considered as feed probiotics as they are helpful in improving growth performance of animals. Lactic acid bacteria utilize endogenous plant sugars and produce lactic acid to bring down pH to near 4, which will be useful for preservation. The main contaminants of forages or silages are clostridia, enterobacteria, listeria, and fungi (D'Mello, 2004). New strategies were proposed to develop improved microbes for silage like the use of more specific lactobacillus to the target crops, addition of heterofermentative lactobacillus that can inhibit volatile fatty acids and yeasts and molds, use of engineered strains that can inhibit specific microorganisms, and cloning and expression of genes that would enable selected lactobacillus strains to utilize polysaccharides of forages (Weinberg and Muck, 1996). Many novel bacterial candidates like mucolytic bacterium *Akkermansia muciniphila* targeting metabolic syndrome have been reported to act by various convergent mechanisms involved in metabolic syndrome (Cani and Van Hul, 2015; Le Barz et al., 2015).

14.5.3 Role of Microorganisms During Stress in Animals

Young animals will be under various forms of stresses like transport, nutrition, climate and handling, and husbandry practices. These stressors will induce imbalances of the host microbiota (Tannock and Savage, 1974). This subsequently leads to increased susceptibility to infections and increased incidences in diarrhea in postweaning animals. There are recent reports targeting the microbiota-gut-brain axis by using various probiotic strains to prevent hypersecretion of cortisol during stressful conditions via the vagal afferent signaling to the brain and reduced stress reactivity at higher levels (Burokas et al., 2017; Takada et al., 2016). The use of probiotics during these conditions is hence helpful in preventing the incidence of disease and results in improved performance of the animals. In intensive farming systems, early weaning of animals to cleaner environment will lead to reduced passive immunity and reduced development of healthy microbiota. There is increasing evidence about the effects of changes in the microbiota in conferring resistance or promoting infection by pathogenic bacteria (Bäumler and Sperandio, 2016). The general approach for the treatment of infections of the gut in animals is the use of antimicrobials. Though this approach clears the infection effectively, it is also associated with altering the native microbiota and subsequent side effects. This often results in antibiotic-induced diarrhea due to the imbalance in the microbiota, mostly due to bacterial overgrowth of existing species in the gut. To prevent this condition, the probiotics are coadministered with antimicrobials in feed (Surawicz, 2008).

14.6 POSSIBLE MECHANISM OF ACTIONS OF PROBIOTICS IN ANIMALS

The proposed mechanism of action of probiotics includes the release of antimicrobial agents and competition between probiotic and pathogen for host cell receptors, essential nutrients, and host immunomodulation. The present studies and information suggest that probiotics can be used as effective agents for animal production in place of antibiotics. This approach is becoming very popular even for nonproduction animals (Di Gioia and Biavati, 2018). Immunomodulation is indicated as one of the major mechanisms of probiotic bacteria as the bacteria and their soluble fractions can activate dendritic cells, macrophages, and monocytes through toll-like receptor recognition and provoking specific Th response and subsequent immunomodulatory effects on intestinal and systemic immunities (Fong et al., 2015). Similar results were reported in poultry, where immunomodulation and decreasing gastrointestinal inflammatory responses with subsequent reduction in colonization by pathogens that require inflammatory environment was noticed. By these processes, these probiotics keep a healthy gut and indirectly help in animal productivity (Vuong et al., 2016). Cartman et al. (2008) have demonstrated the production of antimicrobial agents in the gut by spore probiotics in chicken. In recent studies in mice, probiotics modulated gut microbiota and suppressed hepatocellular carcinoma growth (Li et al., 2016). Epithelial barrier is one of the major etiologic factors of inflammatory bowel disease, and probiotic *Bifidobacterium* species improve intestinal epithelial barrier function via acetate formation (Hsieh et al., 2015) and hence can be effectively used in treating disorders associated with inflammation of the gut. Probiotics may increase the population of beneficial microorganisms like lactobacilli and bifidobacteria that later inhibit harmful microorganisms by producing inhibitory substances like bacteriocins and/or organic acids and by competitive exclusion (Bajagai et al., 2016).

Probiotics are widely used in domestic animals, companion animals, and aquaculture (Akhter et al., 2015), with the aim of reducing disease and improving growth rate. It is well documented that pathogenic microbes at epithelial cells modify the cellular physiology. Recent studies indicate that host microbial interactions in the gut are highly site-specific, and the local cross talk between pathogenic and probiotic microbes determines intestinal function and physiology (Sommer and Bäckhed, 2016). The effect of microbiota on the host health and disease via their metabolites in different invertebrate and vertebrate animal models was reviewed by Lee and Hase (2014). The cellular metabolic change produced by colonization of probiotics in the gut epithelial cell exerts a protective effect on the epithelial surface by preventing colonization of pathogenic strains. The sympathetic and parasympathetic nerves along with hormones will influence the function of the intestinal tissues. Hence stress will play its role by releasing corticosteroids and by direct autonomic stimulation of the intestinal tissues (Matteri et al., 2000; Petrovsky, 2001).

14.7 CONCLUSION

The use of microorganisms is associated with better digestion in monogastric animals and ruminants with alteration in the ruminal fermentation products and subsequent control of ruminal acidosis, immunostimulation or modulation, increased feed conversion ratios, reduced carcass or meat contamination in broilers, improvement in growth rate, increased milk yield, increased egg production and egg quality, and improving the conditions like metabolic syndrome. Therefore, constructive research in using proper microorganisms for the development of livestock is required, and it definitely helps in developing sustainable livestock development approaches.

Acknowledgments

The authors are indebted to all the researchers whom we cited in this review for their significant and valuable research. No funding was received to perform this review.

Conflicts of Interests

The authors declare no relevant competing financial interests to disclose.

References

Akhter, N., Wu, B., Memon, A.M., Mohsin, M., 2015. Probiotics and prebiotics associated with aquaculture: a review. Fish Shellfish Immunol. 45 (2), 733–741.

Arruda, P.H.E., Madson, D.M., Ramirez, A., Rowe, E.W., Songer, J.G., 2016. Bacterial probiotics as an aid in the control of Clostridium difficile disease in neonatal pigs. Can. Vet. J. 57, 183–188.

Bai, S., Wu, A., Ding, X., Lei, Y., Bai, J., Zhang, K., Chio, J., 2013. Effects of probiotic-supplemented diets on growth performance and intestinal immune characteristics of broiler chickens. Poult. Sci. 92 (3), 663–670.

Bajagai, Y.S., Klieve, A.V., Dart, P.J., Bryden, W.L., 2016. In: Makkar, H.P.S. (Ed.), Probiotics in animal nutrition: production, impact and regulation. FAO Animal Production and Health Paper No. 179, Rome.

Bäumler, A.J., Sperandio, V., 2016. Interactions between the microbiota and pathogenic bacteria in the gut. Nature 535 (7610), 85–93.

Bouchard, D., Evan, S., Loir, Y.L., 2016. Lactic acid bacteria in animal production and health. In: Mozzi, F., Raya, R.R., Vignolo, G.M. (Eds.), Biotechnology of Lactic Acid Bacteria: Novel Applications. John Wiley & Sons, United Kingdom.

Burokas, A., Arboleya, S., Moloney, R.D., Peterson, V.L., Murphy, K., et al., 2017. Targeting the microbiota-gut-brain axis: prebiotics have anxiolytic and antidepressant-like effects and reverse the impact of chronic stress in mice. Biol. Psychiatry 82 (7), 472–487.

Cani, P.D., Van Hul, M., 2015. Novel opportunities for next-generation probiotics targeting metabolic syndrome. Curr. Opin. Biotechnol. 32, 21–27.

Cartman, S.T., Ragione, R.M.L., Woodward, M.J., 2008. *Bacillus subtilis* spores germinate in the chicken gastrointestinal tract. Appl. Environ. Microbiol. 74 (16), 5254–5258.

Chaucheyras-Durand, F., Ossa, F., 2014. REVIEW: the rumen microbiome: composition, abundance, diversity, and new investigative tools. Prof. Anim. Sci. 30 (1), 1–12.

Cunningham, J.G., Klein, B.G., 2013. Digestion: the fermentative processes. In: Cunningham's Textbook of Veterinary Physiology. Elsevier/Saunders, India, pp. 320–341. Chapter 31.

Daskiran, M., Onol, A.G., Cengiz, O., Unsal, H., Turkyilmaz, S., Tatli, O., Sevim, O., 2012. Influence of dietary probiotic inclusion on growth performance, blood parameters, and intestinal microflora of male broiler chickens exposed to posthatch holding time. J. Appl. Poult. Res. 21 (3), 612–622.

Dill-McFarland, K.A., Breaker, J.D., Suen, G., 2017. Microbial succession in the gastrointestinal tract of dairy cows from 2 weeks to first lactation. Sci. Rep. 7, 40864.

D'Mello, J.P.F., 2004. Microbiology of Animal Feeds. FAO Animal production and health paper. 89–106.

Elghandour, M.M.Y., Salem, A.Z.M., Castañeda, J.S.M., Camacho, L.M., Kholif, A.E., Chagoyán, J.C.V., 2015. Direct-fed microbes: a tool for improving the utilization of low quality roughages in ruminants. J. Integr. Agric. 14, 526–533.

Ellis, J.L., Bannink, A., Hindrichsen, I.K., Kinley, R.D., Pellikaan, W.F., Milora, N., Dijkstra, J., 2016. The effect of lactic acid bacteria included as a probiotic or silage inoculant on in vitro rumen digestibility, total gas and methane production. Anim. Feed Sci. Technol. 211, 61–74.

FAO/WHO, 2002. Report of a joint FAO/WHO Working Group on Drafting Guidelines for the Evaluation of Probiotics in Food.

Fong, F.L., Shah, N.P., Kirjavainen, P., El-Nezami, H., 2015. Mechanism of action of probiotic bacteria on intestinal and systemic immunities and antigen-presenting cells. Int. Rev. Immunol. 35 (3), 179–188.

Fonty, G., Gouet, P., Jouany, J.P., Senaud, J., 1987. Establishment of the microflora and anaerobic fungi in the rumen of lambs. J. Gen. Microbiol. 133, 1835–1843.

Di Gioia, D., Biavati, B., Di Gioia, D., 2018. Probiotics and prebiotics in animal health and food safety: conclusive remarks and future perspectives. In: Biavati, B. (Ed.), Probiotics and Prebiotics in Animal Health and Food Safety. Springer, Cham. https://doi.org/10.1007/978-3-319-71950-4_11.

Harimurti, S., Hadisaputro, W., 2015. Probiotics in poultry. In: Liong, M.T. (Ed.), Beneficial Microorganisms in Agriculture, Aquaculture and Other Areas, Microbiology Monographs. vol. 29. Springer, Cham.

Herrero, M., Grace, D., Njuki, J., Johnson, N., Enahoro, D., Silvestri, S., Rufino, M.C., 2013. The roles of livestock in developing countries. Animal 7 (Suppl. 1), 3–18.

Hill, C., Guarner, F., Reid, G., Gibson, G.R., Merenstein, D.J., Pot, B., Morelli, L., Canani, R.B., Flint, H.J., Salminen, S., 2014. Expert consensus document: the International Scientific Association for Probiotics and Prebiotics consensus statement on the scope and appropriate use of the term probiotic. Nat. Rev. Gastroenterol. Hepatol. 11, 506–514.

Hou, C., Zeng, X., Yang, F., Liu, H., Qiao, S., 2015. Study and use of the probiotic *Lactobacillus reuteri* in pigs: a review. J. Anim. Sci. Biotechnol. 6 (1), 14.

Hsieh, C.Y., Osaka, T., Moriyama, E., Date, Y., Kikuchi, J., Tsuneda, S., 2015. Strengthening of the intestinal epithelial tight junction by *Bifidobacterium bifidum*. Phys. Rep. 3 (3). pii: e12327.

Jami, E., Israel, A., Kotser, A., Mizrahi, I., 2013. Exploring the bovine rumen bacterial community from birth to adulthood. ISME J. 7 (6), 1069–1079.

Janssen, S.J.C., Porter, C.H., Moore, A.D., Athanasiadis, I.N., Foster, I., Jones, J.W., Antle, J.M., 2017. Towards a new generation of agricultural system data, models and knowledge products: information and communication technology. Agric. Syst. 155, 200–212.

Jernigan, M.A., Miles, R.D., Arafa, A.S., 1985. Probiotics in poultry nutrition—a review. Worlds Poult. Sci. J. 41 (2), 99–107.

Jin, L.Z., Ho, Y.W., Abdullah, N., Jalaludin, S., 1997. Probiotics in poultry: modes of action. Worlds Poult. Sci. J. 53 (4), 351–368.

Kalia, S., Bharti, K.V., Gogoi, D., Giri, A., Kumar, B., 2017. Studies on the growth performance of different broiler strains at high altitude and evaluation of probiotic effect on their survivability. Sci. Rep. 7, 46074.

Khaksar, V., Golian, A., Kermanshahi, H., 2012. Immune response and ileal microflora in broilers fed wheat-based diet with or without enzyme Endofeed W and supplementation of thyme essential oil or probiotic PrimaLac. Afr. J. Biotechnol. 11, 14716–14723.

Khan, R.U., Naz, S., 2013. The applications of probiotics in poultry production. Worlds Poult. Sci. J. 69 (3), 621–632.

Le Barz, M., Anhê, F.F., Varin, T.V., Desjardins, Y., Levy, E., Roy, D., Urdaci, M.C., Marette, A., 2015. Probiotics as complementary treatment for metabolic disorders. Diabetes Metab. J. 39 (4), 291–303.

Lee, W.J., Hase, K., 2014. Gut microbiota-generated metabolites in animal health and disease. Nat. Chem. Biol. 10, 416–424.

Li, J., Sung, C.Y.J., Lee, N., Ni, Y., Pihlajamäki, J., Panagiotou, G., El-Nezami, H., 2016. Probiotics modulated gut microbiota suppresses hepatocellular carcinoma growth in mice. Proc. Natl. Acad. Sci. 113, E1306–E1315.

Link, R., Kováč, G., Pistl, J., 2005. A note on probiotics as an alternative for antibiotics in pigs. J. Anim. Feed Sci. 14 (3), 513–519.

Lücke, F.K., 2000. Utilization of microbes to process and preserve meat. Meat Sci. 56 (2), 105–115.

Martin, S.A., Nisbet, D.J., 1992. Effect of direct-fed microbials on rumen microbial fermentation. J. Dairy Sci. 75 (6), 1736–1744.

Mathieu, F., Jouany, J., Sénaud, J., Bohatier, J., Bertin, G., Mercier, M., 1996. The effect of *Saccharomyces cerevisiae* and *Aspergillus oryzae* on fermentations in the rumen of faunated and defaunated sheep; protozoal and probiotic interactions. Reprod. Nutr. Dev. 36 (3), 271–287.

Matteri, R.L., Carroll, J.A., Dyer, C.J., 2000. Neuroendocrine responses to stress. In: Morberg, G.P., Mench, J.A. (Eds.), The Biology of Animal Stress: Basic Principles and Implications for Animal Welfare. CABI Publishing, New York.

McSweeney, C., Mackie, R., 2012. Micro-organisms and ruminant digestion: state of knowledge, trends and future prospects. Commission on Genetic Resources for Food and Agriculture, FAO. Back Ground Study Paper 61.

Mingmongkolchai, S., Panbangred, W., 2018. Bacillus probiotics: an alternative to antibiotics for livestock production. J. Appl. Microbiol. https://doi.org/10.1111/jam.13690.

Mookiah, S., Sieo, C.C., Ramasamy, K., Abdullah, N., Ho, Y.W., 2014. Effects of dietary prebiotics, probiotic and synbiotics on performance, caecal bacterial populations and caecal fermentation concentrations of broiler chickens. J. Sci. Food Agric. 94 (2), 341348.

Mountzouris, K.C., Tsitrsikos, P., Palamidi, I., Arvaniti, A., Mohnl, M., Schatzmayr, G., Fegeros, K., 2010. Effects of probiotic inclusion levels in broiler nutrition on growth performance, nutrient digestibility, plasma immunoglobulins, and cecal microflora composition. Poult. Sci. 89 (1), 58–67.

Oyarzabal, O.A., Backert, S., 2016. Control of *Campylobacter* spp. in commercial poultry production. In: *Campylobacter* spp. and Related Organisms in Poultry. Springer, Cham, pp. 137–149.

Park, J.H., Kim, I.H., 2014. Supplemental effect of probiotic *Bacillus subtilis* B2A on productivity, organ weight, intestinal Salmonella microflora, and breast meat quality of growing broiler chicks. Poult. Sci. 93, 2054–2059.

Park, Y.H., Hamidon, F., Rajangan, C., Soh, K.P., Gan, C.Y., Lim, T.S., Abdullah, W.N., Liong, M.T., 2016. Application of probiotics for the production of safe and high-quality poultry meat. Korean J. Food Sci. Anim. Resour. 236, 567–576.

Patterson, J.A., Burkholder, K.M., 2003. Application of prebiotics and probiotics in poultry production. Poult. Sci. 82 (4), 627–631.

Pedroso, A.A., Hurley-Bacon, A.L., Zedek, A.S., Kwan, T.W., Jordan, A.P.O., Avellaneda, G., Hofacre, C.L., Oakley, B.B., Collett, S.R., Maurer, J.J., Lee, M.D., 2013. Can probiotics improve the environmental microbiome and resistome of commercial poultry production? Int. J. Environ. Res. Public Health 10, 4534–4559.

Pelicano, E.R., De Souza, P.A., De Souza, H.B.A., Oba, A., Norkus, E.A., Kodawara, L.M., De Lima, T.M.A., 2003. Effect of different probiotics on broiler carcass and meat quality. Rev. Bras. Cienc. Avic. 5 (3), 207–214.

Petrovsky, N., 2001. Towards a unified model of neuroendocrine-immune interaction. Immunol. Cell Biol. 79 (4), 350–357.

Rahman, M., Mustari, A., Salauddin, M., Rahman, M., 2013. Effects of probiotics and enzymes on growth performance and haematobiochemical parameters in broilers. J. Bangladesh Agril. Univ. 11 (1), 111–118.

Roos, T.B., de Moraes, C.M., Sturbelle, R.T., Dummer, L.A., Fischer, G., Leite, F.P.L., 2018. Probiotics *Bacillus toyonensis* and *Saccharomyces boulardii* improve the vaccine immune response to Bovine herpesvirus type 5 in sheep. Res. Vet. Sci. 117, 260–265.

Shim, Y.H., Ingale, S.L., Kim, J.S., Kim, K.H., Seo, D.K., Lee, S.C., Chae, B.J., Kwon, I.K., 2012. A multi-microbe probiotic formulation processed at low and high drying temperatures: effects on growth performance, nutrient retention and caecal microbiology of broilers. Br. Poult. Sci. 53, 482–490.

Simon, O., 2010. An interdisciplinary study on the mode of action of probiotics in pigs. J. Anim. Feed Sci. 19 (2), 230–243.

Sommer, F., Bäckhed, F., 2016. Know your neighbor: microbiota and host epithelial cells interact locally to control intestinal function and physiology. BioEssays 38, 455–464.

Surawicz, C.M., 2008. Role of probiotics in antibiotic-associated diarrhoea, *Clostridium difficile*-associated diarrhoea, and recurrent *Clostridium difficile*-associated diarrhoea. J. Clin. Gastroenterol. 42, S64–S70.

Takada, M., Nishida, K., Kataoka-Kato, A., Gondo, Y., Ishikawa, H., 2016. Probiotic *Lactobacillus casei* strain Shirota relieves stress-associated symptoms by modulating the gut–brain interaction in human and animal models. Neurogastroenterol. Motil. 28 (7), 1027–1036.

Tannock, G.W., Savage, D.C., 1974. Influences of dietary and environmental stress on microbial populations in the murine gastrointestinal tract. Infect. Immun. 9 (3), 591–598.

Thornton, P.K., 2010. Livestock production: recent trends, future prospects. Philos. Trans. R. Soc. Lond. Ser. B Biol. Sci. 365, 2853–2867.

Vuong, C.N., Chou, W.K., Hargis, B.M., Berghman, L.R., Bielke, L.R., 2016. Role of probiotics on immune function and their relationship to antibiotic growth promoters in poultry, a brief review. Int. J. Probiotics Prebiotics 1 (11), 1–6.

Wallace, R.J., Newbold, C.J., 1992. Probiotics for ruminants. In: Probiotics. Springer, Dordrecht.

Weinberg, Z.G., Muck, R.E., 1996. New trends and opportunities in the development and use of inoculants for silage. FEMS Microbiol. Rev. 19 (1), 53–68.

Windsor, P.A., Nampanya, S., Putthana, V., Keonam, K., Johnson, K., Bush, R.D., Khounsy, S., 2018. The endoparasitism challenge in developing countries as goat raising develops from smallholder to commercial production systems: a study from Laos. Vet. Parasitol. 251, 95–100.

Woraprayote, W., Malila, Y., Sorapukdee, S., Swetwiwathana, A., Benjakul, S., Visessanguan, W., 2016. Bacteriocins from lactic acid bacteria and their applications in meat and meat products. Meat Sci. 120, 118–132.

CHAPTER

15

The Status of Research and Application of Biofertilizers and Biopesticides: Global Scenario

Marella Saritha, N.V.K.V. Prasad Tollamadugu

Nanotechnology Laboratory, Institute of Frontier Technology, Regional Agricultural Research Station, Acharya N.G. Ranga Agricultural University, Tirupati, India

OUTLINE

Recent Developments in Applied Microbiology and Biochemistry
https://doi.org/10.1016/B978-0-12-816328-3.00015-5

15.1 INTRODUCTION

Organic farming is nowadays attracting the attention of researchers and the public too with increased health awareness and pollution constraints. The ill effects of agrochemicals that resist pests may accumulate in soil, water, and air, and detrimental effects of the outbreak of secondary pests made it mandatory for the production of biopesticides (Al-Zaidi et al., 2011; Forlani et al., 2014). Biofertilizers are the biologically active products or microbial inoculants (one or more beneficial bacteria of fungal strains) meant for N fixing, phosphate solubilizing, or cellulolytic actions that can promote growth and yield by improving the availability of soil nutrients in its rhizosphere (Mazid et al., 2011a; Arriola et al., 2015). They are usually made of two microbes and a carrier material of which bacteria, blue-green algae, and arbuscular mycorrhiza fungi are the most commonly used biofertilizers. For microbial biopesticides, the best strain used is *Bacillus thuringiensis* (Bt), and almost 90% of the biopesticides are derived from this. Currently, chemical fertilizers and pesticides are being replaced by biofertilizers and biopesticides, respectively, owing to their ease of application, eco-friendly nature, cost effectiveness, and nontoxicity. Demand for biopesticides comes from the increasing awareness of the public towards health and ecology and also hazards in the indiscriminate use of synthetic chemical pesticides. However, only 1%–2% of the total crop protection market is occupied by bioalternatives and that too relied on Bt. Researchers all over the globe are being engaged in developing bioproducts that are easier to use, can overcome the ill effects of phytopathogens, and cover more target crops.

15.2 WHY TO USE BIOPRODUCTS?

In spite of the improved food grain production with the advent of green revolution by the use of inorganic fertilizers, ecological imbalance and depletion of natural resources caused a serious threat to the mankind. The cost of chemical fertilizers is increasing enormously with respect to the deteriorating petroleum-based products and feedstock, and also the energy requirement for producing 1 kg fertilizer is 11.2 kWh for nitrogen (N), 1.1 kWh for phosphorus (P), and 1 kWh for potash. Hence, organic farming became inevitable in order to overcome the limitations.

They also act as supplements to agrochemicals in the production of nutrients prevailing in the soil atmosphere and the eradication of pests that cause damage to crop leading to declined yields. Hence, the promotion of biofertilizers and biopesticides can be admissible through research and development only when demand is created in farmers toward their usage and ultimate production (Malusá and Vassilev, 2014). Biological N fixation is meant to convert elemental N to organic forms, which are readily available for N-fixing microbes in a symbiotic association with plants. These microbes are collectively known as "diazotrophs," and their functions are precisely represented in Fig. 15.1.

15.3 N FIXING INOCULANTS AS BIOFERTILIZERS

The efficacy of plant growth-promoting rhizobacteria (PGPR) over the soil-dwelling bacteria helps in reducing the adverse effects of plant pathogens and enhancing crop

FIG. 15.1 Functional characteristics of biofertilizers and biopesticides.

rotation by uplifting the availability of N, P, and iron (Fe). N, as a component of chlorophyll and protein, is essential for photosynthesis and vegetative and reproductive growth and even determines yield of crop (Marek-Kozaczuk et al., 2014). Azotobacter, a N-fixing bacterium, is best used as biofertilizer for plants like maize, wheat, sorghum, and rice, which aids in the production of plant growth-promoting metabolites; enzymes; and phytohormones like auxins, gibberellins, and cytokinins (Forlain et al., 1998). The use of *Azospirillum brasilense* inoculated maize seeds and also enhanced root and shoot length (Braccini et al., 2012) besides the increased dry weight of the plant (Puente et al., 2009). Farmers' interest in the use of biofertilizers is lacking, especially in some of the North American parts due to price compatibility on application to commodity crops that are of low value. The maximum amount of N-fixing capacity of each inoculum has been graphically represented in Fig. 15.2.

15.3.1 Rhizobium

This N-fixing symbiotic bacterium belongs to the *Rhizobiaceae* family and can fix N at 50–100 kg/ha in pulse crops and hence routinely recommended for all legumes except *Parasponia*, a nonlegume (Saikia and Jain, 2007). Its symbiotic association is visible in the form of root nodules, but stem nodules are also seen in *Sesbania rostrata*. These inoculants can be used as biofertilizers specific to the crop that are mentioned below (Table 15.1).

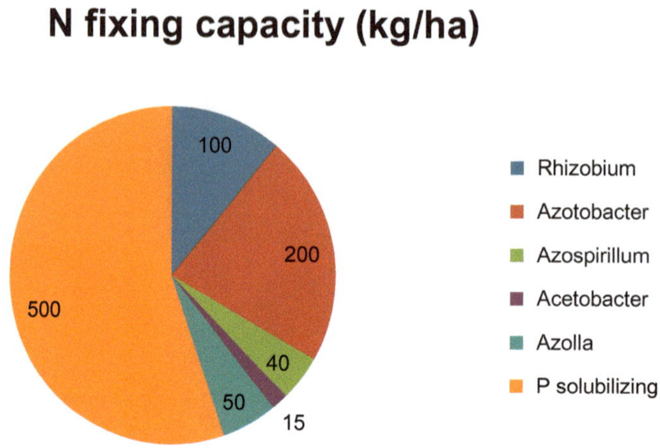

FIG. 15.2 N-fixing capability of microbial species.

TABLE 15.1 Rhizobial Species as Biofertilizer in Various Crops

Rhizobium Inoculant	Crop	N-Fixing Capacity (kg/ha)
Rhizobium trifoli	Berseem	130
Rhizobium meliloti	Lucerne	100–150
Rhizobium phaseoli	Green gram, black gram	80–110
Rhizobium japonicum	Soybean	57–105
Rhizobium leguminosarum	Pea, lentil	62–132
Rhizobium lupine	Chickpea	70–90

15.3.2 Azotobacter

This is a nonsymbiotic, aerobic, and free-living bacterium and belongs to the *Azotobacteriaceae* family, but the root nodules are not visible. It fixes N at 40–200 kg/ha and can afford to 80%–90% of N requirement of the crop. *Azotobacter vinelandii*, *Acinetobacter beijerinckii*, *Agrionoptera insignis*, and *Azomonas macrocytogenes* are the different species occurring in the rhizosphere of crops like rice, maize, sugarcane, bajra, vegetables, and plantation where an increase in the yield is up to 50%. These species help in inhibiting certain root pathogens by seed germination and plant growth as a result of the production of B vitamins and phytohormones (Mazid et al., 2011b).

15.3.3 Herbspirillum

It is a N-fixing symbiont inhabiting roots of sugarcane and enhances uptake of N, K, and P and also promotes growth hormone (gibberellins, auxins, and kinetins) production (Khan et al., 2011a). In a study, researchers investigated the bioprotective effects of inoculating corn grown under drought conditions with two types of PGPR, *Azospirillum brasilense*, strain SP-7, and *Herbaspirillum* sp. The obtained results proved that these bacteria could be used to help plants cope with the negative effects of drought stress conditions (Curá et al., 2017).

15.3.4 Azospirillum

It is a N-fixing microbe also called as *B. polymyxa*. It belongs to the *Spirillaceae* family and is beneficial for nonleguminous plants. The different species under this genus are *Azospirillum amazonense, Azospirillum halopraeferens*, and *Azospirillum brasilense* with a N fixation capacity of 20–40 kg/ha. These are significant in improving leaf area and grain yield and specifically used in maize, sorghum, sugarcane, and pearl millet (C4 plants) as they fix N on salts of organic acids like malic acid and aspartic acid (Brusamarello-Santos et al., 2017).

15.3.5 Acetobacter

It proves best suited for sugarcane crop as it can tolerate the high-sucrose concentration and can be able to fix N up to 15 kg/ha/year.

15.3.6 Azolla

It is a free-floating symbiotic fern belonging to Azollaceae and is used as green manure or as dual crop. The paddy yield is increased when the dry inoculums of *Azolla* is soaked in superphosphate solution and inoculated and is shown to fix N at 45–50 kg/ha. *Azolla pinnata* is the common species used as biofertilizer in paddy crop as it can quickly decompose in soil and make the best availability of N. Its capacity of symbiotic association with blue-green algae favors rice or other crops for dual cropping or green manuring of soil (Ghosh, 2004).

15.3.7 Blue Green Algae (Anabaena, Nostoc, Plectonema)

These are photosynthetic N-fixing free-living organisms and improve soil aeration by adding growth-promoting substances including vitamin B12. They have special structures called heterocysts, which act as micronodules, helping in N fixation. They form symbiotic association with ferns, fungi, liverworts, and flowering plants but in particular with *Azolla* for fixing N.

15.3.8 Mycorrhiza

Mycorrhiza, as is known for crop infection, can also form symbiotic relationship with host plants in exchange of zinc (Zn) and P for carbohydrates. They establish a bridge with soil through their mycelia for the absorption of water and nutrients like P, K, Ca, and N and produce growth promoters like cytokinins further protecting plants from invading pathogens by secreting antimicrobial substances.

15.3.9 Phosphate Solubilizing Bacteria

Pseudomonas, Bacillus, Micrococcus, Aspergillus, Fusarium, etc. are the major phosphate solubilizing bacteria (PSB) in soil. They convert the nonavailable inorganic phosphate to its soluble form, thus increasing the availability of phosphorus by secreting organic acids like citric, oxalic, succinic, tartaric, and malic acids in the soil rhizosphere. They can increase the crop yield up to 200–500 kg/ha. The rhizosphere of sunflower was evidenced by the presence of phosphate-solubilizing bacterial species of *Enterobacter* and *Burkholderia* (Ambrosini et al., 2012).

15.3.10 Potassium Solubilizing Bacteria

Like nitrogen and phosphorus, potassium is one of the primary macronutrients that the plant needs to develop its biological processes. Therefore, potassium-solubilizing activity is an important function of rhizotrophic bacteria. *Aspergillus, Bacillus,* and *Clostridium* genera are efficient in solubilization mobility of K in the soil in different crops (Mohammadi and Sohrabi, 2012).

15.3.11 Bioengineered Microbes

These are microbes encountered with specific functional genes by means of rDNA technology. They protect the crop from pests and degrade biological wastes further enhancing N fixation by application of *nif* genes in *Rhizobium meliloti.*

15.3.12 Liquid Biofertilizers

These are special liquid formulation with desired microbes and their nutrients but endowed with special cell protectants for resting spores/cysts to enhance shelf life and tolerate adverse conditions (Hegde, 2008). They can also overcome the limitations of normal (solid) biofertilizers attributed to their resistance to higher temperatures and ultraviolet radiation (Sumita et al., 2015). They have very high-enzymatic activity that is highly adoptable by farmers and have ease of application like using powder sprayers, hand sprayers, and as basal manure mixed with farmyard manure. Microbial inoculants are of great interest, but the efficacy of microbes other than rhizobium is unpredictable.

15.4 INTERACTION MECHANISM OF PLANT AND MICROBES

Plants defend themselves from the attack of pathogens and also need strategies to take advantage of beneficial microbes. With increasing world population, understanding the interaction between plant and microbe forms a key area of research to maintain and improve crop yield (Fig. 15.3). Changes in ecosystem and physical environment have great impact on beneficial and detrimental plant-microbe interactions. To protect themselves from the attack of pathogens and pests and to avoid crop loss in agriculture, plants have a sophisticated immune system. They also recruit beneficial microbes to their root system by boosting the immune responses further forming a rationale for developing sustainable strategies for disease resistance in next generation crops.

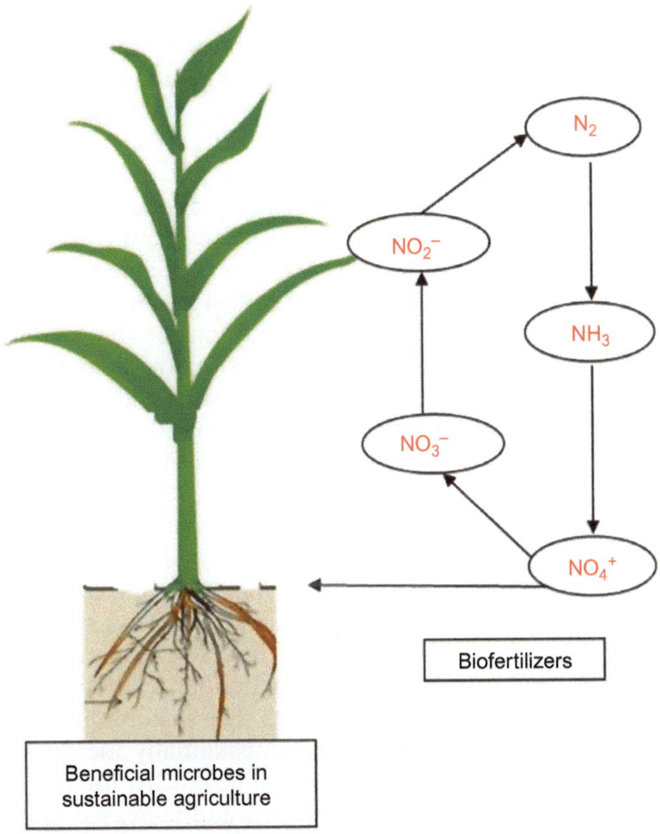

FIG. 15.3 Mechanism of the interaction of plant and biofertilizers.

15.5 GLOBAL MARKET REPORTS OF BIOALTERNATIVES

The biopesticide market is segmented into bioherbicides, bioinsecticides, and biofungicides based on the product type. With respect to ingredients, it is again divided into microbial, plant and biochemical-based biopesticides of which Bt-based microbial pesticides constitute over 50% of the market share followed by fungi, viruses, and nematodes (GBMAIF, 2017).

Production of 3000 tons/year of biopesticides is being seen globally, and this is increasing alarmingly. US market tops the biopesticide use with a sale of 200 products while 60 products in EU (Hubbard et al., 2014). The United States, Canada, and Mexico contribute to 45% of the biopesticide sale, and Asia occupies at least only 5% of the biopesticides sold all over the world (Bailey et al., 2010). Herbal substitutes of biopesticides like beshram, neem, garlic, triphala, and pinus kesiya are being processed and improved to increase consumption by farmers. Further, recombinant DNA (rDNA) technology (Fitches et al., 2004) has been deployed for improving the efficacy of biopesticides. In the year 2015, many biopesticides were introduced in the United States, Canada, Africa, and Australia by companies like Andermatt Biocontrol, Koppert Biological Systems, and Certis USA. Awareness of rhizobium-based fertilizers has been increasing globally especially the Unites States rather than using other biofertilizers that are of low-valued products. Since times are innumerable, rhizobia in particular *Bradyrhizobium japonicum* are being used as inoculants for soybean crops. In Canada, rhizobial strains are most accepted for legume cultivation, but other strains like *Bacillus subtilis, Bradyrhizobium japonicum, Delftia acidovorans*, and *Lactobacillus helveticus* are also being commercialized.

International Plant Nutrition Institute has revealed that the use of biofertilizers in Brazil is about 60–70 K tons/year. In Argentina, Paraguay, Bolivia, and Uruguay of South America, also nearly 70% of 30 million hectares of soybean crops were inoculated with *Bradyrhizobium* sps every year and are considered to be a large soybean market. A difference of 7.5% yield was observed between inoculated and noninoculated crops, whereas *Pseudomonas* sps and *Azospirillum brasilense* are recommended for wheat and maize with an expected increase in yield of 4%–9%. Although countries with advanced R&D and high-end technology like Japan, Taiwan, and Korea are moving more forward toward biofertilizer and biopesticide production, small-scale farmers due to literacy hurdles need to be educated on the comparative use of inorganic and organic constituents (FFTC, 2007).

Asian countries have opportunities for the development and commercialization of bio-alternatives contributing to food safety and environmental concerns. In India, more than 100,000 ha have been occupied by organic farming (Sekhar et al., 2016) supported by nearly 100 biofertilizer producers across the country and are receiving special focus in horticulture, oilseed, and medicinal plant cultivation. In China, 167 million hectares have been cultivated with biofertilizers, and over 300 producers are found all over. However, inoculants other than rhizobium were not found to be efficient. The rhizobium biofertilizer in Japan is Mamezo, and in Taiwan, farmers' economy was substantially increased (USD 27 million) from 1987 to 2006. Vietnam farmers have been using BioGro for rice, which resulted in brighter, less disease-prone, clean and stronger stems and increased grain yield. China and India adopt the larger use of pesticides as proposed by Biopesticides Industry Coalition of American (BPIA) and the European Association for International Producers of Biological Control Products (IBMA). Biopesticide market is emerging as is evidenced with the two top companies AgraQuest and Prophyta adopted by Bayer CropScience and made a significant impact on the biopesticide business sector. Kodiak Concentrate and Serenade SOIL were the products released in 2014, and apart from that, Valent BioSciences had a distribution agreement with North America for a biofungicide product constituting Monsanto and Novozymes (GBMAIF, 2017).

Only 2.8% of the pesticide market is occupied by biopesticides in Korea, and the farmers adopted a method of applying a mixture of egg yolk and cooking oil (EYCO) for pest control. It is cheaper than the chemical pesticide and also eco-friendly. The major biofertilizer industries in Brazil are Embrafós, Instituto de Fosfato Biológico (IFB), Biofosfatos do Brasil, and Liderfós, and IFB is the leading producer in the country. All their products are being applied to beans, carrot, sugarcane, rice, maize, soybean, eucalyptus, citrus, tomatoes, cotton, and forage crops with the aim of lowering soil acidification, having sustainable bioavailability for P for commercial crops, being cost-effective and nature friendly. Biofosfatos do Brasil utilizes sugarcane filter cake (by-product of ethanol and/or sugar production), mixed with PR, as the main raw material. IFB's research aims at new products in plant nutrition among which BioAtivo is the major product promoting biological N fixation by solubilizing phosphate rock (PR). This product has a patent number PI-9401724-7 by the Instituto Nacional de Propriedade Intelectual (Brazilian Institute for Intellectual Property). One of the main objectives is to utilize urban sewage sludge as raw material for the production of biofertilizer (Luis and Jesus, 2005).

15.6 INSIGHTS INTO THE PRODUCT AND APPLICATION IN VARIOUS REGIONS

Forty-five percent of the biopesticides sold was being used by the United States, Canada, and Mexico, whereas Asia is lagging behind with only 4% of the products all over the world (Bailey et al., 2010). The largest product segment was N fixing, which accounts 75% of revenue in 2014 and is encountered majorly by the regional markets of the United States and EU. The global pesticide market (product-incorporated protectants, nematicides, biofungicides, bioherbicides, and bioinsecticides) was over USD 3.3 billion in 2017 and is supposed to be projected to USD 9.5 billion by 2025. North America is leading to reach USD 3667.5 million by 2025 followed by Europe and Asia Pacific with a rapid extension in the coming years. Countries of Southeast Asia especially India and China explore the use of biopesticide potential in fruit production (GBMAIF, 2017). The demand relies on the improvement of soil N yield through microbes in these regions. Fifteen percent of the global revenue in this year was accomplished by phosphate solubilizers as the second largest product segment and is expected to be growing at the fastest rate from 2015 to 2022. The ability of soluble bacteria to hydrolyze organic and inorganic phosphate may be responsible for the growing popularity of the product and boosting in the forecast period. The next segment is occupied by zinc and silicate solubilizers wherein they degrade silicates and aluminum silicates of the soil and enhance metabolism of plant by improving the flow of phosphate ions (Report, 2018).

The largest application accounting for 65% of market in 2014 was seed treatment wherein biofertilizer treated seeds increase the yield besides resisting bacterial and viral attacks. The adaptability in the method of seed treatment, harnessing atmospheric N to soil, may be expected for the driving demand in the forecast period. In developing countries like India, the demand for organic food has been expected to rise by USD 1 billion from 2014 to 2020 (Report, 2018). Fifteen percent of the Asian Pacific market was expected to rise due to biofertilizers, and China is the largest agricultural market of maize, where the organic product manufacturers were exempted from VAT and excise and agricultural tax. North America and Europe together constitute nearly 54% of the global revenue with an increasing awareness of the organic food and the chemical hazards (Report, 2018). "Nitragin" is a laboratory culture of rhizobia meant for value addition to seed with an aim to increase the yield of forage crops like alfalfa, sweet clover, and clover.

In India, maize occupies the third important position after rice and wheat and is known to be the "queen of cereals" attributing to its highest genetic yield potential. This was developed with Nitragin Gold Technology by Canada and is in use only in Canada and the United States. In India, the first study and commercialization on rhizobium was done in 1956 by VN Joshi, but the real effort to promote and popularize the product was set up under the National Project on Development and Use of Biofertilizers (NPDB) by the Ministry of Agriculture under the Ninth Five-Year Plan (Rana et al., 2013). The National Biofertilizer Development Center was developed at Ghaziabad, India, and the marketing was improved in three ways with the following:

– State governments via district level officers and workers at village level
– State marketing federation via cooperative bodies and farmers
– State agro-industry cooperation via the agro service centers

15.7 LIMITATIONS IN PROMOTION OF BIOFERTILIZER AND BIOPESTICIDE CONSUMPTION

1. Resource constraints:
   - The lack of suitable resource constraints/unavailability of suitable strains
   - Ignorance of manufacturer on the quality of product
   - Nonavailability of quality resource/quality control methods
   - Infrastructural constraints like space availability for establishment of laboratory, production, and storage
   - The lack of cold storage equipment for inoculum
2. Market constraints:
   - The lack of farmers' awareness/quality assurance and constraints on marketing
   - Financial constraints related to banks
   - Smaller production units suffer due to less returns on sale with respect to the loan payment
3. Production constraints:
   - The lack of research and development for quality assurance and hence limited production
   - Inability to meet the seasonal demand/requirement due to shortage of skilled personnel
4. Field constraints:
   - Unfavorable environmental factors like soil and climatic factors, microbial factors, and population
   - No immediate visual results as that of inorganic fertilizers
   - The lack of awareness in cropping operations and short span of sowing/plantations in specific locality
5. Technical constraints:
   - Unauthenticated inoculum/microbes, mishandling techniques, and mutations
   - The lack of awareness on the benefits of technology
   - Problems in the adoption of technology due to different methods of inoculation

15.8 POLICIES & PROGRAMS

Many of the governments globally have been investing in sustainable farming pertaining to the ill effects of chemical fertilizers and pesticides. Many policies are being made in recent years to promote organic farming in different countries. A research report by Global Market

Insights has stated that the biofertilizer market share is expected to rise to USD 3 billion by 2024 and the biopesticides market is to reach USD 6.60 billion by 2022. The bionematicide segment is expected to have the highest growth rate from 2017 to 2022. Highly streamlined product registration may be responsible that North America dominates the biopesticide market by 2022, and the ban of neonicotinoids makes Europe the fastest growing biopesticide market by 2022. In European countries, 30% of the budget has been spent toward green payment to farmers under the "Common Agricultural Policy" in promoting bioproducts. The Indian Government has introduced many policies and plans to uplift the market of bioalternatives for sustainable agriculture. A number of schemes like National Mission of Sustainable Development (NMSA)/Paramparagat Krishi Vikas Yojana, Rashtriya Krishi Vikas Yojana (RKVY), National Mission on Oilseeds and Oil Palm (NMOOP), and Indian Council of Agricultural Sciences (ICAR) have been introduced. The "National Organic Program" (NOP) has been implemented by the US Department of Agriculture (USDA) aiming to develop standards in organically grown agricultural products and their growth in the forecast period.

The state and central government agencies in Taiwan encourage extensive use of biofertilizers especially in soybean cultivation, which could be consumed as vegetable, peanut, and red bean so as to support sustainable agriculture. The policy includes the application of biofertilizers involving rhizobial, P-solubilizing microbial inoculants so as to improve N fixation by biological means. The program insists on higher yield and better quality of horticultural crops by conducting awareness programs to farmers regarding production and application of inoculants and their benefits (Young and Chao, 1983). Research has been taken up by Indonesian agricultural sector in response to decreased yield in rice due to diseases. Increasing the productivity by improved soil health by means of composted straw combined with consortia of biofertilizers has significantly increased the grain yield (Simarmata et al., 2016). Product formulation, delivery, and commercialization are still at its early stage, and hence, production by natural means needs to be emphasized. However, biopesticides were largely regulated by the system adopted by chemical pesticides and so barriers to market entry by imposing higher costs (Kumar and Singh, 2014).

15.9 FUTURE PROSPECTS

Global demand for healthier food attributes to promoting awareness in organic farming, and the indiscriminate use of agrochemicals can bring out sustainable agriculture, which can ultimately overcome environmental pollution. Benefits of bioproducts include a high potential for export, less contamination, and longer shelf life; however, care should also be taken to select vital microbe/inoculum, which can best serve the purpose and transfer the technology of biofertilizer and biopesticide formulation to the industry. For this, a regulatory body has to be formed with the aim of focusing on the research and development and also marketing strategies (Safir, 1994), and the governments should come up with a policy for safeguarding of the ecosystem as of the US Environmental Protection Agency (Kumar and Singh, 2015).

15.10 CONCLUSION

The major drawback in the promotion of biopesticides is the lack of their profile. Awareness has to be created in farmers on the use of biofertilizers and biopesticides by making them part of workshops and seminars, and however, the inorganic chemical fertilizers used have been reduced to 30%–50%. One way of improving plant growth and yield of crop is to use a mixture of biofertilizers so as to enhance the benefits of synergistic effect of the components. However, biofertilizers with multiple strains may be avoided for compatibility issues (Khan et al., 2011b).

References

Al-Zaidi, A.A., Elhag, E.A., Al-Otaibi, S.H., Baig, M.B., 2011. Negative effects of pesticides on the environment and the farmers awareness in Saudi Arabia: a case study. J. Anim. Plant Sci. 21, 605–611.

Ambrosini, A., Beneduzi, A., Stefanski, T., Pinheiro, F.G., Vargas, L.K., Passaglia, L.M., 2012. Screening of plant growth promoting rhizobacteria isolated from sunflower (*Helianthus annuus* L.). Plant Soil 356, 245–264.

Arriola, K.G., Queiroz, O.C., Romero, J.J., Casper, D., Muniz, E., Hamie, J., Adesogan, A.T., 2015. Effect of microbial inoculants on the quality and aerobic stability of bermudagrass round-bale haylage. J. Dairy Sci. 98, 478–485.

Bailey, K.I., Boyetchko, S.M., Langle, T., 2010. Social and economic drivers shaping the future of biological control: a Canadian perspective on the factors affecting the development and use of microbial biopesticides. Biol. Control 52, 221–229.

Braccini, A.L., Demoraes Dan, L.G., Piccinin, G.G., Albrecht, L.P., Barbosa, M.C., Ortiz, A.H.T., 2012. Seed inoculation with *Azospirillum brasilense* associated with the use of bioregulators in maize. Rev. Caatinga 25, 58–64.

Brusamarello-Santos, L.C., Gilard, F., Brulé, L., Quilleré, I., Gourion, B., Ratet, P., Maltempi de Souza, E., Lea, P.J., Hirel, B., 2017. Metabolic profiling of two maize (Zea mays L.) inbred lines inoculated with the nitrogen fixing plant-interacting bacteria *Herbaspirillum seropedicae* and *Azospirillum brasilense*. PLoS One 12. e0174576.

Curá, J.A., Franz, D.R., Filosofía, J.E., Balestrasse, K.B., Burgueño, L.E., 2017. Inoculation with *Azospirillum* sp. and *Herbaspirillum* sp. bacteria increases the tolerance of maize to drought stress. Microorganisms 5 (3). pii: E41.

FFTC, 2007. Appropriate use of bio-fertilizers and bio-pesticides for small-scale farmers in Asia. Annual Report.

Fitches, E., Edwards, M.G., Mee, C., Grishin, E., Gatehouse, A.M.R., et al., 2004. Fusion proteins containing insect-specific toxins as pest control agents: snowdrop lectin delivers fused insecticidal spider venom toxin to insect haemolymph following oral ingestion. J. Insect Physiol. 50, 61–71.

Forlain, G., Pastorelli, R., Branzoni, M., Favilli, F., 1998. Root colonization efficiency, plant growth promoting activity and potentially related properties in plant associated bacteria. J. Genet. Breed. 49, 343–351.

Forlani, L., Juárez, M.P., Lavarías, S., Pedrini, N., 2014. Toxicological and biochemical response of the entomopathogenic fungus *Beauveria bassiana* after exposure to deltamethrin. Pest Manag. Sci. 70, 751–756.

GBMAIF, 2017. Global Biopesticides Market Analysis and Industry Forecasts 2017–2022—A $7.62 Billion Opportunity. News provided by Research and Markets, Dublin, September 22, 2017 07:00 ET.

Ghosh, N., 2004. Promoting biofertilisers in Indian agriculture. Econ. Polit. Wkly. 5, 5617–5625.

Hegde, S.V., 2008. Liquid bio-fertilizers in Indian agriculture. Bio-Fertilizer News Letter, 17–22.

Hubbard, M., Hynes, R.K., Erlandson, M., Bailey, K.L., 2014. The biochemistry behind biopesticide efficacy. Sustain. Chem. Process. 2, 18.

Khan, T.A., Mazid, M., Mohammad, F., 2011a. Role of ascorbic acid against pathogenesis in plants. J. Stress Physiol. Biochem. 7, 222–234.

Khan, T.A., Mazid, M., Mohammad, F., 2011b. Ascorbic acid: an enigmatic molecule to developmental and environmental stress in plant. Int. J. Appl. Biol. Pharm. Technol. 2, 468–483.

Kumar, S., Singh, A., 2014. Biopesticides for integrated crop management: environmental and regulatory aspects. J. Biofertil. Biopestici. 5. e121.

Kumar, S., Singh, A., 2015. Biopesticides: present status and the future prospects. J. Biofertil. Biopestici. 6, 2.

Luis, E.F.R., Jesus, C.M., 2005. Bacterial biofertilizers. In: Siddiqui, Z.A. (Ed.), PGPR: Biocontrol and Biofertilization. Springer, Dordrecht, pp. 143–172.

Malusá, E., Vassilev, N., 2014. A contribution to set a legal framework for biofertilisers. Appl. Microbiol. Biotechnol. 98, 6599–6607.

Marek-Kozaczuk, M., Wielbo, J., Pawlik, A., Skorupska, A., 2014. Nodulation competitiveness of *Ensifer meliloti* alfalfa nodule isolates and their potential for application as inoculants. Pol. J. Microbiol. 63, 375–386.

Mazid, M., Khan, T.A., Mohammad, F., 2011a. Potential of NO and H_2O_2 as signalling molecules in tolerance to abiotic stress in plants. J. Ind. Res. Technol. 1, 56–68.

Mazid, M., Khan, T.A., Mohammad, F., 2011b. Cytokinins, A classical multifaceted hormone in plant system. J. Stress Physiol. Biochem. 7 (4), 347–368.

Mohammadi, K., Sohrabi, Y., 2012. Bacterial biofertilizers for sustainable crop production: a review. J. Agric. Biol. Sci. 7, 307–316.

Puente, M.L., García, J.E., Alejandro, P., 2009. Effect of the bacterial concentration of *Azospirillum brasilense* in the inoculum and its plant growth regulator compounds on crop yield of corn (*Zea mays* L.) in the field. World J. Agric. Sci. 5 (5), 604–608.

Rana, R., Ramesh, Kapoor, P., 2013. Biofertilizers and their role in agriculture. Pop. Kheti 1, 56–61.

Report, 2018. Biofertilizers Market Size, Share & Trends Analysis Report by Product (Nitrogen Fixing, Phosphate Solubilizing), by Application (Seed Treatment, Soil Treatment), and Segment Forecasts, 2012–2022. ID: 978-1-68038-038-5.

Safir, G.R., 1994. Involvement of cropping systems, plant produced compounds and inoculum production in the functioning of VAM fungi. In: Pfleger, F.L., Linderman, R.G. (Eds.), Mycorrhizae and Plant Health. APS Press, St. Paul, MN, pp. 239–259.

Saikia, S.P., Jain, V., 2007. Biological nitrogen fixation with non-legumes: an achievable target or a dogma. Curr. Sci. 92, 317–322.

Sekhar, M., Riotte, J., Ruiz, L., Jouquet, P., Braun, J.J., 2016. Influences of climate and agriculture on water and biogeochemical cycles: Kabini critical zone observatory. Proc. Indian Natl. Sci. Acad. 82, 833–846.

Simarmata, T., Hersanti, Turmuktini, T., Fitriatin, B.N., Setiawati, M.R., Purwanto, 2016. Application of bioameliorant and biofertilizers to increase the soil health and rice productivity. HAYATI J. Biosci. 23, 181–184.

Sumita, P., Singh, H.B., Farooqui, A., Rakshit, A., 2015. Fungal biofertilizers in Indian agriculture: perception, demand and promotion. J. Eco-friendly Agri. 10, 101–113.

Young, C.C., Chao, C.C., 1983. In: Legume production and application of Rhizobium in Taiwan. Proceedings, ROC-JAPAN Seminar on Promoting Nitrogen Fixation in Agriculture, pp. 39–44.

CHAPTER

16

Role of Plant Growth-Promoting Microorganisms as a Tool for Environmental Sustainability

Shameer Syed, N.V.K.V. Prasad Tollamadugu

Nanotechnology Laboratory, Institute of Frontier Technology, Regional Agricultural Research Station, Acharya N.G. Ranga Agricultural University, Tirupati, India

OUTLINE

16.1 INTRODUCTION

Environmental sustainability is the rates of renewable resource harvest, pollution creation, and nonrenewable resource depletion that can be continued indefinitely (Daly, 1990; Morelli, 2011) and is the most vital for the survival of living organisms (Díaz et al., 2006). Human activities like industrialization, deforestation, and overexploitation of natural resources by exploding population, depriving agricultural practices, and much more have led the environment to the brink of collapse driving world nations to arms to prevent further deterioration of the environment and restore its stability (Hopwood et al., 2005; Tilman and Clark, 2014). Many approaches like waste recycling and effluent treatment systems have been tried to alleviate the situation but only met with little or no success. Agriculture, in the beginning, was an eco-friendly activity, but with the

https://doi.org/10.1016/B978-0-12-816328-3.00016-7

advent of the green revolution, chemical fertilizers and pesticides dominated and increased its negative footprint on the environment (Horrigan et al., 2002). As a remedy for this, microbial agents capable of plant growth promotion and protection collectively defined as plant-growth-promoting microorganisms have been proposed, studied, and developed ever since the human activities showed visible impact on the environment (Bass and Dalal-Clayton, 2012). PGPMs have been the most extensively studied group of beneficial microorganisms (in agriculture), but they have acquired popularity recently owing to the increasing awareness toward the aftermath of environmental pollution and increasing research revealing versatile applications of PGPMs toward agriculture and the environment alike, thus decreasing the impact of agriculture on the environment (Garnett et al., 2013).

PGPMs have such multipurpose appliance owing to the definition, which allows a vast group of microorganisms comprising bacteria, fungi, and algae to be termed PGPMs (Schäfer and Adams, 2015). Being composed of such diversity, the PGPMs can have a multitude of potential and yet to be tapped (Saharan and Nehra, 2011). As summarized in Fig. 16.1, PGPMs have several potential applications for agroenvironmental sustainability. But still, the PGPMs are not far reached in utilization due to inconsistency due to limited colonization ability or

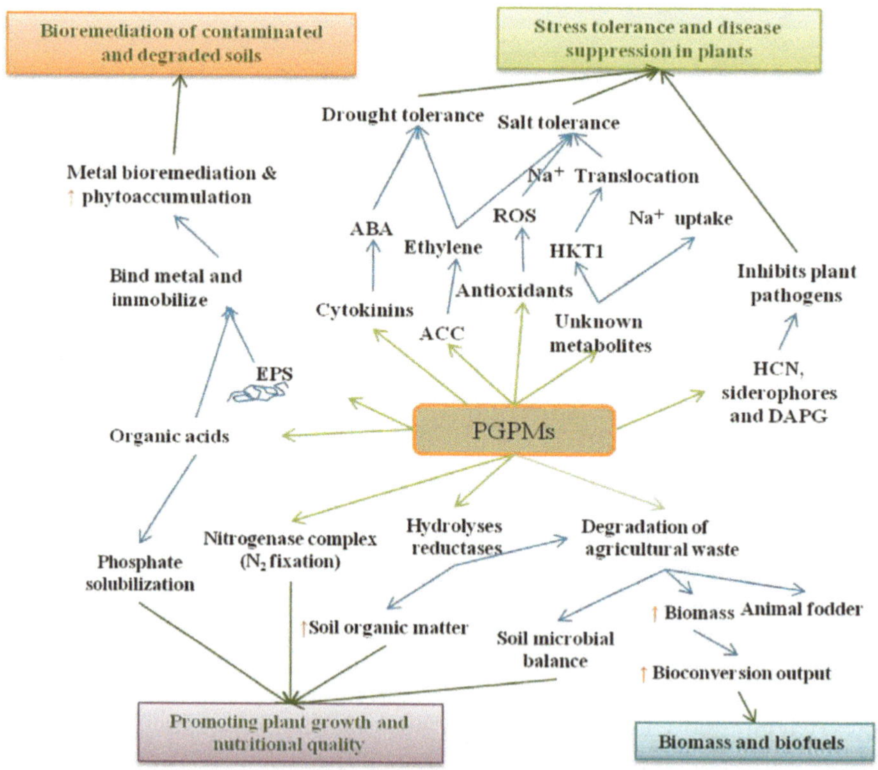

FIG. 16.1 Potential applications of plant-growth-promoting microorganisms for agroenvironmental sustainability.

specificity in colonization, the lack of awareness, socioeconomic hindrances, the desired level of PGPM's activity set too high, etc. These blockades are being slowly overcome through creating awareness toward the benefits of PGPMs and damages of chemical fertilizers along with the development of PGPMs to cater the needs of the modern agriculture by combining state-of-the-art technology with basic traits (plant growth promotion and protection) of the microbial world to create balanced products that are affordable and more effective, in the form of a consortium of different genera of PGPMs with different characteristic plant-promoting traits and complimenting each other (Sachs, 2012; Sarma et al., 2015). The modern technology can be employed to improve the production (genetic engineering) or isolate the effector compounds/molecules and directly apply it to the crop plants, eliminating the microbial agent that sometimes is an opportunistic pathogen and adding more recent approaches like the application of nanofertilizers or encapsulated nanopesticides (Fulekar et al., 2014; Grillo et al., 2016).

16.2 PGPMs IN AGRICULTURE

PGPMs are a vast number of bacteria and fungi that live in close association with the plant, particularly in the root system, and have a positive influence on the plant, including *Rhizobia, Bacillus, Pseudomonads, Acetobacter, Nitrosomonas, Nitrobacter, Bradyrhizobium, Mesorhizobium, Serratia, Azotobacter, Brevibacillus, Enterobacter, Azospirillum, Serratia, Paenibacillus,* and *Stenotrophomonas* (Rajkumar et al., 2009; Kumar et al., 2013; Weselowski et al., 2016). Along with these, there are also fungal species (*Trichoderma, Penicillium,* and *Streptomyces*) that have a beneficial action on the plant and are well established (Pellegrino and Bedini, 2014). The PGPMs have multifaceted applications in agriculture broadly sectioned into direct and indirect. These activities include enhancing the availability of essential nutrients (N_2 fixation and P and K solubilization), growth regulation via phytohormones (auxins, gibberellins, and cytokinins) coming under direct mechanisms exhibited by the PGPMs toward agricultural crop plants (Dubey et al., 2016).

Production of antibiotics against plant pathogens (butyrolactones, zwittermicin A, kanosamine, oligomycin A, oomycin A, phenazine-1-carboxylic acid, pyoluteorin, pyrrolnitrin, viscosinamide, xanthobaccin, phenazines, 2,4-diacetylphloroglucinol, pyrrole compounds, and cyclic lipopeptides) (Keswani et al., 2014; Lee et al., 2016), induced stress resistance, and protection from abiotic agents (drought, salinity, and temperature) through the production of primary and secondary metabolites (trehalose, antifreeze proteins, and IAA and ACC deaminase) (Kaushal and Wani, 2016) are considered indirect activities of PGPMs for plant protection. Many researchers have reported that with the use of PGPMs in agriculture be it in fertile, well-irrigated soils or not, the demand and input quantity of fertilizers and allied chemical additives can be reduced to half the original when PGPMs are employed (Yobo et al., 2009), reducing the environmental footprint of agriculture. The PGPMs reduce the drought stress on plants and therefore lower the water utilization and allow waterlogging required crops to be grown in the low-water-level area also (Kaushal and Wani, 2016). As stated in Table 16.1, there are several commercially available microbial (biocontrol) products useful for plant disease management worldwide.

TABLE 16.1 Commercial Biocontrol Products Used for Plant Disease Management Worldwide

Trade Name	Biocontrol Agent	Target Pathogen/ Disease/Crop	Type of Formulation	Mode of Action	Delivery/ Application	Manufacturer
AQ10 Biofungicide	*Ampelomyces quisqualis* M-10	Powdery mildew	Water-dispersible granules	Primarily by antibiosis (production of antibiotic like products)	Spray	Ecogen, Inc., 2005 Cabot Blvd., West Israel
Bio-fungus	*Trichoderma* sp.	*Sclerotinia*, *Phytophthara*, *Rhizoctonia solani*, *Pythium* sp., *Fusarium* sp.	Granular wettable powder, sticks, and crumbles	Antagonism and competitive inhibition of pathogen	After fumigation, incorporation into soil by spraying or injection	De Ceuster, Meststoffen NV, Belgium
Aspire	*Candida oleophila* I-182	*Botrytis* sp., *Penicillium* sp.	Wettable powder	Colonizes fruit surfaces and avoids establishment of saprophytes and other yeast	Postharvest to fruit as drench, drip, or spray	Ecogen, Inc., 2005 Cabot Blvd., West Israel
Binab T	*Trichoderma harzianum* (ATCC 20476) and *T. polysporum* (ATCC 20475)	Wilt, root rot, stem rots of woody plants	Wettable powder and pellets	Antagonism is the primary mode of pathogen inhibition	Spray, inserting pellets in wounds, painting on wounds	Bio-Innovation AB, Sweden and Henry Doubleday research, United Kingdom
Biofox C	*Fusarium oxysporum* (nonpathogenic)	*Fusarium oxysporum*, *F. moniliforme*	Dust or alginate granules	Competitive inhibition and antagonistic effect of secondary metabolites	Seed treatment or soil incorporation	SIAPA, Via Vittorio Veneto 1 Galliera, Italy
Bio-save 110	*Pseudomonas syringae* ESC-11	*Botrytis cinerea*, *Penicillium* sp., *Mucor piriformis*, *Geotrichum candidum*	Frozen cell concentrated pellets	Inhibits the colonization of fungi and molds through competition and secondary metabolites	Pellets, postharvest, to fruit as drench, dip, or spray	EcoScience Corp., United States
BlightBan A506	*Pseudomonas fluorescens* A506	Frost damage, *Erwinia amylovora*, and rust-inducing bacteria	Wettable powder	Competitive inhibitor of other bacteria and biofilm prevents frost damage	Spray during bloom time on flower and fruits	Plant Health Technologies, United States
Cedomon	*Pseudomonas chlororaphis*	Leaf stripe, net blotch, *Fusarium* spp., spot blotch, leaf spot	Seed treatment	Competition for nutrients and space, growth stimulation, induced resistance, and antibiosis	Seed dressing	BioAgri AB, Sweden

Companion	*Bacillus subtilis* GB03	*Rhizoctonia*, *Pythium*, *Fusarium*, and *Phytophthora* sp.	Liquid	Competitive inhibition pathogen and ISR in host	Drench at time of seeding and transplanting or spray on turf	Growth Products, United States
Conquer	*Pseudomonas fluorescens*	*Pseudomonas tolaasii*	Liquid	Antagonism	Spray	Mauri Foods, Australia
Contans	*Coniothyrium minitans*	*Sclerotinia sclerotiorum* and *S. minor*	Water-dispersible granules	Competitive destruction of pathogen	Spray	Prophyta Biologischer Pflanzenschutz, Germany
Deny	*Burkholderia cepacia*	*Rhizoctonia*, *Pythium*, *Fusarium* sp.	Peat-based solid biomass from solid fermentation and liquid suspension	Produces antagonistic metabolites and plant hormone production	Applied to seed in a sticking box, aqueous suspension for drip irrigation	Stine Microbial Products, United States
Epic	*Bacillus subtilis*	*Rhizoctonia solani*, *Fusarium* spp., *Alternaria* sp., *Aspergillus* sp. that attack roots	Dry powder	Fungicide and antagonist, produces antifungal antibiotics	Added to a slurry; mix with a chemical fungicide for commercial seed treatment	Gustafson, Inc., United States
Fusaclean	*Fusarium oxysporum* (nonpathogenic)	*Fusarium oxysporum*	Spore, microgranules	Competition and antibiosis	In drip to rock wool, in potting mix	Natural Plant Protection, France
Galltrol-A	*Agrobacterium radiobacter* strain 84	Crown gall caused by *A. tumefaciens*	Petri plates with pure culture grown on agar plates	Competitive inhibition and antibiosis	Bacterial mass from one plate to 1 gal of water, seed drenching, seedlings, cuttings, roots	AgBioChem, Inc., United States
Intercept	*Pseudomonas cepacia*	*Rhizoctonia solani*, *Fusarium* sp., *Pythium* sp.	Liquid suspension	Antagonism and competitive suppression of pathogens	Spray, drench, or through irrigation	Soil Technologies Corporation, United States
Kodiak, Kodiak HB, Kodiak AT	*Bacillus subtilis*	*Rhizoctonia solani*, *Fusarium* spp., *Alternaria* sp., *Aspergillus* sp. that attack roots	Dry powder, usually applied with chemical fungicide	Growth promotion, antagonistic to pathogens and imparts stress tolerance	Added to a slurry for seed treatments	Bayer Crop Science, United States

Continued

TABLE 16.1 Commercial Biocontrol Products Used for Plant Disease Management Worldwide—cont'd

Trade Name	Biocontrol Agent	Target Pathogen/ Disease/Crop	Type of Formulation	Mode of Action	Delivery/ Application	Manufacturer
Ecological niche	*Coniothyrium minitans*	*Sclerotinia sclerotiorum* and *S. minor*	Granules	Converts soil matter and influences soil parameters	Incorporated into soil or soilless mix granules	BIOVED, Ltd., Hungary
Monitor SD	*Trichoderma* sp.	Soilborne plant pathogens	Seed dresser	Parasitizes fungal pathogens, solubilizes nutrients, and promotes plant growth	Seed dressing	M/s Agriland Biotech Pvt. Ltd., India
Mycostop	*Streptomyces griseoviridis* strain K61	*Fusarium* spp., *Alternaria brassicicola*, *Phomopsis* sp., *Botrytis* sp., *Pythium* sp., and *Phytophthora* sp. that cause seed, stem, and root rot and wilt disease	Powder	Competitive inhibition, plant hormone production, and parasitizes	Drench, spray, or through irrigation system	Kemira Agro Oy, Finland
Nogall, Diegall	*Agrobacterium radiobacter*	*Agrobacterium tumefaciens*	Washed plates; culture suspensions	Competitive colonization and antagonism against *A. tumefaciens*	Root dips	Bio-Care Technology Pvt. Ltd., Australia
Norbac 84C	*Agrobacterium radiobacter* strain K84	Crown gall caused by *A. tumefaciens*	Aqueous suspension containing bacterial cells, methyl cellulose, and phosphate buffer	Competitive colonization and antagonism against *A. tumefaciens*	Root, stem, cutting dip, or spray	New BioProducts Inc., United States
Paecil	*Paecilomyces lilacinus*	Various nematode sp.	Dry spore concentrate	Antagonistic in nature to nematodes	Seedling or soil drench	Shakti Biotech, India
Polygandron	*Pythium oligandrum*	*Pythium ultimum*	Granule or powder		Seed treatment or soil incorporation	Výskumny ustav rastlinnej výroby, Slovak Republic
Primastop	*Gliocladium catenulatum*	*Pythium* sp., *R. solani* sp., *Botrytis* sp., *Didymella* sp.	Wettable powder		Drench and incorporation	Kemira Agro Oyj, Finland

Protus WG	*Talaromyces flavus*, isolate V117b	*Verticillium dahliae*, *V. alboatrum*, and *R. solani*	Water-dispersible powder containing ascospores	Parasitizes the pathogen upon contact	Soil or seed treatment, soil drench, and root application	Prophyta Biologischer Pflanzenschutz GmbH, Germany
FZB24	*Bacillus subtilis* FZB24	*Rhizoctonia solani*, *Fusarium* sp., *Alternaria* sp., *Sclerotinia*, *Verticillium*, *Streptomyces scabies*	Water-dispersible granules	Produces plant hormones and exhibits antibiotic-based antagonism	As suspension for seed treatment, soil drench, drip and addition to nutrient solutions	KFZB Biotechnik, Germany
Root Pro	*Trichoderma harzianum*	*Rhizoctonia solani*, *Pythium* sp., *Fusarium* sp., and *Sclerotium rolfsii*	Fungal spores mixed with peat and other organic materials	Parasitizes the pathogen upon contact	Mixed into growing media at the time of seedling or transplanting	Mycontrol Ltd., Israel
RootShield	*Trichoderma harzianum* Rifai strain KRL-AG2	*Pythium* sp., *R. solani*, *Fusarium* sp.	Granule or wettable powder	Pathogen inhibition by lytic enzymes effecting pathogens	Granules mixed with potting medium, mixed with water, and added as soil drench	Bioworks Inc., Geneva
Rotstop, PG suspension	*Phlebia gigantea*	*Heterobasidion annosum*	Spores in inert powder	Selective antagonism	Spray	Kemira Agro Oyj, Finland
Serenade ASO	*Bacillus subtilis*	Powdery mildew, cercospora leaf spot, early blight, late blight	Wettable powder	Promotes systemic responses and promotes plant growth and stress resistance	Spray	AgraQuest, Inc., United States
Shakti SD	Bacterial sp.	Soilborne diseases of field crops	Powder	Produce antibiotics and exhibits antagonism and plant growth promotion	Seed dressing	Sangam Biodynamic, India
Spot-less	*Pseudomonas aureofaciens* strain Tx-1	*Pythium aphanidermatum*, anthracnose	Liquid	Antagonistic and produces lytic enzymes targeting pathogens	Overhead irrigation	Eco Soil Systems, Inc., United States
SoilGard	*Gliocladium virens* GL-21	Damping off and root rot pathogens especially *R. solani* and *Pythium* sp.	Granules	Parasitizes the pathogen and inhibits the pathogen establishment	Incorporated into soil or soilless media prior to seeding	Thermo Trilogy, United States

Continued

TABLE 16.1 Commercial Biocontrol Products Used for Plant Disease Management Worldwide—cont'd

Trade Name	Biocontrol Agent	Target Pathogen/ Disease/Crop	Type of Formulation	Mode of Action	Delivery/ Application	Manufacturer
Star-T WP	*Trichoderma* sp.	Soilborne pathogens of field crops, wilt of sugar cane, gummosis of citrus	Wettable powder	Competitive colonization and parasitic toward other fungi	Seed and soil	Sangam Biodynamic, India
Supresivit	*Trichoderma harzianum*	Various fungi	Powder and granules	Competitive antagonism by parasitizing the pathogen	Dip of cuttings and spray on seeds	Borregaard BioPlant, Denmark
System 3	*Bacillus subtilis* GB03 and chemical pesticides	Seedling pathogens	Dust	Antibiosis and competitive colonization	Seed treatment in planter box	Helena Chemical Co., United States
Trieco	*Trichoderma viride*	*Rhizoctonia* spp., *Pythium* sp., *Fusarium* sp.	Powder	Competitive antagonism by parasitizing the pathogen	Wet or dry seed dressing or soil drench	Ecosense Labs (I) Pvt. Ltd., India
Trichodex	*Trichoderma harzianum*	Primarily *B. cinerea*, also *Colletotrichum* spp., *Fulvia fulva*, *Monilinia laxa*, *Plasmopara viticola*, *Pseudoperonospora cubensis*, *Rhizopus stolonifer*, *S. sclerotiorum*	Wettable powder	Antagonistic suppression of pathogens and parasitism of the pathogen	Spray	Makhteshim Chemical works, Israel
Trichopel	*Trichoderma harzianum* and *T. viride*	*Armillaria*, *Botryosphaeria*, *Chondrosternum*, *Fusarium*, *Nectria*, *Phytophthora*, *Pythium*, *Rhizoctonia*	Wettable powder	Competitive colonizer exhibiting antagonism by parasitizing the pathogen	Soil incorporation and mix with potting medium	Agrimm Technologies, Ltd., New Zealand
Victus	*Pseudomonas fluorescens* NCIB 12089	Soilborne pathogens	Aqueous suspension of fermentation broth	Pathogen suppression by antibiosis and hydrolytic enzymes	Spray or seed drenching and soil addition	Sylvan Spawn, United States

Subtiliex	*Bacillus subtilis* isolated from *Vicia faba* phylloplane	*Fusarium*, *Rhizoctonia*, *Alternaria*, *Aspergillus*	Aqueous suspension	Nutrient solubilization and antagonism	Soil incorporation, seed dressing, and spray or drip	Becker Underwood Saskatoon, Canada
RhizoVital 42 l	*Bacillus amyloliquefaciens* strain FZB24	Cereals, pulses, corn, vegetables (potatoes, carrots, lettuce, etc.)	Liquid formulation	Production of phytohormones and phytase enzyme	Spray and drenching	ABiTEP GmBh, Germany
Bisolbisan, F	*Bacillus subtilis*, strain Ch-13 + metabolites on silica micro granules	Wheat, potatoes, carrots, leafy vegetables		Nutrient solubilization (organic acid production), enzymes (phosphatases, phytases)		Biosolbi microbial preparations, Russia
Vitormone	*Azotobacter* sp.	Grapes, pomegranate, onion, and vegetables	Liquid formulation	N_2 fixation along with antifungal metabolites, amino acids, and "B" group vitamins that are absorbed easily by the leaves	Foliar application	Kan Biosys, India
TeamBio-3	Consortia of *Azotobacter* (NFB), *B. polymyxa* (PSB), and *Bacillus* sp. (KMB)	Grapes, pomegranate, vegetables in polyhouses	Liquid, wettable powder, and granules	Nutrient augmentation (N, P, and K), competitive colonization leading to disease resistance	Spray or soil incorporation	Kan Biosys, India
Tricho-Shield COMBAT	*Trichoderma viride* spores	Cowpea, mung bean, urd bean, chili seedlings, cowpea, sesamum, and pigeon pea	Wettable powder	Antifungal activity through competition and antagonistic activity	Root zone drenching, soil, and foliar application	Kan Biosys, India
Sudo	*Pseudomonas fluorescens*	Rice, chili seedlings, banana, tomato, chickpea, black gram, sesamum, groundnut, and sugarcane	Powder	Antagonistic to fungal pathogens via primary and secondary metabolites	Soil incorporation	Kan Biosys, India

Continued

TABLE 16.1 Commercial Biocontrol Products Used for Plant Disease Management Worldwide—cont'd

Trade Name	Biocontrol Agent	Target Pathogen/ Disease/Crop	Type of Formulation	Mode of Action	Delivery/ Application	Manufacturer
Speed Kompost	*Aspergillus, Penicillium, Chaetomium, Trichoderma, Paecilomyces, Streptomyces, Bacillus, Azotobacter* dynamic formulation	Organic-carbon-depleted soils	Powder	Cellulolytic, amylolytic, proteolytic enzyme production to increase the organic matter in soils	Addition to organic material before soil application	Kan Biosys, India
ABTEC Pseudo Chitinase	*Pseudomonas fluorescens* (chitinase induced)	Ginger, turmeric, cardamom, vegetables	Liquid formulation	Chitinolysis-mediated suppression of fungal pathogens *Phytophthora* sp., *Pythium* sp., *Sclerotium* sp., *Rhizoctonia* sp., and *Fusarium*	Seed dressing, soil incorporation, and foliar spray	Agro Bio-Tech Research Centre Ltd., India
ABTEC Btk	*Bacillus thuringiensis* var. *kurstaki*	Insecticide for different crops	Wettable powder	Production of delta endotoxin	Foliar spray	Agro Bio-Tech Research Centre Ltd., India
PSEUDO-GUN	*Pseudomonas* sp.	Leaf blight, sheath blight, damping off, panama wilt, seedling rot, and other diseases of plant	Powder and granules	Produce phenazine type antibiotics and induce systemic resistance in host plant	Soil incorporation	Maharashtra biofertilizers, India
BIO-PHORCE NO. 1	Phosphate-solubilizing bacteria	Biofertilizer for pulses and legume crops	Wettable powder	Solubilize phosphates and produce growth hormones like IAA and GA	Seed dressing	Maharashtra biofertilizers, India
SANJIVANI-N3	*Azospirillum* sp.	Biofertilizer for pulses and legumes and ornamental plants	Powder and granules	Nitrogen fixation, production of (IAA), disease resistance, and drought tolerance	Band placement/ side dress application at the time of planting or during early stages of the cropping	Sushila Biofertilizer Company Ltd., India

16.3 PGPMs IN BIOREMEDIATION

Soil serves as a major component of environmental sustainability as it is the only route to obtain nutrition for the primary food producers on earth, that is, plants, on which all others depend directly or indirectly. Hence, soil contamination with a multitude of compounds, including organic (polychlorinated dibenzo-*p*-dioxins, polychlorinated biphenyls, and polycyclic aromatic hydrocarbons) and inorganic (Cd, Cr, Ni, Pb, Cu, and As), along with a vast number of compounds comprising both organic and inorganic components (pesticides, herbicides, untreated industrial waste, construction waste, etc.) is a reason for great concern (Banwart, 2011). To some extent, agricultural practices have added to the menace of soil contamination, by means of excessive use of pesticides, fertilizers, etc. This escalating soil pollution has been a challenge, and the measures to solve it were nowhere near to calm the situation as methods to remediate others like water bodies won't work here because of the physical dynamics of the soil. Microbiological agents like bacteria can form/participate in microhabitats and become persistent, leading to the formation of habitat niche that can immobilize metals, break down organic pollutants, etc. (Khan, 2005; Khan et al., 2009; Abhilash et al., 2012). Among the microorganisms, PGPMs are most potential in alleviating soil pollution, thereby restoring environmental sustainability (Burd et al., 2000; Kong and Glick, 2017). PGPMs employed in agriculture hold the immense potential for environmental bioremediation as they reduce the fertilizer dose and eliminate the use of pesticides without compromising on crop growth and yield (Hrynkiewicz and Baum, 2012). The majority of PGPMs are able to alter the mobility of contaminants, particularly heavy metals (biosorbing, bioaccumulation, precipitation, and chelation), thereby restraining the metal toxicity effects in plants (Ma et al., 2016; Khan et al., 2017). In another mode of action, the PGPMs alter metal itself so that they become less toxic to the plant by producing metal reductases, chelating complexes, and biomethylation (Kang et al., 2016; Dąbrowska et al., 2017). This mechanism of action along with the association of the plant root system enhances the metal accumulation in different parts of the plant through the roots and reduces the cumulative availability of the metal in the environment (Ullah et al., 2015; Moreira et al., 2016; Sobariu et al., 2016). Certain PGP fungi belonging to arbuscular mycorrhizal fungi (AMF) have both plant growth promotion and heavy-metal detoxification by accumulating the heavy metal in their hyphae (Harms et al., 2011). Along with PGPMs themselves operating as heavy-metal remediating agents when added to soils, they also increase the metal accumulation by the plants and augment phytoremediation of the metal-contaminated soils.

16.4 PGPMs IN BIOENERGY CROP/BIOMASS FOR ENERGY PRODUCTION

The bioenergy crops, that is, the crops solely grown for the production of biofuels, may contradict with the crops for food, as the land available is limited, requiring the role of PGPMs to enhance the bioenergy crop biomass even in unsuitable/contaminated soils allowing the fertile soils to be exploited for food crop cultivation (Contreras-Cornejo et al., 2009; Babu et al., 2014; Abhilash et al., 2016). It should be understood that PGPMs themselves might not be used as biofuel producers but they have tremendous potential in enhancing the biomass of bioenergy crops like miscanthus (*Miscanthus x giganteus*—a perennial grass), reed canary grass (*Phalaris arundinacea*), switch grass (*Panicum virgatum*), ryegrass (*Lolium perenne*), silver birch (*Betula pendula*), and duckweed (*Lemna trisulca*) (Mei and Flinn, 2010; Li et al., 2013; Zeilinger et al., 2016; Toyama et al., 2017). Even food crops like maize, sorghum, wheat, sugar beets, and oilseeds like canola can be produced in less fertile/contaminated soils through the assistance of PGPMs, thereby playing an indirect but vital role in the biofuel production (Sheng et al., 2012; Owen et al., 2015; Adak et al., 2016). The ability of PGPMs to enhance overall biomass has been demonstrated by many researchers (Weyens et al., 2009; Labbe et al., 2016). The fundamental trait that is essential in this facet is the production of phytohormones that induce growth and increase total biomass, thereby enhancing bioenergy potential (del Rosario Cappellari et al., 2017). Another aspect functional in increasing the biomass production is the ability to reduce the toxicity of different contaminants (heavy metals, organic pollutants, industrial effluents, etc.) on the plant, thereby protecting the crop from stress induced by them leading to a higher biomass production. The ability to accumulate/immobilize heavy metals is an imperative feature for bioenergy crop production in less fertile soils achieving a dual goal of producing a sustainable fuel source and remediation of the contaminated soil at once, balancing the challenge of food crop and bioenergy crop cultivation (Pandey et al., 2016).

16.5 FUTURE PROSPECTS

The future developments in the use of plant-growth-promoting microorganisms would be to overcome the fundamental hurdle faced, that is, acceptability among the farmers arising due to pathogenic interactions during handling of some of the opportunistic pathogens we employ as PGPMs. To this end, technologies like selective molecule application, instead of whole microbe, can be done if we are able to economically produce the target molecule on a large scale especially in the area of plant protection. The PGPMs should also be persistent; slow degrading materials can be employed to be a carrier for the PGPM formulation so that it serves in both as energy source enhancing the survival of the PGPM and matrix to which PGPM bind and thrive.

Acknowledgments

The authors are thankful to DST-SERB N-PDF division for providing the necessary financial funding under sanction order DST-SERB N-PDF/2015/000777 and Institute of Frontier Technology, Regional Agricultural Research Station, Acharya N.G. Ranga Agricultural University, Tirupati, for providing the necessary research facilities.

Conflict of Interest

We declare that there is no conflict of interest regarding the publication of this article.

References

Abhilash, P.C., Powell, J.R., Singh, H.B., Singh, B.K., 2012. Plant–microbe interactions: novel applications for exploitation in multipurpose remediation technologies. Trends Biotechnol. 30, 416–420.

Abhilash, P.C., Tripathi, V., Edrisi, S.A., Dubey, R.K., Bakshi, M., Dubey, P.K., Singh, H.B., Ebbs, S.D., 2016. Sustainability of crop production from polluted lands. Energy Ecol. Environ. 1, 54–65.

Adak, A., Prasanna, R., Babu, S., Bidyarani, N., Verma, S., Pal, M., Shivay, Y.S., Nain, L., 2016. Micronutrient enrichment mediated by plant-microbe interactions and rice cultivation practices. J. Plant Nutr. 39, 1216–1232.

Babu, A.G., Shim, J., Bang, K.S., Shea, P.J., Oh, B.T., 2014. *Trichoderma virens* PDR-28: a heavy metal-tolerant and plant growth-promoting fungus for remediation and bioenergy crop production on mine tailing soil. J. Environ. Manag. 132, 129–134.

Banwart, S., 2011. Save our soils. Nature 474, 151–152.

Bass, S., Dalal-Clayton, B., 2012. Sustainable Development Strategies: A Resource Book. Routledge, London.

Burd, G.I., Dixon, D.G., Glick, B.R., 2000. Plant growth-promoting bacteria that decrease heavy metal toxicity in plants. Can. J. Microbiol. 46 (3), 237–245.

Contreras-Cornejo, H.A., Macías-Rodríguez, L., Cortés-Penagos, C., López-Bucio, J., 2009. *Trichoderma virens*, a plant beneficial fungus, enhances biomass production and promotes lateral root growth through an auxin-dependent mechanism in Arabidopsis. Plant Physiol. 149, 1579–1592.

Dąbrowska, G., Hrynkiewicz, K., Trejgell, A., Baum, C., 2017. The effect of plant growth-promoting rhizobacteria on the phytoextraction of Cd and Zn by *Brassica napus* L. Int. J. Phytoremediation 19, 597–604.

Daly, H.E., 1990. Toward some operational principles of sustainable development. Ecol. Econ. 2, 1–6.

del Rosario Cappellari, L., Chiappero, J., Santoro, M.V., Giordano, W., Banchio, E., 2017. Inducing phenolic production and volatile organic compounds emission by inoculating *Mentha piperita* with plant growth-promoting rhizobacteria. Sci. Hortic. 220, 193–198.

Díaz, S., Fargione, J., Chapin III, F.S., Tilman, D., 2006. Biodiversity loss threatens human well-being. PLoS Biol. 4 (8). e277.

Dubey, R.K., Tripathi, V., Dubey, P.K., Singh, H.B., Abhilash, P.C., 2016. Exploring rhizospheric interactions for agricultural sustainability: the need of integrative research on multi-trophic interactions. J. Clean. Prod. 115, 362–365.

Fulekar, M.H., Pathak, B., Kale, R.K., 2014. Nanotechnology: perspective for environmental sustainability. In: Environment and Sustainable Development. Springer, India, pp. 87–114.

Garnett, T., Appleby, M.C., Balmford, A., Bateman, I.J., Benton, T.G., Bloomer, P., Burlingame, B., Dawkins, M., Dolan, L., Fraser, D., Herrero, M., Hoffmann, I., Smith, P., Thornton, P.K., Toulmin, C., Vermeulen, S.J., Godfray, H.C., 2013. Sustainable intensification in agriculture: premises and policies. Science 341, 33–34.

Grillo, R., Abhilash, P.C., Fraceto, L.F., 2016. Nanotechnology applied to bio-encapsulation of pesticides. J. Nanosci. Nanotechnol. 16, 1231–1234.

Harms, H., Schlosser, D., Wick, L.Y., 2011. Untapped potential: exploiting fungi in bioremediation of hazardous chemicals. Nat. Rev. Microbiol. 9, 177–192.

Hopwood, B., Mellor, M., O'Brien, G., 2005. Sustainable development: mapping different approaches. Sustain. Dev. 13 (1), 38–52.

Horrigan, L., Lawrence, R.S., Walker, P., 2002. How sustainable agriculture can address the environmental and human health harms of industrial agriculture. Environ. Health Perspect. 110, 445–456.

Hrynkiewicz, K., Baum, C., 2012. The potential of rhizosphere microorganisms to promote the plant growth in disturbed soils. In: Environmental Protection Strategies for Sustainable Development. Springer, Dordrecht, pp. 35–64.

Kang, C.H., Kwon, Y.J., So, J.S., 2016. Bioremediation of heavy metals by using bacterial mixtures. Ecol. Eng. 89, 64–69.

Kaushal, M., Wani, S.P., 2016. Plant-growth-promoting rhizobacteria: drought stress alleviators to ameliorate crop production in dry lands. Ann. Microbiol. 66, 35–42.

Keswani, C., Mishra, S., Sarma, B.K., Singh, S.P., Singh, H.B., 2014. Unraveling the efficient applications of secondary metabolites of various *Trichoderma* spp. Appl. Microbiol. Biotechnol. 98, 533–544.

Khan, A.G., 2005. Role of soil microbes in the rhizospheres of plants growing on trace metal contaminated soils in phytoremediation. J. Trace Elem. Med. Biol. 18, 355–364.

Khan, M.S., Zaidi, A., Wani, P.A., Oves, M., 2009. Role of plant growth promoting rhizobacteria in the remediation of metal contaminated soils. Environ. Chem. Lett. 7, 1–19.

Khan, A.R., Ullah, I., Waqas, M., Park, G.S., Khan, A.L., Hong, S.J., Ullah, R., Jung, B.K., Park, C.E., Ur-Rehman, S., Lee, I.J., Shin, J.H., 2017. Host plant growth promotion and cadmium detoxification in *Solanum nigrum*, mediated by endophytic fungi. Ecotoxicol. Environ. Saf. 136, 180–188.

Kong, Z., Glick, B.R., 2017. The role of plant growth-promoting bacteria in metal phytoremediation. Adv. Microb. Physiol. 71, 97–132.

Kumar, A., Munder, A., Aravind, R., Eapen, S.J., Tümmler, B., Raaijmakers, J.M., 2013. Friend or foe: genetic and functional characterization of plant endophytic *Pseudomonas aeruginosa*. Environ. Microbiol. 15, 764–779.

Labbe JL, Muchero W, et al. (2016) Complex of mutualistic microbes designed to increase plant productivity. U.S. Patent Application 15/334,793.

Lee, S., Yap, M., Behringer, G., Hung, R., Bennett, J.W., 2016. Volatile organic compounds emitted by *Trichoderma* species mediate plant growth. Fungal Biol. Biotechnol. 3, 7.

Li, G.X., Wu, X.Q., Ye, J.R., 2013. Biosafety and colonization of *Burkholderia multivorans* WS-FJ9 and its growth-promoting effects on poplars. Appl. Microbiol. Biotechnol. 97, 10489–10498.

Ma, Y., Rajkumar, M., Zhang, C., Freitas, H., 2016. Beneficial role of bacterial endophytes in heavy metal phytoremediation. J. Environ. Manag. 174, 14–25.

Mei, C., Flinn, B.S., 2010. The use of beneficial microbial endophytes for plant biomass and stress tolerance improvement. Recent Pat. Biotechnol. 4 (1), 81–95.

Moreira, H., Pereira, S.I., Marques, A.P., Rangel, A.O., Castro, P.M., 2016. Mine land valorization through energy maize production enhanced by the application of plant growth-promoting rhizobacteria and arbuscular mycorrhizal fungi. Environ. Sci. Pollut. Res. Int. 23, 6940–6950.

Morelli, J., 2011. Environmental sustainability: a definition for environmental professionals. J. Environ. Sustain. 1 (1), 2.

Owen, D., Williams, A.P., Griffith, G.W., Withers, P.J.A., 2015. Use of commercial bio-inoculants to increase agricultural production through improved phosphrous acquisition. Appl. Soil Ecol. 86, 41–54.

Pandey, V.C., Bajpai, O., Singh, N., 2016. Energy crops in sustainable phytoremediation. Renew. Sust. Energ. Rev. 54, 58–73.

Pellegrino, E., Bedini, S., 2014. Enhancing ecosystem services in sustainable agriculture: biofertilization and biofortification of chickpea (*Cicer arietinum* L.) by arbuscular mycorrhizal fungi. Soil Biol. Biochem. 68, 429–439.

Rajkumar, M., Ae, N., Freitas, H., 2009. Endophytic bacteria and their potential to enhance heavy metal phytoextraction. Chemosphere 77, 153–160.

Sachs, J.D., 2012. From millennium development goals to sustainable development goals. Lancet 379, 2206–2211.

Saharan, B.S., Nehra, V., 2011. Plant growth promoting rhizobacteria: a critical review. Life Sci. Med. Res. 21, 30.

Sarma, B.K., Yadav, S.K., Singh, S., Singh, H.B., 2015. Microbial consortium-mediated plant defense against phytopathogens: readdressing for enhancing efficacy. Soil Biol. Biochem. 87, 25–33.

Schäfer, T., Adams, T., 2015. The importance of microbiology in sustainable agriculture. In: Principles of Plant-Microbe Interactions. Springer International Publishing, Switzerland, pp. 5–6.

Sheng, X., Sun, L., Huang, Z., He, L., Zhang, W., Chen, Z., 2012. Promotion of growth and Cu accumulation of bio-energy crop (*Zea mays*) by bacteria: implications for energy plant biomass production and phytoremediation. J. Environ. Manag. 103, 58–64.

Sobariu, D.L., Fertu, D.I.T., Diaconu, M., Pavel, L.V., Hlihor, R.M., Drăgoi, E.N., Curteanu, S., Lenz, M., Corvini, P.F., Gavrilescu, M., 2016. Rhizobacteria and plant symbiosis in heavy metal uptake and its implications for soil bioremediation. New Biotechnol. 39 (Pt A), 125–134.

Tilman, D., Clark, M., 2014. Global diets link environmental sustainability and human health. Nature 515, 518–522.

Toyama, T., Kuroda, M., Ogata, Y., Hachiya, Y., Quach, A., Tokura, K., Tanaka, Y., Mori, K., Morikawa, M., Ike, M., 2017. Enhanced biomass production of duckweeds by inoculating a plant growth-promoting bacterium, *Acinetobacter calcoaceticus* P23, in sterile medium and non-sterile environmental waters. Water Sci. Technol. 76, 1418–1428.

Ullah, A., Heng, S., Munis, M.F.H., Fahad, S., Yang, X., 2015. Phytoremediation of heavy metals assisted by plant growth promoting (PGP) bacteria: a review. Environ. Exp. Bot. 117, 28–40.

Weselowski, B., Nathoo, N., Eastman, A.W., MacDonald, J., Yuan, Z.C., 2016. Isolation, identification and characterization of *Paenibacillus polymyxa* CR1 with potentials for biopesticide, biofertilization, biomass degradation and biofuel production. BMC Microbiol. 16, 244.

Weyens, N., van der Lelie, D., Taghavi, S., Newman, L., Vangronsveld, J., 2009. Exploiting plant–microbe partnerships to improve biomass production and remediation. Trends Biotechnol. 27, 591–598.

Yobo, K.S., Laing, M.D., Hunter, C.H., 2009. Effects of single and dual applications of selected *Trichoderma* and *Bacillus* isolates on performance of dry bean seedlings grown in composted pine bark growth medium under shade house conditions. J. Plant Nutr. 32, 1271–1289.

Zeilinger, S., Gupta, V.K., Dahms, T.E., Silva, R.N., Singh, H.B., Upadhyay, R.S., Gomes, E.V., Tsui, C.K., Nayak, S.C., 2016. Friends or foes? Emerging insights from fungal interactions with plants. FEMS Microbiol. Rev. 40, 182–207.

CHAPTER

17

Phosphate-Solubilizing Microorganisms and Their Emerging Role in Sustainable Agriculture

*Mokula Mohammed Rafi**,†, M.S. Krishnaveni*†, *P.B.B.N. Charyulu*†

*Department of Natural Resource Management, College of Agriculture and Rural Transformation, University of Gondar, Gondar, Ethiopia †Department of Microbiology, Sri Krishnadevaraya University, Anantapuramu, India

OUTLINE

Recent Developments in Applied Microbiology and Biochemistry
https://doi.org/10.1016/B978-0-12-816328-3.00017-9

17.1 INTRODUCTION

The existence of life on earth is favored by the cycling of biological elements wherein the complex biological systems after their decay are converted into simpler forms. These cycles like carbon, nitrogen, phosphorus, and sulfur are essential for agriculture, and the role played by microorganisms in these cycles is very important. Phosphorus (P) is second only to nitrogen as one of the most essential macronutrients required by plants. P is essential for biological growth and proper plant development. Organic matter derived from dead and decaying plant debris is rich in organic source of P. The deficiency of P may occur in crop plants growing in soils containing even adequate phosphates, which may be partly due to the fact that the plants are able to absorb P only in available form. Most of the P (95%–99%) present in the soil is in the insoluble form and hence cannot be utilized by the plants due to chemical fixation in the soil and its interaction with other metallic elements that are present in the rhizosphere area. Soil phosphates are rendered available either by plant roots or by soil microorganisms through their secretion of organic acids. Therefore, phosphate-solubilizing microorganisms (PSMs) in soils play a vital role in correcting P deficiency of crop plants. These microorganisms may also release soluble inorganic phosphates into soil through decomposition of phosphate-rich organic compounds. The recent trend of nutrient supply approach for improving agricultural production laid greater emphasis on the use of biofertilizers along with chemical fertilizers. Organic P constitutes 20%–40% of total P. Several soil bacteria, particularly *Bacillus* and *Pseudomonas*, and fungi possess the ability to solubilize insoluble phosphates into soluble forms. Phosphate-solubilizing bacteria (PSB) and nitrogen-fixing bacteria attracted the attention of many workers for their commercial utilization in agriculture. One way to correct the deficiency of P in plants is to inoculate seeds or soils with PSMs along with phosphatic fertilizers. Microorganisms contribute directly and indirectly to the soil health through their advantageous or harmful activities.

17.2 PHOSPHATE-SOLUBILIZING MICROORGANISMS (PSMs)

PSMs, which include many bacteria, fungi, and actinomycetes, play an important role in correcting P deficiency of plants. They assimilate not only P but also a large portion of soluble phosphate that is released in amounts of more than what is required by them. In addition, they may release soluble inorganic phosphate into soil through the decomposition of phosphate affluent organic compounds, thus playing an important role in soil by solubilizing P and making it available to plants. The mineralization of organic P relies largely upon the number and the activity of PSMs in soil. Bacteria are predominant among them and proved to be more effective in P solubilization than fungi. Rhizosphere helps large populations of soil microorganisms by accommodating a wide range of plant and microbial exudates and metabolites. In natural soil systems, PSMs consist of a broad class of bacteria and fungi that interact in soil especially in the extreme microenvironments found around plant roots called the rhizosphere.

PSB are present in almost all the soils, although their number varies depending upon the soil and climatic conditions. An extensively high concentration of PSB is usually found in the rhizosphere in comparison with nonrhizosphere soils. In a study, 76 PSB were isolated from chickpea that showed phosphate-solubilizing efficiency from 6% to 118% on solid and 22 to

248 μg mL^{-1} in liquid medium (Anu and Kundu, 2005). A majority of the PSB isolated from the rhizosphere exhibited moderate P-solubilizing capacity, while others showed high to very high efficiency (Oteino et al., 2015). Since these bacteria can play an important role in plant nutrition through an increase in P uptake by plants, they are also essential for sustainable agriculture in order to encourage the growth and yield of plants and supply of plant nutrients, thereby reducing the use of chemical fertilizers.

Hence, these bacteria could be a promising source of plant growth-promoting agent in agriculture and can be used to improve the soil health (Dhankhar et al., 2013). Ranjan et al. (2013) isolated PSB strains from rhizosphere soils of maize, onion, jasmine, and tomato, which released high amounts of P from calcium phosphate. Oligonucleotide primers in RAPD characterization showed genetic variation among the bacterial strains. Amplified fragments showed 40% shared and 20% unique polymorphic bands and the rest being monomorphic bands. The salt-tolerant or halophilic soil microorganisms that also demonstrate the ability to solubilize insoluble P facilitate the expansion of saline-alkali soil-based agriculture. Recently, PSB were isolated from rhizosphere soils of okra, chili, tomato, cotton, and eggplant, which were identified as *Bacillus megaterium*, *Pseudomonas putida*, and *P. fluorescens* (Baliah et al., 2016). Among PSMs, phosphate-solubilizing fungi also contribute much to plant growth. These fungi have been reported to be capable to traverse long distances within the soil and may be more important in the solubilization of inorganic phosphate in soils as they secrete more acids, such as gluconic, citric, lactic, 2-ketogluconic, oxalic, tartaric, and acetic acid, than bacteria.

17.3 MECHANISM OF PHOSPHATE SOLUBILIZATION

Solubilization of insoluble P by microorganisms was reported by Pikovskaya in the year 1948. When P is applied to soil, it interrelates with other metallic elements such as Fe, Al, and Ca ions, which make the P unavailable to plants through the formation of ferrous phosphate, aluminum phosphate, calcium phosphate, etc., and the release of organic acids by PSMs leads the chelation reaction, and because of this, the bound P to other metallic elements is released and becomes available to plants. As summarized in Fig. 17.1, the principal mechanisms for phosphate solubilization include the production of organic acids and acid phosphatases.

17.3.1 Production of Organic Acids

A key mechanism for mineralization of phosphates in soil is through microbial secretion of low-molecular-weight organic acids. The PSB have the ability to reduce the pH of the surroundings by the production of organic acids. These organic acids either can dissolve phosphates as a result of anion exchange or can chelate Ca, Fe, or Al ions associated with the phosphates. However, soil microorganisms vary considerably in their ability to secrete organic acids, thereby solubilizing mineral phosphates to different extent. Behera et al. (2017) isolated a phosphate-solubilizing bacterium, PSB-37, from a mangrove soil of the Mahanadi River Delta and identified the strain as *Serratia* sp. The maximum phosphate-solubilizing activity of the strain was 44.84 μg/mL, and also the pH of the growth medium decreased from 7.0 to 3.15. HPLC analysis of the culture broth revealed the presence of various organic acids such as malic acid (237 mg/L), lactic acid (599.5 mg/L), and acetic acid (5.0 mg/L).

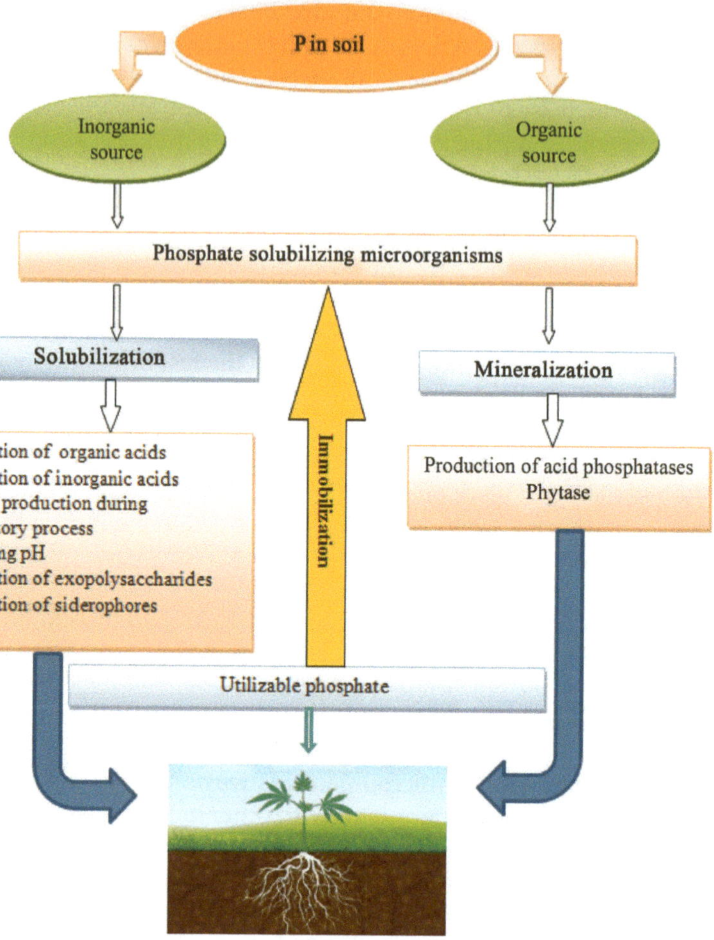

FIG. 17.1 Mechanism of solubilization of phosphate in soil.

17.3.2 Production of Acid Phosphatases

The mineralization of P compounds is carried out by the action of several phosphatases (also called phosphor hydrolases), present in a wide variety of soil microorganisms that play a significant role in the assimilation of phosphate from organic compounds by plants and microorganisms. It involves the hydrolysis of phosphor ester or phosphor anhydride bonds. The major mechanism of mineral phosphate solubilization is the action of organic acids by soil microorganisms. Phosphatases produced by PSMs play a key role in the solubilization of insoluble P into plant-available inorganic forms. They are produced by plant roots, microorganisms, and soil fungi.

17.4 PSB AS PLANT GROWTH PROMOTING RHIZOBACTERIA

Plant growth-promoting rhizobacteria (PGPR) are soil and rhizosphere bacteria that can benefit plant growth by different mechanisms. The bacteria that establish positive interaction with plant roots are called plant growth-promoting rhizobacteria, which play an important role in phytostimulation, phytoremediation, and biofertilization. Important characteristics of PGPR include production of exopolysaccharides, plant hormones, and siderophores; solubilization of P and Ca; and resistance of antibiotics. The PSB facilitate the growth of plants by stimulating the efficiency of nitrogen fixation; accelerating the accessibility of other trace elements; synthesizing important growth-promoting phytohormones like indoleacetic acid (IAA), gibberellic acid (GA), siderophores, and antibiotics; and providing protection to plants against soilborne pathogens. The benefits of PGPR addition to plants include increase in germination rate, root growth, yield, leaf area, biocontrol, and chlorophyll content; tolerance to drought; and increase in shoot and root weight. There are a number of reports on plant growth promotion by bacteria that have the ability to solubilize inorganic and/or organic P in soil after their inoculation in soil or plant seeds. It was reported that phytohormones produced by microorganisms can alter root branching and root hair development and thus have an impact on P acquisition. *Achromobacter* could enhance root hair number and length in oilseed rape (*Brassica napus*) and increase NO_3 and K uptake and consequently shoot and root dry weight by 22%–33% and 6%–21%, respectively (Bertrand et al., 2001).

Most of the agricultural soils were deficient in readily available P to the plants and hence require the application of P fertilizers to make the plants productive. Even though PSB occur in soil, their numbers were typically not sufficient to compete with other bacteria usually established in rhizosphere. Thus, the quantity of P liberated by them was usually not adequate for a substantial increase in in situ plant growth. Consequently, inoculation of plants with a target microorganism at a higher concentration than that usually found in soil was indispensable for the solubilization of phosphate to increase plant yield. PSMs act as biofertilizers by making otherwise unavailable P available to growing plants. PSB may also aid the growth of plants by stimulating the efficiency of biological nitrogen fixation, synthesizing phytohormones, and enhancing the availability of some trace elements such as Zn and Fe. The PSB displaying multiple plant growth-promoting traits on soil-plant system are required to uncover their efficacy as effective bioinoculants. The inoculation of PSB and PGPR collectively could decrease 50% of P fertilizer application without any considerable decrease in crop yield.

17.5 PLANT RESPONSES TO PSB INOCULATION

The use of PSB as bioinoculants plays a vital role in maintaining the soil nutrient status and structure and opens up a new horizon for better crop productivity and for greater yield performance (Ingle and Padole, 2017). The use of efficient PSMs opens up a new horizon for better crop productivity and for greater yield performance without affecting the soil health. Increase in the incidence of PSB in natural and alkaline soils is essential to increase and maintain the supply of available P. In a study, Chendrayan et al. (2003) conducted field experiments in rice, finger millet, bajra, black gram, red gram, groundnut, and soybean to find the effect of various biofertilizers and reported that PSB increased the grain yield in black gram over control. An increase in plant height, total dry matter production, and seed weight was noticed in soybean when inoculated with PSB over control.

The use of P solubilizers as inoculants increased growth, yield, and P uptake by many crop plants. The field and pot trials with PSMs with or without phosphatic fertilizers showed increase in yield and P uptake ranging from 10% to 27%. The effects of these microorganisms are due to P solubilization and other factors like phytohormones, nitrogen fixation, mineralization, mobilization of other nutrients, and promotion of PGPR microorganisms. Egamberdiyeva (2005) isolated bacteria from the rhizosphere of wheat, alfalfa, cotton, and tomato grown in fields of semiarid regions of Uzbekistan and identified them as *Bacillus laevolacticus, B. amyloliquefaciens, Pseudomonas denitrificans, P. rathonis, and Arthrobacter simplex*. All the isolated strains of bacteria increased the plant growth of wheat and maize in pot experiments. Seed inoculation of phosphate-solubilizing *Bacillus* sp. along with phosphate resulted in significant increase in grain yield of maize, French beans, amaranth, and finger millet (Pal, 1998). These bacteria enhanced the growth and grain yield of different plants. Moreover, PSB and NPK had a positive effect on germination and seed quality of some plants like radish that directly improved vigor index (Lamo et al., 2012). The benefits of adopting microbial management of the rhizosphere for sustainable agriculture production include enhancing the bioavailability of phosphate to crops, stimulated root and shoot growth, and improved root and shoot length, as well as improved fresh and dry shoot weight, P-labeled phosphate uptake, and significant enhancement in grain and dry matter yields. Walpola and Yoon (2013) studied the effect of two PSB strains, *Pantoea agglomerans* and *Burkholderia anthina*, which under greenhouse conditions enhanced shoot and root length, shoot and root dry matter, and P uptake of mung bean plants. Coinoculation of both PSB strains and TCP resulted in the highest growth performances and P uptake of mung bean plants. Panhwar et al. (2013) reported that the application of PSB along with organic acids enhanced soluble P in the soil solution, improved root growth, and increased plant biomass of aerobic rice seedlings. Kolekar et al. (2017) reported an improvement in the growth and yield of *Vigna radiata* plants when inoculated with a mixture of yeast, bacteria, and fungi. The authors attributed the improvement to synergistic effect of PSM. Phosphate-solubilizing strains like *Serratia marcescens* EB 67 and *Pseudomonas* sp. CDB 35 promoted the growth and plant biomass of maize with an increase of 66% and 51%, respectively (Hameeda et al., 2008). Phosphate-solubilizing endophytic *Pseudomonas* isolates stimulated the growth of *Pisum sativum* L. by producing gluconic acid that solubilized phosphate (Oteino et al., 2015). Gunes et al. (2009) reported an enhancement in the yield of strawberry by P-solubilizing *Bacillus* FS-3 and *Aspergillus* FS9 in P-deficient calcareous soils of Turkey. Inoculation of *Lysinibacillus fusiformis* strain 31MZR, a PSB, together with biochar showed positive effects on maize plant height and nutrient concentration (Rafique et al., 2017). Similarly, inoculation of maize seeds with PSB resulted in significant increase in yield and yield components over control plots under semiarid conditions (Amanullah and Khan, 2015). A field experiment was conducted to assess the effects of the inoculation of *Sinorhizobium ciceri* and PSB with 18/20 kg NP ha^{-1} as urea and DAP on the performance of chickpea and observed improved nodule number, nodule dry weight, nodule volume, and dry matter over uninoculated control at mid-flowering stage (Messele and Pant, 2012). Kaur and Sudhakara (2014) reported a significant increase in growth parameters, grain yield, total P uptake, and soil fertility in the context of organic carbon, available P, total P, and enzyme activities in maize and wheat when inoculated with *Pantoea cypripedii* and *Pseudomonas plecoglossicida* along with RP.

17.6 PLANT RESPONSES TO CO-INOCULATION OF PSB WITH DIAZOTROPHIC BACTERIA

Microbial inoculants that supply nitrogen and P have an important role in supplementing the nutrient needs of crops and agroforestry in dry lands as the farmers here are resource poor and cannot invest on chemical fertilizers. Dry lands are also identified as potential areas for organic production of oilseeds, medicinal herbs, and nutritious cereals. Biofertilizers are key components of organic farming. A large number of microbial inoculants containing *Bradyrhizobium*, *Azotobacter*, *Azospirillum*, *Pseudomonas*, and *Bacillus* were developed and used extensively in pulses, oilseeds, and coarse cereals. Trends of nutrient supply approach for indicating agricultural production laid greater emphasis on the use of biofertilizers along with chemical fertilizers. Bioinoculants may be helpful in reducing fertilizer input. They are cheap, balanced, and environmentally safe than chemical fertilizers. *Azotobacter*, *Azospirillum*, and PSB inoculants can be expected to improve plant growth if the inoculated bacteria are established and grow well in the rhizosphere. The P deficiency of the soil can be corrected by inoculating seeds with PSB along with phosphatic fertilizers (Bakhshandeh et al., 2015).

Synergistic effect of *Azospirillum* and PSB inoculation on biometric, nutritional, and yield parameters in pearl millet was observed (Mane et al., 2000). In the experiments conducted in the fields of black soil with low N (0.032%) and available P by inoculating seeds of pearl millet with *A. brasilense* and PSB along with superphosphate, increase in dry weight of roots and shoots at 75 days after sowing (DAS) in all treatments over control was observed that though was not significant, but maximum increase was observed with dual inoculum containing *Azospirillum* and PSB that might be due to more availability of P made available by PSB coupled with better nitrogen fixation by *Azospirillum* in the rhizosphere. Synergistic effect of a gradual-growing nitrogen-fixing bacterium *Phyllobacterium* sp. and a fast-growing PSB *B. licheniformis*, both isolated from the rhizosphere of mangrove, was reported by Rojas et al. (2001). Prabakaran et al. (2005) reported that combined inoculation of *Azospirillum* and PSB showed better effect on growth and yield of rice when compared with individual inoculation of *Azospirillum* and PSB. Increase in straw yield, grain yield, and weight of 1000 grain weight (g) was observed. Chendrayan et al. (2003) studied the effect of combined inoculation of various biofertilizers on the yield of different crops. In CO 45 rice variety, yield increase due to rhizobacterial inoculation ranged from 0.8% to 20% over control and dual inoculation of *A. lipoferum* and *P. fluorescens* resulted in higher grain yield (6996 kg/ha) than other treatments. In finger millet and bajra, the dual inoculation of *Azospirillum* and AMF led to 25.7% and 28.5% increase in yield, respectively. Guggari and Kalaghatagi (2005) studied the effect of biofertilizers on pearl millet and pigeon pea intercropping system under rainfed conditions and reported that combined inoculation of *Azospirillum* and PSB increased plant height, number of tillers, length of ear, straw yield, and grain yield in both the plants over control. Biofertilizers like *A. amazonense* A10 and *B. megaterium* P7 alone and in combination increased the grain yield of rice in an autoclaved soil by 103%–256% over control (Khan et al., 2003). In pot experiments,

P mobilization in wheat inoculated with *B. megaterium* or *A. lipoferum* 137 as single or mixed inocula was studied in the presence of tricalcium phosphate. Wheat plants inoculated with mixed inocula exhibited higher shoot dry weight and total N yield, while shoot P content was increased by 37% and 53% compared with the plants inoculated with *A. lipoferum* and uninoculated ones used as control, respectively (El-Komy, 2005). Ramakrishnaiah and Vijaya (2013) conducted experiments to evaluate the influence of different combinations of microbial inoculants (VAM+*Azotobacter*+PSB) on soil phosphatase activity and nutrients (N, P, K, Cu, Zn, Fe, Cu, and Mn) of *Stevia rebaudiana* plant rhizosphere. The study proved that the effect of combined application of microbial inoculants on soil alkaline and acid phosphatase activity was maximum at 60th day in triple inoculated plants than dual inoculated and control. Inoculation of foxtail millet with two strains of *Azospirillum* sp. and one strain of a PSB individually and in combination significantly increased plant height and dry weight of shoot and root over control plants. The increase in yield was due to synergistic effect of *Azospirillum* and PSB inoculation (Rafi and Charyulu, 2016). Phosphate-solubilizing microorganisms exhibited considerable synergistic effect on the growth and development of crops. Besides solubilizing P, some PSMs also demonstrated their potential as biocontrol agents against some plant pathogens. PSMs manage the pathogens by producing antifungal compounds, siderophores, antibiotics, hydrogen cyanide, and lytic enzymes all of which enhance inhibition of the growth of plant pathogens (Alori et al., 2017).

17.7 VESICULAR ARBUSCULAR MYCORRHIZAE (VAM) IN PHOSPHATE SOLUBILIZATION

Soil inoculation with *Bacillus circulans* (PSB) along with VAM increased P uptake and shoot and root weight over control in finger millet when superphosphate and tricalcium phosphate were added (Raj et al., 1981). Results on experiments conducted to determine the influence of VAM fungi (*Glomus fasciculatum* and *G. mosseae*), *A. brasilense*, and phosphobacterial (*B. megaterium*) inoculation, either individually or in combination on seedling growth and nutrient uptake of four shola tree species, revealed that the seedlings that received the combined inoculum of VAM fungi, *Azospirillum*, and phosphobacteria increased plant height and total dry weight. When *Azospirillum* was coinoculated with phosphobacteria, the seedling height of shola tree species at 180 days was recorded as 26.04, 32.14, and 26.98 cm in *E. munroii*, *G. obtusa*, and *L. roxburghii*, respectively, which was higher than the uninoculated control. Similarly, total dry weight (g/plant) also increased over uninoculated control (Kennedy and Chellapillai, 1998). The interactive effect of arbuscular mycorrhizal fungi (AMF) and rhizobacteria on the growth and nutrient uptake of *Sorghum bicolor* was studied in acid and low-availability phosphate soil (Widada et al., 2007). Combined inoculation of AMF and PSB supported improved uptake of native P from the soil and P coming from the phosphatic rock and improved plant growth by solubilizing P from different fractions of soil (Ahmed et al., 2008). The interactive effect of PSB and AMF on the use of soil P sources of low bioavailability by onion plant was evaluated. Dual inoculation with an AMF (*G. intraradices*) and two PSB (*Enterobacter* sp. and *B. subtilis*) increased biomass and N and P accumulation significantly in onion plant tissues (Toro et al., 1997). In the interactive effects of AMF *G. fasciculatum* and *Acaulospora laevis* with *Pseudomonas striata*, PSB at different doses of rock phosphate (0, 15, 30, and 45 kg/ha) on two varieties of *Sesamum indicum* were studied, which resulted in significant increase in shoot growth, root growth, and biomass and content of shoot P (Sabannavar and Lakshman, 2009). Coinoculation of wheat plants with AMF and PSB increased shoot dry matter yield, spike number, and grain yield by 52%, 19%, and 26%, respectively (Yousefi et al., 2011). Zhang et al. (2011) investigated the effect of interaction between two AMF, *G. aggregatum* and *G. mosseae*, and a P-solubilizing fungus (*Mortierella* sp.) on the growth of *Kosteletzkya virginica*. Dual inoculation of *Mortierella* sp. with AMF had significant positive effects on shoot and root dry weight, available P, pH values, electric conductivities of rhizosphere, and bulk soils under salt stress. A positive synergistic effect of an AMF, *G. etunicatum*, and a PSB strain, *Burkholderia cepacia* BAM-6, was observed on wheat plants grown in soil with low available P, which resulted in plant growth promotion and nutrient uptake by the plants (Saxena et al., 2014). The combined effect of an AMF and a phosphate-solubilizing fungus on salt tolerance of beach plum (*Prunus maritima*) showed enhanced salt tolerance (Xueming et al., 2014).

17.8 CONCLUSIONS

The benefits that accrued to crop plants by PSMs in sustainable agriculture are indisputable. Likewise, synergistic and beneficial effects of inoculation of PSMs with other microorganisms like diazotrophs to crop plants are also evident. However, bioinoculation experiments on long-term basis are warranted to confirm the synergistic benefits to crop plants. Similarly, efforts are also needed to augment the potential of efficient PSMs by a better understanding of the mechanisms of P solubilization through genetic engineering.

References

Ahmed, M.F., Kennedy, I.R., Choudhury, A.T.M.A., Kecskes, M.L., Deker, R., 2008. Phosphorus adsorption in some Australian soils and influence of bacteria on the desorption of phosphorus. Commun. Soil Sci. Plant Anal. 39, 1269–1294.

Alori, E.T., Glick, B.R., Babalola, O.O., 2017. Microbial phosphorus solubilization and its potential for use in sustainable agriculture. Front. Microbiol. 8, 971.

Amanullah, Khan, A., 2015. Phosphorus and compost management influence maize (*Zea mays*) productivity under semiarid condition with and without phosphate solubilizing bacteria. Front. Plant Sci. 6, 1083.

Anu, Kundu, B.S., 2005. Development and characterization of *lacZ* marked strains of phosphate solubilizing bacteria. Indian J. Microbiol. 45, 261–263.

Bakhshandeh, E., Rahimian, H., Pirdashti, H., Nematzadeh, G.A., 2015. Evaluation of phosphate-solubilizing bacteria on the growth and grain yield of rice (Oryza sativa L.) cropped in northern Iran. J. Appl. Microbiol. 119, 1371–1382.

Baliah, N.T., Pandiarajan, G., Kumar, B.M., 2016. Isolation, identification and characterization of phosphate solubilizing bacteria from different crop soils of Srivilliputtur Taluk, Virudhunagar District, Tamil Nadu. Trop. Ecol. 57 (3), 465–474.

Behera, B.C., Yadav, H., Singh, S.K., Mishra, R.R., Sethi, B.K., Dutta, S.K., Thatoi, H.N., 2017. Phosphate solubilization and acid phosphatase activity of *Serratia* sp. isolated from mangrove soil of Mahanadi river delta, Odisha. Indian J. Genet. Eng. Biotechnol. 15, 169–178.

Bertrand, H., Nalin, R., Bally, R., Cleyet-Marel, J.C., 2001. Isolation and identification of the most efficient plant growth promoting bacteria associated with canola (*Brassica napus*). Biol. Fertil. Soils 33, 152–156.

Chendrayan, K., Natarajan, T., Umamaheswari, T., 2003. Combined inoculation of biofertilizers for increasing crop production. Biofertil. Newsl. 11, 24–26.

Dhankhar, R., Sheoran, S., Dhaka, A., Soni, R., 2013. The role of phosphorus solubilizing bacteria (PSB) in soil management an overview. Internat. J. Develop. Res. 3 (9), 31–36.

Egamberdiyeva, D., 2005. Plant-growth-promoting rhizobacteria isolated from a Calcisol in a semi-arid region of Uzbekistan: biochemical characterization and effectiveness. J. Plant Nutr. Soil Sci. 168, 94–99.

El-Komy, H.M.A., 2005. Coimmobilization of *Azospirillum lipoferum* and *Bacillus megaterium* for successful phosphorus and nitrogen nutrition of wheat plants. Food Technol. Biotechnol. 43, 19–27.

Guggari, A.K., Kalaghatagi, S.B., 2005. Effect of fertilizer and biofertilizer on pearl millet (*Pennisetum glaucum*) and pigeonpea (*Cajanus cajan*) intercropping system under rainfed conditions. Indian J. Agron. 50, 24–26.

Gunes, A., Ataoglu, N., Turan, M., Eşitken, A., Ketterings, Q.M., 2009. Effects of phosphate-solubilizing microorganisms on strawberry yield and nutrient concentrations. J. Plant Nutr. Soil Sci. 172, 385–392.

Hameeda, B., Harini, G., Rupela, O.P., Wani, S.P., Reddy, G., 2008. Growth promotion of maize by phosphate-solubilizing bacteria isolated from composts and macrofauna. Microbial Res. 163, 234–242.

Ingle, K.P., Padole, D.A., 2017. Phosphate solubilizing microbes: an overview. Int. J. Curr. Microbiol. App. Sci. 6, 844–852.

Kaur, G., Sudhakara, R.M., 2014. Role of phosphate-solubilizing bacteria in improving the soil fertility and crop productivity in organic farming. Arch. Agron. Soil Sci. 60, 549–564.

Kennedy, R.W., Chellapillai, K.L., 1998. Synergistic effect of VAM, *Azospirillum* and phosphobacteria on growth response and nutrient uptake of shola tree species. Indian J. For. 21, 308–312.

Khan, M.R., Talukdar, N.C., Thakuria, D., 2003. Detection of *Azospirillum* and PSB in rice rhizosphere soil by protein and antibiotic resistance profile and their effect on grain yield of rice. Indian J. Biotechnol. 2, 246–250.

Kolekar, S.S., Desai, P.D., Panchal, H.K., Shah, K.B., 2017. Study of phosphate solubilizing microorganisms with biofertilizer potential. Int. J. Pharm. Bio. Sci. 8, 751–757.

Lamo, K., Korla, B.N., Shukla, Y.R., 2012. Effect of different organic and inorganic nutrient sources on seed production of Radish (*Raphanus sativus CV. Chinese pink*). Life Sci. Leaflets 2, 38–44.

Mane, S.S., Hadgaonkar, A.K., Surya, W.A.P., Salunke, S.D., 2000. Response of pearl millet to inoculation of phosphate solubilizing bacteria and *Azospirillum*. J. Indian Soc. Soil Sci. 48, 617–619.

Messele, B., Pant, L.M., 2012. Effects of inoculation of *Sinorhizobium ciceri* and phosphate solubilizing bacteria on nodulation, yield and nitrogen and phosphorus uptake of chickpea (*Cicer arietinum* L.) in Shoa Robit area. J. Biofertil. Biopestici. 3, 129.

Oteino, N., Lally, R.D., Kiwanuka, S., Lloyd, A., Ryan, D., Germaine, K.J., Dowling, D.N., 2015. Plant growth promotion induced by phosphate solubilizing endophytic *Pseudomonas* isolates. Front. Microbiol. 6, 745.

Pal, S.S., 1998. Interaction of an acid tolerant strain of phosphate solubilizing bacteria with a few acid tolerant crops. Plant Soil 198, 169–177.

Panhwar, Q.A., Jusop, S., Naher, U.A., Othman, R., Razi, M.A., 2013. Application of potential phosphate-solubilizing bacteria and organic acids on phosphate solubilization from phosphate rock in Aerobic Rice. Sci. World J. 2013. 272409.

Prabakaran, J., Kumar, K., Rajan, B.G., Paulsamy, A., Rajah, V.G., 2005. In: Significance of Azophos and silicate solubilizing bacteria on low land rice. Proc. National Symposium on Microbial Technology for Productive Agriculture, TNAU, Coimbatore, pp. 130–131.

Rafi, M.M., Charyulu, P.B.B.N., 2016. Synergistic effect of *Azospirillum* and PSB inoculation on growth and yield of foxtail millet. Int. J. Plant Anim. Environ. Sci. 6 (1), 138–147.

Rafique, M., Sultan, T., Ortas, I., Chaudhary, H.J., 2017. Enhancement of maize plant growth with inoculation of phosphate-solubilizing bacteria and biochar amendment in soil. Soil Sci. Plant Nutr. 63, 460–469.

Raj, J., Bagyaraj, D.J., Manjunath, A., 1981. Influence of soil inoculation with vesicular-arbuscular mycorrhiza and a phosphate-dissolving bacterium on plant growth and ^{32}P-uptake. Soil Biol. Biochem. 13, 105–108.

Ramakrishnaiah, G., Vijaya, T., 2013. Influence of VAM fungi, *Azotobacter sp.* and PSB on soil phosphatase activity and nutrients (N, P, K, Cu, Zn, Fe and Mn) status in the rhizosphere of *Stevia rebaudiana* (Bert.) plants. Amer. J. Plant Sci. 4, 1443–1447.

Ranjan, A., Mahalakshmi, M.R., Sridevi, M., 2013. Isolation and characterization of phosphate-solubilizing bacterial species from different crop fields of Salem, Tamil Nadu, India. Int. J. Nutr. Pharmacol. Neurol. Dis. 3 (1), 29–33.

Rojas, A., Holguin, G., Glick, B.R., Bashan, Y., 2001. Synergism between *Phyllobacterium* sp. (N_2-fixer) and *Bacillus licheniformis* (P-solubilizer), both from a semiarid mangrove rhizosphere. FEMS Microbial Ecol. 35, 181–187.

Sabannavar, S.J., Lakshman, H.C., 2009. Effect of rock phosphate solubilization using mycorrhizal fungi and phosphobacteria on two high yielding varieties of *Sesamum indicum* L. World J. Agric. Sci. 5 (4), 470–479.

Saxena, J., Minaxi, Jha, A., 2014. Impact of a phosphate solubilizing bacterium and an arbuscular mycorrhizal fungus (*Glomus etunicatum*) on growth, yield and P concentration in wheat plants. Clean: Soil, Air, Water 42, 1248–1252.

Toro, M., Azcon, R., Barea, J.M., 1997. Improvement of arbuscular mycorrhiza development by inoculation of soil with phosphate-solubilizing rhizobacteria to improve rock phosphate bioavailability (^{32}P) and nutrient cycling. Appl. Environ. Microbiol. 63, 4408–4412.

Walpola, B.C., Yoon, M., 2013. Phosphate solubilizing bacteria: assessment of their effect on growth promotion and phosphorous uptake of mung bean (*Vigna radiata* [L.] R. Wilczek). Chilean J. Agric. Res. 73, 275–281.

Widada, J., Damarjaya, D.I., Kabirun, S., 2007. The interactive effects of arbuscular mycorrhizal fungi and rhizobacteria on the growth and nutrients uptake of sorghum in acid soil. In: Rodriguez-Barrueco, C., Velazquez, E. (Eds.), First International Meeting on Microbial Phosphate Solubilization. Springer, pp. 173–177.

Xueming, Z., Zhenping, H., Yu, Z., Huanshi, Z., Pei, Q., 2014. Arbuscular mycorrhizal fungi (AMF) and phosphate-solubilizing fungus (PSF) on tolerance of beach plum (*Prunus maritima*) under salt stress. Aust. J. Crop. Sci. 8, 945–950.

Yousefi, A.A., Khavazi, K., Moezi, A.M., Rejali, F., Nadian, H.A., 2011. Phosphate solubilizing bacteria and arbuscular mycorrhizal fungi impacts on inorganic phosphorus fractions and wheat growth. World Appl. Sci. J. 15, 1310–1318.

Zhang, H., Wu, X., Li, G., Qin, P., 2011. Interactions between arbuscular mycorrhizal fungi and phosphate-solubilizing fungus (*Mortierella* sp.) and their effects on *Kosteletzkya virginica* growth and enzyme activities of rhizosphere and bulk soils at different salinities. Biol. Fertil. Soils 47, 543–554.

Further Reading

Istina, I.N., Widiastuti, H., Joy, B., Antralina, M., 2015. Phosphate-solubilizing microbe from Saprists peat soil and their potency to enhance oil palm growth and P uptake. Proc. Food Sci. 3, 426–435.

Sonboir, H.L., Sarawgi, S.K., 1998. Effect of phosphorus, bacterial cultures and micronutrients on growth, yield and balance sheet of N and P in chickpea. Ann. Agric. Res. 19, 488–491.

CHAPTER

18

Microbes in the Generation of Genetically Engineered Plants for Disease Resistance

Shameer Syed, N.V.K.V. Prasad Tollamadugu

Nanotechnology Laboratory, Institute of Frontier Technology, Regional Agricultural Research Station, Acharya N.G. Ranga Agricultural University, Tirupati, India

OUTLINE

18.1 INTRODUCTION

As the planet's human populace swells from the current 7.3 billion to an anticipated 9.7 billion by 2050, there will be an escalating requirement to efficiently produce and distribute food, leading to a sharp increase in food demand from 50% to 90% by 2050. The rapid growth in human population and related intensification of demand for food leave us with an inescapable adversary, i.e., increase in agricultural yield under the stresses like diseases and pests,

https://doi.org/10.1016/B978-0-12-816328-3.00018-0

alongside drastically changing environmental and climatic conditions, parallelly maintains environmental harmony in doing so. Plant diseases caused by many diverse agents ranging from fungi, bacteria, viruses, nematodes, insects, pests, etc. alone cause up to 40% of economic loss (including pre- and postharvest) to the agriculture sector. Therefore, this pressing situation requires revamping our agricultural practices both traditional and contemporary with modern ones, concerning the molecular level regulation or control mechanisms to enhance the crops' ability to resist, evade, and tolerate pathogens and their effector proteins and/or products, decreasing economic and yield burdens of the agricultural sector. Keeping in view of the essentiality to reduce the agricultural footprint on the environment, the genetic engineering or genome editing (GE) technologies would seem to be go to ones to many in this stance, as they can manipulate the crop at genome level, in which the changes become permanent and transferable to future generations in a short period. GE technologies rely on the fact that the optimal functionality of living species is the functional outcome of its DNA, more precisely the functional genes. By changing (making it overexpressive or underexpressive) and/or eliminating and inserting a whole or part of a new gene (transgenesis), leading to gene changes even to single nucleotide level, the outcome will be surprisingly different. Thus, GE can be defined as the targeted modification of very specific functional or nonfunctional DNA segment (gene/s) within a living organism, leading to the development of new traits or enhancement of old ones, conferring the desired quality of the crop.

The GE technology-based applications in agriculture include and majorly focus on altering plant structure or function to develop a useful characteristic or confer disease or stress resistance or both to the crop plant. The selection of target DNA segment for GE technologies is based on either insertion or modification of foreign gene or existing gene segment in the plant to impart a new trait or modify the previous one. It is easier to select a gene responsible for the pathogen resistance in microbial domain than to try out changing the genes in the plant randomly to identify the gene, which augments disease resistance, and modify it to improve its related disease-resistant function. From the beginning, it's been a major hurdle for modern agriculture to balance environmental stability and plant disease control, as the extensive use of chemical agents of disease control is still a foremost approach even as biological and disease-resistant varieties are being gradually adapted globally (Table 18.1). The evolution of new virulent plant pathogens is overshadowing the traditional plant breeding system as it takes long periods and is expensive to produce disease-resistant varieties that are too prone to environmental conditions (Rojas et al., 2010). The technological growth in regard to identification, seclusion, and transfer of genes enhancing disease resistance in plants without affecting the plant native properties has seen tremendous growth. With the inception of the first *Agrobacterium*-mediated transformation more than a decade ago, a range of advantageous qualities have been transferred to major crops using different techniques (Baltes et al., 2017). Many instances saw endogenous plant genes being altered to improve the crop feature or resistance, such as the introduction of sense and antisense configurations to modify ripening and senescence processes (Fray and Grierson, 1993). In the attempts to develop phytopathogen resistance or to study the mechanisms of pathogenicity via constitutive and inducible promoters, several genes from acceptor or other species were used (Silva and Klessig, 1998).

Much research was put up to identify and isolate genes from diverse sources, including viruses, bacteria, fungi, animals, and humans, useful for transfer to plants with the potential to

TABLE 18.1 Global GE Crop Cultivation in 1996 and 2016 (ISAAA, 2016)

	1996		2016	
Country	**Crop**	**GE Crop Area (Mha)**	**Crop**	**GE Crop Area (Mha)**
The United States	Tomato, cotton, soybean, maize, canola, potato, squash	1.5	Cotton, soybean, maize, canola, potato, squash, papaya, alfalfa, sugar beet	72.9
Argentina	Soybean	0.1	Soybean, maize, cotton	23.8
Brazil	–	–	Soybean, cotton, maize	49.1
Canada	Canola, maize	0.1	Canola, maize, soybean, sugar beet	11.6
India	–	–	Cotton	10.8
China	Tobacco, tomato	Trace	Cotton, tomato, poplar, petunia, papaya, sweet pepper	2.8
Paraguay		–	Soybean	3.6
South Africa		–	Maize, soybean, cotton	2.7
Uruguay		–	Soybean, maize	1.3
Philippines		–	Maize	0.8
Australia	Cotton	<0.05	Cotton, canola, carnation	0.9
Spain	–	–	Maize	0.1
Mexico	Cotton, tomato	<0.05	Cotton, soybean	0.1
Colombia		–	Cotton, carnation	0.1
Chile		–	Maize, soybean, canola	<0.05
Honduras		–	Maize	<0.05
Czech Republic		–	Maize	<0.05
Portugal		–	Maize	<0.05
Slovakia		–	Maize	<0.05
Bolivia		–	Soybean, maize, cotton	1.2
Total	Soybean, maize, tobacco, cotton, canola, tomato, and potato	1.7	Soybean, maize, cotton, canola, squash, papaya, alfalfa, carnation, tomato, poplar, petunia, sweet pepper, sugar beet	185.1

Countries like Poland, Egypt, Romania, Germany, and Burkina Faso have averted from GE crops from recent years (2015) due to stringent regulations and political policies against GE crops.

(i) augment resistance to microorganisms and insects (Düring, 1996), herbicides and chemicals (Powles, 2018), or environmental strain and nutrient scarcity (Cabello et al., 2014; Marco et al., 2015) and (ii) study plant/pathogen interactions and other physiological or genetic processes (Nejat et al., 2017). As a result of the tremendous thrust to search and employ nonplant genes resulting in successful endeavors in improving plant disease resistance, it may not be far to have any gene prospectively useful in improving plant disease resistance that is now or will soon be available, leading to novel transgenic plants (Butler et al., 2018). The expression of transgenes in plants is not without exertion and defeat, commonly due to GE plants were futile to bring into being the expected gene product to level desired or lacked functionality significantly than expected (Araki and Ishii, 2015). GE of crop plants is based on two major modes of transgene functioning in the receptor plant that can be seen as either producing the transgenic metabolite that develops the antimicrobial and insecticidal ability of the plant or expressing the pathogen-derived transgene in inducing plant defense response or endowing resistance to toxins or enzymes of invading phytopathogens (Silva et al., 2018). In this regard, the potential of microbial genes (fungi and bacteria) in the genetic engineering of plants to develop or create new and novel crop varieties possessing enhanced disease resistance is still at an early stage and needs to be encouraged as a major research area (Lorito and Scala, 1999; Mittler and Blumwald, 2010; Naqvi et al., 2010).

18.2 MICROBIAL SOURCES OF TRANSGENES FOR PLANT GENETIC ENGINEERING

The sources of genes for genetic engineering plants to improve disease resistance are diverse, and at times, even animal genes too were employed (Punja, 2001; Mittler and Blumwald, 2010). Proteins imparting antifungal and antibacterial characteristics are the primary tools of protection and contravention toward diverse bacterial and fungal groups and are usually effective on a broad range of targets, even influencing synergistically with biologically active substances (i.e., antibiotics). Many bacterial and fungal genes have been studied in vitro, but most never went into the transgenesis phase, to be transferred into plants. The bacterium *Serratia marcescens*'s chitinase gene *chi*A was the first gene to be transferred into plants to enhance disease resistance against phytopathogens (*Alternaria longipes* and *Rhizoctonia solani*) and was studied extensively for its effectiveness and stable expression of the desired protein, in this case chitinase enzyme (Kamble et al., 2016). Similarly, another chitinase gene from *Streptomyces* sp. strain N174 showed tremendous antimicrobial potential in in vitro studies against *Fusarium oxysporum*, *Verticillium albo-atrum*, and other fungi. The most effective chitinase gene *chit42* from *Trichoderma harzianum* has rapid lysis of many chitin structures like chlamydospores, mature hyphae, and sclerotia, inhibits a variety of fungal pathogens, and has shown 400-fold increase in chitinase activity in comparison with controls, with no perceptible influence on the plant growth parameters (Zhang et al., 2016). Similar incorporation of *Trichoderma* gene into potato showed absolute resistance toward airborne pathogens (*Alternaria alternata*, *Alternaria solani*, and *Botrytis cinerea*) and soilborne pathogens (*R. solani* and *Sclerotinia sclerotiorum*). Following this success of *chit42* in potato, the gene has been transferred into many crops (apple, petunia, chickpea, pea, brassica, wheat, grape, alfalfa, and tomato) with successful disease resistance enhancement against *Venturia inaequalis* (apple), *Uncinula necator* (grape), and *R. solani* (petunia) (Lorito and Scala, 1999). Along with *T. harzianum* chitinase, many specific hydrolytic enzyme genes from species like *Erwinia carotovora* (pectate lyase (*PL3*)), *Schizosaccharomyces pombe* (ribonuclease (*pac1*)), *Streptomyces* sp. (chitosanase), *Rhizopus oligosporus* (chitinase (*chi1*)), and *E. carotovora* (oligogalacturonide lyase) (Zhang et al., 2016) have been used in the transgenesis of crop plants.

Apart from the fungal genes, many bacterial genes for the production of hydrolytic enzymes and other metabolites have been genetically introduced into the plant system and established to express the desired gene or gene product, enhancing the disease tolerance and/or resistance from partial to absolute (van der Biezen, 2001; Nejat et al., 2017). The most famous of them all is the δ-endotoxin-producing gene from *Bacillus thuringiensis*, capable of controlling many plant insects and thereby protecting the crop and yield, but that is not without setbacks completely. Many bacterial genes have been studied for similar applications like *Vibrio cholerae* (cholera toxin, subunit A), *Pseudomonas syringae* pv. *tabaci* (tabtoxin resistance), *Phytophthora cryptogea* (b-cryptogein), *Bacillus amyloliquefaciens* (barnase (RNase)), *P. syringae* (avirulence (*avrB*), avirulence (*avrPto*), and avirulence (*avrRpt2*)), and *E. carotovora* (pectate lyase (*PL3*) and oligogalacturonide lyase) (Table 18.2). The source of novel genes for the transgenesis of crops was limited to culturable microflora in the past due to growth limitations imposed by the nonculturable microbes. Due to the advent of modern technologies and improvements leading to the cost-effectiveness of many protocols, the sources have become far more extensive recently (Zhang et al., 2008). Metagenomics and metabolomics combined together along with restriction mapping techniques (RAPD, RFLP, AmpFLP, and STR) give large data in a short period about the potential genes and metabolites without the need to grow and culture the entire microbiome. With the emergence of new species every now and then, the possibility of these new species possessing novel antimicrobial properties is not entirely questionable, which can be utilized for genetic engineering of the crop plants (Figs. 18.1 and 18.2).

TABLE 18.2 Microbial Genes Expressed in Plants That Enhance Defense Response to Pathogens, Activate Plant Defense Responses, or Protect Them From Pathogen Toxins and Purposes Other Than to Increase Disease Resistance

Plant	Gene Encoding	Source	Activity/Effect
Tobacco	Chitinase (*chiA*)	*Serratia marcescens*	Antifungal chitinase. Resistance to *A. longipes* and *R. solani*
	Chitosanase	*Streptomyces* sp.	Antifungal chitosanase
	Chitinase (*chi1*)	*Rhizopus oligosporus* chitinase involved in autolysis	Resistance to *S. sclerotiorum* and *B. cinerea*
	Ribonuclease (*pac1*)	*Schizosaccharomyces pombe*	Degradation of dsRNA. Resistance to viruses
	Trichodiene synthase (*Tri5*)	*Fusarium sporotrichioides*	Produce sesquiterpenoid phytoalexin. Resistance to microbes and insects
	Killer toxin	*Ustilago maydis* (mycovirus)	Toxic to closely related *Ustilago* species. Resistance to *U. maydis*

Continued

TABLE 18.2 Microbial Genes Expressed in Plants That Enhance Defense Response to Pathogens, Activate Plant Defense Responses, or Protect Them From Pathogen Toxins and Purposes Other Than to Increase Disease Resistance—cont'd

Plant	Gene Encoding	Source	Activity/Effect
	b0	*Halobacterium halobium*	Resistance to *P. syringae* pv. *tabaci*
	Cholera toxin, subunit A	*Vibrio cholerae*	Antibacterial. Resistance to *P. syringae* pv. *tabaci*
	FLP recombinase (FLP/FRT)	*Saccharomyces cerevisiae*	Site specific recombination. Study effect on development
	Mitotic inducer (*cdc25*)	*Schizosaccharomyces pombe*	Alter mitosis
	Presequence of mitochondrial ATPse (b-subunit)	*Saccharomyces cerevisiae*	Target proteins to mitochondria
	Tabtoxin resistance	*Pseudomonas syringae* pv. *tabaci*	Inactivate tabtoxin. Resistance to *P. syringae* pv. *tabaci*
	Fe(III) reductase (FRE1, FRE2)	*Saccharomyces cerevisiae*	Improve Fe nutrition. Tolerance to Fe deficiency
	Lignin-degrading enzymes (*LiP, Mnp*)	*Phanerochaete chrysosporium*	Alter growth and development. Study effect of pathogen enzymes on plant
	TL-DNA (various ORFs)	*Agrobacterium rhizogenes*	Alter morphogenesis and response to hormones
	Inorganic pyrophosphatase (*ppa*) and invertase (*suc2*)	*E. coli, S. cerevisiae*	Alter photoassimilate partitioning
	Sarcotoxin peptide	*Sarcophaga peregrina*	Enhanced seedling survival following inoculation with *R. solani, Pythium aphanidermatum*, and *P. nicotianae*
	polO and polA (polyoxins)	*Streptomyces cacaoi*	Fungicide and insecticide
	Glucose oxidase	*Aspergillus niger*	Delayed lesion development due to *Phytophthora infestans*, reduced disease development due to *A. solani* and *V. dahliae*
	Chitinase	*Serratia marcescens*	Reduced disease incidence due to *R. solani* on seedlings
	Aminolevulinate synthase (*ALA-S*)	*Saccharomyces cerevisiae*	Synthesis of ALA. Tolerance to inhibitors of the C5 pathway
Tomato	Virulence (*Ecp2*)	*Cladosporium fulvum*	Elicit HR. Resistance to *C. fulvum*
	Avirulence (*avrB*)	*Pseudomonas syringae*	Elicit genotype-specific HR

TABLE 18.2 Microbial Genes Expressed in Plants That Enhance Defense Response to Pathogens, Activate Plant Defense Responses, or Protect Them From Pathogen Toxins and Purposes Other Than to Increase Disease Resistance—cont'd

Plant	Gene Encoding	Source	Activity/Effect
Arabidopsis thaliana	Acyltransferase (*SLC1-1*)	*Saccharomyces cerevisiae*	Increase seed oil content
	Bt toxin	*Bacillus thuringiensis*	Insecticidal. Resistance to caterpillars, looper, bollworms, hornworms, armyworm, beetles, and other insects
Corn, potato, cotton, tomato, rice, carrot, etc.	Pectate lyase (*PL3*)	*Erwinia carotovora*	Activate disease response by producing elicitors from plant tissue. Resistance to *E. carotovora*
Potato	Oligogalacturonide lyase	*Erwinia carotovora*	Degrade sugar oligomers into inactive products. Interfere with signaling and suppress pathogenicity in the potato/*E. carotovora* interaction
	expl	*Erwinia carotovora*	Resistance to *E. carotovora*
	aiiA	*Erwinia carotovora*	Resistance to *E. carotovora*
	Avirulence (*avr9*, *infl*)	*Cladosporium fulvum*, *P. infestans*	Elicit HR. Resistance to potato virus X
	B-Cryptogein	*Phytophthora cryptogea*	Elicit defense response. Broad-spectrum disease resistance
	Bacterio-opsin proton pump (*bO*)	*Halobacterium halobium*	Activate local and systemic response. Resistance to viruses, *P. syringae* pv. *tabaci* and *P. infestans*
	Barnase (RNase)	*Bacillus amyloliquefaciens*	Delayed sporulation and reduced sporangia production by *P. infestans*
	Ornithine carbamoyltransferase (*argK*)	*Pseudomonas syringae* pv. *phaseolicola*	Resistance to bacteria phaseolotoxin. Repress symptom formation
Bean	Antifungal protein	*Aspergillus giganteus*	Reduced development of colonies of *Blumeria graminis* and *Puccinia recondita*
	Xylanase (*xynD*)	*Clostridium thermocellum* and *Ruminococcus flavefaciens*	Xylanase production
Grape	Endochitinase	*Trichoderma harzianum*	Reduction of *B. cinerea* development in preliminary tests

From Broglie, R., Broglie, K., Roby, D., Chet, I., 1993. Production of transgenic plants with enhanced resistance to microbial pathogens. In: Transgenic Plants, vol. 1, pp. 265–276; Lorito, M., Scala, F., 1999. Microbial genes expressed in transgenic plants to improve disease resistance. J. Plant Pathol. 1999, 73–88.

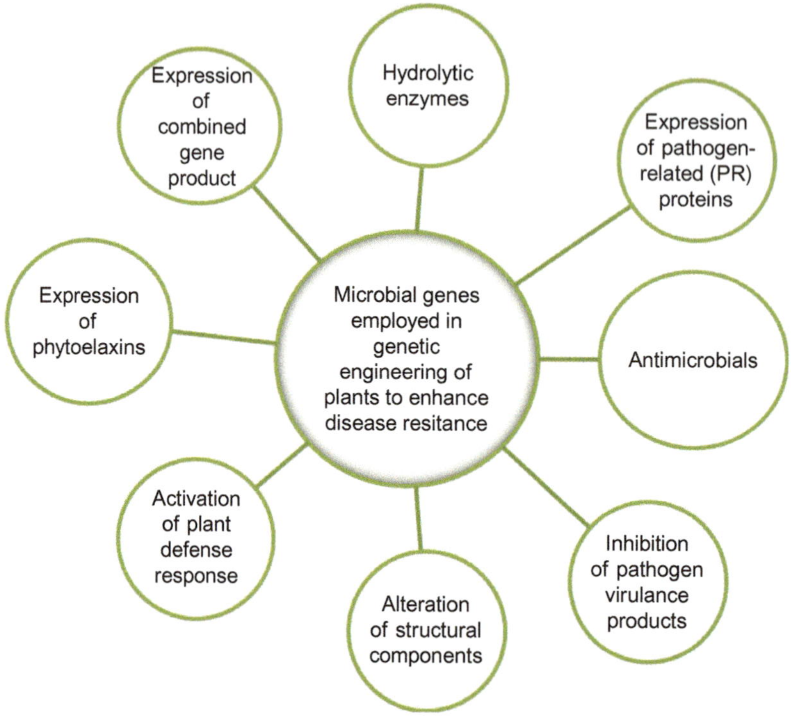

FIG. 18.1 Strategies used in genetic engineering of plant species to enhance disease resistance against invading phytopathogens.

18.3 ATTRIBUTES FLAUNTED BY THE GENETICALLY ENGINEERED CROP PLANTS

The basic strategy involved in genetic engineering of crops to enhance disease resistance is to impart the genes that either inhibit phytopathogens or render the plant insusceptible to the pathogen (van der Biezen, 2001). This can be achieved by many means; the most common ones include genes responsible for the synthesis of antimicrobials (antibiotics, antimicrobial peptides, and hydrolytic enzymes-chitinases), metabolites (ethylene, salicylic acid, and protease inhibitors), and induced systemic resistance (structural alterations and pathogen-related (PR) elicitors) (Mourgues et al., 1998; Ho et al., 2006). Most of the time, the transgene works synergistically with other cellular metabolites (phytoalexins and defensins) in plants while conferring disease resistance.

18.3.1 Antimicrobials

Antimicrobials are the pathogen inhibitory compounds produced by the transgene in the host plant capable of complete inhibition of pathogen growth and proliferation in host plant

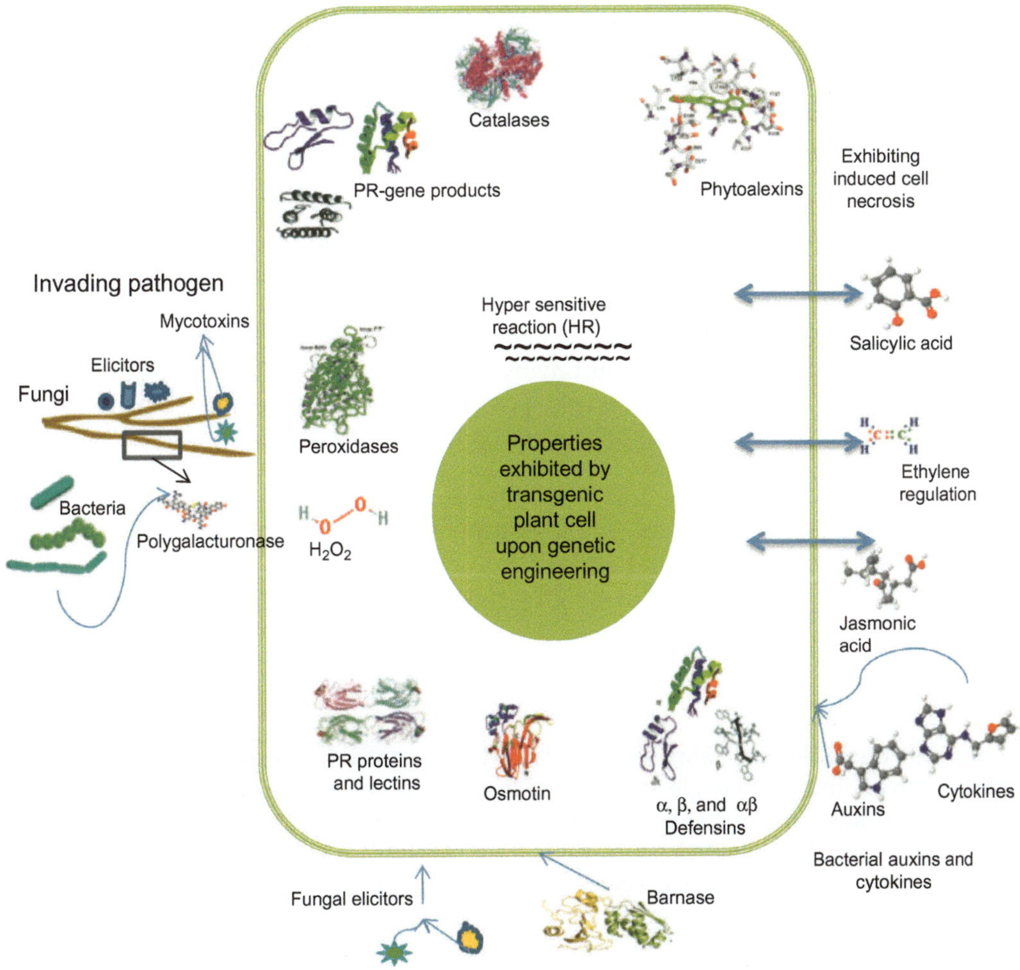

FIG. 18.2 Transgenic plants with enhanced disease resistance have been engineered to express gene products to counterattack invading pathogen through the expression of transgene products inside of cell or outside of cell (membrane attached).

upon entry. Antimicrobials vary depending on the source and can be of different types based on the core functional molecule (lipopeptide-lipid chain linked to cyclic protein and antimicrobial peptides—short sequences of amino acids) (Eulgem et al., 2017).

18.3.2 Antimicrobial Peptides (AMPs) Antibiotics

Antimicrobial peptides (AMPs), also known as host defense peptides (HDPs), are α-helical structures of short amino acid sequences found in many organisms, including microbes

(Ilyas et al., 2017). These AMPs play a major role in innate immunity of the host as they are present on the surface of the cell membrane and cause lipid bilayer to disintegrate and trigger autolysis of the cell via internal hydrolytic enzymes. For the insertion of AMP production, gene chloroplast is a better choice over nuclear transformation, as there is a lesser risk of transfer of the transgene into weeds or other invasive plants. Chloroplast expression gives a high copy number required to inhibit the pathogen; the insertion of AMP gene as inducible gene is more stable and devoid of resistance development risk (Lin et al., 2010). Microorganisms like *Bacillus brevis* (gramicidins), *Bacillus licheniformis* (bacitracins), *Paenibacillus polymyxa* (fusaricidins), *Pseudomonas fluorescens* (viscosins), *P. syringae* (syringopeptins), *P. putida* (putisolvins), *P. tolaasii* (tolaasins), and *B. subtilis* (rhizocticin and bacilysin) produce short peptide sequences with certain modifications (lipid attachment or cyclization of amino acids) and many more from fungal species that are available (Montesinos, 2007). The transformation of crops with AMP genes has been shown to provide disease resistance against many phytopathogens like *B. cinerea*, *V. albo-atrum*, *Phytophthora parasitica*, *Clavibacter michiganensis*, *and P. syringae* pv. *tabaci* in crops like tobacco, tomato, potato, rice, poplar, and canola. Most of the literature suggests the major portion of AMPs be utilized in genetic engineering, employing plant or animal analogue encoding genes only, and microbial AMP gene sources are barely utilized. This makes the bacterial AMPs a potential source of antibiotic genes for crop transformation to enhance disease resistance (Montesinos, 2007).

18.3.3 Enzymes

Enzymes, the biological catalysts, are also effective pathogen suppressors by hydrolyzing the cell wall, hyphae, etc., leading to the lysis of the cell and preventing its subsequent infection. The most effective enzyme class in this category is chitinases, as they are capable of degrading fungal hyphae where chitin is the core structural element. Chitinase production genes from many sources (*S. marcescens*, *Streptomyces* sp., and *R. oligosporus*) have been studied and introduced into crops including tobacco and potato exhibiting exceptional pathogen resistance. The most effective chitinase without question is endochitinase from *T. harzianum*, which showed around 400 times increased production of chitinase and has been successfully incorporated into crops like tobacco, potato, apple, petunia, and grape, demonstrating resistance toward pathogens like *A. alternata*, *A. solani*, *B. cinerea*, and *R. solani* (Cao et al., 2017; Singh et al., 2018). In addition to chitinases, many microbial catalysts exhibit pathogen suppression potential, including *B. amyloliquefaciens* (barnase and barstar, potato), *Aspergillus niger* (glucose oxidase, potato), *S. pombe* (ribonuclease, tobacco), and *Fusarium sporotrichioides* (trichodiene synthase, tobacco) (Park et al., 2001). Apart from the hydrolytic activity of these enzymes, they also produce reactive oxygen species (ROS) like H_2O_2 and reactive oxygen, which disrupt the pathogen virulence factors and elicitation of hypersensitive reaction, leading to necrosis of the infected area and further leading to the suppression of pathogenicity and reducing the spread of the infection. Enzymes like barnase and barstar counteract each other and are coincorporated into crop plants to reduce the barnase activity (a ribonuclease) on healthy tissue, and when challenged with *Phytophthora infestans*, no effect was observed on uninfected sites; similarly, *S. pombe* gene *pac1* was effective against viral dsDNA.

18.4 INDUCED SYSTEMIC RESISTANCE

Many transgenes producing compounds capable of inducing hypersensitive reaction leading to induced systemic resistance toward pathogen have been employed in genetically engineering disease resistance (Jain et al., 2018). Particularly, genes producing compounds that in turn synergistically function with host cellular machinery and induce a hypersensitive reaction or pathogen-related avirulence (Avr) and resistance (R) genes that govern the plant susceptibility toward a particular class of pathogens have been the subject of extensive research interest and exploited for transforming the crop varieties (Cao et al., 1998). The functions of these genes can be seen as varying either structural elements by HR reaction or the interaction between avirulence and resistance gene products or their elicitors.

18.4.1 Structural Alterations

Plant-pathogen interaction mainly depends on the recognition and binding of the pathogen receptors on plant membranes at the cellular level initiating infection. Altering the structural characteristics would directly affect the infection process, thus intervening many genes coding for enzymes, toxins creating necrotic lesions, and changes in pathogens binding to the cell membrane is a potential route for protecting crops (Stefani and Hamelin, 2010). With the potato plant, the incorporation of bacterio-opsin (bO) gene into plants that regulate the proton pump conferred local and systemic resistance even in the absence of pathogen. The expression of bO gene in tobacco saw elevated levels of salicylic-acid-induced resistance of many viral pathogens and absolute resistance to *P. syringae* pv. *tabaci*. The argK gene encoding ornithine transcarbamylase rendered tobacco plant insensitive to the phaseolotoxin; similarly, transgenic beans were completely resistant to *P. syringae* pv. *phaseolicola* (Wani et al., 2010). Unsaturated oligogalacturonates, formed from pectic degradation present in the plant cell wall, are thought to bring about defense reaction; based on this pectate lyase, similar enzymes in cell membranes will enhance the plant disease resistance by inducing systemic resistance.

18.4.2 Pathogen Related (PR) Elicitors

Genes that may protect plants by activating defense responses or inhibiting pathogen virulence factors have also been exploited in engineering resistance. The development of Avr and R genes against each other is the result of an arms race between pathogen and plant through evolution. The resistance genes or gene cluster in some cases of bacteria and fungi can be utilized for genetically improving crops to either increase pathogen resistance or eliminate the interacting receptors from the tissues of the plant (Di et al., 2010). Most of the cases involve the insertion of a pathogen-inducible promoter to function as a regulator and avoid the generalization of HR reaction, finally leading to the plant death. The genes inf1 and Avr9 from *P. infestans* and *Cladosporium fulvum* showed inducible HR reaction in tomato and tobacco against potato virus X (PVX). The *ecp2* gene from *C. fulvum* was found to be effective in tomato through induction of HR by *ecp2* protein expressed by the host transgene. These results signify that HR may be a very efficient means to shield plants against different pathogens independent of the causal agent. Disease confrontation reactions triggered by elicitors apart from avirulence gene products have been also exploited to augment plant resistance to pathogens. Cryptogein, an elicitor protein from *P. cryptogea*, imparted resistance to *P. parasitica* var. *nicotianae* infection in tobacco by liberating upon the pathogen at infection site, creating an HR reaction and later leading to resistance toward *Thielaviopsis basicola*, *Erysiphe cichoracearum*, and *B. cinerea*.

18.5 CONCLUSION AND FUTURE PROSPECTS

The genetic engineering of crops may not be an entirely new concept, but right now, it is at a deciding junction as the need and yield from agriculture are hardly balancing. The continuous growth in demand for food requires robust but sustainable approaches, leaving only genetic engineering as a viable option as it can utilize many diverse sources of genes to impart disease resistance to crops, contrasting to conventional plant breeding where gene sources are limited for a said trait. Being said so, the transgenes employed and expressed in plants to enhance disease resistance need to be assayed and demonstrated to function optimally over long periods of time under diverse environmental and soil conditions before utilizing on a large scale. The lack of proper awareness, involvement of foreign genes, stringent regulations, and certain failures in transgenic varieties at field level have led to a conflict of opinion between transgenic crop producers and others. To address these concerns, protocols that assess the efficacy of transgenic crop varieties should be established based on source, target species, host, and residual impact on the microflora of the rhizosphere. The selection of source for genetic engineering of plants should be supported by the stability of the gene product, efficacy, impact on consumers, and the environment. The transformation of crops should have inducible transgenes instead of constitutive as the inducible genes express only when triggered by a stimuli (pathogen). The choice of a microbial gene for transgenic expression in the plant must seriously consider the consequences of such a transgene on native microbial populations. A responsible use of fungal and bacterial genes for genetic plant improvement should produce a minimal effect on the ecosystem, have very low chances of horizontal transfer possibility, and contribute to more productive and less chemically dependent agriculture.

Acknowledgments

The authors are thankful to DST-SERB N-PDF division for providing the necessary financial funding under sanction order DST-SERB N-PDF/2015/000777 and Institute of Frontier Technology, Regional Agricultural Research Station, Acharya NG Ranga Agricultural University, Tirupati, for providing the necessary research facilities.

Conflict of Interest

We declare that there is no conflict of interest regarding the publication of this chapter.

References

Araki, M., Ishii, T., 2015. Towards social acceptance of plant breeding by genome editing. Trends Plant Sci. 20, 145–149.

Baltes, N.J., Gil-Humanes, J., Voytas, D.F., 2017. Genome engineering and agriculture: opportunities and challenges. In: Progress in Molecular Biology and Translational Science. vol. 149. Academic Press, pp. 1–26.

Butler, N.M., Jiang, J., Stupar, R.M., 2018. Crop improvement using genome editing. Plant Breed. Rev. 41, 55–101.
Cabello, J.V., Lodeyro, A.F., Zurbriggen, M.D., 2014. Novel perspectives for the engineering of abiotic stress tolerance in plants. Curr. Opin. Biotechnol. 26, 62–70.
Cao, H., Li, X., Dong, X., 1998. Generation of broad-spectrum disease resistance by overexpression of an essential regulatory gene in systemic acquired resistance. Proc. Natl. Acad. Sci. U. S. A. 95, 6531–6536.
Cao, Y., Hinkle, G.J., Slater, S.C., Chen, X., Goldman, B.S., Monsanto Technology LLC, 2017. Expression of microbial proteins in plants for production of plants with improved properties. U.S. Patent Application 15/612,580.
Di, R., Blechl, A., Dill-Macky, R., Tortora, A., Tumer, N.E., 2010. Expression of a truncated form of yeast ribosomal protein L3 in transgenic wheat improves resistance to Fusarium head blight. Plant Sci. 178, 374–380.
Düring, K., 1996. Genetic engineering for resistance to bacteria in transgenic plants by introduction of foreign genes. Mol. Breed. 2, 297–305.
Eulgem, T., Rodriguez-Salus, M., Knoth, C., 2017. Molecules that induce disease resistance in plants. U.S. Patent Application 15/453,759.
Fray, R.G., Grierson, D., 1993. Identification and genetic analysis of normal and mutant phytoene synthase genes of tomato by sequencing, complementation and co-suppression. Plant Mol. Biol. 22, 589–602.
Ho, N.H., Baisakh, N., Oliva, N., Datta, K., Frutos, R., Datta, S.K., 2006. Translational fusion hybrid Bt genes confer resistance against yellow stem borer in transgenic elite Vietnamese rice (*Oryza sativa* L.) cultivars. Crop Sci. 46, 781–789.
Ilyas, H., Datta, A., Bhunia, A., 2017. An approach towards structure based antimicrobial peptide design for use in development of transgenic plants: a strategy for plant disease management. Curr. Med. Chem. 24, 1350–1364.
ISAAA, 2016. Global Status of Commercialized Biotech/GM Crops: 2016. ISAAA Brief No. 52. ISAAA, Ithaca, NY.
Jain, S., Choudhary, D.K., Sharma, K.P., Aggarwal, R., 2018. Bacterial mediated plant protection: induced systemic resistance in soybean. In: Microbial Biotechnology. Springer, Singapore, pp. 193–206.
Kamble, S., Mukherjee, P.K., Eapen, S., 2016. Expression of an endochitinase gene from Trichoderma virens confers enhanced tolerance to Alternaria blight in transgenic *Brassica juncea* (L.) czern and coss lines. Physiol. Mol. Biol. Plants 22, 69–76.
Lin, Y.H., Huang, H.E., Wu, F.S., Ger, M.J., Liao, P.L., et al., 2010. Plant ferredoxin-like protein (PFLP) outside chloroplast in Arabidopsis enhances disease resistance against bacterial pathogens. Plant Sci. 179, 450–458.
Lorito, M., Scala, F., 1999. Microbial genes expressed in transgenic plants to improve disease resistance. J. Plant Pathol. 1999, 73–88.
Marco, F., Bitrián, M., Carrasco, P., Rajam, M.V., Alcázar, R., Tiburcio, A.F., 2015. Genetic engineering strategies for abiotic stress tolerance in plants. In: Plant Biology and Biotechnology. Springer, New Delhi, pp. 579–609.
Mittler, R., Blumwald, E., 2010. Genetic engineering for modern agriculture: challenges and perspectives. Annu. Rev. Plant Biol. 61, 443–462.
Montesinos, E., 2007. Antimicrobial peptides and plant disease control. FEMS Microbiol. Lett. 270, 1–11.
Mourgues, F., Brisset, M.N., Chevreau, E., 1998. Strategies to improve plant resistance to bacterial diseases through genetic engineering. Trends Biotechnol. 16, 203–210.
Naqvi, S., Farré, G., Sanahuja, G., Capell, T., Zhu, C., Christou, P., 2010. When more is better: multigene engineering in plants. Trends Plant Sci. 15, 48–56.
Nejat, N., Rookes, J., Mantri, N.L., Cahill, D.M., 2017. Plant–pathogen interactions: toward development of next-generation disease-resistant plants. Crit. Rev. Biotechnol. 37, 229–237.
Park, J.M., Park, C.J., Lee, S.B., Ham, B.K., Shin, R., Paek, K.H., 2001. Overexpression of the tobacco Tsi1 gene encoding an EREBP/AP2-type transcription factor enhances resistance against pathogen attack and osmotic stress in tobacco. Plant Cell 13, 1035–1046.
Powles, S.B., 2018. Herbicide Resistance in Plants: Biology and Biochemistry. CRC Press.
Punja, Z.K., 2001. Genetic engineering of plants to enhance resistance to fungal pathogens—a review of progress and future prospects. Can. J. Plant Pathol. 23, 216–235.
Rojas, C.A., Hemerly, A.S., Ferreira, P.C.G., 2010. Genetically modified crops for biomass increase. Genes and strategies. GM Crops 1, 137–142.
Silva, H., Klessig, D.F., 1998. Engineering disease and pest resistance in plants. Trends Microbiol. 6, 54–61.
Silva, M.S., Arraes, F.B.M., de Araújo Campos, M., Grossi-de-Sa, M., Fernandez, D., et al., 2018. Potential biotechnological assets related to plant immunity modulation applicable in engineering disease-resistant crops. Plant Sci. 270, 72–84.
Singh, H.R., Hazarika, P., Agarwala, N., Bhattacharyya, N., Bhagawati, P., et al., 2018. Transgenic tea over-expressing *Solanum tuberosum* endo-1, 3-beta-d-glucanase gene conferred resistance against blister blight disease. Plant Mol. Biol. Report., 1–16.

Stefani, F.O.P., Hamelin, R.C., 2010. Current state of genetically modified plant impact on target and non-target fungi. Environ. Rev. 18, 441–475.

van der Biezen, E.A., 2001. Quest for antimicrobial genes to engineer disease-resistant crops. Trends Plant Sci. 6, 89–91.

Wani, S.H., Haider, N., Kumar, H., Singh, N.B., 2010. Plant plastid engineering. Curr. Genomics 11, 500–512.

Zhang, Y., Liu, T., Meyer, C.A., Eeckhoute, J., Johnson, D.S., et al., 2008. Model-based analysis of ChIP-Seq (MACS). Genome Biol. 9, R137.

Zhang, F., Ruan, X., Wang, X., Liu, Z., Hu, L., Li, C., 2016. Overexpression of a chitinase gene from Trichoderma asperellum increases disease resistance in transgenic soybean. Appl. Biochem. Biotechnol. 180, 1542–1558.

Further Reading

Broglie, R., Broglie, K., Roby, D., Chet, I., 1993. Production of transgenic plants with enhanced resistance to microbial pathogens. In: Transgenic Plants. vol. 1. Academic Press, New York, pp. 265–276.

SECTION 4

ENVIRONMENTAL MICROBIOLOGY

CHAPTER

19

Recent Approaches in the Production of Novel Enzymes From Environmental Samples by Enrichment Culture and Metagenomic Approach

Ravuri Jaya Madhuri, Maddu Saraswathi, Krishnamoorthy Gowthami, Maram Bhargavi, Yaramolu Divya, Velluri Deepika

Department of Applied Microbiology, Sri Padmavati Mahila Visvavidyalayam, Tirupati, India

OUTLINE

Recent Developments in Applied Microbiology and Biochemistry
https://doi.org/10.1016/B978-0-12-816328-3.00019-2

19.1 INTRODUCTION

Microorganisms are found everywhere in nature, making soil, water, and air as their habitat. Microbes thrive in diverse environmental conditions such as different degrees of temperature, radiation, pressure, gravity, vacuum, desiccation, salinity, pH, oxygen tension, and chemical extremes and overcome these limitations by specific evolutionary adaptations. In these diverse environments, microbes, out of inevitability to survive, develop an array of communities that are adapted for given environmental conditions and/or limitations (Cowan et al., 2015). The conservative estimate suggests that the number of characterized and studied organisms is only 0.1%–1.0% of the total living biota on the planet (Mora et al., 2011). Certain extreme environments, mentioned previously, have a very rich species diversity, particularly of microorganisms as revealed by recent DNA-based studies, and the discovery of new species is ever increasing thanks to the modern sampling and culturing tools and methods (Gascon et al., 2015). One such habitat of such rich biodiversity is marine environments where multiple factors control and limit the biological activities of the inhabitants, and recent literature suggest that marine environments are potential habitat for inimitable metabolite (enzymes, antibiotics, biosurfactants, and antiviral and anticancer compounds)-producing microorganisms (Li et al., 2012; Cragg and Newman, 2013; Laurila-Pant et al., 2015). The major hurdle to optimally utilize the marine habitat's potential is the inability to grow the majority of the microbial species in artificial culture medium due to the very specific growth requirements (Zengler et al., 2002).

Therefore, the endeavors undertaken to further understand the untapped marine environments are essential to utilize and understand the metabolic pathways and pave the way for new methods of drug synthesis becoming available via the discovery of novel enzymes and metabolites (Laurila-Pant et al., 2015). Enzymes are the primary example of natural compounds with a rapidly increasing range of sources and applications (Li et al., 2012). The enzymes occurring biologically are industrially important as they mediate reactions that are difficult or expensive to complete through chemical synthesis (Bisswanger, 2014). These biological catalysts have advantages over nonbiological catalysts in that they are enantioselective and are capable of acting on a wide range of substrates (Wallace and Balskus, 2014). Microbial enzymes are known to play a critical role as metabolic catalysts, leading to their use in various industries and applications (Gurung et al., 2013). Currently, the applications of enzymes are mainly focused in industries like pulp and paper, leather, detergents, textiles, pharmaceuticals, chemical, food and beverages, biofuels, and animal feed (Adrio and Demain, 2014). Today, there is a need for new, improved, or/and more versatile enzymes in order to develop more precise, sustainable, and economically competitive industrial production processes. In addition, there is a global political drive to promote white biotechnology as a central feature of the sustainable economic future of modern industrialized societies, which makes the development of distinctive enzymes, unique processes, exceptional products, and inimitable applications imperative.

Modern molecular techniques such as "metagenomics" that are primarily culture-independent technologies directly assess the availability of specific gene(s) responsible for the production of unique metabolites, including novel enzymes in given environmental sample promises to provide new molecules with diverse functions, but ultimately, expression systems are required for any new enzymes and bioactive molecules to become an economic success (Berini et al., 2017). It is disappointing to know that the uncultivable microorganisms from extreme environments are the principal resources for these exceptional enzymes, isolated from different environments through following different strategies of isolation and selection procedures. Isolation and selection by enrichment cultivation methods were employed for selecting different bacterial species involved in xenobiotic degradation through enzymes such as azoreductases and depolymerases (*Bacillus* spp., *Pseudomonas* spp., *Nocardia* spp., *Streptomyces* spp., *Aspergillus* spp., and *Alternaria* spp.) (Jain et al., 2011). Similarly, speciality enzymes with multiple application potential like chitosanases, depolymerases, anticancer enzymes, and fibrinolytic enzymes are majorly isolated from fungal, bacterial, algal, and actinomycete species of marine sources, and they hold more potential of such enzymes and molecules (Mane and Tak, 2015; Bhargavi and Jaya Madhuri, 2016; Bhargavi and Jaya Madhuri, 2017; Parte et al., 2017; Velumani, 2017; Mohammadipanah and Momenilandi, 2018).

19.2 PRODUCTION OF NOVEL ENZYMES FROM ENVIRONMENTAL SAMPLES BY ENRICHMENT CULTURE

Enrichment culture is basically an isolation technique designed to make conditions of growth very favorable for an organism of interest while having an unfavorable environment for any competition. It is the use of certain growth media to favor the growth of a particular microorganism over others, enriching a sample for the microorganism of interest. This enrichment is generally done by introducing nutrients or environmental conditions that only allow the growth of an organism of interest. Enrichment culture techniques are used to increase a small number of the desired organisms to detectable levels. This allows for the detection and identification of microorganisms with a variety of nutritional needs (Liu et al., 2016). Enrichment culturing is often employed for isolating microbial species of interest from soil and marine habitats. The enrichment culturing includes two methods depending on type of media utilized: One is submerged (liquid medium), which is easy to control and optimize environmental and nutritional parameters, and the other is solid-state fermentation in which solid materials (raw or processed) are used as substrates (Mahitha and Madhuri, 2016). Enrichment-dependent novel enzyme production in liquid fermentation method commonly involves the utilization of physical parameters like temperature, pH, incubation time, static and shaking conditions, and/or chemical properties like carbon and nitrogen sources and substrate and inoculum concentration as determining factors for the enzyme-producing isolates, thus providing the chance to grow and proliferate only the adapting species to the determining physical and/or chemical factor (Shah and Patel, 2014). If the case is solid media utilization, certain factors have no influence as they are not maintained like static and stirring of the fermentation medium, and some new factors come into play as the substrate type (raw or processed) (Figs. 19.1 and 19.2).

The enrichment culture technique primarily employs substrates that are inexpensive (Jambul seeds; agrowastes for enzymes like depolymerase, glutaminase, asparaginase, and bagasse; distillery effluent—asparaginase; dairy waste whey; and fermented food waste for fibrinokinase) as the overall isolation process should be economical and repeatable until the

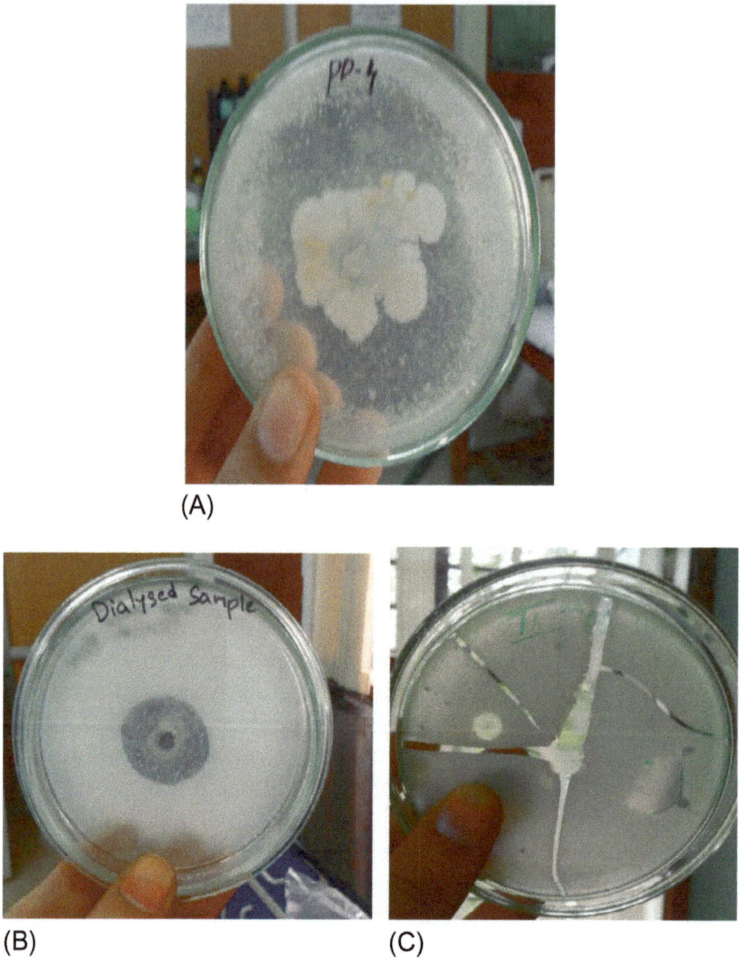

FIG. 19.1 (A) Depolymerase activity indicated by formation of halo zone around the colony utilizing plastic as a nutrient source at the end of the seventh week. (B) Depolymerase activity in purified sample at the end of 1-week incubation.

target species is isolated (Siddiqui et al., 2015; Vijayaraghavan et al., 2017). The enrichment technique also differs in the isolation procedure based on enzyme type like extracellular or intracellular in nature; if extracellular, the presence of a soluble substrate is a universally utilized method as it gives the indication of enzyme activity through halo or clear-zone development that is easy to distinguish when compared with intracellular where it often requires destruction of the cell, addition of multiple reagents, and separate addition of substrate or secondary substrate to assess whether the target activity is present or not. The enrichment method of isolation was a powerful technique utilizing in many applications like selective enrichment, mimicking habitat to enrich microbes from extreme habitations in the search for many metabolites, antibiotics, and enzymes that are crucial in many sectors. The prospective effectiveness of the methods was multiplied manyfold by adding features like optimization

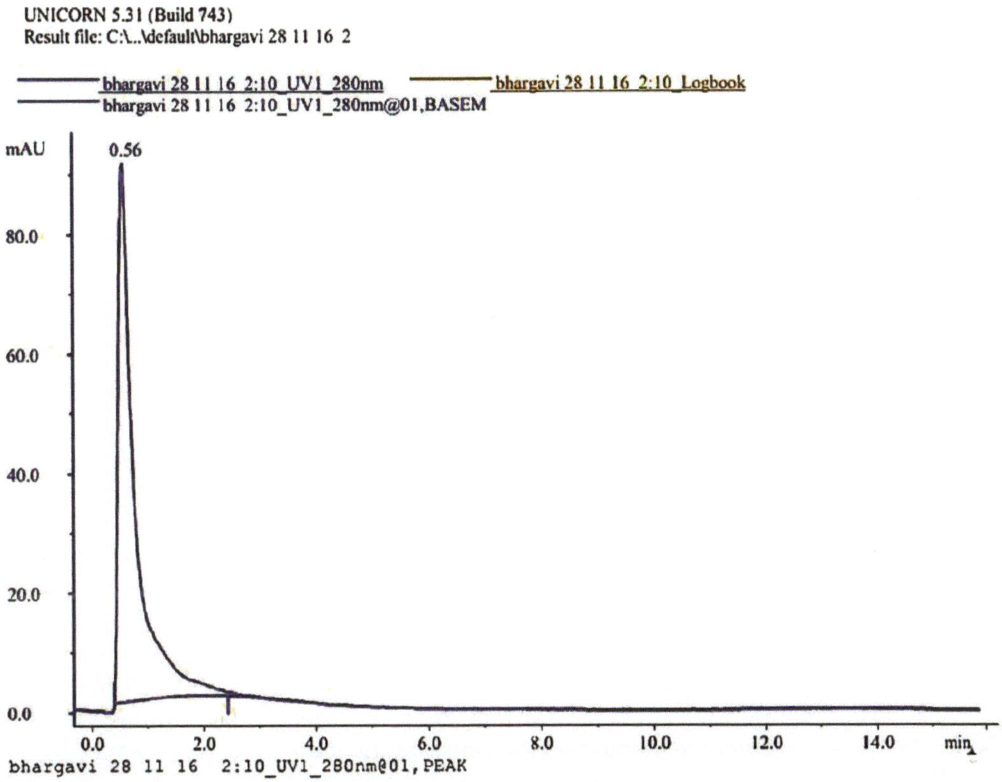

No	Ret min	Area mAU*min	Height mAU
1	0.56	31.1446	90.598

Total number of detected peaks = 158
Total area = 32.0620 mAU*min
Area in evaluated peaks = 31.1446 mAU*min
Ratio peak area / total area = 0.971386
Total peak width = 2.04 min
Calculated from : bhargavi 28 11 16 2:10_UV1_280nm
Baseline : bhargavi 28 11 16 2:10_UV1_280nm@01,BASEM
Peak rejection on:
Maximum number of peaks: 20

FIG. 19.2 Purification of asparaginase enzyme by gel-filtration chromatography.

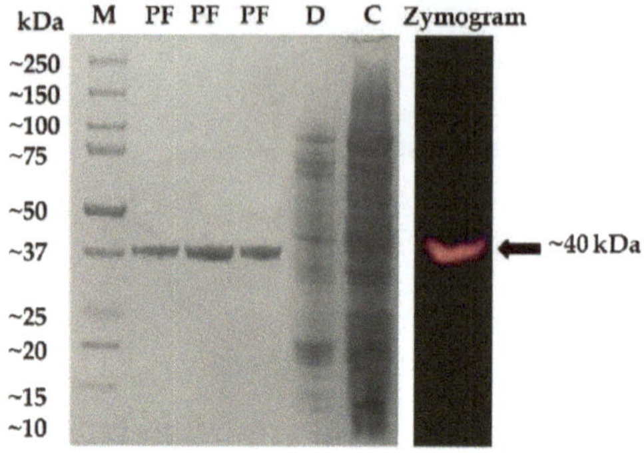

M - Protein marker (Precision Plus Protein™ Dual Color Standards- Biorad)
PF - Positive fraction
D - Dialysate
C - Crude culture filtrate

FIG. 19.3 Characterization of purified azoreductase enzyme by SDS-PAGE zymogram.

of selecting method, innovations in fermentation technology, and secondary modifications of the initially selected organisms. The final active compound's functionality can also be enhanced by increasing the purity of the compound by ammonium sulfate precipitation, dialysis, and chromatographic separation coupled with the determination of molecular weight using SDS-PAGE method. But, having said that, the fundamental disadvantage it encompasses is the dependency of the method on the cultivability of the microorganism being studied; sadly, only a fraction (<1%) of total microbial diversity in a given habitat is culturable, leaving out the rest being untapped (Fig. 19.3).

19.3 METAGENOMIC APPROACHES FOR DETECTION OF NOVEL ENZYMES

The rapid growth in industrial biotechnology in manufacturing of functional metabolites (antibiotics, enzymes, anticancer therapeutics, antiviral agents, surfactants, etc.) warranted the search for metabolites with unique, exceptional, and incomparable properties to mediate the synthesis process of new compounds and molecules that otherwise could not materialize. The potential source for such agents without doubt is microbial domain, but the culture-dependent methods are not efficient enough to study the diversity where a vast majority (>99%) of species are uncultivable. In order to tap into the nonculturable microbial diversity in the environment, DNA-based methods were developed that directly assess the presence of a gene for the novel metabolites (enzymes) irrespective of the cultivability of the organisms, and at the same time, the complete diversity can be assayed in a short time period. Application of metagenomics promises the success in the search for new molecules and metabolites with diverse functions; we can target microorganisms that are recalcitrant to isolation on common laboratory medium (Natarajan, 2018).

19.4 METAGENOMIC STRATEGIES FOR NOVEL MARINE ENZYME DISCOVERY

The marine biodiversity is an excellent source for acquiring an inventory of enzymes that can be used for a variety of biotech applications. This diversity has, to date, not been fully exploited. Extensive quantity of literature has emerged in recent times on isolating lipase enzymes of microbial origin, due to its multiple applications in detergent and fabric maintenance industry. DNA mined from environmental samples can be cloned into plasmids, fosmids, bacterial artificial chromosomes, or cosmids for proliferation in an appropriate heterologous host organism, such as *Escherichia coli*, and then be screened for catalytic activities. The rapid progress of efficient selection of metagenomic libraries to uncover a new enzymatic activity has indicated the significance of microbial multiplicity in the novel enzyme recognition. DNA purity from marine samples has also been a problem because of the intricate and varied nature of the marine environments and the role of coextracted substances, such as humic acids, that restrain biochemical reactions (Corinaldesi et al., 2005).

Presently, the most preferred and the best host system for the construction of heterologous protein and for screening in metagenomic libraries is *E. coli* as the judgmental factors are dependent on the nature of target protein, such as thermotolerance, and also, *E. coli* is a well-studied model organism and most used organism for heterologous protein construction. Thus, the *E. coli* strains like *E. coli* BL21 and K12 are the most used hosts for industrial protein construction. Other substitute bacterial hosts were also used as the number of heterologous proteins expressed in a given host is limited, like *Bacillus brevis*, *B. megaterium*, and *Bacillus subtilis*. *B. subtilis* and other *Bacillus* strains were suggested to be the most well-known microbes for the metagenomic library screening and heterologous protein construction. *Bacillus* species boast additional benefits in the field of protein fabrication and selection in metagenomic libraries for industrial purposes, due to the deficiency of lipopolysaccharides, and they are known endotoxin promoting antigens (Zweers et al., 2008). While metagenomics may aid progress in our understanding of microbial physiology, genetics, and community ecology (even provides insights into the uncultivable communities of different habitats, especially extreme ones), it can be an advanced and dominant tool for finding out a novel enzyme that is useful for biotechnology application and industrialization.

19.4.1 Sequence-Based Approach

In order to recover a gene encoding for a novel biocatalyst from genomic DNA (gDNA) or environmental DNA (eDNA), a common methodology is the targeted amplification of the gene employing primers to similar known enzymes by PCR or sequencing of the entire genome using the next-generation sequencing technology (shotgun or DNA library sequencing) and mining the target while compared with the available metagenomic database comparing its sequences against genomic databases. The latter method was previously expensive and laborious (long segment sequencing—Sanger's method), but due to second (e.g., 454, MiSeq Illumina, and Ion Torrent)- and third (e.g., PacBio)-generation sequencing methods, the technology is economical and rapid (Faure and Joly, 2015). The sequences of the target gene and the known gene sequences should be similar at least in order to find novel enzymes; it is very difficult and rare to find completely novel enzymes or enzyme families employing a

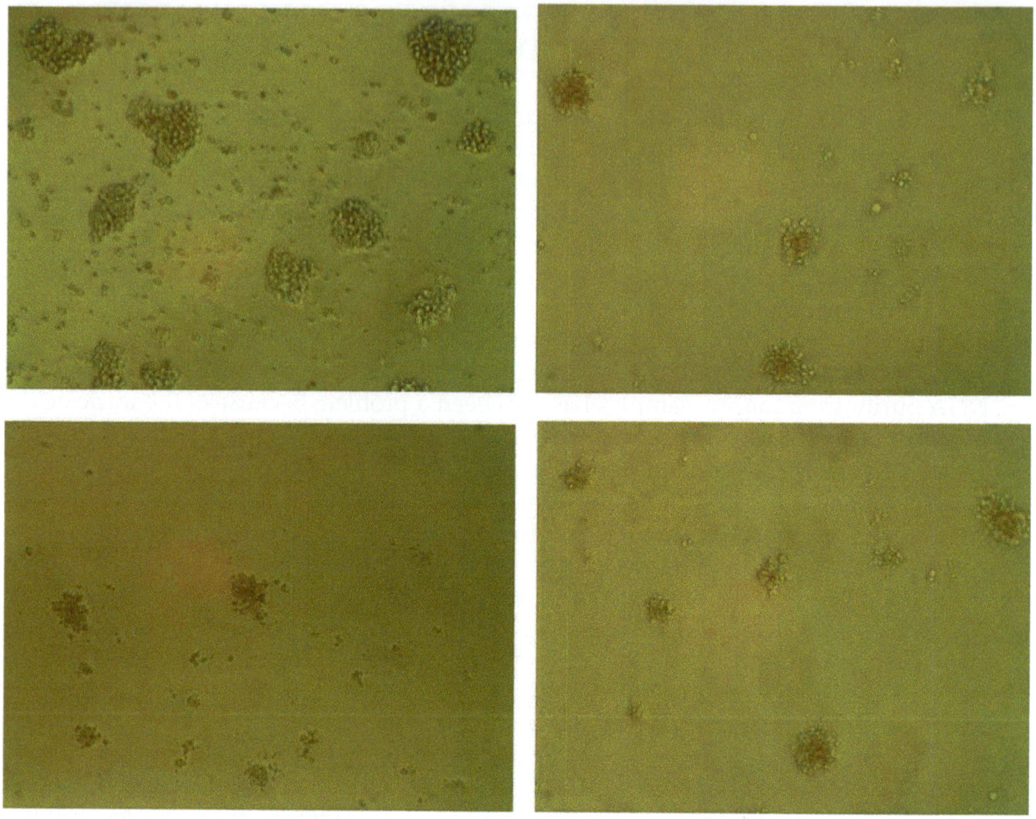

FIG. 19.4 Anticancer activity of glutaminase and asparaginase on acute lymphoblastic leukemia cell lines Jurkat J_6 cell lines.

sequence-based prospecting of metagenomic data. Sequence-based methods are preferred, commonly to investigate the microbial multiplicity of an environment or habitat by sequence exploration of 16S or 18S rRNA genes or to study the gene sequential arrangement in the genome of a species. Even with the simplification of sequence analysis and advancements in bioinformatic tools, it might still be difficult to specifically find a sequence that might be responsible for a novel enzyme production from vast data produced or to predict its characteristics (activity or heterologous expression) (Fig. 19.4).

19.4.2 Function-Based Approach

The genomic DNA from a target microorganism or the eDNA from a premeditated milieu might be used to manufacture (meta) genomic libraries into the cultivable host cell and employ them to detect a concerned biological activity. Functional (meta) genomics relies exclusively on gene function rather than similarities in the obtained sequences and demonstrates the substantial advantage when applied to unique strains, species, genera, or unknown bacteria, as it has a high likelihood of yielding genes encoding novel enzymes. Specific enzymatic activities can also screen by controlling physical and chemical conditions like temperature, pH, and substrate concentration. Extraordinary novel enzymes from diverse habitats from around the world have been discovered by functional (meta) genomics termed as "hypothetical proteins" in different databases (Ferrer et al., 2016). The functional screening is very time-consuming and laborious in nature as large quantities of sequences have to be assayed and yields are too low due to detrimental factors like heterologous expression in the chosen host cell, substrate affinity, or a missing cofactor (Ekkers et al., 2012). Functional metagenomics are primarily recommended when working with low-density uncultivable microorganisms. Introduction of robotized high-throughput screening will significantly improve the quantity of screened clones and resulting positive matches. The development of culture techniques involving the approaches like media mimicking the natural habitat intricately, selective enrichment (addition of specific nutrients), and underexplored sample and habitat exploration will inarguably increase the chances of novel biomolecule discoveries (Kamagata, 2015). Emergences like cycling culture (varying growth and culture conditions), utilizing microtiter plates with phenotypic microarrays for growth and novel growth media, have led to the unearthing of exhilarating novel enzymes and sources.

19.5 LIMITATION OF METAGENOMIC APPROACHES OF NOVEL BIOMOLECULE PROSPECTING

The DNA-based metagenomic strategy has its challenges too, the most fundamental of which is the isolation of environmental DNA (eDNA) from a habitat; as the number of copies of the DNA segment or a gene is very low, the chances of isolating a target gene is a very ambitious task to take up (Corinaldesi et al., 2005). In most of the metagenomic studies, the eDNA isolated is fragmented into small segments, leading to the generation of thousands of readouts from DNA sequencing, requiring the assembly of these segments employing bioinformatic tools, making it virtually impossible to sequence complete genome of a microbial species to reconstruct, but development is seen in recent times in this direction but a long way from perfection (Henson et al., 2012). These hindrances can be overcome by advanced technologies like single-molecule real-time sequencing (SMRT) technology, a next-generation sequencing method that will increase current sequencing and readouts of genes (Rhoads and Au, 2015). Another approach is annotation of gene readouts without genome reassembly that is efficient to reveal gene functions, but the short-read sequences generated are again a hurdle making it inadequate to predict gene-product data (Teeling and Glockner, 2012). To enhance the accuracy and lower the time consumption in the generation of data from microbial metagenomics, both amplicon-based method (amplification of rRNA genes as targets) and shotgun sequencing (that generates huge numbers of short sequences from all genes not confining to rRNA genes) can be utilized to assess the communal diversity, thereby envisaging the type of adaptations arising so that the enzymes involved can be predicted and overall microbial diversity with much relevant proportions can be calculated (Woese and Fox, 1970). The success rate of enzyme prospecting could be significantly increased in shorter times when metagenomic sequences obtained by direct sequencing are selected and subjected to high-throughput expression for subsequent functional characterization (Kimura, 2018). The expression host, the most crucial component after the eDNA in metagenomic studies, should also be a

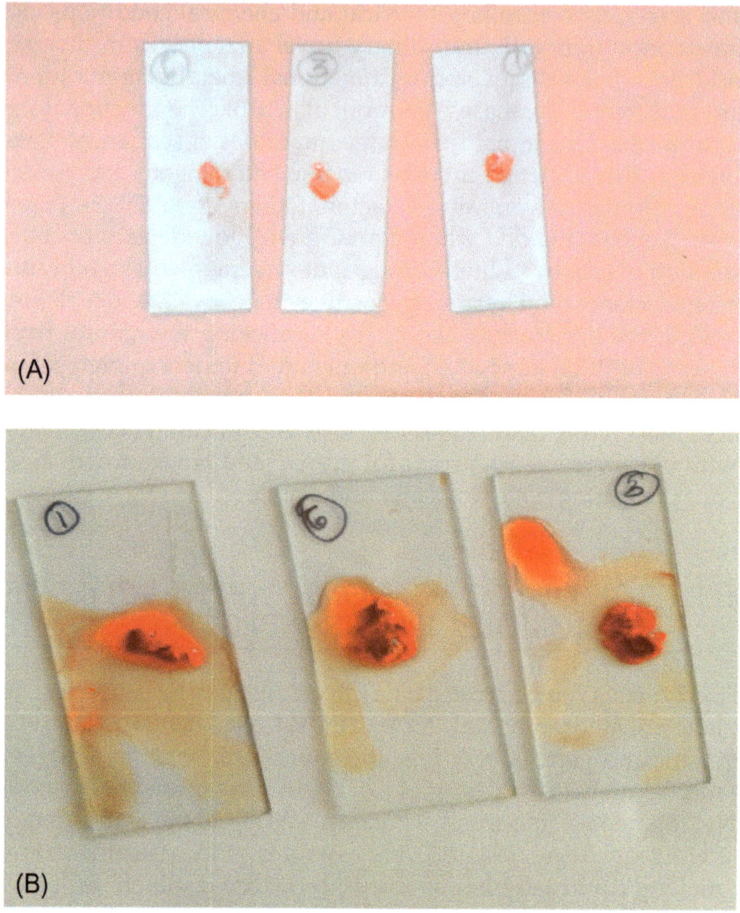

FIG. 19.5 Fibrinolytic activity of GPJ3 on blood clot in slide method: (A) Blood clot as control. (B) Liquefaction of blood clot by positive isolates.

compatible one as it governs the expression of the clone with the target gene (used to characterize the function of the gene isolated from a prospective habitat) (Fig. 19.5).

19.6 CONCLUSIONS

There is a huge demand for the novel enzymes on the industrial market that can be used in industrial bioprocesses. As the classical methods like culture-dependent methods have limitations, exploring the uncharted tertiary of actual microbial biodiversity where the majority of the populations are uncultivable metagenomics offers a promising tool for discovering novel enzymes that have multiple applications significantly influencing many sectors simultaneously. Metagenomics presents the possibility of discovering novel biocatalysts from microbial communities that either confounded cultivation or failed to yield new culture isolates upon repeated attempts. Yet, there are still limitations to overcome, reading vast data, sequence generation limit (NSG gives 200–1000bp lengths), predicting the enzyme nature or properties from metagenomic studies, etc. The combination of function- and sequence-based approaches will improve the output from metagenomic studies to a great extent. Nevertheless, metagenomic analysis of eDNA or gDNA is still the best way to understand the whole habitat diversity and characterize the potential biomolecules from the same.

Acknowledgments

The authors are indebted to all the researchers whom we cited in this review for their significant and valuable research. No funding was received to perform this review.

Conflicts of Interests

The authors declare no relevant competing financial interests to disclose.

References

Adrio, J.L., Demain, A.L., 2014. Microbial enzymes: tools for biotechnological processes. Biomol. Ther. 4, 117–139.

Berini, F., Casciello, C., Marcone, G.L., Marinelli, F., 2017. Metagenomics: novel enzymes from non-culturable microbes. FEMS Microbiol. Lett. 364 (21), https://doi.org/10.1093/femsle/fnx211.

Bhargavi, M., Jaya Madhuri, R., 2016. Anticancer enzyme from marine bacteria. Am. J. Marine Sci. 3 (1), 1–3.

Bhargavi, M., Jaya Madhuri, R., 2017. Evaluation of cytotoxic activity of asparaginase from marine fungi. Int. J. Pharm. Biosci. 21, 43–50.

Bisswanger, H., 2014. Enzyme assays. Perspect. Sci. 1, 41–55.

Corinaldesi, C., Danovaro, R., Dell'Anno, A., 2005. Simultaneous recovery of extracellular and intracellular DNA suitable for molecular studies from marine sediments. Appl. Environ. Microbiol. 71, 46–50.

Cowan, D.A., Ramond, J.B., Makhalanyane, T.P., De Maayer, P., 2015. Metagenomics of extreme environments. Curr. Opin. Microbiol. 25, 97–102.

Cragg, G.M., Newman, D.J., 2013. Natural products: a continuing source of novel drug leads. Biochim. Biophys. Acta 1830, 3670–3695.

Ekkers, D.M., Cretoiu, M.S., Kielak, A.M., Elsas, J.D., 2012. The great screen anomaly—a new frontier in product discovery through functional metagenomics. Appl. Microbiol. Biotechnol. 93, 1005–1020.

Faure, D., Joly, D., 2015. Next-generation sequencing as a powerful motor for advances in the biological and environmental sciences. Genetica 143, 129–132.

Ferrer, M., et al., 2016. Estimating the success of enzyme bioprospecting through metagenomics: current status and future trends. Microb. Biotechnol. 9, 22–34.

Gascon, C., Brooks, T.M., Contreras-MacBeath, T., Heard, N., Konstant, W., et al., 2015. The importance and benefits of species. Curr. Biol. 25, R431–R438.

Gurung, N., Ray, S., Bose, S., Rai, V., 2013. A broader view: microbial enzymes and their relevance in industries, medicine, and beyond. Biomed. Res. Int. 2013, 329121.

Henson, J., Tischler, G., Ning, Z., 2012. Next-generation sequencing and large genome assemblies. Pharmacogenomics 13, 901–915.

Jain, P.K., Gupta, V.K., Gaur, R.K., Lowry, M., Jaroli, D.P., Chauhan, U.K., 2011. Bioremediation of petroleum oil contaminated soil and water. Res. J. Environ. Toxicol. 5, 1–26.

Kamagata, Y., 2015. Keys to cultivating uncultured microbes: elaborate enrichment strategies and resuscitation of dormant cells. Microbes Environ. 30, 289–290.

Kimura, N., 2018. Novel biological resources screened from uncultured bacterial resources by metagenomics methods. In: Nagarajan, M. (Ed.), Metagenomics Perspectives, Methods, and Applications, pp. 273–288.

Laurila-Pant, M., Lehikoinen, A., Uusitalo, L., Venesjärvi, R., 2015. How to value biodiversity in environmental management? Ecol. Indic. 55, 1–11.

Li, S., Yang, X., Yang, S., Zhu, M., Wang, X., 2012. Technology prospecting on enzymes: application, marketing and engineering. Comput. Struct. Biotechnol. J. 2. e201209017.

Liu, X.J., Zhang, J., Hui, H.P., Li, Z.J., 2016. Microbial production and characterization of poly-3-hydroxybutyrate by *Neptunomonas* Antarctica. PeerJ 4, e2291.

Mahitha, G., Madhuri, J., 2016. Microbial polyhydroxybutyrate production by using cheap raw materials as substrates. Indian J. Pharm. Biol. Res. 4, 57–62.

Mane, P., Tak, V., 2015. Overview of microbial therapeutic enzymes. Int. J. Curr. Microbiol. App. Sci. 4, 17–26.

Mohammadipanah, F., Momenilandi, M., 2018. Potential of rare actinomycetes in the production of metabolites against multiple oxidant agents. Pharm. Biol. 56, 51–59.

Mora, C., Tittensor, D.P., Adl, S., Simpson, A.G., Worm, B., 2011. How many species are there on Earth and in the ocean? PLoS Biol. 9. e1001127.

Natarajan, M., 2018. Metagenomics. A paradigm shift in microbiology. In: Metagenomics. Perspectives, Methods, and Applications. Elsevier Publication, pp. 1–13. (Chapter I). ISBN No. 978-08-122068-6.

Parte, S., Sirisha, V.L., Souza, J.S.D., 2017. Biotechnological applications of novel marine enzymes from algae, fungi, bacteria and sponges. Adv. Food Nutr. Res. 80, 75–106.

Rhoads, A., Au, K.F., 2015. PacBio sequencing and its applications. Genomics Proteomics Bioinformatics 13, 278–289.

Shah, M.P., Patel, K.A., 2014. Microbial decolorization and degradation of remazol Black & Mordant Orange by microbial consortia isolated from common effluent treatment plant. Int. J. Environ. Bioremed. Biodegrad. 2, 117–124.

Siddiqui, H.N., Karkun, A., Jangir, S., 2015. Isolation and identification of microorganism to study of their potential to degrade harmful azo dyes by the enzyme azoreductase produced by the microorganism. Scholars Acad. J. Biosci. 3 (6), 541–557.

Teeling, H., Glockner, F.O., 2012. Current opportunities and challenges in microbial metagenome analysis—a bioinformatic perspective. Brief. Bioinform. 13, 728–742.

Velumani, S., 2017. Isolation, screening, characterization and production of fibrinolytics from marine microorganisms. IJARIIE 2 (2), 1632–1642.

Vijayaraghavan, P., Rajendran, P., Prakash Vincent, S.G., Arun, A., Abdullah Al-Dhabi, N., Valan Arasu, M., Young Kwon, O., Kim, Y.O., 2017. Novel sequential screening and enhanced production of fibrinolytic enzyme by Bacillus sp. IND12 using response surface methodology in solid-state fermentation. Biomed. Res. 22, 112–118.

Wallace, S., Balskus, E.P., 2014. Opportunities for merging chemical and biological synthesis. Curr. Opin. Biotechnol. 30, 1–8.

Woese, C.R., Fox, G.E., 1970. Phylogenetic structure of the prokaryotic domain: the primary kingdoms. Proc. Natl. Acad. Sci. U. S. A. 74, 5088–5090.

Zengler, K., Toledo, G., Rappe, M., Elkins, J., Mathur, E.J., Short, J.M., Keller, M., 2002. Cultivating the uncultured. Proc. Natl. Acad. Sci. U. S. A. 99, 15681–15686.

Zweers, J.C., Barak, I., Becher, D., Driessen, A.J., Hecker, M., Kontinen, V.P., 2008. Towards the development of Bacillus subtilis as a cell factory for membrane proteins and protein complexes. Microb. Cell Factories 7, 10.

Further Reading

Li, H., Fei, Z., Gong, J., Yang, T., Xu, Z., Shi, J., 2015. Screening and characterization of a highly active chitosanase based on metagenomic technology. J. Mol. Catal. B Enzym. 111, 29–35.

CHAPTER

20

Development in the Treatment of Municipal and Industrial Wastewater by Microorganism

*Mekapogu Madakka**, *Nadimikeri Jayaraju*[†], *Nambi Rajesh**, *Muni Ramanna Gari Subhosh Chandra*[‡]

***Department of Biotechnology & Bioinformatics, Yogi Vemana University, Kadapa, India**
[†]Department of Geology, Yogi Vemana University, Kadapa, India [‡]Department of Microbiology, Yogi Vemana University, Kadapa, India

OUTLINE

20.1 INTRODUCTION

Safe drinking water and proper sanitation have constantly been recognized as indispensable factors to sustain life. Nevertheless, despite remarkable global progress to improve access to drinking water facilities, currently, there are 884 million and an additional 2.5 billion

Recent Developments in Applied Microbiology and Biochemistry
https://doi.org/10.1016/B978-0-12-816328-3.00020-9

people lacking improved water sources and sanitation, respectively (Naidoo and Olaniran, 2013). This crisis is further compounded by factors such as increasing poverty, accelerated population growth, and rapid urbanization coupled with hydrologic variability and climate change. These socioeconomic and environmental factors place even further stress on the deteriorating water and sanitation infrastructure, more so in developing regions, where billions are still at risk of water, sanitation, and hygiene (WaSH)-related diseases. Despite meeting the Millennium Development Goals regarding access to potable water, the depletion of existing finite water resources still continues to be a major problem, with projections that approximately 605 million people will still lack access to improved drinking water. In addition, the lack of access to potable water is estimated to cost countries between 1% and 7% of their annual GDP, with slow water- and sanitation-related progress further impeding national economic growth (Naidoo and Olaniran, 2014). This together with the abovenamed factors serves as the major driving force behind the increased use of wastewater, surrounding surface water, and gray water for various recreational, agricultural, and aquaculture activities. Over the last few years, the quantity of municipal wastewater produced has drastically increased due to the constant increase in population numbers together with an increased dependence on diminishing water resources. The release of inefficiently treated wastewater into surrounding surface water serves as a threat not only to micro- and macroflora but also to the prerequisite of good-quality water required for all socioeconomic functions. Thus, the constant monitoring of the operational status of existing wastewater treatment plants (WWTPs) and the increasing emphasis on environmental and water resource health have become key factors in determining the quantity and quality of wastewater generated by respective municipalities.

In addition to municipal wastewater, treatment of industrial wastewater becomes increasingly essential to (i) recovering freshwater from industrial wastewater for beneficial uses that is an ever-increasing imperative due to ever-increasing freshwater demand (irrigation water being the largest consumer) due to demographic and economic growth and (ii) reducing environmental impacts. Thereby, the metalloid arsenic (As) is of principal concern, since it is highly toxic to environment and humans and present in many industrial wastewater streams such as those from (i) mining and related activities (ore processing, smelters, etc.); (ii) wood processing industry and wood combustion; (iii) carbon and petroleum exploitation, processing, refining, and combustion; (iv) geothermal exploitation; (v) agrochemical industry (e.g., related to impurities in pesticides) and agricultural wastewater due to their applications; (vi) water treatment sludge; and (vi) municipal solid waste incineration. The available conventional techniques for industrial wastewater (physicochemical) treatments include processes that can be used alone or in combination, such as oxidation, coprecipitation and adsorption onto coagulated flocs, lime treatment, and use of ion exchange resins. Adsorption onto suitable surfaces such as iron oxide (Shih et al., 2015), iron-oxide-coated sand (Yang et al., 2015), manganese oxide (Yu et al., 2015), magnesite (Masindi and Gitari, 2016), granular-activated carbon (GAC) (Podder and Majumder, 2016), zerovalent iron, biochar (Vithanage et al., 2017), and other materials as pollutant removal option. Electrochemical methods and plasmotechnologies are also applied. Another group of technologies is membrane-based applications, particularly reverse osmosis, electrodialysis, electrodialysis reversal, and nanofiltration, as well as new membrane-based applications such as forward osmosis and membrane distillation (Marino and Figoli, 2015) and capillary zone electrophoresis. All of these conventional techniques have the disadvantage that they produce toxic residual waste that requires expensive management and often have limited efficiency, operational difficulty, and high operational cost. Recently, microbial remediation tends to be a promising technology to recover and remediate arsenic in industrial wastewater in an environmentally and economically sound way (Fazi et al., 2015). Such microorganisms are termed as "eco-friendly nanofactories" by some environmentalist. This review highlights the developments in the treatment of municipal and industrial wastewater treatments by using microorganisms.

20.2 TYPES OF BACTERIAL COMMUNITIES IN THE TREATMENT OF MUNICIPAL AND INDUSTRIAL WASTEWATER

In WWTPs, microorganisms are organized into species-rich structures that biodegrade a wider range of substrates than pure cultures. Bacteria constitute from 5% to 20% of organics in flocs; the rest are extracellular polymeric substances (EPS). The size, density, shape, and content of filamentous bacteria determine floc stability and settling. Environmental stress may lead to floc fragmentation and increase the suspended solid concentration in the effluent. Activated sludge and biofilm differ in EPS composition within the same systems (Mahendran et al., 2012). In floc-derived EPS, proteins predominate, whereas in biofilm-derived EPS, proteins and polysaccharides are present in similar proportions. The granule size is correlated with organic load; too high load causes disintegration of granules. The increase in the granule diameter limits the diffusion of nutrients; starving cells derive energy from EPS, which impairs the mechanical strength of granules. The outer layer of granule is inhabited by aerobic microorganisms, while in deeper layers, facultative and obligatory anaerobes develop, which enables the removal of carbon and nutrients in one-stage system.

20.2.1 Selection of Microalgal Species for Wastewater Treatment

Using microalgal cultivation as a tertiary wastewater treatment process has started since 1970. The most desired characteristics of algae for the use in wastewater treatment and biofuel production include higher growth rate, high lipid content and productivity, higher tolerance to the possible pollutants—metal ions and toxic compounds present in the treated wastewater—high NH_4^+ tolerance, high O_2 generation rates, high CO_2 sinking capacity, and robust growth properties with improved tolerance for varied environmental conditions. These criteria are of prime importance, and algal growth has been reported as the limiting factor in nutrient and pollutant removal efficiency of an algae-bacteria consortium. The selection of microalgae for a particular treatment option can be based on the knowledge about the indigenous species in such wastewaters, making use of their characteristics for our advantage. Several microalgal species such as *Chlorella*, *Scenedesmus*, *Desmodesmus*, *Neochloris*, *Chlamydomonas*, *Nitzschia*, and *Cosmarium* have been applied to various types of wastewater treatments coupled with biofuel production under sterilized or nonsterilized conditions. Among them, species of the genera *Chlorella*, *Scenedesmus*, and some *Cyanobacteria* are the most employed species in various wastewater treatments due to their high growth rate, high environmental tolerance, and high lipid/starch accumulation potential (Kim et al., 2016).

In another study, Zhou et al. (2011) isolated microalgal strains from various wastewater treatment sites and found that five strains (of the genera *Chlorella, Hindakia, Scenedesmus,* and *Auxenochlorella*) had higher biomass and lipid productivity in addition to wastewater treatment. *Botryococcus braunii*, a green freshwater microalga, is known for its unique ability to constitutively synthesize and store high quantities of a wide variety of lipids along with wastewater treatment (Chinnasamy et al., 2010). The marine microalga, *Nannochloropsis gaditana,* is useful in treating centrate (concentrated municipal wastewater generated during sludge centrifuge) along with biomass production for lipid yield. *Nannochloropsis gaditana* efficiently removes N and P, being useful for wastewater treatment (Ledda et al., 2015). In addition to microalgae, some brown macroalgae (*Ascophyllum nodosum, Fucus spiralis, Laminaria hyperborea,* and *Pelvetia canaliculata*) are successful in removing transition metal ions from petrochemical wastewaters (Cechinel et al., 2016). These observations can be used as a guideline to choose the microalgal strain for wastewater treatment with additional applications.

20.2.2 Microalgae-Bacteria Consortium for Wastewater Treatment

Utilization of microalgae-bacteria consortium in wastewater treatment can represent a practical choice for sustainable wastewater treatment and support for algal productivity for their potential applications in industry (Tricolici et al., 2014). Microalgae and bacteria grown symbiotically could improve the relevance of microalga-based approaches for the treatment of wastewater (Ryu et al., 2017). In this mutual effort, the oxygen released by algal photosynthesis is utilized by bacteria (aerobic-heterotrophic) to mineralize organic compounds, and bacterial respiration provides CO_2 as a carbon source to the algae (Fig. 20.1). As a result, the advantages are cost-efficient aeration, efficient removal of pollutants, and mitigation of greenhouse gas emissions (Su et al., 2012). Hence, the associated bacteria reduce photosynthetic oxygen tension of phototrophic microalgae by consuming oxygen as their electron acceptor. It is also hypothesized that the algae-bacteria symbiosis is beneficial for algae as they can supply essential nutrients like vitamins and other compounds, as many algae are cobalamin auxotrophs.

Additionally, siderophores present in some bacteria can promote microalgal growth under iron-deficient conditions. In a study, it was shown that the associated bacteria plays a profound role in improving flocculation of *Chlorella vulgaris*, by increasing the floc size resulting in the sedimentation of microalgae (Lee et al., 2013). The cocultivation of microalgae and bacteria not only reduces high costs but also decreases the spatial distance for O_2 and CO_2 exchange, when compared with separate culture units. The formation of microalgae-bacteria consortium is dynamic and is usually divided into four stages:

- *Stage 1*: Adsorption of microalgae onto the surface of flocculated sludge, probably due to extracellular polymeric substance (EPS) bridging and large superficial area of the activated sludge flocs (Salim et al., 2014)
- *Stage 2*: Attachment of the nascent bacteria onto the surface of microalgae, mainly to the phycosphere on the surface of algal cell and its immediate environs (Eigemann et al., 2013)
- *Stage 3*: Uniform distribution of microalgae and bacteria in the pellet, accompanied by the growth of both of the constituent biota (Su et al., 2011)

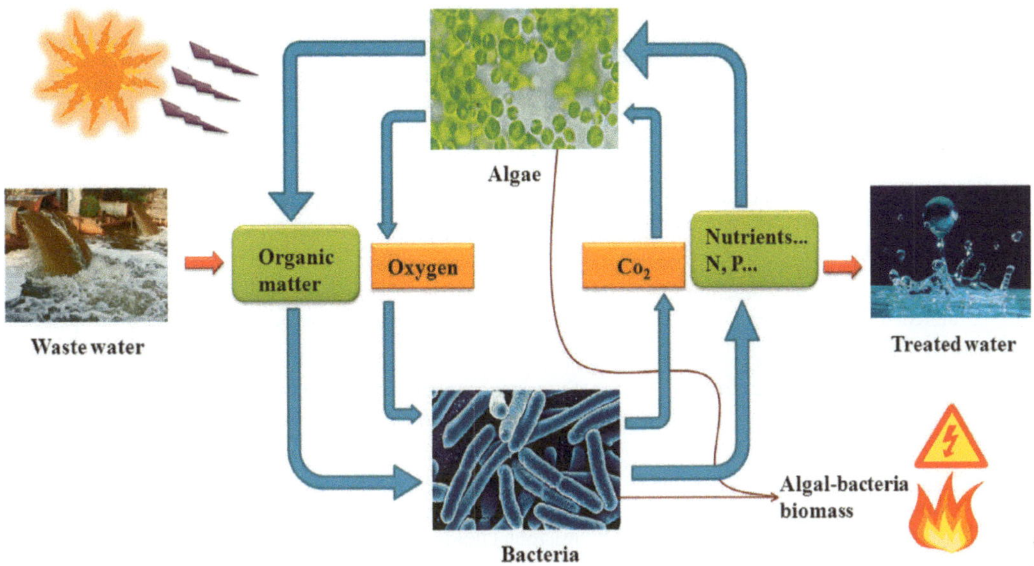

FIG. 20.1 Schematic representation of the mechanism of microalgae-bacteria consortium for wastewater treatment.

- *Stage 4*: Formation of mature consortium (i.e., dynamic balance between biomass attachment and detachment), which is considered to be synergistic between algae and bacteria

Several studies have combined microalgal photosynthesis with conventional wastewater treatment. Indirect adhesion of bacterial symbionts on the sheath and the direct adhesion onto the algal cell surface may reduce diffusion distance and permit rapid and efficient exchange of substrates (Park et al., 2008). The three main algae-bacteria consortia currently in use are as follows: (i) algae plus wastewater bacteria, (2) algae plus activated sludge, and (3) coculture of algae and assigned bacteria (such as nitrogen-fixing bacteria *Azospirillum*). Major bacteria that usually exist in initial wastewater were identified as members of the following genera: *Arcobacter*, *Clostridium*, *Desulfobulbus*, *Flavobacterium*, *Acidovorax*, *Bacteroides*, *Propionivibrio*, *Dechloromonas*, *Flexibacter*, *Pseudomonas*, *Rhodobacter*, *Enhydrobacter*, *Sphingobacterium*, *Paludibacter*, *Smithella*, *Pedobacter*, and *Zoogloea*. The abundance of *Proteobacteria*, *Flavobacteria*, and *Sphingobacteria* in algal cultures has been previously reported (Handschuh et al., 2017). Depending on the algal species in the algae-bacteria consortia and the immediate environments during the various growth stages of algae, the associated bacterial community structure can be affected in a species-specific manner. Cocultures of bacteria and algae have been examined, including both bacteria-algae consortia in nature and artificially induced symbioses (Tate et al., 2013). In a study, researchers identified the stimulation of microalgal growth by associated bacteria and coined the term "microalgal-growth-promoting bacteria" (MGPB). In this preliminary study, they demonstrated that coimmobilization of *Chlorella* sp. and *Azospirillum brasilense* was successful in efficient removal of nutrients from wastewater (De-Bashan et al., 2004). Table 20.1 is a summary of recently used algae-bacteria consortium systems in a successful manner in the wastewater treatment.

TABLE 20.1 List of Recently Used Algae-Bacteria Consortium Systems in Wastewater Treatment

S. No.	Name of Algae-Bacteria Consortium	Advantages	References
1.	*Chlorella vulgaris-Bacillus licheniformis* and *Microcystis aeruginosa-Bacillus licheniformis* consortia	Helps to eliminate total dissolved nitrogen, total dissolved phosphorus, and soluble chemical oxygen demand	Ji et al. (2018)
2.	Heterotrophic bacteria, autotrophs, algae, and phosphate-accumulating organisms	Successful in removing complete ammonia from wastewater	Yang et al. (2018)
3.	Algal-sludge bacterial-membrane bioreactor	Helps in effective wastewater treatment and low membrane fouling system	Sun et al. (2018)
4.	*Chlorella* sp. + cyanobacteria	Successful in almost complete removal of 17β-estradiol within 12 h in urban wastewater	Parladé et al. (2018)
5.	*Scenedesmus quadricauda* + activated sludge	Helps in treating coke wastewater with concomitant production of microbial lipids	Ryu et al. (2017)
6.	Scenedesmus sp. + heterotrophic cyanobacteria	C, N, P nutrients were more efficiently removed under the two-phase photoperiod	Lee et al. (2016)
7.	*Chlorella vulgaris* + activated sludge	Successful in removing organic carbon, ammonium nitrogen, and phosphate from wastewater	Xu et al. (2016)
8.	Microalgae + aerobic bacteria + *Navicula pelliculosa*	Helps in enhancing the detoxification process and in reducing toxicity	
9.	*Chlorella vulgaris* and bacteria	Helps in the removal of nitrogen, phosphorus, and organic matter from wastewaters	He et al. (2013)

20.3 BIOLOGICAL TREATMENTS OF DIFFERENT WASTEWATERS USING MICROORGANISMS

20.3.1 Pharmaceutical Wastewater

Pharmaceutical wastewater is usually distinguished by elevated toxicity and the incidence of refractory compounds that limit its biodegradability, making it a potential hazard to the natural environment and to wastewater treatment plants, if not handled properly (Lefebvre et al., 2014). Pharmaceutically active compounds (PhACs), mainly present in aquatic environments, are emerging as a severe risk for both wildlife and humans. Many PhACs have been systematically detected in wastewater over past decades. Recently, Vasiliadou's group assessed the toxicity and inhibition of three pharmaceutical compounds (carbamazepine, caffeine, and sulfamethoxazole) on two cultures of microbial consortia enriched from wastewater aerobic activated sludge (Vasiliadou et al., 2018). In this study, elevated values of toxicity and inhibition of toxic pharmaceutical compounds were observed for the control culture as compared with the acclimated one. In another study, researchers have scrutinized the connection between pharmaceutical biotransformation and biodiversity in wastewater treatment plant bioreactor microbial communities and recognize taxa and functional genes that were strongly associated with biotransformation (Stadler et al., 2017). These strategies will help in designing biological wastewater treatment plants for better removal of chemicals that negatively impact environmental health. Similarly, Aydin's group examined the occurrence and variation in the antibiotic resistance genes found during the 1-year operation of the anaerobic sequencing batch reactors used to treat pharmaceutical wastewater that contained combinations of sulfamethoxazole-tetracycline-erythromycin and sulfamethoxazole-tetracycline (Aydin et al., 2015). The anaerobic treatment procedures indicate that anaerobic treatment procedures executed in the occurrence and dissemination of antibiotic resistance genes revealed the presence of high nonlethal concentrations of sulfamethoxazole-tetracycline-erythromycin and sulfamethoxazole-tetracycline combinations. There are several successful strategies of pharmaceutical wastewater treatment by microorganism (Prasertkulsak et al., 2016; Popa et al., 2014; Ng et al., 2014; Gadipelly et al., 2014).

20.3.2 Dye-Containing Wastewater

Textile industries commonly release a vast amount of wastewater, which contains various fabric dyes as the main constituent. Textile industrial wastewater is characterized by strong color, high salinity, high temperature, variable pH, and high COD. In general, most textile industries handle a huge quantity of synthetic dyes, sodium sulfide, Glauber's salt (in dye bath solution), and hydrogen peroxide (as oxidizing agent) (Vijayaraghavan and Shanthakumar, 2015). Microorganisms have been used to remove or to neutralize these colorful dyes through biosorption or reductive mechanisms (Chen et al., 2017; Fan et al., 2018). Treatment of textile dye wastewater was carried out by Selvakumar et al. (2013) in a batch reactor using *Ganoderma lucidum* and concluded that *G. lucidum* could be utilized for the effective treatment of textile dye wastewater. It was reported that more than 30 azo dyes could be decomposed by species *Chlorella* and *Oscillatoria* to aromatic amines that were further completely metabolized by bacterial cultures. Marungrueng and Pavasant (2007) also proved that the *Caulerpa lentillifera* can effectively remove textile dyes by biosorption. Several studies have provided some exciting ideas for alga-based dye wastewater treatment (Wang et al., 2016; Logroño et al., 2017). There are three standard advantages of alga-based dye wastewater treatment: First, these processes can be carried out in situ at the contaminated site; second, bioprocess approaches are usually environmentally benign; and finally, they are cost-effective (Vijayaraghavan and Yun, 2008).

20.3.3 Metal-Containing Wastewater

There is an increased consciousness of the possible dangers of environmental pollution by heavy metals. With the authoritarian statutory limits imposed by international agencies, the conventional physicochemical methods for the treatment of metal-containing wastewaters are proving to be expensive and also inadequate to meet the required standards. This technoeconomic impasse has led to an intensive search for new technologies, and among the new technologies, the treatment of metal-containing wastewater by microorganisms is one with several applications (Mekuto et al., 2016; Mosier et al., 2015). It was reported that copper cations could be removed more efficiently by an algae-bacteria consortium compared with individual constituent organism (Boivin et al., 2007), suggesting that metal removal by the consortium is a synergistic effect. Future research direction for algae-bacteria consortium-based heavy-metal removal can involve the following points to be addressed: (i) The mechanism of algae-bacteria consortium for heavy-metal removal, such as biosorption, bioconvention, or bioaccumulation, should be systematically investigated; (ii) the synergistic effects of algae and bacteria need more elucidation; (iii) ecological cation exchanger can be developed, since dried biomass of algae and bacteria possesses the ability of ion exchange due to charged cell walls.

20.3.4 Agro-Industrial Wastewater

Agro-industrial wastewaters include potato-processing wastewater, swine wastewater, livestock wastewater, dairy wastewater, slaughterhouse wastewater, fish-farm wastewater, and some digestion effluent. One common characteristic of agro-industrial wastewater is high ammonium concentration, which is highly correlated to eutrophication. Among the microorganisms, microalgae can grow well under "nutrient-rich" environment and rapidly convert the nutrients (e.g., ammonium used as a nitrogen source) contained in agro-industrial wastewaters to biomass. The microalgae *Neochloris oleoabundans*, *C. vulgaris*, and *S. obliquus* were cultivated using agrozootechnical digestate and compared for their ability in nitrogen removal and lipid production (Franchino et al., 2013). However, some agro-industrial wastewaters are characteristic of high COD levels (i.e., 20,180 $mg\,L^{-1}$), and thus, it may not be easy to directly treat the original wastewaters by microalgae. Under optimal conditions, the ammonium removal efficiency by microalgae can often reach higher than 90%. *Cyanobacteria* are also very efficient in nitrogen removal. Some *Cyanobacteria*, such as species of *Oscillatoria*, *Anabaena*, and *Spirulina*, are capable of utilizing elemental nitrogen as their sole nitrogen source, by the reduction of N_2 to ammonium. High temperatures favor the formation of free ammonia. Although free ammonia is generally toxic to photosynthetic organisms, the toxicity appears to be reduced in alkalophilic species such as *S. platensis*. Hernández's group is successful in treating agro-industrial wastewater using microalgae-bacteria consortium combined with anaerobic digestion of the produced biomass (Hernández et al., 2013).

20.4 CONCLUSION

Microorganisms play an essential role in environmental fate with a multiplicity of bioremediation mechanisms affecting its transformation from aqueous to volatile and from more toxic to less toxic. Recent advances on bioremediation make us confident that the development of desired industrial-scale remediation technology is achievable. Exploration of this area on a broader spectrum could thus further lead to a better environment. These studies will add to the current knowledge on the potential of cost-effective technologies for cleaner production and sustainable management and reuse of contaminated industrial wastewater. The application of microbial systems in the treatment of wastewater requires little energy and results in low operation and maintenance costs. Microalgae and microalgae-bacteria consortium can be successfully applied for the treatment of nutrient-rich wastewater, such as agro-industrial wastewater, municipal wastewater, pharmaceutical wastewater, and textile dye wastewaters by biosorption or bioconversion pathways. Several advantages are clearly observed in an algae-bacteria consortium system with the recent developments in industrial and municipal wastewater treatment. Nevertheless, mechanisms of pollutant removal and bioconversion are not clearly understood but require further investigations.

Acknowledgments

The authors are indebted to all the researchers whom we cited in this review for their significant and valuable research. No funding was received to perform this review.

Conflict of Interest

The authors declare no relevant competing financial interests to disclose.

References

Aydin, S., Ince, B., Ince, O., 2015. Development of antibiotic resistance genes in microbial communities during long-term operation of anaerobic reactors in the treatment of pharmaceutical wastewater. Water Res. 83, 337–344.

Boivin, M.E., Greve, G.D., Garcia-Meza, J.V., Massieux, B., Sprenger, W., Kraak, M.H., Breure, A.M., Rutgers, M., Admiraal, W., 2007. Algal-bacterial interactions in metal contaminated floodplain sediments. Environ. Pollut. 145, 884–894.

Cechinel, M.A., Mayer, D.A., Pozdniakova, T.A., Mazur, L.P., Boaventura, R.A., de Souza, A.A.U., de Souza, S.M.G.U., Vilar, V.J., 2016. Removal of metal ions from a petrochemical wastewater using brown macro-algae as natural cation exchangers. Chem. Eng. J. 286, 1–15.

Chen, Z., Li, Z., Liu, P., Liu, Y., Wang, Y., Li, Q., He, N., 2017. Characterization of a novel bioflocculant from a marine bacterium and its application in dye wastewater treatment. BMC Biotechnol. 17 (1), 84.

Chinnasamy, S., Bhatnagar, A., Hunt, R.W., Das, K.C., 2010. Microalgae cultivation in a wastewater dominated by carpet mill effluents for biofuel applications. Bioresour. Technol. 101, 3097–3105.

De-Bashan, L.E., Hernandez, J.P., Morey, T., Bashan, Y., 2004. Microalgae growth promoting bacteria as "helpers" for microalgae: a novel approach for removing ammonium and phosphorus from municipal wastewater. Water Res. 38 (2), 466–474.

Eigemann, F., Hilt, S., Salka, I., Grossart, H.P., 2013. Bacterial community composition associated with freshwater algae: species specificity vs. dependency on environmental conditions and source community. FEMS Microbiol. Ecol. 83 (3), 650–663.

Fan, J., Chen, D., Li, N., Xu, Q., Li, H., He, J., Lu, J., 2018. Adsorption and biodegradation of dye in wastewater with Fe_3O_4@MIL-100 (Fe) core-shell bio-nanocomposites. Chemosphere 191, 315–323.

Fazi, S., Amalfitano, S., Casentini, B., Davolos, D., Pietrangeli, B., Crognale, S., Lotti, F., Rossetti, S., 2015. Arsenic removal from naturally contaminated waters: a review of methods combining chemical and biological treatments. Rendiconti Lincei 27 (1), 51–58.

Franchino, M., Comino, E., Bona, F., Riggio, V.A., 2013. Growth of three microalgae strains and nutrient removal from an agro-zootechnical digestate. Chemosphere 92 (6), 738–744.

Gadipelly, C., Pérez-González, A., Yadav, G.D., Ortiz, I., Ibáñez, R., Rathod, V.K., Marathe, K.V., 2014. Pharmaceutical industry wastewater: review of the technologies for water treatment and reuse. Ind. Eng. Chem. Res. 53 (29), 11571–11592.

Handschuh, H., Ryan, M.P., O'Dwyer, J., Adley, C.C., 2017. Assessment of the bacterial diversity of aircraft water: identification of the frequent fliers. PLoS One 12 (1), e0170567.

He, P.J., Mao, B., Lü, F., Shao, L.M., Lee, D.J., Chang, J.S., 2013. The combined effect of bacteria and *Chlorella vulgaris* on the treatment of municipal wastewaters. Bioresour. Technol. 146, 562–568.

Hernández, D., Riaño, B., Coca, M., García-González, M.C., 2013. Treatment of agro-industrial wastewater using microalgae-bacteria consortium combined with anaerobic digestion of the produced biomass. Bioresour. Technol. 135, 598–603.

Ji, X., Jiang, M., Zhang, J., Jiang, X., Zheng, Z., 2018. The interactions of algae-bacteria symbiotic system and its effects on nutrients removal from synthetic wastewater. Bioresour. Technol. 247, 44–50.

Kim, H.C., Choi, W.J., Chae, A.N., Park, J., Kim, H.J., Song, K.G., 2016. Evaluating integrated strategies for robust treatment of high saline piggery wastewater. Water Res. 89, 222–231.

Ledda, C., Villegas, G.R., Adani, F., Fernández, F.A., Grima, E.M., 2015. Utilization of centrate from wastewater treatment for the outdoor production of *Nannochloropsis gaditana* biomass at pilot-scale. Algal Res. 12, 17–25.

Lee, J., Cho, D.H., Ramanan, R., Kim, B.H., Oh, H.M., Kim, H.S., 2013. Microalgae associated bacteria play a key role in the flocculation of *Chlorella vulgaris*. Bioresour. Technol. 131, 195–201.

Lee, C.S., Oh, H.S., Oh, H.M., Kim, H.S., Ahn, C.Y., 2016. Two-phase photoperiodic cultivation of algal-bacterial consortia for high biomass production and efficient nutrient removal from municipal wastewater. Bioresour. Technol. 200, 867–875.

Lefebvre, O., Shi, X., Wu, C.H., Ng, H.Y., 2014. Biological treatment of pharmaceutical wastewater from the antibiotics industry. Water Sci. Technol. 69 (4), 855–861.

Logroño, W., Pérez, M., Urquizo, G., Kadier, A., Echeverría, M., Recalde, C., Rákhely, G., 2017. Single chamber microbial fuel cell (SCMFC) with a cathodic microalgal biofilm: a preliminary assessment of the generation of bioelectricity and biodegradation of real dye textile wastewater. Chemosphere 176, 378–388.

Mahendran, B., Lishman, L., Liss, S.N., 2012. Structural, physicochemical and microbial properties of flocs and biofilms in integrated fixed-film activated sludge (IFFAS) systems. Water Res. 46, 5085–5101.

Marino, T., Figoli, A., 2015. Arsenic removal by liquid membranes. Membranes 5, 150–167.

Marungrueng, K., Pavasant, P., 2007. High performance biosorbent (*Caulerpa lentillifera*) for basic dye removal. Bioresour. Technol. 98 (8), 1567–1572.

Masindi, V., Gitari, W.M., 2016. Removal of arsenic from wastewaters by cryptocrystalline magnesite: complimenting experimental results with modelling. J. Clean. Prod. 113, 318–324.

Mekuto, L., Ntwampe, S.K., Akcil, A., 2016. An integrated biological approach for treatment of cyanidation wastewater. Sci. Total Environ. 571, 711–720.

Mosier, A.P., Behnke, J., Jin, E.T., Cady, N.C., 2015. Microbial biofilms for the removal of Cu^{2+} from CMP wastewater. J. Environ. Manag. 160, 67–72.

Naidoo, S., Olaniran, A.O., 2013. Treated wastewater effluent as a source of microbial pollution of surface water resources. Int. J. Environ. Res. Public Health 11, 249–270.

Naidoo, S., Olaniran, A.O., 2014. Prevalence of antimicrobial resistant and virulent *Salmonella* spp. in treated effluent and receiving aquatic milieu of wastewater treatment plants in Durban, South Africa. Int. J. Environ. Res. Public Health 12, 9692–9713.

Ng, K.K., Shi, X., Yao, Y., Ng, H.Y., 2014. Bio-entrapped membrane reactor and salt marsh sediment membrane bioreactor for the treatment of pharmaceutical wastewater: treatment performance and microbial communities. Bioresour. Technol. 171, 265–273.

Park, Y., Je, K.W., Lee, K., Jung, S.E., Choi, T.J., 2008. Growth promotion of *Chlorella ellipsoidea* by co-inoculation with Brevundimonas sp. isolated from the microalga. Hydrobiologia 598, 219–228.

Parladé, E., Hom-Diaz, A., Blanquez, P., Martínez-Alonso, M., Vicent, T., Gaju, N., 2018. Effect of cultivation conditions on β-estradiol removal in laboratory and pilot-plant photobioreactors by an algal-bacterial consortium treating urban wastewater. Water Res. https://doi.org/10.1016/j.watres.2018.02.060.

Podder, M., Majumder, C., 2016. Application of granular activated carbon/$MnFe_2O_4$ composite immobilized on *C. glutamicum* MTCC 2745 to remove as (III) and as (V): kinetic, mechanistic and thermodynamic studies. Spectrochim. Acta A: Mol. Biomol. Spectrosc. 153, 298–314.

Popa, C., Favier, L., Dinica, R., Semrany, S., Djelal, H., Amrane, A., Bahrim, G., 2014. Potential of newly isolated wild Streptomyces strains as agents for the biodegradation of a recalcitrant pharmaceutical, carbamazepine. Environ. Technol. 35 (21–24), 3082–3091.

Prasertkulsak, S., Chiemchaisri, C., Chiemchaisri, W., Itonaga, T., Yamamoto, K., 2016. Removals of pharmaceutical compounds from hospital wastewater in membrane bioreactor operated under short hydraulic retention time. Chemosphere 150, 624–631.

Ryu, B.G., Kim, J., Han, J.I., Yang, J.W., 2017. Feasibility of using a microalgal-bacterial consortium for treatment of toxic coke wastewater with concomitant production of microbial lipids. Bioresour. Technol. 225, 58–66.

Salim, S., Kosterink, N.R., Tchetkoua Wacka, N.D., Vermue, M.H., Wijffels, R.H., 2014. Mechanism behind autoflocculation of unicellular green microalgae *Ettlia texensis*. J. Biotechnol. 174, 34–38.

Selvakumar, S., Manivasagan, R., Chinnappan, K., 2013. Biodegradation and decolourization of textile dye wastewater using *Ganoderma lucidum*. 3 Biotech 3 (1), 71–79.

Shih, Y.-J., Huang, R.-L., Huang, Y.-H., 2015. Adsorptive removal of arsenic using a novel akhtenskite coated waste goethite. J. Clean. Prod. 87, 897–905.

Stadler, L.B., Delgado Vela, J., Jain, S., Dick, G.J., Love, N.G., 2017. Elucidating the impact of microbial community biodiversity on pharmaceutical biotransformation during wastewater treatment. Microb. Biotechnol. https://doi.org/10.1111/1751-7915.12870.

Su, Y., Mennerich, A., Urban, B., 2011. Municipal wastewater treatment and biomass accumulation with a wastewater-born and settleable algal-bacterial culture. Water Res. 45 (11), 3351–3358.

Su, Y., Mennerich, A., Urban, B., 2012. Synergistic cooperation between wastewater-born algae and activated sludge for wastewater treatment: influence of algae and sludge inoculation ratios. Bioresour. Technol. 105, 67–73.

Sun, L., Tian, Y., Zhang, J., Li, L., Zhang, J., Li, J., 2018. A novel membrane bioreactor inoculated with symbiotic sludge bacteria and algae: performance and microbial community analysis. Bioresour. Technol. 251, 311–319.

Tate, J.J., Gutierrez-Wing, M.T., Rusch, K.A., Benton, M.G., 2013. The effects of plant growth substances and mixed cultures on growth and metabolite production of green algae *Chlorella* sp.: a review. J. Plant Growth Regul. 32 (2), 417–428.

Tricolici, O., Bumbac, C., Postolache, C., 2014. Microalgae–bacteria system for biological wastewater treatment. J. Environ. Prot. Ecol. 15, 268–276.

Vasiliadou, I.A., Molina, R., Martinez, F., Melero, J.A., Stathopoulou, P.M., Tsiamis, G., 2018. Toxicity assessment of pharmaceutical compounds on mixed culture from activated sludge using respirometric technique: the role of microbial community structure. Sci. Total Environ. 630, 809–819.

Vijayaraghavan, G., Shanthakumar, S., 2015. Removal of sulphur black dye from its aqueous solution using alginate from Sargassum sp. (brown algae) as a coagulant. Environ. Prog. Sustain. Energy 34 (5), 1427–1434.

Vijayaraghavan, K., Yun, Y.S., 2008. Bacterial biosorbents and biosorption. Biotechnol. Adv. 6 (3), 266–291.

Vithanage, M., Herath, I., Joseph, S., Bundschuh, J., Bolan, N., Ok, Y.S., Kirkham, M.B., Rinklebe, J., 2017. Interaction of arsenic with biochar in soil and water: a critical review. Carbon 113, 219–230.

Wang, Y., Ho, S.H., Cheng, C.L., Guo, W.Q., Nagarajan, D., Ren, N.Q., Lee, D.J., Chang, J.S., 2016. Perspectives on the feasibility of using microalgae for industrial wastewater treatment. Bioresour. Technol. 222, 485–497.

Xu, Y.J., Wang, Y., Yang, Y., Zhou, D.D., 2016. The role of starvation in biomass harvesting and lipid accumulation: co-culture of microalgae-bacteria in synthetic wastewater. Environ. Prog. Sustain. Energy 35 (1), 103–109.

Yang, J.S., Kwon, M.J., Park, Y.T., Choi, J., 2015. Adsorption of arsenic from aqueous solutions by iron oxide coated sand fabricated with acid mine drainage. Sep. Sci. Technol. 50, 267–275.

Yang, J., Gou, Y., Fang, F., Guo, J., Lu, L., Zhou, Y., Ma, H., 2018. Potential of wastewater treatment using a concentrated and suspended algal-bacterial consortium in a photo membrane bioreactor. Chem. Eng. J. 335, 154–160.

Yu, Z., Zhou, L., Huang, Y., Song, Z., Qiu, W., 2015. Effects of a manganese oxide-modified biochar composite on adsorption of arsenic in red soil. J. Environ. Manag. 163, 155–162.

Zhou, W., Li, Y., Min, M., Hu, B., Chen, P., Ruan, R., 2011. Local bioprospecting for high-lipid producing microalgal strains to be grown on concentrated municipal wastewater for biofuel production. Bioresour. Technol. 102 (13), 6909–6919.

CHAPTER

21

Emerging Trends of Microorganism in the Production of Alternative Energy

*Golla Ramanjaneyulu**[,†], *Bontha Rajasekhar Reddy**

***Department of Microbiology, Sri Krishnadevaraya University, Anantapur, India [†]Council of Scientific and Industrial Research—Central Institute of Medicinal and Aromatic Plants Research Centre, Bangalore, India**

OUTLINE

21.1 INTRODUCTION

The rapid utilization of petroleum and its products has resulted in drastic environmental change. The greenhouse gases (GHG), basically CO_2 discharged chiefly because of transportation, are expected to reach 2.7 billion tons by 2030. The economy of most developed and developing countries is reliant, as it were, on oil and its subsidiaries, and thus, any disturbance in the oil supply either due to geopolitical unrest or otherwise will have a huge impact not only on the economy but also in national security. We have witnessed few vitality changes lately, for example, the Deepwater Horizon oil slick (the BP oil slick in the Gulf of Mexico) of 2010, the current Fukushima atomic reactor mishap in Japan (2011), and the oil price volatility and uncertainty in petroleum product supply due to colossal uprisings in the Arab

Recent Developments in Applied Microbiology and Biochemistry
https://doi.org/10.1016/B978-0-12-816328-3.00021-0

world. The interest for oil and other sources of energy is growing at extremely great and quick pace, and global energy utilization is anticipated to increase by approximately 36% by the year 2030. This necessitates the development of alternative sources of energy to avert future energy crises. There is a need to devise a plan of action designed for protecting the environment, in the meantime sufficiently meeting the ever-growing energy needs of the world. Energy and environment are interlinked. From one perspective, there is a need to reasonably harness the accessible energy sources and on the other side to create novel sustainable energy development advancements for addressing cutting-edge needs. Nevertheless, there is an urgent need to protect the environment while meeting the worldwide energy prerequisites—an extreme challenge indeed. In the last few decades, energy utilization has expanded exponentially worldwide, and it is obvious that energy security is indispensable to the future financial success and ecological well-being. The United States with only 4.5% of the total populace is responsible for about 25% of worldwide energy utilization and 25% of worldwide CO_2 emissions (Yat et al., 2008). Endeavors have been made to anticipate future energy needs and at the same time to harness different sources of energy. Harnessing the power and importance of microorganisms in alternative and renewable energy is one step toward finding cheap, green, and sustainable solution to the issues of energy and environment.

21.2 BIOFUELS

Worldwide biofuel production has increased close to 137 billion L (3.3 EJ) in 2016. Conventional biofuels would ahead course to meet 2°C scenario (2DS) target for 2025; however, accelerated generation of advanced biofuels will be important to meet 2DS needs for transportation sector decarbonization. In the United States, ethanol output is anticipated to stabilize because of lower investment in new capacity and reaching the corn ethanol limit within the renewable fuel standard. To meet Brazil's 2030 commitments, a 43% reduction in greenhouse gas emissions (vs 2005 levels) and a 45% share of renewable energy in the country's energy matrix with biofuels would equate to over 50 billion L of fuel ethanol demand representing 18% share of sustainable biofuels, but accelerated production growth will be required if this goal is to be met. Biodiesel policy support stays strong in both countries (Brazil and the United States), for production growth needed. In the European Union, proposals to the revised Renewable Energy Directive (RED) coating 2020–30 incorporated a scale-down of the top food-crop-based biofuels from 7% to 3.8% (by energy) of the 2030 renewable energy target. Conversely, many petroleum-importing countries in Asia have improved policy approach for domestically produced biofuels, boosting businesses markets for ethanol (e.g., India and Thailand) and biodiesel (e.g., Indonesia and Malaysia) (https://www.iea.org/etp/tracking2017/transportbiofuels/).

21.3 SUSTAINABLE ENERGY FROM MICROBES

21.3.1 Alcohols as Biofuels

Generation for alternate, economical, and renewable energy by microorganisms utilizing renewable resources is an important alternative to petroleum-based fuels. Bioenergy is a critical part of renewable and alternate energy solution for today's energy crisis that threatens world economic growth. Corn ethanol has been growing rapidly in the past few years. Policy makers and researchers have reported that corn ethanol (first generation) possesses serious drawbacks. It adversely impacts the food prices and is harsh on soil fertility. Lignocellulosic ethanol on the other hand uses abundant lignocellulosic biomass, which contains different concentrations of cellulose and hemicellulose and is a viable option for grain ethanol. Transformation of abundant lignocellulosic biomass to biofuels as transportation fuels will be a feasible alternative for enhancing energy security. Such approach also will reduce greenhouse gas emissions. They would use cleaner-burning over conventional fossil fuels. Another advantage is the short cycle required for plant growth (Table 21.1).

21.3.2 Ethanol

Ethanol produced from lignocellulosic biomass is a potential biofuel, which generates four to five times more energy and has been reported to reduce greenhouse gas emissions by 115% (Limayem and Ricke, 2012). Lignocellulosic materials, for example, agriculture residues

TABLE 21.1 Composition of Different Agricultural and Other Lignocellulosic Substrates

Material	Cellulose[a]	Hemicellulose	Lignin	Ash	Extractives
Algae (green)	20–40	20–50	–	–	–
Cotton, flax, etc.	80–95	5–20	–	–	–
Grasses	25–40	25–50	10–30	–	–
Hardwoods	45 ± 2	30 ± 5	20 ± 4	0.6 ± 0.2	5 ± 3
Hardwood barks	22–40	20–38	30–55	0.8 ± 0.2	6 ± 2
Softwoods	42 ± 2	27 ± 2	28 ± 3	0.5 ± 0.1	3 ± 2
Softwood barks	18–38	15–33	30–60	0.8 ± 0.2	–
Cornstalk	39–47	26–31	3–5	12–16	–
Wheat straw	37–41	27–32	13–15	11–14	–
Newspaper	40–55	25–40	18–30	–	–
Chemical pulp	60–80	20–30	2–10	–	–
Sorghum stalks	27	25	11	–	–
Corn stover	38–40	28	7–21	3.6–7.0	–
Coir	36–43	0.15–0.25	41–45	2.7–10.2	–
Bagasse	32–48	19–24	23–32	1.5–5	–
Rice straw	28–36	23–28	12–14	14–20	–
Wheat straw	33–38	26–32	17–19	6–8	–
Barley straw	31–45	27–38	14–19	2–7	–
Sorghum straw	32	24	13	12	–
Sweet sorghum Bagasse	34–45	18–28	14–22	–	–

[a] *Composition represented in %wt on dry matter basis (Kumar et al., 2014).*

(wheat straw, sugarcane bagasse, and corn stover), forestry waste (hardwood and softwood), and also dedicated crops (switchgrass and *Salix*), are renewable resources of energy. As the raw materials based on the above products are available in plenty, it could be used for ethanol production (Yat et al., 2008). With respect to alcohol production to be used as biofuels, ethanol production using lignocellulosic material as substrate has been well studied. Recent investigations have been centering on metabolic engineering of hosts such as *Saccharomyces cerevisiae*, *Escherichia coli*, *Corynebacterium glutamicum*, and *Zymomonas mobilis* for enhanced ethanol production. Increasing microbial tolerance for inhibitory compounds present in lignocellulosic hydrolysates will be another significant area of research in this area. A variety of inhibitory compounds are formed during hydrolysis of lignocellulosic substrates. These inhibitors, such as furans, phenolics, and weak acids, have been reported to reduce ethanol yield and productivity. Additionally, at higher concentrations, ethanol itself has an inhibitory effect (Almeida et al., 2007). Metabolic, genetic, and evolutionary engineering strategies could be employed to improve tolerance for microorganisms against such inhibitors that could ultimately increase their accumulation (Almeida et al., 2007) (Table 21.2).

TABLE 21.2 Different Pretreatment Processes of Lignocellulosic Biomass

Pretreatment Process	Advantages	Limitations and Disadvantages
Mechanical comminution	Reduces cellulose crystallinity	Power consumption usually higher than inherent biomass energy
Steam explosion	Causes hemicellulose degradation and lignin transformation, cost-effective	Destruction of a portion of the xylan fraction, incomplete disruption of the lignin-carbohydrate matrix, generation of compounds inhibitory to microorganisms
AFEX	Increases accessible surface area, removes lignin and hemicellulose to an extent, does not produce inhibitors for downstream processes	Not efficient for biomass with high lignin content
CO_2 explosion	Increases accessible surface area, cost-effective, does not cause formation of inhibitory compounds	Does not modify lignin or hemicelluloses
Ozonolysis	Reduces lignin content, does not produce toxic residues	Large amount of ozone required, expensive
Acid hydrolysis	Hydrolyzes hemicellulose to xylose and other sugars, alters lignin structure	High cost, equipment corrosion, formation of toxic substances
Organosolv	Hydrolyzes lignin and hemicelluloses	Solvents need to be drained from the reactor, evaporated, condensed, and recycled; high cost
Pyrolysis	Produces gas and liquid products	High temperature, ash production
Pulsed electric field	Ambient conditions, disrupts plant cells, simple equipment	Simple equipment Process needs more research
Biological	Degrades lignin and hemicelluloses, low-energy requirements	Rate of hydrolysis is very low

Data from Kumar et al. (2009).

21.3.2.1 Utilization of Xylan

Xylan is the most abundant polysaccharide present in hemicellulose after cellulose in lignocellulosic biomass, which gives a pentose sugar xylose after hydrolysis. Fermentation of both pentose and hexose sugars is important to enhance the economic feasibility of the lignocellulose-to-ethanol conversion process. However, most of the ethanol-producing microorganisms cannot ferment xylose because of the lack of an endogenous xylose assimilation pathway especially pentose transporters. To this end, *S. cerevisiae* and *Z. mobilis* have been engineered with heterologous xylose assimilation pathways. The heterologous xylose reductase (XR)/xylitol dehydrogenase (XDH) or xylose isomerase (XI) pathway has been expressed in fungi and bacteria to convert xylose to xylulose-5-P, which is further metabolized through those pentose phosphate pathway (PPP) and glycolysis (Moysés et al., 2016). Rational methodologies to improve xylose fermentation focused primarily on the identification of more efficient enzymes for its transport and utilization (Lee et al., 2016), the maintenance of the redox balance, the flux through the PPP, and the elimination of the by-product xylitol (Moysés et al., 2016). Despite the above attempts, the ethanol yield and productivity of xylose fermentation are not on par with glucose fermentation. It has been proposed that endogenous regulation of xylose metabolism in the recombinant strains could be the possible reason for low yield and productivity (Table 21.3).

TABLE 21.3 Biological Pretreatment of Different Lignocellulosic Substrates

Feedstock	Fungal Strain	Type of Process	Mediator	Delignification %	Glucose Yield	Ethanol Yield
Wood pulp	*Trametes villosa*	Enzymatic	HBT	50%	–	–
		Laccase	HPA	39%		
Wood pulp	*Aspergillus fumigatus*	Enzymatic	HBT	14%	–	–
		Laccase				
Hardwood chips	*Trametes versicolor*	Enzymatic	VA	54.2%	–	–
		Laccase				
Sugarcane bagasse	*Ceriporiopsis subvermispora*	Fungal	–	20%	–	–
		MnP, LiP,				
		Laccase				
Eucalyptus globulus	*Myceliophthora thermophila*	Enzymatic	MS	>25%	–	–
		Laccase				
		Enzymatic	AC	21%	–	–
			SA	25%		

Continued

TABLE 21.3 Biological Pretreatment of Different Lignocellulosic Substrates—cont'd

Feedstock	Fungal Strain	Type of Process	Mediator	Delignification %	Glucose Yield	Ethanol Yield
		Laccase	CA	−9%		
			HBT	40%		
Corn stover	*Ceriporiopsis subvermispora*	Fungal Laccase	–	>25%	56% 37%	–
Switchgrass				>25%		
Ricinus communis		Enzymatic	–	85.69%	288.83 mg/g	–
		Laccase				
Eucalyptus globulus	*Pycnoporus cinnabarinus*	Enzymatic	HBT	27%	–	–
		Laccase				
Wheat straw	*Isolated eucalyptus*	Fungal	–	35%	70%	–
	Irpex lacteus	Laccase		35%	70%	
Wheat straw	*Ceriporiopsis subvermispora*	Fungal	–	34%	69%	90%
		MnP,		39%		
	Irpex lacteus	LiP Laccase				
Corn stover	*Ceriporiopsis subvermispora* 1	Fungal	–	39.2%	66.9%	–
		MnP Laccase				
Corn stover (cs)	*Pycnoporus sanguineus* H275	Fungal laccase	–	27% (cs)	–	–
Wheat straw (ws)			–	25% (ws)		
			–	71% (cs)		
			VA	72% (ws)		
Elephant grass (eg)	*Trametes villosa*	LMS	HBT	36% (eg)	70.7% (eg)	16.2% (eg) 12.3% (ew)
Eucalyptus wood (ew)				58% (ew)	57.8% (ew)	
Corn stover	*Trametes versicolor*	LMS	HBT	–	31.7%	–
Wheat straw	*Sclerotium* sp.	LMS	ABTS	–	82.29%	–

TABLE 21.3 Biological Pretreatment of Different Lignocellulosic Substrates—cont'd

Feedstock	Fungal Strain	Type of Process	Mediator	Delignification %	Glucose Yield	Ethanol Yield
Cotton gin trash	*Trametes versicolor*	LMS	PrimaGreen	15%	23%	31.6%
Cotton gin trash	*Trametes versicolor*	LMS	PrimaGreen	27%	41%	64%
Sugarcane bagasse	*Pleurotus ostreatus* IBL-02	Enzymatic	–	33.6%	72.9%	16.3 g/L
	MnP, LiP, laccase					
Wheat straw	*Trametes versicolor*	Fungal	–	–	68.33%	–
		MnP				
	Irpex lacteus	LiP			56.26%	
		Laccase				
Sugarcane bagasse	*Phlebia* sp. MG-60	IFF	–	30%	–	45%
Sugarcane bagasse	*Phanerochaete*	Fungal	–	64%	–	–
	Chrysosporium					
	Lentinula edodes	MnP				
	Pleurotus ostreatus					
Oil palm empty fruit	*Pleurotus floridanus*	Fungal	–	5%	15.4%	27.9%
bunches						
Corn stover	*Coprinus comatus*	IFF	–	66.5%	82%	
Wheat straw	*Irpex lacteus*	Fungal	–	–	84%	74%
		MnP Laccase				
Corn stover	*Phanerochaete chrysosporium*	Enzymatic	–	–	323 mg/g	–
		MnP				
	Coridus versicolor	Laccase				
Bamboo culms	*Punctularia* sp.	Fungal	–	–	10%	–
Poplar wood	*Trametes velutina*	Fungal	–	–	43.6%	22.2%

Continued

TABLE 21.3 Biological Pretreatment of Different Lignocellulosic Substrates—cont'd

Feedstock	Fungal Strain	Type of Process	Mediator	Delignification %	Glucose Yield	Ethanol Yield
Sugarcane bagasse	*Pleurotus florida*	Fungal	–	7.91%	303.33 mg/g	–
		MnP Laccase				
Corn stover	*Irpex lacteus*	Fungal	–	25.48%	308.98 mg/g	144.03 mg/g
		MnP				
Sugarcane bagasse	*Phlebia* sp. MG-60	IFF	–	–	–	44%
Hardwood kraft pulp	*Phlebia* sp. MG-60	IFF	–	–	–	71.8% 51.1%
Waste newspaper						
Sugarcane bagasse	*Phlebia* sp. MG-60	IFF	–	–	–	65.7%
Oakwood	*Phlebia* sp. MG-60	IFF	–	–	–	43.9%

21.3.2.2 *Utilization of Mixed-Sugars*

Inhibition of uptake and utilization of xylose in the vicinity for glucose have been observed in the recombinant *S. cerevisiae* strains. Similarly, higher concentration of glucose inhibits xylose metabolism in *E. coli* (Vinuselvi and Lee, 2012). Moreover, *S. cerevisiae* fails to offer particular pentose transporters, in spite of the fact that it encodes a large number of hexose transporters with different substrate specificity and natural inclination. Pentoses cannot compete viably with glucose because of that higher affinity for transporters for glucose (Shin et al., 2015). Different pathways for lignocellulosic biomass utilization by engineered *S. cerevisiae* are shown in Fig. 21.1. Evolutionary engineering of xylose-utilizing *E. coli* and *S. cerevisiae* exhibits carbon catabolite repression and enhanced the resistance of xylose metabolism to glucose (Vinuselvi and Lee, 2012). Glucose and galactose are the prevalent sugars in the hydrolysates of marine biomass. Galactose metabolic system of *S. cerevisiae* is under glucose repression and basically linked to respiration. *Saccharomyces cerevisiae* strains that could use galactose even under anaerobic conditions are highly needed from the industrial point of view (Lee et al., 2015a,b; Quarterman et al., 2016). Coutilization of sugars is hampered throughout the later phases of fermentation because of the accumulation of high concentration of ethanol. At higher ethanol concentrations, xylose utilization was reduced when the concentration of glucose became low in the media (Zhang et al., 2016). Molecular engineering of microorganisms to improve biofuel production is represented in Fig. 21.2 by mixed-sugar utilization approach.

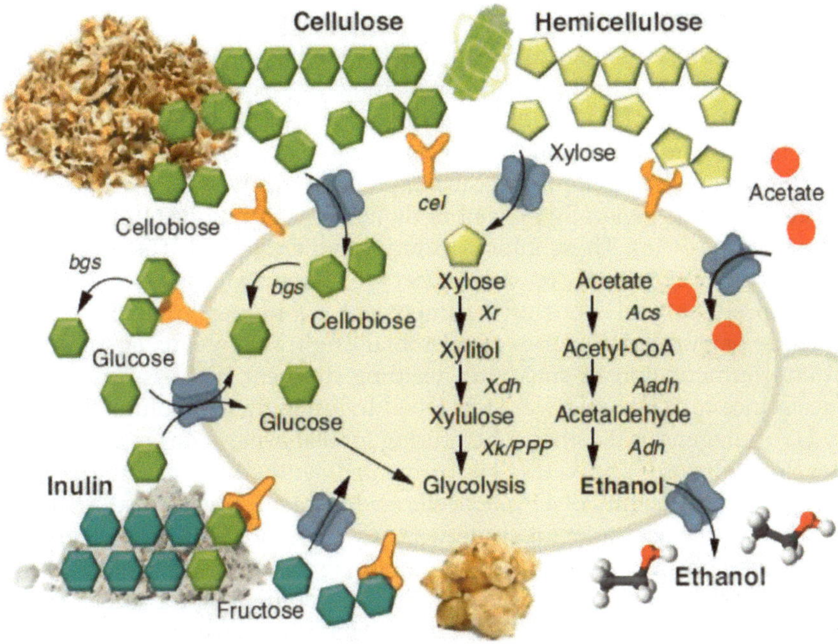

FIG. 21.1 Utilization of lignocellulosic biomass pathways by engineered *S. cerevisiae*.

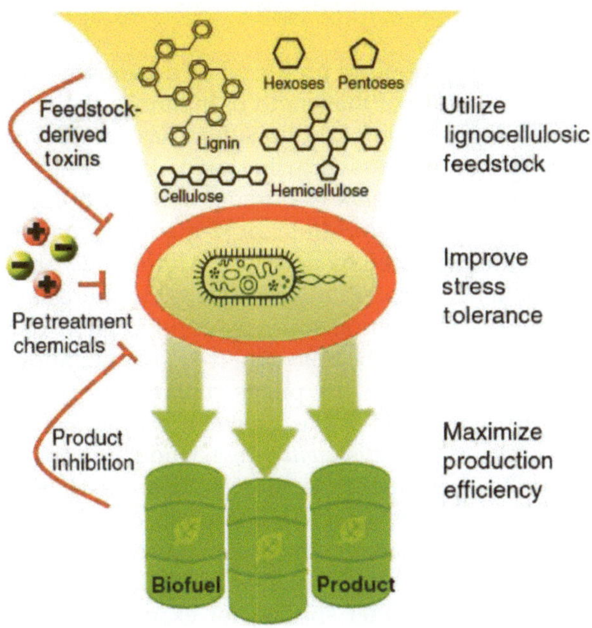

FIG. 21.2 Molecular engineering of microorganisms to improve mixed-sugar utilization from lignocellulosic biomass.

21.3.2.3 *Inhibitor Stress*

Lignocellulosic hydrolysates have different kinds of inhibitors in varying amounts, depending upon the raw material and the pretreatment method. Lignocellulose-derived inhibitors are basically grouped as furan derivatives, weak acids, and phenolic compounds. Lignocellulosic hydrolysates contain significant amounts of weak acids, such as acetic acid, formic acid, and levulinic acid, produced as a result of solubilization and hydrolysis of hemicellulose portion. In particular, acetic acid is generally present at higher concentrations (1–10 g/L) in hydrolysates (Hasunuma et al., 2016). These inhibitors reduce ethanol yield. Therefore, the lignocellulosic ethanol production depends on strains that are tolerant to these inhibitors (Ling et al., 2014). Biological pretreatment is one of the important techniques in which the substrate is treated with crude or purified enzymes of microbial origin that results in simultaneous saccharification and fermentation of substrates yielding different biofuels. *Spathaspora passalidarum* with resistance to acetic acid was developed by utilizing an evolutionary engineering technique by UV mutagenesis. Further subculturing for 380 generations resulted in increasing acetic acid levels in glucose- and xylose-supplemented synthetic medium. The ethanol productivity and yield in the vicinity of 4.5 g/L acetic acid were improved by seven- and twofold, respectively. When acidic acid-tolerant strain was grown in inhibitor-rich *Eucalyptus globulus* autohydrolysate, significant sugar utilization without lag phase was observed (Morales et al., 2017). Implication of the above or improved strain development could accelerate bioethanol production to use as an alternative fuel. Besides, metabolic engineering in yeast or other microbial strains can enhance ethanol production, such as sugar utilization, inhibitor resistance, and ethanol tolerance if engineered appropriately (Fig. 21.3).

21.3.3 Butanol

Butanol is another important compound produced by microorganisms that can be utilized as a fuel or fuel component. Because of the length of butanol's chain, it is easier to mix with higher hydrocarbons, including gasoline. Additionally, it has another advantage with respect to ethanol that butanol is substantially less corrosive and volatile and also has lower water solubility. Among the four isomers of butanol (n-butanol (butan-1-ol), *sec*-butanol (butan-2-ol), isobutanol (2-methylpropan-1-ol), and *tert*-butanol (2-methylpropan-2-ol)), only n-butanol, *sec*-butanol, and isobutanol are produced by microbes. Although *tert*-butanol is only received in refinery, others can be obtained through microbial fermentation. Only some of the strains are reported to ferment biomass to butanol, and yeast is not capable of producing butanol.

In biological procedures, butanol can be delivered from an indistinguishable biomass from ethanol. As a substrate, it can be utilized by plants containing carbohydrates, for instance, sugarcane, sugar beets, corn, and wheat. More interest is centered on feedstocks that do not compete for food, for example, *Miscanthus*, *switchgrass*, wood and crop waste, algae biomass, and food processing waste. The principal procedure that was employed to get n-butanol was acetone-butanol-ethanol (ABE) fermentation. The products of ABE maturation are solvents, for example, acetone, butanol, and ethanol in the proportion of 3:6:1. In this procedure, strains of anaerobic microbes from the class *Clostridium* are included. The best-studied and generally utilized species is *C. acetobutylicum*; other species, for example, *C. beijerinckii*, *C. aurantibutyricum*, or *C. tetanomorphum*, likewise have been previously reported to give higher butanol yield (Jones and Woods, 1986).

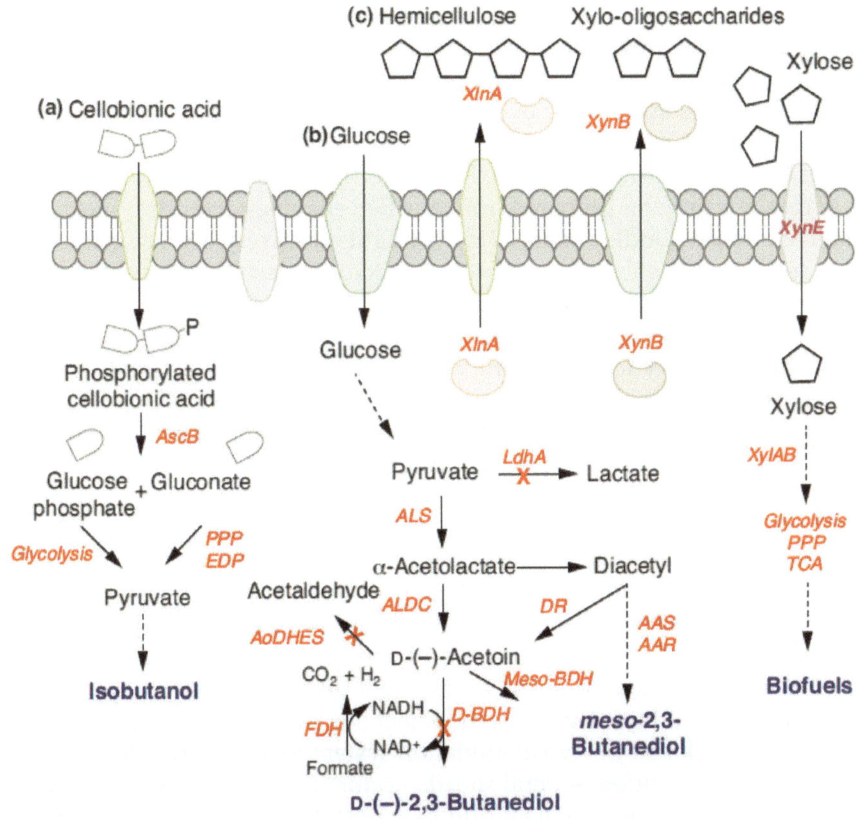

FIG. 21.3 Metabolic engineering strategies for microbial biofuel production under adverse conditions.

The ABE aging procedure is more intricate than a generation of ethanol. There are two fundamental phases of fermentation. In the first phase (acidogenesis stage), metabolites like acetone, butyrate, hydrogen, and carbon dioxide are major products produced. This reduces the pH of the growth medium. In the next stage (solventogenesis), microbial metabolism is altered to generate butanol, acetone, ethanol, H_2, and CO_2; interestingly, when the concentration of glucose is less than some threshold levels, *Clostridium* strains produced just acids. Other than glycolytic reactions in the metabolic pathway for the production of acids and solvents, the reaction between pyruvate and butyryl-CoA is predominant. During acidogenesis, acetone is delivered from acetyl-CoA and butyrate from butyryl-CoA. However, during solventogenesis, both acetyl-CoA and butyryl-CoA are intermediates for the production of ethanol and butanol. Acetyl-CoA is the key intermediate for the synthesis of acetone. A few strains of *Clostridia*, for instance, *C. beijerinckii* and *C. aurantibutyricum*, reduce acetone to isopropanol at later stages (Leja et al., 2014). Concentration of alcohol cannot surpass 12 g for each liter of a fermentation medium. Higher amount of n-butanol inhibits alcohol production by bacterial cells. Nonetheless, little enhancements have been made for n-butanol levels to reach up to 20 g/L of medium (Nigam and Singh, 2011). The formation of by-products, primarily acetone in ABE fermentation, essentially influences the solvent yield and downstream separation process. Wang et al. (2018) improved isopropanol-butanol-ethanol mixture production through the manipulation of intracellular NAD(P)H levels in recombinant *C. acetobutylicum* XY16.

One of the potential solutions for overcoming these impediments is to develop a simplified pathway in vitro utilizing a limited number of enzymes through an approach known as cell-free metabolic engineering (CFME). CFME is a cell-free biosystem that employs in vitro ensembles of catalytic proteins prepared from purified enzymes or crude lysates of cells to produce product of interest (Dudley et al., 2015). It might effectively dispose cell-associated process barriers, for example, substrate or product toxicity, intracellular flux balance that brings about low target product yield and unwanted by-products, and product excretion constraints by intracellular transport barriers (Dudley et al., 2015). Zhang et al. (2017) designed a totally simulated reaction pathway for upgrading ethanol to acetoin, 2,3-butanediol, and 2-butanol in a cell-free biosystem composed of ethanol dehydrogenase, formolase, 2,3-butanediol dehydrogenase, dioldehydratase, and NADH oxidase. This strategy resulted in the production of acetoin, 2,3-butanediol, and 2-butanol at 88.78% and 88.28% under optimized conditions and 27.25% of the theoretical yield from 100 mM ethanol, respectively. A few investigations on 2-butanol production have been led by extending the terminal product of 2,3-butanediol or acetolactate utilizing metabolic engineering techniques (Oh et al., 2014; Chen et al., 2015). The final accumulation of 2-butanol might have been still very much lower due to metabolic flux limitations and its toxicity on microbial cells. Extensive metabolic engineering for solventogenic *Clostridia* could be accomplished by means of productive genetic tools, for example, an in vivo methylation system to avoid an inherent genetic restriction system of *Clostridia* to segregate stable plasmids and viable gene knockout (KO) system (Lütke-Eversloh, 2014). Lesiak et al. (2014) developed in vivo methylation system for *C. acetobutylicum* ATCC 824 and *C. saccharobutylicum*. In addition, several shuttle vectors that can reproduce in both *Clostridia* and *E. coli* have been developed (Leblanc and Lee, 1984; Purdy et al., 2002; Yu et al., 2012; Lee et al., 2015a,b). Among them, plasmids such as pIM1, pMTL500E, and pLK1-MCS holding pIM13, pAMβ1, and pUB110 replicons, respectively, stably reproduce in *C. acetobutylicum* (Lee et al., 2015a,b).

After successful introduction of secondary alcohol dehydrogenase into *C. acetobutylicum* XY16, the recombinant XY16 could completely eliminate acetone formation and convert it into isopropanol, demonstrating a great possibility for industrial-scale production for IBE mixtures. Especially, pH could significantly improve final solvent titer through regulation of NADH and NADPH levels in vivo. At optimum pH level of 4.8, total IBE production was significantly increased from 3.88 to 16.09 g/L with a final concentration of 9.97, 4.98, and 1.14 g/L of butanol, isopropanol, and ethanol, respectively. Meanwhile, NADH and NADPH levels were maintained at optimal levels for IBE formation compared with the control one without pH adjustment. In view of this, the possibility of modulating NAD(P)H is a productive approach to enhance IBE production. Chen et al. (2018) established a hybrid procedure that integrated fermentation, pervaporation, and esterification pointing to improve the economic feasibility of the conventional ABE fermentation process by using *Candida* sp. 99–125 as a fuel cell catalyst. The novel integration methodology gives a promising technique to in situ upgrading ABE products.

21.3.4 Methane

Methane is a potent greenhouse gas that is discharged during the burning of fossil fuels and also generated by microbial activity. It has been estimated that not less than 1 billion tons of methane is being formed and consumed by microorganisms annually (Zhang et al., 2018). Biogas (methane) is a gaseous biofuel that is delivered through the digestion of organic materials without oxygen and comprises for the most part of methane and CO_2. Methane is a noteworthy part in natural gas, the production of which has experienced an exponential increase phase in the last two decades. The other compounds that are formed during the digestion include sulfur compounds (hydrogen sulfide and alkylthiols), ammonia, alcohols, carbonyl compounds (such as aldehydes and ketones), carboxylic acids, (e.g., formic acid and acetic acid), terpenes, and aromatic compounds (Papurello et al., 2012). The uses of biogas include its utilization as vehicle fuel (after enriching its composition in methane) and for the production of electricity and heat through burning in a special vessel called combined heat and power. The utilization of methane as a vehicle fuel introduces a few advantages as it does not require a broad change of the vehicle, and it is a technology that has already been proved in practice (Aslam et al., 2006). The procedure of anaerobic digestion is a complex method that includes different stages, such as hydrolysis of complex molecules to its monomers, which is followed by acidogenesis, acetogenesis, and methanogenesis. Each stage is catalyzed by various consortia of microorganisms. Products of specific stages turn into a substrate for the following groups of microorganisms (bacteria). Methane is a bacterial metabolic waste of the last fermentation step. This highlights an urgent need to scan for more productive carbon source. Low-throughput methane from a landfill or natural gas wells that is generally flared can be utilized specifically by methanotrophs to produce fuels, or it can be converted to methanol (CH_3OH) and, in the long run, used by methylotrophs for fuel generation (Liao et al., 2016).

Lignocellulosic material (LCM) shows an attractive choice as biomass feedstock for second-generation biofuels explicitly hydrogen, biogas, bio-oils, bioethanol, and biodiesel. In developing nations, extensive measures of agricultural, forest, and agroindustrial deposits are generated every year as waste which are not used alternatively and are either left in the fields or incinerated. Thus, these leftovers could be harnessed as potential alternative raw material to produce a variety of industrially significant bioproducts, specifically biofuels in an environmentally friendly way. In addition, the crop residues from agriculture have a short harvesting period that presents them more reliably accessible resources of biofuel production (Batista et al., 2014, 2018). Complex methanogenesis pathways used by archaea are the main route for bioconversion of carbon dioxide to methane in nature (Brauer et al., 2006; Thauer et al., 2008; Yvon-Durocher et al., 2014). Zhang et al. (2018) reported that wild-type iron-iron (Fe-only) nitrogenase from bacterium *Rhodopseudomonas palustris* reduces CO_2 simultaneously with nitrogen gas (N_2) and protons to yield CH_4, ammonia (NH_3), and hydrogen gas (H_2) in a single enzymatic step. The amount of CH_4 generated by purified Fe-only nitrogenase was low when compared with its other by-products. However, CH_4 production by this enzyme in *R. palustris* might have been sufficient to support the growth of an obligate CH_4-utilizing *Methylomonas* strain when the two microorganisms were coculture, with oxygen (O_2) added at intervals. In freshwater environment, degradation of complex organic materials by microorganisms involves food chain of microbes that culminates in the production of CH_4 by methanogenic archaea (Ferry, 2011). In aerobic oceans, marine microorganisms can produce CH_4 as a by-product by utilizing the phosphate source methylphosphonate, which is generated by *Nitrosopumilus maritimus* (Metcalf et al., 2012), and small amounts of CH_4 are produced by bacteria and archaea that utilize the Wood-Ljungdahl pathway either to fix CO_2 or to oxidize acetyl-CoA to CO_2 (Vorholt et al., 1995).

Till date, methanogenic archaea are the only known microorganisms that are capable of performing biological CH_4 production in the anoxic conditions (Taubner et al., 2015). Taubner et al. (2018) investigated methanogens under Enceladus-like conditions; three thermophilic and also methanogenic strains, *Methanothermococcus okinawensis* (65°C) (Takai et al., 2002), *Methanothermobacter marburgensis* (65°C) (Schönheit et al., 1980), and *Methanococcus villosus* (80°C) (Bellack et al., 2011), were able to fix carbon and produce energy through the reduction of CO_2 with H_2 to form CH_4 and were studied regarding growth and biological CH_4 generation under distinctive headspace gas compositions on H_2/CO_2, H_2/CO, H_2, blend 1 (H_2, CO_2, CO, CH_4, and N_2), and blend 2 (H_2, CO_2, CO, CH_4, N_2, and C_2H_4). These methanogens were prioritized because of their capability to grow in a wide temperature range characteristic in the vicinity of hydrothermal vents (Martin et al., 2008), in a chemically defined medium (Taubner and Rittmann, 2016), and at low partial pressures of H_2 (Thauer et al., 2008). In a study, Taubner et al. (2018) showed that the methanogenic strain *M. okinawensis* has the ability to propagate and/or to produce CH_4 under putative Enceladus-like conditions.

Anaerobic digestion of microalgae to produce biogas is another alternative for energy generation apart from the biofuel production. However, the high amount of proteins of microalgae may repress the anaerobic bacteria, resulting in the lower biogas production. Anaerobic codigestion of microalgae with high carbon cosubstrates could balance the carbon to nitrogen proportion and in future could be promising for methane production. Indeed, the anaerobic biodegradability of aromatic compounds has been extensively studied, but less attention has been paid to the relationship of aromatic compound structure and their toxic impacts on the anaerobic bacterial community. Information about the impact of aromatic compound structure on the inhibition of biogas biosynthesis is vital in foreseeing the effect of xenobiotics with respect to anaerobic waste and wastewater treatments, thereby potentially preventing the cost of treatment plant operations. A better knowledge of the structure-toxicity relationship will attain the provision of anaerobic technologies to waste and wastewater containing aromatic compounds (Donlon et al., 1995; Debord, 2012).

Pant and Rai (2018) studied the impact of chlorpyrifos (CP), its degradation product trichloropyridinol (TCP), and N-substituted aromatic compounds (nitrophenol and amino phenol) on methane production from organic solid by coculture of facultative anaerobe *Pseudomonas aeruginosa* and anaerobe *Methanosarcina mazei* in a concentration range of 0.5–5.0 g/L. The study reported that the generation of methane in the reaction system might have been influenced by the addition of CP and its degradation product TCP and of N-substituted aromatic compounds (nitrophenol and aminophenol). There was a slight increase in methane generation at lower concentrations (<0.5 g/L) of CP and TCP. However, higher concentrations of these products resulted in inhibitory effects, and they also inferred with N-substituted aromatic compounds.

21.3.5 Hydrogen

Hydrogen (H_2) is a gaseous fuel that is perceived as another promising alternative to fossil fuels. Hydrogen is an eco-friendly form of energy because it will not produce GHG during fuel combustion; instead, it produces only water, and also it can be used directly for electricity generation through hydrogen fuel cells or other type of fuel cells such as SOFC (La Licata et al., 2011; Matsakas et al., 2017). Hydrogen is mostly (approximately 95% globally) produced from fossil-derived materials and has a high energy yield (122 kJ/g, which is 2.75 times higher over that of hydrocarbons) (Kim et al., 2012). To minimize the reliance on fossil sources, it is essential to switch the processing to microbial transformation of renewable resources. Considering the need for cleaner energy carriers and a reduction in carbon dioxide emissions, the production of hydrogen from renewable sources is a promising alternative for the future hydrogen-based carbon-free fuel (Batista et al., 2018).

Hydrogen generation from complex substrates composed of simple sugars has the potential to mitigate future worldwide energy demand (Sagır et al., 2017). However, presently, hydrogen is produced mainly from fossil fuels (e.g., natural gas steam reforming and coal gasification) or water (e.g., electrolysis and photolysis) through energy-intensive processes. It is therefore essential to develop environmentally friendly and optimize more energy-efficient and sustainable hydrogen production processes, through biological biomass fermentation. Biological production of H_2 can be classified as fermentation (photofermentation or dark fermentation) or photosynthetic (direct biophotolysis of water by algae and cyanobacteria or bioelectrohydrogenesis in microbial fuel cells) processes (Vijayaraghavan and Mohd Soom, 2006).

Fermentation can be carried out by pure cultures (such as *Clostridium* sp.) or mixed cultures (Dong et al., 2009). In mixed cultures, hydrogen levels are decreased due to the potential utilization of hydrogen by hydrogenotrophic bacteria, and for this reason, heat treatment of the sludge is widely used to suppress phenomenon (Kiran et al., 2014). *Clostridium acetobutylicum* is one of the highest hydrogen-producing bacteria isolated so far. The bacteria are reported to generate 185.9 mL H_2/kg TS/d, followed by *Serratia marcescens* with a production of 171.2 mL H_2/kg TS/d. Different authors have likewise utilized FW from different sources, for example, dining halls, for the production of hydrogen (Shin et al., 2004), achieving high hydrogen yields (91. 5 mL H_2/g VS), consequently underpinning the vitality of the FW fraction in the MSW for the biohydrogen production (Matsakas et al., 2017).

Hydrogen production through anaerobic fermentation is simple and can utilize a wide range of substrates, including waste products (Das and Veziroglu, 2008) and microalgae, for example, *Scenedesmus obliquus* (Batista et al., 2014), *Nannochloropsis* sp. (Nobre et al., 2013), *Chlorella vulgaris* (Lakaniemi et al., 2011), *Dunaliella tertiolecta* (Lakaniemi et al., 2011), and *Spirogyra* sp. (Pacheco et al., 2015). *Enterobacter aerogenes* is known as a proficient biohydrogen production bacterium, as it is one of the most widely studied among all model strains (Converti et al., 2003; Uma et al., 2017). Moreover, it is a facultative anaerobic bacterium, which has the ability to grow in the vicinity of oxygen, and subsequently makes the manipulation in bioprocesses much easier compared with the use of strict anaerobes (Zhang et al., 2011). *Enterobacter aerogenes* has proved the ability to convert microalgae *Anabaena* sp. (Ferreira et al., 2012), *Nannochloropsis* sp. (Nobre et al., 2013; Ferreira et al., 2012), *S. obliquus* (Batista et al. 2014, 2015), and *C. vulgaris* (Batista et al., 2015).

Sagır et al. (2017) demonstrated the improvement of hydrogen yield by the sequential microaerobic dark fermentation and photofermentation by immobilized *Rhodobacter capsulatus* JP91 using glucose as substrate. The results indicated that hydrogen production is determined by inoculum concentration, oxygen, and glucose levels. The maximum hydrogen production (21 ± 0.25 mmol H_2/L) and yield (7.8 ± 0.1 mol H_2/mol glucose) were obtained at 6 mM glucose, 4.5% oxygen, and 62.5 v/v% inoculum concentration in a sequential microaerobic dark fermentative and photofermentative systems by the use of immobilized *R. capsulatus* JP91 demonstrating the production of enhanced hydrogen in a sequential system by immobilized *R. capsulatus* JP91.

Substrates holding glucose has been used and studied widely for biohydrogen generation by means of single-stage (Kumar and Das, 2000; Ghosh and Hallenbeck, 2010), sequential (Hitit et al., 2017a) cocultures (Xie et al., 2010; Hitit et al., 2017b) of different dark fermentative and photofermentative bacteria. However, these methods need to be improved with increment hydrogen yields, downstream purification, and processing cost associated with it.

21.3.6 Biodiesel/Microbial Lipids

Biodiesel refers to a renewable diesel fuel made from oil crops, animal fats, and food waste oil that can substitute petroleum diesel and has environmental protection, safety, renewability, and other advantages. During 2006–13, global biodiesel market output grew rapidly from 6.5 billion liters in 2006 to 24.6 billion liters in 2013, at a CAGR of 20.9%. Global biodiesel market is expected to produce 45,291 million liters by 2020, representing a CAGR of 10.1% during 2009–20. Biodiesel demonstrates an animal fat-based or vegetable oil diesel fuel, including long-chain alkyl (methyl, ethyl, or propyl) esters. Biodiesel is generally made by esterifying lipids (e.g., soybean oil, vegetable oil, and animal fat (tallow)) with an alcohol generating fatty acid esters. Biodiesel is suggested to be utilized in standard diesel engines and is thus well defined from the vegetable and waste oils used to fuel converted diesel engines. Biodiesel can be used singly or blended with gasoline in any proportions. Biodiesel blends can also be utilized as heating oil. Oleaginous microorganisms when grown under nitrogen-limited conditions can accumulate 20%–80% about their dry weight in the form of lipids (triacylglycerols) that are considered as an alternative substrate for biodiesel processing (Zhang et al., 2011; Annamalai et al., 2018). Similar fatty acid profile of oils from agricultural oil crops and microbial lipids or single cell oils makes them generally suitable as a substitute for biodiesel (Ratledge, 2010). However, currently, the production of vegetable oil is cheap compared with that of microbial oil production due to the carbon sources used for production. Carbon source is estimated to be about 80% of the total medium cost, but there are many methods to drastically improve the technoeconomics of microbial oil production processes (Ahmad et al., 2017). Lignocellulosic biomass consisting of 55%–65% carbohydrate is widely considered as a promising feedstock for biofuels due to its sustainability and abundance (Ahmad et al., 2017). Microbial lipid production from lignocellulosic biomass is considered as a potential alternative feedstock for large-scale production of biodiesel at low cost and also to prevent shortage of feedstock (Ruan et al., 2013; Harde et al., 2016; Park et al., 2017). The majority of yeasts and filamentous fungi are oleaginous in nature since they have the proficiency to accumulate high amounts of triacylglycerols (TAGs) that are the significant components of lipids. TAGs are rarely found in bacteria as they contain polyhydroxyalkanoates for storage purposes; nevertheless, bacteria belonging to the actinobacterial group have been reported to produce lipids, for example, *Mycobacterium*, *Rhodococcus*, *Streptomyces*, and *Nocardia* (Alvarez and Steinbuchel, 2002; Lohit et al., 2017).

Yeast represents an important class of microbiota that are prevalent in all ecosystems (e.g., fresh and marine waters, oceans, and soil) and are even found in extreme environments (Butinar et al., 2007). Most of the lipid-producing yeasts belong to the phylum *Basidiomycetes* (*Rhodotorula* sp., *Cryptococcus* sp., *Rhodosporidium* sp., *Sporidiobolus* sp., *Trichosporan* sp., etc.) and a few are ascomycetes (*Yarrowia lipolytica*, *Candida* sp., etc.) (Ageitos et al., 2011; Polburee et al., 2015). A list of different potential microbial sources of lipids for biodiesel production is summarized in a recent article by Gujjala et al. (2017). Well-known yeast *Cryptococcus curvatus* is a potential strain to produce lipids, the hydrocarbon precursor with high energy density, from cellulose (Zhang et al., 2011; Seo et al., 2014). Nowadays, filamentous fungi such as *Mortierella alpina* (Holdsworth and Ratledge, 1991), *Mucor circinelloides* (Zheng et al., 2012), *Umbelopsis isabellina* (Li et al., 2011), and *Aspergillus oryzae* (Meeuwse et al., 2011) have also been reported to assimilate intracellular lipids. Hence, they are being tapped for biodiesel applications. Different biochemical investigations have suggested that oleaginous fungi can accumulate lipids since they contain a malic enzyme that generates nicotinamide adenine dinucleotide phosphate (NADPH), which is capable of reducing acetyl units and forming the backbone of fatty acids (Gutierrez et al., 2011). Different biochemical and metabolic strategies are being used for improving lipid accumulation. Induction of lipogenesis mainly depends on nutrient-limiting conditions such as the limitation of nitrogen, phosphorus, magnesium, zinc, iron, etc. (Ratledge and Wynn, 2002) in the presence of carbon in excess quantity. Among these, limitation of nitrogen is the most important since it yields maximum lipid accumulation and results in better conversion of substrate (Gutierrez et al., 2011). Various investigations have been carried out to optimize the media constituents for maximum lipid accumulation, which reveal that a high C/N ratio (80–350 mol/mol) (Rossi et al., 2011) leads to improved lipid accumulation within the biomass. A few investigations recommended that microbial lipid generation by oleaginous yeasts was exceptionally influenced by the C/N proportion of the culture medium (Seo et al., 2013; Chang et al., 2015). Metabolic engineering is a promising method that is being explored for enhancing the yield of many biological products. This method includes the genetic manipulation of vital genes that are controlling different regulatory process. In the connection of maximizing lipid yield, the final step of the TAG biosynthetic pathway is crucial and has been reported to be regulated for enhanced lipid production at both the transcriptional and the translational level (Ratledge, 2002). Modifications of existing fermentative pathways can also result in the production of different classes of microbial biofuels. For example, biopropane production pathways of two *E. coli* were recently described, utilizing CoA intermediates derived from either a clostridial-like fermentative butanol pathway (Menon et al., 2015) or modifications of the fatty acid biosynthetic route (Kallio et al., 2014) (Table 21.4).

21.3.7 Bisabolene

Bisabolene has been identified as a potential precursor and a monocyclic sesquiterpene to D2 diesel fuel bisabolane. Utilizing metabolic engineering methods, different microbial resources have been explored for nonnatural production of bisabolene in the past

TABLE 21.4 Microbial Sources of Lipids for Biodiesel Production

S. No.	Genus	Species	Lipid (%w/w)
	Yeast		
1.	*Myxozyma*	*melibiosi, udenii*	23.00, 20.30
2.	*Lipomyces*	*lipofer, tetrasporus, starkeyi, doorenjongii, kockii*	43.00, 66.50, 62.00, 72.30, 77.80
3.	*Kodamaea*	*ohmeri*	53.28
4.	*Metschnikowia*	*pulcherrima, gruessii*	30.00, 34.00
5.	*Cyberlindnera*	*jadinii, saturnus 1*	20.00–30.00, 25.00
6.	*Candida*	*freyschussii*	32.00
7.	*Wickerhamomyces*	*ciferrii*	22.00
8.	*Schwanniomyces*	*occidentalis*	23.00
9.	*Kurtzmaniella*	*cleridarum*	33.00
10.	*Candida*	*diddensiae, tenuis, tropicalis*	37.00, 20.46–56.58, 23.00
11.	*Geotrichum*	*fermentans*	19.50
12.	*Magnusiomyces*	*magnusii*	27.00
13.	*Galactomyces*	*pseudocandidus*	28.04
14	*Torulaspora*	*delbrueckii*	40.00
15.	*Trigonopsis*	*variabilis*	20.00–43.70
16.	*Yarrowia*	*lipolytica*	37.60–54.80
17.	*Rhodotorula*	*minuta, mucilaginosa, colostri, graminis, glutinis, babjevae, diobovatum, toruloides, fluvial, glacialis, terpenoidalis*	24.62, 28.00, 26.89, 41.10, 35.40, 65.00, 20.00–41.00, 19.00–51.10, 25.00, 68.00, 27.00
18.	*Leucosporidiella*	*creatinivora*	61.00–62.00
19.	*Rhodosporidium*	*paludigenum, sphaerocarpum*	31.00, 43.00
20.	*Cryptococcus*	*terreus, terricola, aerius, adeliensis, albidus, wieringae, oeirensis*	51.00, 69.30, 63.30, 32.10, 40.10, 23.00–53.00, 25.00
21.	*Trichosporon*	*coremiiforme, domesticum, montevideense, brassicae*	37.80, 35.16, 25.60, 20.34
	Filamentous fungi		
22.	*Aspergillus*	*niger, terreus*	9.60, 37.40
23.	*Cunninghamella*	*elegans*	33.60
24.	*Mucor*	*circinelloides, plumbeus*	23.80, 20.60
25.	*Neosartorya*	*fischeri*	8.80
26.	*Rhizopus*	*oryzae*	34.80

TABLE 21.4 Microbial Sources of Lipids for Biodiesel Production—cont'd

S. No.	Genus	Species	Lipid (%w/w)
27.	*Thermomyces*	*lanuginosus*	21.00
28.	*Mortierella*	*isabellina, vinacea*	67.00, 51.90
29.	*Chaetomium*	*globosum*	5.10
	Microalgae		
30.	*Botryococcus*	*braunii*	25.00–75.00
31.	*Chaetoceros*	*muelleri, calcitrans*	33.60, 14.60–16.40
32.	*Chlorella*	*emersonii, protothecoides, sorokiniana, vulgaris, pyrenoidosa*	25.00–63.00, 14.60–57.80, 19.00–22.00, 5.00–58.00, 2.00
33.	*Crypthecodinium*	*cohnii*	20.00–51.10
34.	*Dunaliella*	*salina, primolecta, tertiolecta*	6.00–25.00, 23.10, 16.70–71.00
35.	*Euglena*	*gracilis*	14.00–20.00
36.	*Haematococcus*	*pluvialis*	25.00

Data from Gujjala, L.K.S., Jeevan Kumar, S.P., Talukdar, B., Dash, A., Kumar S., et al., 2017. Biodiesel from oleaginous microbes: opportunities and challenges. Biofuels. doi:10.1080/17597269.2017.1402587.

5 years (Davies et al., 2014; Phelan et al., 2014). Bisabolene synthase genes from *Arabidopsis thaliana, Pseudotsuga menziesii, Abies grandis,* and *Picea abies* have been screened in *E. coli* (carrying entire MVA pathway in a single plasmid) and *S. cerevisiae* (Peralta-Yahya et al., 2011) to obtain higher titers of bisabolene. How high the levels in these two platforms is included. Bioengineering and synthetic biology approaches have revolutionized the field of biotechnology, enabling the introduction of nonnative and *de novo* pathways for biofuel production. This reprogramming of microorganisms is also applied to the utilization of mixed carbon components derived from lignocellulosic biomass, a major technical barrier for the development of economically feasible fermentations. To generate high-value drop-in alcohol biofuels, metabolic engineering of microorganisms has enabled the incorporation of nonnative pathway genes in different hosts. These fuels include branched and linear higher alcohols, such as propanol and butanol, which have higher energy content and lower hygroscopy.

For third-generation biodiesel production, microalgae have been extensively explored as a feedstock in recent years. There may be an extensive possibility that they can play a considerable role in solving the problem between the production of biofuels and food. Cyanobacteria can use light energy to fix CO_2 into biomass that makes them exceedingly useful hosts for the production of bioalternatives to petroleum-based products. A variety of chemicals have been produced in cyanobacteria, but many of them suffer from low titers and productivities (Case and Atsumi, 2016). The recent studies have used knowledge gained from mass-spectrometry-based metabolomic analysis to improve the efficiency of engineered pathways, enabling utilization of lignocellulosic sugars and increasing stress tolerance in biofuel-producing microbes (Martien and Amador-Noguez, 2017).

21.3.8 Microbial Fuel Cells

Transformation of chemical energy to electric energy is known from eighteenth century of Volta, the inventor of voltaic pile and who was the contemporary of Luigi Galvani who initially observed animal electricity. In the presence of biological catalysts like enzymes (enzymatic fuel cells) and microorganisms (microbial fuel cells, MFCs), the chemical energy accessible in biomass surrounding us can be harnessed. Many researchers have started focusing on the ability of microorganisms to produce electric energy in biological systems (Potter, 1910). In 1911, Potter observed that a maximum voltage of 0.3–0.5 V could be generated with glucose as a substrate and Pt (platinum) as electrode by the *S. cerevisiae*. Presently, for almost a century, research is continuously progressing on MFCs by the oxidation of organic matter to produce electric energy providing a great scope toward alternate energy (Pant et al., 2012). Bioelectrochemical cells (BEC) have gained significant interest in the production of bioenergy from natural biomass and wastewaters. In particular, microbial fuel cells (MFCs) and microbial electrolysis cells (MECs) have been extensively exploited for bioelectricity and biohydrogen generation (Logan et al., 2015). From a biological perspective, both kinds of fuel cells work on a similar principle; consequently, common microorganisms can be deployed in these fuel cells in bioenergy production. The distinctive character of these microorganisms (referred as exoelectrogens or electricigens) in BEC is the display of particular molecular machinery that helps exchange the electrons from microbial outer membrane to the conductive surfaces (Kumar and Kumar, 2017). Hence, the electrons can be utilized to produce electricity and hydrogen. MFCs are one of the widely studied technologies that have potential for waste valorization into energy in the form of bioelectricity production (Koók et al., 2016). In spite of critical progression occurring in this field with respect to microbiology, materials science, chemistry, electrochemistry, etc., process economization and process sustainability were observed to be the most essential elements to move the field to the next level (Mohanakrishna et al., 2012). Microorganisms present in MFCs as catalysts to drive to the anodic and cathodic reaction to generate electricity. By exoelectrogenic microorganisms, biodegradable substances containing chemical energy can be converted into electricity. At the anode compartment, electrons and protons are produced by the oxidation of organic compounds by certain microbes. The electrons reached to the cathode compartment through the external circuit where it reduces the electron acceptor present, and in the meantime, protons produced at the anode are exchanged to the cathode through a membrane separator or through the electrolyte in a membraneless cell.

The anodic reaction that is based on microbial oxidation is similar for all types of MFC applications. However, different cathodic reactions can be employed in MFCs to generate electric energy if the overall reaction is thermodynamically favored. Because of unlimited availability and positive redox potential, oxygen is mostly widely considered as the favorable electron acceptor for practical applications. The attainability of utilizing other electron acceptors with a high redox potential, for example, nitrate, sulfate, and some other contaminants in the environment with high redox potential, which are electrochemically or naturally reducible in the cathode chamber, can also be considered (Berchmans, 2018). Additionally, to increase the voltage of the cell, permanganate, dichromate, peroxide, and ferricyanide are being used as a part of MFCs in light of their high redox potential (Yang et al., 2011). Microorganisms that can reoxidize reduced metal oxides and Fe^{2+} species like *Acidithiobacillus ferrooxidans* and *Thiobacillus ferrooxidans* can also be utilized as cathodic biocatalysts (Kumar et al., 2015).

In MFCs, the anode and cathode are isolated by an ion-exchange membrane, and solutions comprising biomass and microorganisms are used as fuel (Logan and Regan, 2006; Lal, 2013):

Anode: $C_6H_{12}O_6 + 6H_2O \rightarrow 6CO_2 + 24H^+ + 24e^-$

Cathode: $6O_2 + 24H^+ + 24e^- \rightarrow 12H_2O$

$C_6H_{12}O_6 + 6O_2 \rightarrow 6CO_2 + 6H_2O +$ Electric Energy

However, the outputs of energy from MFCs and MECs are inadequate for industrial-level applications and, therefore, not feasible for commercialization. Hypothetically, an MFC can create a maximum voltage of 1.2 V, and the optimum hydrogen generation yields in MEC would be 3.4 mol H_2/mol acetate (Logan et al., 2015). Initially, the research on MFC was focused on wastewater treatment; however, over a period of time, the field of MFCs has developed into a much more diverse field of research called bioelectrochemical systems (BES) because of the advent of several related technologies such as microbial electrolysis, microbial desalination, microbial electrosynthesis, and photomicrobial cells. Latest studies by the use of distinctive electrode materials and also MFC reactor designs in the scope of 200 mL to a couple of hundred liters were assessed toward the advancement of pilot-scale MFC systems (Janicek et al., 2014). Transforming this technology from lab scale to industrial scale will bring it a step toward the realization of commercial application of bioelectricity generation. Among the few materials used for the electrodes, carbon-based materials, like graphite and carbon with metal-based impregnations or coatings, were observed to be economical and proficient for bioelectricity production (Xie et al., 2010; Mohanakrishna et al., 2012).

If MFCs are benchmarked against anaerobic digestion, two distinct differences are to be observed. MFCs cannot deal with suspended and particulate organic material, though anaerobic assimilation is capable of dealing with them. Using MFCs other than electricity, many other value-added products can be obtained that is otherwise not possible with anaerobic digestion. However, the rates of energy conversions are lower in MFC technology. Thus, all the technological challenges need to be clearly understood to make the MFC technology more viable. Proper power management systems should be evolved to maximize the power output derived from MFCs and to integrate with MFC. Synthetic biology may help in developing robust exoelectrogens with perfect electron-exchange properties. MFC, as energy-saving technology, may well wean for us far from the dwindling oil assets. Before that, there are many technical challenges that must be considered for sustainable and renewable energy generation. Regardless, the technology may open the way to new method for renewable and sustainable energy products. In the near future, MFCs might be developed to such a phase, to the point that they can give a sensible and usable power yield per unit the MFC volume. In such scenario, a larger battery size could be ignored, provided the maintenance is simple and has a green and safe label. This eco-friendly fuel cell will then lead to several groundbreaking applications. As the amount of low-power devices implanted in the human body increases, the long term, stable power source used may well be the MFC (Table 21.5).

21.4 CONCLUSIONS

Cofermentation of hexose and pentose and inhibitor and stress tolerance ability of the microbes are main obstacles in economic conversion of lignocellulosic biomass to biofuels. Bioengineering techniques have revolutionized the field of biotechnology, expanding both the range of compounds produced and types of organisms utilized as cell factories. From the

TABLE 21.5 Different Applications of Metabolomic-Based Analyses to Biofuel

Area	Major Findings	Metabolomics Methods	Bioproduct(s)	Organism
Pathway engineering	Metabolite quantification detected accumulation of isopentenyl pyrophosphate, indicating that NudB was a bottleneck enzyme in engineered heterologous MVA pathway. Increased NudB expression resulted in a 60% increase in methyl butanol production	LC-MS	Isoprene-derived alcohols, 3-methyl-2-butenol, 3-methyl-3-butenol, and 3-methyl-1-butanol	*Escherichia coli*
	Time-dependent metabolite profiling was used to verify enzymatic activity of engineered xylose to butanediol pathway before regulation of the pathway was optimized	GC and GC-MS	1,4-Butanediol	*Escherichia coli*
	Integrated omics was used to design and optimize a succinyl-CoA to butanediol pathway. Dynamic labeling showed that aldehyde dehydrogenase was a rate-limiting step, guiding targeted enzyme engineering that resulted in a 20% increase in titer. Competing TCA cycle reactions were identified using targeted transcriptomics, directed by isotopic labeling	LC-MS, dynamic isotopic labeling	1,4-Butanediol	*Escherichia coli*
	The mevalonate (MVA) and methylerythritol phosphate (MEP) pathways were found to be synergistic in isoprene production. Dual overexpression of MAV and MEP pathways resulted in enhanced productivity beyond the expected additive effect of individual overexpression, with a final titer of 24 g/L	LC-MS, steady-state isotopic labeling, 13C MFA	Isoprene	*Escherichia coli*
	Medium-chain hydrocarbon nonatetraene was found to be synthesized via a polyketide-synthesis-related pathway whereby head-to-tail condensation of acetate is followed by decarboxylation. Candidate polyketide synthases were identified	GC-MS, steady-state isotopic labeling	(3E,5E,7E)-Nona-1,3,5,7-tetraene	*Nigrograna mackinonnii*

TABLE 21.5 Different Applications of Metabolomic-Based Analyses to Biofuel—cont'd

Area	Major Findings	Metabolomics Methods	Bioproduct(s)	Organism
Redox balancing	Yeast responded to increased NADPH demand by increasing acetate production and rerouting flux toward the PPP. These responses were metabolically or transcriptionally controlled depending on these varieties of NADPH demand	GC-MS and IE-MS/MS, steady-state isotopic labeling, 13C MFA	Any	*Saccharomyces cerevisiae*
	Constitutive expression of phosphoglucomutase and transaldolase increased ethanol yield. Metabolomic analysis identified increased NADPH availability leading to reduced acetate production as a major source of improvement	GC-MS	Ethanol	*Fusarium oxysporum*
	Increased carbon and redox demands of fatty acid over production resulted in increased flux through the oxidative PPP and increased conversion of NADH to NADPH by transhydrogenases. ED glycolysis increased only lightly despite its ability to generate both NADPH and acetyl-coA for fatty acid synthesis, making ED glycolysis a potential target for metabolic improvement	GC-MS, steady-state isotopic labeling, 13C MFA	Fatty acids	*Escherichia coli*
	Increased carbon and redox demands of mevalonate over production were met by conversion of NADH to NADPH via transhydrogenase. This study did not see increases in oxidative PPP flux, suggesting that the moderate NADPH demands for mevalonate production could be met by transhydrogenase reactions alone	GC-MS, steady-state isotopic labeling, 13C MFA	Mevalonate	*Escherichia coli*
	A correlation was observed between increased PPP flux and increased expression of recombinant protein. Combined overexpression of glucose-6-phosphate dehydrogenase and 6-phosphogluconolactone resulted in the highest PPP flux and the highest expression levels of recombinant protein	GC-MS, steady-state isotopic labeling, 13C MFA	Any	*Pichia pastor*

Continued

TABLE 21.5 Different Applications of Metabolomic-Based Analyses to Biofuel—cont'd

Area	Major Findings	Metabolomics Methods	Bioproduct(s)	Organism
Central metabolism	Flux modeling of central carbon metabolism verified the absence of ED glycolysis and oxidative PPP and showed high TCA cycle flux	GC-MS, steady-state isotopic labeling, 13C MFA	Any	*Thermus thermophilus*
	Flux modeling of central carbon metabolism revealed noncanonical TCA cycle reactions, generation of C1 from pyruvate, and isoleucine production via citramalate synthase	GC-MS, parallel steady-state isotopic labeling, 13C MFA	Butanol, ethanol, and acetone	*Clostridium acetobutylicum*
	Flux modeling of central carbon metabolism showed that the TCA cycle and oxidative PPP are responsible for NADPH production during growth on xylose	GC-MS, parallel steady-state isotopic labeling, 13C MFA	Any	*Geobacillus* sp.
	13C fingerprinting based on labeling patterns of only a few amino acids was used to assess the metabolic activity of EMP and ED glycolysis, gluconeogenesis, glyoxylate shunt, anaplerotic pathways, and amino acid synthesis in a nonmodel organism	GC-MS, parallel steady-state isotopic labeling, 13C fingerprinting	Any	*Rhodococcus opacus PD630*
Xylose utilization	Expression of heterologous xylose reductase (XR), xylitol dehydrogenase (XDH), and xylulose kinase enzymes led to increased flux through the oxidative PPP and TCA cycle to meet increased NADPH and energy demands, limiting ethanol production	GC-MS, steady-state isotopic labeling, 13C MFA	Ethanol	*Saccharomyces cerevisiae (XR-XDH xylose assimilation)*
	Yeast strain with xylose isomerase (XI)-based xylose assimilation did not exhibit high flux through oxidative PPP suggesting that XI ameliorates the redox imbalances seen in XR-HDH strains. However, bottlenecks in lower glycolysis limit ethanol production	GC-MS, steady-state isotopic labeling, 13C MFA	Ethanol	*Saccharomyces cerevisiae (XI xylose assimilation)*
	During sugar coutilization, hexoses were assimilated via glycolysis, while pentoses were incorporated in to the PPP. The phosphoketolase pathway plays an important role in pentose metabolism and could be targeted for strain improvement	GC-MS, parallel steady-state isotopic labeling, 13C MFA	Butanol, ethanol, and acetone	*Clostridium acetobutylicum BOH3*

TABLE 21.5 Different Applications of Metabolomic-Based Analyses to Biofuel—cont'd

Area	Major Findings	Metabolomics Methods	Bioproduct(s)	Organism
	In xylose-utilizing strain developed via directed evolution, NADPH production was identified as a limiting factor during growth on xylose, suggesting that expression of heterologous oxidative PPP enzymes may improve strain performance	GC-MS, parallel steady-state isotopic labeling, 13C MFA	Any	*Thermus thermophilus developed via adaptive evolution of strain HB8*
Stress tolerance	Acetic acid was found to inhibit xylose fermentation due to an accumulation of intermediates of the nonoxidative PPP. Overexpression of transaldolase relieved this bottleneck and improved ethanol yields	GC-MS and CE-MS	Ethanol	*Saccharomyces cerevisiae (XR-XDH xylose assimilation)*
	Proline and myoinositol were identified as key metabolites in tolerance to furfural, acetic acid, and phenol. Accumulation or supplementation of these metabolites mitigates growth inhibition	GC-MS, regression modeling (PLS-DA)	Any	*Saccharomyces cerevisiae*
	Phenolic amides inhibit nucleotide biosynthesis via competitive inhibition of glutamine amidotransferase. External supplementation with nucleosides mitigates growth inhibition	LC-MS, dynamic isotopic labeling	Any	*Escherichia coli*
	Threonine was identified as a key metabolite contributing to butanol tolerance based on metabolomics-based regression modeling. Gene deletions aimed at increasing threonine accumulation resulted in improved butanol tolerance, providing a proof of concept for semirational engineering based on metabolomics data	GC-MS, regression modeling (OPLS)	1-Butanol	*Saccharomyces cerevisiae*
	Directed evolution for improved butanol tolerance resulted in increased abundance of disaccharides and saturated fatty acids and decreased levels of carotenoids and carotenoid precursors, suggesting that membrane fluidity and osmotic control are important factors in butanol tolerance	GC-MS and LC-MS	1-Butanol	*Methylobacterium extorquens AM1*

Data from Martien, J.I., Amador-Noguez, D., 2017. Recent applications of metabolomics to advance microbial biofuel production. Curr. Opin. Biotechnol. 43, 118–126.

past few years, successful application of metabolic engineering of yeast capable of utilizing the pentose sugars (especially xylose and arabinose) as sole carbon source has been generated. However, still, many major obstacles remain for engineering to meet simultaneous, exogenous sugar metabolism and improve product yield. Beyond catabolic pathways, the attention should be paid more toward nontraditional aspects of cellular engineering such as host molecular transport capability, catabolite sensing, and stress response mechanisms. Traditional use of biomass that is abundant in nature, yet challenging to access sugars, is a key target to minimize the usage of land for food versus fuel and to minimize ethical problems. Exploiting photosynthetic organisms for biofuel generation has shown exciting potential. It has been demonstrated that the availability of boundless raw materials (sunlight, water, and CO_2) has been used to produce isopropanol in engineered *Cyanobacteria*. Indeed, rapid advances in genetic engineering techniques and owing to its potential have made several breakthroughs in this area. In addition, synthetic biology approaches are likely to be used in the near future for biofuel production. Nevertheless, the high cost of processing at industrial levels limits its commercial production of biofuels. Hence, further research should be focused on the exploration of new technologies in improving yields of biofuels.

Acknowledgments

GR is thankful to the Science and Engineering Research Board (SERB), Govt. of India, for their financial assistance under SERB-National Postdoctoral fellowship (PDF/2016/001667).

Conflicts of interests

The authors declare no relevant competing financial interests to disclose.

References

Ageitos, J.M., Vallejo, J.A., Veiga-Crespo, P., et al., 2011. Oily yeasts as oleaginous cell factories. Appl. Microbiol. Biotechnol. 90 (4), 1219–1227.

Ahmad, F.B., Zhang, Z., Doherty, W.O.S., Teo, V.S.J., Hara, I.M.O., 2017. Improved microbial oil production from oil palm empty fruit bunch by *Mucor plumbeus*. Fuel 194, 180–187.

Almeida, J.R.M., Modig, T., Petersson, A., Hahn-Hagerdal, B., Liden, G., Gorwa-Grauslund, M.F., 2007. Increased tolerance and conversion of inhibitors in lignocellulosic hydrolysates by *Saccharomyces cerevisiae*. J. Chem. Technol. Biotechnol. 82, 340–349.

Alvarez, H.M., Steinbuchel, A., 2002. Triacylglycerols in prokaryotic microorganisms. Appl. Microbiol. Biotechnol. 60 (4), 367–376.

Annamalai, N., Sivakumara, N., Oleskowicz-Popielc, P., 2018. Enhanced production of microbial lipids from waste office paper by the oleaginous yeast *Cryptococcus curvatus*. Fuel 217, 420–426.

Aslam, M.U., Masjuki, H.H., Kalam, M.A., Abdesselam, H., Mahlia, T.M.I., Amalina, M.A., 2006. An experimental investigation of CNG as an alternative fuel for a retrofitted gasoline vehicle. Fuel 85, 717–724.

Batista, A.P., Ambrosano, L., Graça, S., Sousa, C., Marques, P.A.A.S., Ribeiro, B., et al., 2015. Combining urban wastewater treatment with biohydrogen production e an integrated microalgae-based approach. Bioresour. Technol. 184, 230–235.

Batista, A.P., Gouveia, L., Marques, P.A.S.S., 2018. Fermentative hydrogen production from microalgal biomass by a single strain of bacterium *Enterobacter aerogenes*: effect of operational conditions and fermentation kinetics. Renew. Energy 119, 203–209.

Batista, A.P., Moura, P., Marques, P.A.S.S., Ortigueira, J., Alves, L., Gouveia, L., 2014. *Scenedesmus obliquus* as feedstock for biohydrogen production by *Enterobacter aerogenes* and *Clostridium butyricum*. Fuel 117, 537–543.

Bellack, A., Huber, H., Rachel, R., Wanner, G., Wirth, R., 2011. *Methanocaldococcus villosus* sp. nov., a heavily flagellated archaeon that adheres to surfaces and forms cell–cell contacts. Int. J. Syst. Evol. Microbiol. 61, 1239–1245.

Berchmans, S., 2018. Microbial fuel cell as alternate power tool: potential and challenges. In: Das, D. (Ed.), Microbial Fuel Cell. Springer, Cham, pp. 403–419.

Brauer, S.L., Cadillo-Quiroz, H., Yashiro, E., Yavitt, J.B., Zinder, S.H., 2006. Isolation of a novel acidiphilic methanogen from an acidic peat bog. Nature 442, 192–194.

Butinar, L., Spencer-Martins, I., Gunde-Cimerman, N., 2007. Yeasts in high Arctic glaciers: the discovery of a new habitat for eukaryotic microorganisms. Antonie Van Leeuwenhoek 91, 277–289.

Case, A.E., Atsumi, S., 2016. *Cyanobacterial* chemical production. J. Biotechnol. 231, 106–114.

Chang, Y.H., Chang, K.S., Lee, C.F., Hsu, C.L., Huang, C.W., Jang, H.D., 2015. Microbial lipid production by oleaginous yeast *Cryptococcus* sp. in the batch cultures using corncob hydrolysate as carbon source. Biomass Bioenergy 72, 95–103.

Chen, C., Cai, D., Qin, P., Chen, B., Wang, Z., Tan, T., 2018. Bio-plasticizer production by hybrid acetone-butanol-ethanol fermentation with full cell catalysis of *Candida sp*. Bioresour. Technol. 257, 217–222.

Chen, Z., Wu, Y., Huang, J.H., Liu, D.H., 2015. Metabolic engineering of *Klebsiella pneumoniae* for the de novo production of 2-butanol as a potential biofuel. Bioresour. Technol. 197, 260–265.

Converti, A., Perego, P., Del Borghi, M., 2003. Effect of specific oxygen uptake rate on *Enterobacter aerogenes* energetics: carbon and reduction degree balances in batch cultivations. Biotechnol. Bioeng. 82, 370–377.

Das, D., Veziroglu, T.N., 2008. Advances in biological hydrogen production processes. Int. J. Hydrogen Energy 33, 6046–6057.

Davies, F.K., Work, V.H., Beliaev, A.S., Posewitz, M.C., 2014. Engineering limonene and bisabolene production in wild type and a glycogen-deficient mutant of *Synechococcus* sp. PCC 7002. Front. Bioeng. Biotechnol. 2, 21.

Debord, J., 2012. Coefficient de partage. Available: http://eduscol.education.fr/rnchimie/chi_org/eb/coefficient_de_partage.pdf. [Accessed March 14, 2018].

Dong, L., Zhenhong, Y., Yongming, S., Xiaoying, K., Yu, Z., 2009. Hydrogen production characteristics of the organic fraction of municipal solid wastes by anaerobic mixed culture fermentation. Int. J. Hydrog. Energy 34, 812–820.

Donlon, B., Razo-Flores, E., Field, J., et al., 1995. Toxicity of N-substituted aromatics to acetoclastic methanogenic activity in granular sludge. Appl. Environ. Microbiol. 61, 3889–3893.

Dudley, Q.M., Karim, A.S., Jewett, M.C., 2015. Cell-free metabolic engineering: biomanufacturing beyond the cell. Biotechnol. J. 10, 69–82.

Ferreira, A.F., Marques, A.C., Batista, A.P., Marques, P.A.S.S., Gouveia, L., Silva, C., 2012. Biological hydrogen production by Anabaena sp. e yield, energy and CO_2 analysis including fermentative biomass recovery. Int. J. Hydrogen Energy 7, 179–190.

Ferry, J.G., 2011. Fundamentals of methanogenic pathways that are key to the biomethanation of complex biomass. Curr. Opin. Biotechnol. 22, 351–357.

Ghosh, D., Hallenbeck, P.C., 2010. Response surface methodology for process parameter optimization of hydrogen yield by the metabolically engineered strain *Escherichia coli* DJT135. Bioresour. Technol. 101, 1820–1825.

Gujjala, L.K.S., Jeevan Kumar, S.P., Talukdar, B., Dash, A., Kumar, S., et al., 2017. Biodiesel from oleaginous microbes: opportunities and challenges. Biofuels https://doi.org/10.1080/17597269.2017.1402587.

Gutierrez, A., Lopez-Garcıa, S., Garre, V., 2011. High reliability transformation of the basal fungus *Mucor circinelloides* by electroporation. J. Microbiol. Methods 84, 442–446.

Harde, S.M., Wang, Z., Horne, M., Zhu, J.Y., Pan, X., 2016. Microbial lipid production from SPORL-pretreated Douglas fir by *Mortierella isabellina*. Fuel 175, 64–74.

Hasunuma, T., Sakamoto, T., Kondo, A., 2016. Inverse metabolic engineering based on transient acclimation of yeast improves acid-containing xylose fermentation and tolerance to formic and acetic acids. Appl. Microbiol. Biotechnol. 100, 1027–1038.

Hitit, Z.Y., Zampol, C., Hallenbeck, P.C., 2017a. Increased hydrogen yield and COD removal from starch/glucose based medium by sequential dark and photo-fermentation using *Clostridium butyricum* and *Rhodopseudomonas palustris*. Int. J. Hydrog. Energy 42, 18832–18843.

Hitit, Z.Y., Zampol, C., Hallenbeck, P.C., 2017b. Hydrogen production by co-cultures of *Clostridium butyricum* and *Rhodospeudomonas palustris*: Optimization of yield using 19 response surface methodology. Int. J. Hydrog. Energy 42 (10), 6578–6589.

Holdsworth, J.E., Ratledge, C., 1991. Triacylglycerol synthesis in the oleaginous yeast *Candida curvata* D. Lipids 26, 111–118.

Janicek, A., Fan, Y., Liu, H., 2014. Design of microbial fuel cells for practical application: a review and analysis of scale-up studies. Biofuels 5, 79–92.

Jones, D., Woods, D., 1986. Acetone-butanol fermentation revisited. Microbiol. Rev. 50, 484–524.

Kallio, P., Pásztor, A., Thiel, K., Akhtar, M.K., Jones, P.R., 2014. An engineered pathway for the biosynthesis of renewable propane. Nat. Commun. 5, 4731.

Kim, H., Moon, S., Abug, A., Choi, S.C., Zhang, R., Oh, Y.S., 2012. Effect of fermentation conditions on biohydrogen production from lipid-rich food material. Int. J. Hydrog. Energy 37, 15062–15069.

Kiran, U.E., Trzcinski, A.P., NgWJ, L.Y., 2014. Bioconversion of food waste to energy: a review. Fuel 134, 389–399.

Koók, L., Rózsenberszki, T., Nemestóthy, N., Bélafi-Bakó, K., Bakonyi, P., 2016. Bioelectrochemical treatment of municipal waste liquor in microbial fuel cells for energy valorization. J. Clean. Prod. 112, 4406–4412.

Kumar, P., Barrett, D.M., Delwiche, M.J., Stroeve, P., 2009. Methods for pretreatment of lignocellulosic biomass for efficient hydrolysis and biofuel production. Indus Eng. Chem. Res. 48 (8), 3713–3729.

Kumar, N., Das, D., 2000. Enhancement of hydrogen production by *Enterobacter cloacae* IIT- BT 08. Process Biochem. 35, 589–593.

Kumar, P., Pant, D.C., Mehariya, S., Sharma, R., Kansal, A., Kalia, V.C., 2014. Ecobiotechnological strategy to enhance efficiency of bioconversion of wastes into hydrogen and methane. Indian J. Microbiol. 54 (3), 262–267.

Kumar, R., Kumar, P., 2017. Future microbial applications for bioenergy production: a perspective. Front. Microbiol. 8, 450.

Kumar, R., Singh, L., Wahid, Z.A., Din, M.F.M., 2015. Exoelectrogens in microbial fuel cells toward bioelectricity generation: a review. Int. J. Energy Res. 39, 1048–1067.

La Licata, B., Sagnelli, F., Boulanger, A., Lanzini, A., Leone, P., Zitella, P., et al., 2011. Biohydrogen production from organic wastes in a pilot plant reactor and its use in a SOFC. Int. J. Hydrog. Energy 36, 7861–7865.

Lakaniemi, A.M., Hulatt, C.J., Thomas, D.N., Tuovinen, O.H., Puhakka, J.A., 2011. Biogenic hydrogen and methane production from *Chlorella vulgaris* and *Dunaliella tertiolecta* biomass. Biotechnol. Biofuels 4, 34.

Lal, D., 2013. Microbes to generate electricity. Indian J. Microbiol. 53, 120–122.

Leblanc, D.J., Lee, L.N., 1984. Physical and genetic analyses of streptococcal plasmid pAMb1 and cloning of its replication region. J. Bacteriol. 157, 445–453.

Lee, H.J., Kim, S.J., Yoon, J.J., Kim, K.H., Seo, J.H., Park, Y.C., 2015a. Evolutionary engineering of *Saccharomyces cerevisiae* for efficient conversion of red algal biosugars to bioethanol. Bioresour. Technol. 191, 445–451.

Lee, S.H., Kwon, M.A., Choi, S., Kim, S., et al., 2015b. A new shuttle plasmid that stably replicates in *Clostridium acetobutylicum*. J. Microbiol. Biotechnol. 25, 1702–1708.

Lee, S.M., Jellison, T., Alper, H.S., 2016. Bioprospecting and evolving alternative xylose and arabinose pathway enzymes for use in *Saccharomyces cerevisiae*. Appl. Microbiol. Biotechnol. 100, 2487–2498.

Leja, K., et al., 2014. Przemysłowe wykorzystanie bakterii z rodzaju *Clostridium*. Postępy Mikrobiologii 53, 15–24.

Lesiak, J.M., Liebl, W., Ehrenreich, A., 2014. Development of an *in vivo* methylation system for the solventogen *Clostridium saccharobutylicum* NCP 262 and analysis of two endonuclease mutants. J. Biotechnol. 188, 97–99.

Li, C.H., Cervantes, M., Springer, D.J., et al., 2011. Sporangiospore size dimorphism is linked to virulence of *Mucor circinelloides*. PLoS Pathog. 7, e1002086.

Liao, J.C., Mi, L., Pontrelli, S., Luo, S., 2016. Fuelling the future: microbial engineering for the production of sustainable biofuels. Nat. Rev. Microbiol. 14, 288–304.

Limayem, A., Ricke, S.C., 2012. Lignocellulosic biomass for bioethanol production: current perspectives, potential issues and future prospects. Prog. Energy Combust. Sci. 38, 449–467.

Ling, H., Teo, W., Chen, B., Leong, S.S., Chang, M.W., 2014. Microbial tolerance engineering toward biochemical production: from lignocellulose to products. Curr. Opin. Biotechnol. 29, 99–106.

Logan, B.E., Regan, J.M., 2006. Feature article: microbial fuel cells—challenges and applications. Environ. Sci. Technol. 40, 5172–5180.

Logan, B.E., Wallack, M.J., Kim, K.Y., He, W., Feng, Y., Saikaly, P.E., 2015. Assessment of microbial fuel cell configurations and power densities. Environ. Sci. Technol. Lett. 2, 206–214.

Lohit, K.S., Gujjala, S.P., Kumar, J., Talukdar, B., Dash, A., Kumar, S., et al., 2017. Biodiesel from oleaginous microbes: opportunities and challenges. Biofuels https://doi.org/10.1080/17597269.2017.1402587.

Lütke-Eversloh, T., 2014. Application of new metabolic engineering tools for *Clostridium acetobutylicum*. Appl. Microbiol. Biotechnol. 98, 5823–5837.

Martien, J.I., Amador-Noguez, D., 2017. Recent applications of metabolomics to advance microbial biofuel production. Curr. Opin. Biotechnol. 43, 118–126.

Martin, W., Baross, J., Kelley, D., Russell, M.J., 2008. Hydrothermal vents and the origin of life. Nat. Rev. Microbiol. 6, 805–814.

Matsakas, L., Gao, Q., Jansson, S., Rova, U., Christakopoulos, P., 2017. Green conversion of municipal solid wastes into fuels and chemicals. Electron. J. Biotechnol. 26, 69–83.

Meeuwse, P., Tramper, J., Rinzema, A., 2011. Modeling lipid accumulation in oleaginous fungi in chemostat cultures: I. Development and validation of a chemostat model for *Umbelopsis isabellina*. Bioprocess Biosyst. Eng. 34 (8), 939–949.

Menon, N., Pásztor, A., Menon, B.R., Kallio, P., Fisher, K., Akhtar, M.K., Leys, D., Jones, P.R., Scrutton, N.S., 2015. A microbial platform for renewable propane synthesis based on a fermentative butanol pathway. Biotechnol. Biofuels 8, 12–61.

Metcalf, W.W., et al., 2012. Synthesis of methylphosphonic acid by marine microbes: a source for methane in the aerobic ocean. Science 337, 1104–1107.

Mohanakrishna, G., Mohan, S.K., Mohan, S.V., 2012. Carbon based nanotubes and nanopowder as impregnated electrode structures for enhanced power generation: evaluation with real field wastewater. Appl. Energy 95, 31–37.

Morales, P., Gentina, J.C., Aroca, G., Mussatto, S.I., 2017. Development of an acetic acid tolerant *Spathaspora passalidarum* strain through evolutionary engineering with resistance to inhibitor compounds of autohydrolysate of *Eucalyptus globulus*. Ind. Crop. Prod. 106, 5–11.

Moysés, D.N., Reis, V.C., Almeida, J.R., Moraes, L.M., Torres, F.A., 2016. Xylose fermentation by *Saccharomyces cerevisiae*: challenges and prospects. Int. J. Mol. Sci. 17, 207.

Nigam, P., Singh, A., 2011. Production of liquid biofuels from renewable resources. Prog. Energy Combust. Sci. 37, 52–68.

Nobre, B.P., Villalobos, F., Barrag, B.E., Oliveira, A.C., Batista, A.P., Marques, P.A.S.S., 2013. A biorefinery from *Nannochloropsis* sp. microalga e extraction of oils and pigments. Production of biohydrogen from the leftover biomass. Bioresour. Technol. 135, 128–136.

Oh, B.R., Heo, S.Y., Lee, S.M., Hong, W.K., Park, J.M., Jung, Y.R., Kim, D.H., Sohn, J.H., Seo, J.W., Kim, C.H., 2014. Production of 2-butanol from crude glycerol by a genetically-engineered *Klebsiella pneumoniae* strain. Biotechnol. Lett. 36 (1), 57–62.

Pacheco, R., Ferreira, A.F., Pinto, T., Nobre, B., Loureiro, D., Moura, P., et al., 2015. The production of pigments & hydrogen through a *Spirogyra* sp. biorefinery. Energ. Convers. Manage. 89, 789–797.

Pant, A., Rai, J.P.N., 2018. Impact of chlorpyrifos, TCP and N-substituted aromatic compounds on methane production from organic solid waste (OSW) using co-culture of *Pseudomonas aeruginosa* and *Methanosarcina mazei*. Biofuels https://doi.org/10.1080/17597269.2018.1432276.

Pant, D., Singh, A., Van Bogaert, G., Irving Olsen, S., Singh Nigam, P., Diels, L., Vanbroekhoven, K., 2012. Bioelectrochemical systems (BES) for sustainable energy production and product recovery from organic wastes and industrial wastewaters. RSC Adv. 2, 1248–1263.

Papurello, D., Soukoulis, C., Schuhfried, E., Cappellin, L., Gasperi, F., Silvestri, S., et al., 2012. Monitoring of volatile compound emissions during dry anaerobic digestion of the organic fraction of municipal solid waste by proton transfer reaction time-of-flight mass spectrometry. Bioresour. Technol. 126, 254–265.

Park, G.W., Chang, H.N., Jung, K., Seo, C., Kim, Y.C., Choi, J.H., et al., 2017. Production of microbial lipid by *Cryptococcus curvatus* on rice straw hydrolysates. Process Biochem. 56, 147–153.

Peralta-Yahya, P.P., Ouellet, M., Chan, R., Mukhopadhyay, A., Keasling, J.D., Lee, T.S., 2011. Identification and microbial production of a terpene-based advanced biofuel. Nat. Commun. 2, 483.

Phelan, R.M., Sekurova, O.N., Keasling, J.D., Zotchev, S.B., 2014. Engineering terpene biosynthesis in *Streptomyces* for production of the advanced biofuel precursor bisabolene. ACS Synth. Biol. 4, 93–399.

Polburee, P., Yongmanitchai, W., Lertwattanasakul, N., et al., 2015. Characterization of oleaginous yeasts accumulating high levels of lipid when cultivated in glycerol and their potential for lipid production from biodiesel-derived crude glycerol. Fungal Biol. 119, 1194–1204.

Potter, M.C., 1910. In: On the difference of potential due to the vital activity of microorganisms. Proceedings of the University of Durham Philosophical Society. pp. 245–249.

Purdy, D., O'Keeffe, T.A., Elmore, M., Herbert, M., et al., 2002. Conjugative transfer of *clostridial* shuttle vectors from *Escherichia coli* to *Clostridium difficile* through circumvention of the restriction barrier. Mol. Microbiol. 46, 439–452.

Quarterman, J., Skerker, J.M., Feng, X., Liu, I.Y., Zhao, H., Arkin, A.P., Jin, Y.S., 2016. Rapid and efficient galactose fermentation by engineered *Saccharomyces cerevisiae*. J. Biotechnol. 229, 13–21.

Ratledge, C., 2002. Regulation of lipid accumulation in oleaginous micro-organisms. Biochem. Soc. Trans. 30, 1047–1050.

Ratledge, C., 2010. Single cell oils for the 21st century. In: Cohen, Z., Ratledge, R. (Eds.), Single Cell Oils-Microbial and Algal Oils, second ed. AOCS Press, Urbana; IL, pp. 3–26.

Ratledge, C., Wynn, J.P., 2002. The biochemistry and molecular biology of lipid accumulation in oleaginous microorganisms. Adv. Appl. Microbiol. 51, 1–44.

Rossi, M., Amaretti, A., Raimondi, S., et al., 2011. Getting lipids for biodiesel production from oleaginous fungi. In: Stoytcheva, M., Montero, G. (Eds.), Biodiesel – Feedstocks and Processing Technologies. InTech Publishers, Krautzeka, Croatia.

Ruan, Z.H., Zanotti, M., Zhong, Y., Liao, W., Ducey, C., Liu, Y., 2013. Co-hydrolysis of lignocellulosic biomass for microbial lipid accumulation. Biotechnol. Bioeng. 110, 1039–1049.

Sagır, E., Yucel, M., Hallenbeck, P.C., 2017. Demonstration and optimization of sequential microaerobic dark- and photo-fermentation biohydrogen production by immobilized *Rhodobacter capsulatus* JP91. Bioresour. Technol. 250, 43–52.

Schönheit, P., Moll, J., Thauer, R.K., 1980. Growth parameters (Ks, μmax, Ys) of *Methanobacterium thermoautotrophicum*. Arch. Microbiol. 127, 59–65.

Seo, Y., Lee, I., Han, J., 2013. Cultivation and lipid production of yeast *Cryptococcus curvatus* using pretreated waste active sludge supernatant. Bioresour. Technol. 135, 304–308.

Seo, Y.H., Lee, I., Jeon, S.H., Han, J.I., 2014. Efficient conversion from cheese whey to lipid using *Cryptococcus curvatus*. Biochem. Eng. J. 90, 149–153.

Shin, H.S., Youn, J.H., Kim, S.H., 2004. Hydrogen production from food waste in anaerobic mesophilic and thermophilic acidogenesis. Int. J. Hydrog. Energy 29, 1355–1363.

Shin, H.Y., Nijland, J.G., de Waal, P.P., de Jong, R.M., Klaassen, P., Driessen, A.J.M., 2015. An engineered cryptic Hxt11 sugar transporter facilitates glucose-xylose co-consumption in *Saccharomyces cerevisiae*. Biotechnol. Biofuels 8, 176.

Takai, K., Inoue, A., Horikoshi, K., 2002. *Methanothermococcus okinawensis* sp. nov., a thermophilic, methane-producing archaeon isolated from a western Pacific deep-sea hydrothermal vent system. Int. J. Syst. Evol. Microbiol. 52, 1089–1095.

Taubner, R.S., Pappenreiter, P., Zwicker, J., Smrzka, D., Pruckner, C., Kolar, P., et al., 2018. Biological methane production under putative Enceladus-like conditions. Nat. Commun. 9, 748.

Taubner, R.S., Rittmann, S.K.M.R., 2016. Method for indirect quantification of CH4 production *via* H2O production using hydrogenotrophic methanogens. Front. Microbiol. 7, 532.

Taubner, R.S., Schleper, C., Firneis, M.G., Rittmann, S.K.M.R., 2015. Assessing the ecophysiology of methanogens in the context of recent astrobiological and planetological studies. Life 5, 1652–1686.

Thauer, R.K., Kaster, A.K., Seedorf, H., Buckel, W., Hedderich, R., 2008. Methanogenic archaea: ecologically relevant differences in energy conservation. Nat. Rev. Microbiol. 6, 579–591.

Uma, J., Kim, D.G., Jung, M.Y., Saratale, G.D., Oha, M.K., 2017. Metabolic engineering of *Enterobacter aerogenes* for 2,3-butanediol production from sugarcane bagasse hydrolysate. Bioresour. Technol. 1245, 1567–1574.

Vijayaraghavan, K., Mohd Soom, M.A., 2006. Trends in bio-hydrogen generation: a review. Environ. Sci. 3, 255–271.

Vinuselvi, P., Lee, S.K., 2012. Engineered *Escherichia coli* capable of co-utilization of cellobiose and xylose. Enzym. Microb. Technol. 50, 1–4.

Vorholt, J., Kunow, J., Stetter, K.O., Thauer, R.K., 1995. Enzymes and coenzymes of the carbon monoxide dehydrogenase pathway for autotrophic CO_2 fixation in *Archaeoglobus lithotrophicus* and the lack of carbon monoxide dehydrogenase in the heterotrophic *A. profundus*. Arch. Microbiol. 163, 112–118.

Wang, C., Xin, F., Kong, X., Zhao, J., Dong, W., Zhang, W., et al., 2018. Enhanced isopropanol–butanol–ethanol mixture production through manipulation of intracellular NAD(P)H level in the recombinant *Clostridium acetobutylicum* XY16. Biotechnol. Biofuels 11, 1–10.

Xie, X., Hu, L., Pasta, M., Wells, G.F., Kong, D., Criddle, C.S., Cui, Y., 2010. Three-dimensional carbon nanotube–textile anode for high-performance microbial fuel cells. Nano Lett. 11, 291–296.

Yang, Y., Sun, G., Xu, M., 2011. Microbial fuel cells come of age. J. Chem. Technol. Biotechnol. 86, 625–632.

Yat, S.C., Berger, A., Shonnard, D.R., 2008. Kinetic characterization of dilute surface acid hydrolysis of timber varieties and switchgrass. Bioresour. Technol. 99, 3855–3863.

Yu, M., Du, Y., Jiang, W., Chang, W.L., et al., 2012. Effects of different replicons in conjugative plasmids on transformation efficiency, plasmid stability, gene expression and *n*-butanol biosynthesis in *Clostridium tyrobutyricum*. Appl. Microbiol. Biotechnol. 93, 881–889.

Yvon-Durocher, G., et al., 2014. Methane fluxes show consistent temperature dependence across microbial to ecosystem scales. Nature 507, 488–491.
Zhang, B., Sun, H.B., Li, J., Wan, Y.H., Li, Y., Zhang, Y.P., 2016. High-titer-ethanol production from cellulosic hydrolysate by an engineered strain of *Saccharomyces cerevisiae* during an *in situ* removal process reducing the inhibition of ethanol on xylose metabolism. Process Biochem. 51, 967–972.
Zhang, J., Fang, X., Zhu, X.L., Li, Y., Xu, H.P., Zhao, B.F., et al., 2011. Microbial lipid production by the oleaginous yeast *Cryptococcus curvatus* O3 grown in fed-batch culture. Biomass Bioenergy 35, 1906–1911.
Zhang, L., Singh, R., Sivakumar, D., Guo, Z., Li, J., Chen, F., He, Y., et al., 2017. An artificial synthetic pathway for acetoin, 2,3-butanediol, and 2-butanol production from ethanol using cell free multi-enzyme catalysis. Green Chem. 20, 230–242. https://doi.org/10.1039/c7gc02898a.
Zhang, Y., Paul, S., Caldwell, G., 2018. Renewable energy-assessment of co-digestion strategies to optimize methane production from microalgae. Int. J. Biotechnol. Bioeng. 12 (5).
Zheng, Y., Yu, X., Zeng, J., et al., 2012. Feasibility of filamentous fungi for biofuel production using hydrolysate from dilute sulphuric acid pretreatment of wheat straw. Biotechnol. Biofuel 5, 50.

Further Reading

Fei, Q., Chang, H.N., Shang, L., Choi, J., Kim, N.J., Kang, J.W., 2011. The effect of volatile fatty acids as a sole carbon source on lipid accumulation by *Cryptococcus albidus* for biodiesel production. Bioresour. Technol. 102, 2695–2701.
Ferreira, A.F., Ribeiro, L.A., Batista, A.P., Marques, P.A.S.S., Nobre, B.P., Palavra, A.M.F., et al., 2013. A biorefinery from *Nannochloropsis* sp. microalga e energy and CO_2 emission and economic analyses. Bioresour. Technol. 138, 235–244.
Guo, K., Hassett, D.J., Gu, T., 2012. Microbial fuel cells: electricity generation from organic wastes by microbes. In: Arora, R. (Ed.), Microbial Biotechnology: Energy and Environment. CAB International, Oxon, United Kingdom.
Khan, M.Z., Singh, S., Sreekrishnan, T.R., Ahammad, S.Z., 2014. Feasibility study on anaerobic biodegradation of *azo* dye reactive orange 16. RSC Adv. 4, 46851–46859.
Kumar, G., Mudhoo, A., Sivagurunathan, P., Nagarajan, D., 2016. Recent insights into the cell immobilization technology applied for dark fermentative hydrogen production. Bioresour. Technol. 219, 725–737.
Oh, S.T., Kim, J.R., Premier, G.C., et al., 2010. Sustainable wastewater treatment: how might microbial fuel cells contribute. Biotechnol. Adv. 28, 871–881.
Pandey, A., Chang, J.S., Hallenbeck, P.C.L.C. (Eds.), 2013. Biohydrogen. Elsevier, London, pp. 185–221.
Potter, M.C., 1911. Electrical effects accompanying the decomposition of organic compounds. Proc. R. Soc. Lond. B 260–276.

CHAPTER

22

Present Status and Future Perspectives of Marine Actinobacterial Metabolites

Gurushankara Hunasanahally Puttaswamygowda, Shilpa Olakkaran, Anet Antony, Anupama Kizhakke Purayil

Department of Animal Science, School of Biological Sciences, Tejaswini Hills, Periye, Kasaragod, India

OUTLINE

22.1 INTRODUCTION

Actinobacteria are ubiquitous gram-positive bacteria with high guanine and cytosine contents in DNA, having a characteristic filamentous morphology (Dhakal et al., 2017). *Actinobacteria* have a number of important functions, including decomposition of all sorts of organic substances. These filamentous bacteria have evolved with the wealth of biosynthetic gene clusters and thereby showed their unsurpassed capacity for the

Recent Developments in Applied Microbiology and Biochemistry
https://doi.org/10.1016/B978-0-12-816328-3.00022-2

production of various bioactive secondary metabolites (Hassan et al., 2017). As per *Bergey's Manual of Systematic Bacteriology*, the phylum *Actinobacteria* is divided into six classes, namely, *Actinobacteria, Acidimicrobiia, Coriobacteriia, Nitriliruptoria, Rubrobacteria,* and *Thermoleophilia*. The class *Actinobacteria* is further divided into 16 orders that are *Actinopolysporales, Actinomycetales, Bifidobacteriales, Catenulisporales, Corynebacteriales, Frankiales, Glycomycetales, Jiangellales, Kineosporiales, Micrococcales, Micromonosporales, Propionibacteriales, Pseudonocardiales, Streptomycetales, Streptosporangiales,* and *Incertae sedis*. The *Actinomycetales* members are commonly referred to as actinomycetes (Goodfellow et al., 2012; http://www.bacterio.net/).

Marine actinobacteria are considered as a treasure house of secondary metabolites, and the large numbers of these bioactive metabolites belong to the family *Actinomycetaceae* including the genera of *Streptomyces, Actinobaculum,* and *Acanobacterium*, and others are commercially available due to their capability to produce novel bioactive molecules (Hassan et al., 2017). Every strain of *Actinobacteria* is believed to have the genetic potential for the production of 15–25 secondary metabolites (Lam, 2006). About 10,000 antibiotics have been isolated from *Actinobacteria*, which is 45% of all bioactive microbial metabolites discovered (Jackson et al., 2018). The actinomycin and a number of other antibiotics have been discovered from *Actinobacteria*, especially from the genus *Streptomyces*. The genus *Streptomyces* only describes 80% of the richest drug-prolific family in all kingdoms, producing therapeutic compounds (van Keulen and Dyson, 2014). The first report of streptomycin by Selman Waksman and associates in the 1940s and the golden era (1950–1970) of antibiotic discovery was evidenced by the commercialization of several lifesaving antibiotics such as vancomycin and rifamycin (Jose and Jha, 2016). The subsequent modern approach on the recovery of marine bacteria from diverse habitat samples and screening of bioactivity encouraged the research on *Actinobacteria* being continued by several research groups. *Actinobacteria* are widely distributed in marine environments such as sponges, fish, mollusks, mangroves, and seaweeds, besides seawater and sediments (Ward and Bora, 2006; Gogineni and Hamann, 2018). These organisms are obtaining importance not only for their taxonomic and ecological perceptions but also for their invention of unique bioactive compounds like antibiotics; antioxidants; and cytotoxic, antitumor, immunosuppressive, and cardiovascular agents with their unique carbon skeletons that also provide a strong base for the synthesis of therapeutics (Hassan et al., 2017). The *Actinobacteria* group has a vast pharmacological potential that remains unopposed among other microbial groups. The massive diversity, along with its underutilization, is attracting the researchers toward determining novel bioactive metabolites. This chapter gives an overview about the microbes in marine environment, actinobacterial diversity, present status, and future perspectives of marine actinobacterial metabolites and their applications.

22.2 MARINE MICROBES—TREASURE HOUSE OF BIOACTIVE MOLECULES

Marine ecosystem, covering over 70% of the Earth's area, harbors most of the planet's biodiversity. Marine microbes are essentially contributing to global biomass, biogeochemical cycling, and biodiversity, since the origin of the Earth. Oceans provide a variety of distinct biotopes, which offers many habitats for microorganisms. Marine microorganisms inhabit all available niches from the polar ice to hydrothermal vents, from the deep biosphere to mangrove forests, and from the oligotrophic open ocean waters to polluted coastal waters and sandy beaches. The surface of macroorganisms such as algae, sponges, fish, and corals has been particularly attractive as an ecological niche for microorganisms. In many cases, bacteria live in a habitat-specific close association with higher organisms and form mutualistic or symbiotic relationships. In the variety of habitats, the composition of marine microbial communities may vary in different ecosystems. This might have contributed the oceans an abode of unbelievable diversity of microorganisms. Yet, so far, microbial distribution patterns remain unknown in most marine ecosystems. The major environmental determinant of microbial community composition and diversity is salinity, temperature, pH, or other physical and chemical factors (Lozupone and Knight, 2007). It is estimated that, in over 1.2 million species already cataloged taxonomically in a central database, some 91% of species in the ocean still await description (Mora et al., 2011). Less than 0.1%, probably only 0.01% of all microbes in the oceans is known (Simon and Daniel, 2009). Systematic research of marine bacteria with respect to their secondary metabolite profiles revealed a large biosynthetic potential (Gurgui and Piel, 2010). Marine wealth is aptly described as "blue gold"; more than 2.30 lakhs marine species and isolation of more than 28,000 structurally unique bioactive natural compounds have been documented over the past 50 years (Blunt et al., 2016). However, the immense microbial diversity of the marine environment is unexplored. Molecular approaches on the analysis of marine metagenomes have revealed a remarkable number of newly described bacterial taxa of marine origin. Hence, it can be assumed that biosynthetic potential of the unknown microbes is still hidden in the oceans. Considering the incredible biodiversity of marine microorganisms and the gap in our knowledge, particularly regarding their potential of natural product biosynthesis, they are expected to represent affluent opportunity for the discovery of many bioactive compounds.

22.3 ACTINOBACTERIA IN MARINE ENVIRONMENT

The phylum *Actinobacteria* has been reported to be common or even abundant in deep marine sediments; however, knowledge about their diversity, distribution, and function is scanty. *Actinobacteria* found in various marine habitats are spread globally across oceanic realms, separated geographically and influenced by varying geophysical parameters of temperature, salinity, underlying geochemistry, and ocean currents. Salt marshes and wetlands, estuaries, continental shelves, open oceans, and deep-sea ecosystems influence the presence of *Actinobacteria* (Ward and Bora, 2006). The marine actinobacterial diversity has been investigated by 16S phylogenetic diversity inventories and culture-dependent methods. Thus, deep-sea sediments were found to contain >1300 different actinobacterial operational taxonomic units, a great proportion of which is predicted to represent novel species and genera (Bull and Stach, 2007). Among newly described marine genera are *Salinispora* and *Demequina*, and others are awaiting formal taxonomic description (Jensen et al., 2015). New species of known actinobacterial genera are being described on a regular base; the serious limitation in this endeavor is the taxonomic identification.

22.4 SECONDARY METABOLITES FROM MARINE ACTINOBACTERIA

Actinobacteria are prolific producers of secondary metabolites with biological activities (Goodfellow and Fiedler, 2010). Secondary metabolites are metabolic products that are not necessary for vegetative growth of the producing organisms, but they are considered as differentiating compounds conferring adaptive roles, by functioning as defense compounds or signaling molecules in ecological interactions. *Actinobacteria* produce the major fraction of bioactive compounds among the different microbial phyla in marine ecosystems. These diverse bioactivities are mediated by secondary metabolites usually several classes of chemical moieties such as polyketides, alkaloids, fatty acids, peptides, terpenes, and sugars. Biological synthesis of secondary metabolite is catalyzed by various enzymes, encoded by a cluster of genes. The gene cluster contains all the necessary genes for the synthesis of a particular secondary metabolite. This comprises the genes that encode the biosynthetic enzymes, genes for resistance to the toxic action of secondary metabolites, regulatory proteins, and genes for secretion of the metabolites. Enzymes such as polyketide synthase and nonribosomal peptide synthetase are involved in the synthesis of secondary metabolites (Donadio et al., 2007). The complete process of the production and transportation of secondary metabolites is rigorously regulated by transcriptional regulators and transporters. The size of the gene cluster responsible for the synthesis of each secondary metabolite is usually between 10 and 100 kb (Ichikawa et al., 2013). Genome mining for novel candidate secondary metabolic pathways based on clustering and coexpression was confirmed to be a highly successful approach in microbes (Osbourn, 2010). This helps to predict the types of metabolite one might expect to find after extraction and purification. With the rising number of genome nucleotide sequence information in the GenBank and the advent of next-generation sequencing, it will be possible to search for candidate secondary metabolite gene cluster in a wide range of actinobacterial species. The evolution of microbial natural product collections and the development of high-throughput screening methods have attracted many workers to the use of natural product libraries in drug discoveries.

22.5 MARINE ACTINOBACTERIA AS A NOVEL SOURCE OF BIOACTIVE COMPOUNDS

Marine environment provides characteristics such as physical, chemical, and biological parameters, which may have given rise to the evolution of metabolic pathways producing novel chemical skeletons. Marine actinobacterial natural products exhibit a wide range of bioactivities that include antimicrobial, antituberculosis, antiviral, antiparasitic, antihelminthic, antimalarial, antiprotozoal, anticoagulant, antiplatelet, antiinflammatory, antidiabetic, and antitumor effects. They may also affect the cardiovascular, immune, and nervous systems (Imhoff et al., 2011; Gogineni and Hamann, 2018). Some examples of novel metabolites produced by marine actinobacteria showing various biological activities are provided in the Tables 22.1–22.4.

TABLE 22.1 Examples of Novel Antibacterial Metabolites Produced by Marine Actinobacteria

Compound	Species	References
Lutoside	*Micrococcus luteus*	Bultel-Ponce et al. (1998)
Bonactin	*Streptomyces* sp.	Schumacher et al. (2003)
Himalomycins A, B	*Streptomyces* sp.	Maskey et al. (2003a)
Chandrananimycin	*Actinomadura* sp.	Maskey et al. (2003b)
Diazepinomicin	*Micromonosproa* sp.	Charan et al. (2004)
Abyssomicin	*Verrucosispora* sp.	Riedlinger et al. (2004)
Gutingimycin	*Streptomyces* sp.	Maskey et al. (2004a)
Helquinoline	*Janibacter limosus*	Asolkar et al. (2004)
Trioxacarcin	*Streptomyces* sp.	Maskey et al. (2004b)
Chloro-dihydroquinones	*Novel actinomycete*	Soria-Mercado et al. (2005)
Glaciapyrroles	*Streptomyces* sp.	Macherla et al. (2005)
Frigocyclinone	*Streptomyces griseus*	Bruntner et al. (2005)
Lajollamycin	*Streptomyces nodosus*	Manam et al. (2005)
1-Hydroxy-1-norresistomycin	*Streptomyces chinaensis*	Gorajana et al. (2005)
Resistoflavin methyl ether	*Streptomyces* sp.	Kock et al. (2005)
Glyciapyrroles A–C	*Streptomyces* sp.	Macherla et al. (2005)
Manumycins	*Streptomyces* sp.	Li et al. (2005)
Marinomycins	*Marinispora* sp.	Kwon et al. (2006)
Bisanthraquinone	*Streptomyces* sp.	Socha et al. (2006)
Lincomycin	*Streptomyces lincolnensis*	Peschke et al. (2006)
Lipoxazolidinones A and B	*Marinispora* sp.	Macherla et al. (2007)
Proximicins	*Verrucosispora* sp.	Fiedler et al. (2008)
Essramycin	*Streptomyces* sp.	El-Gendy et al. (2008)
Lynamicins	*Marinispora* sp.	McArthur et al. (2008)
Marinopyrroles	*Streptomyces* sp.	Hughes et al. (2008)
Streptophenazines A–H (phenazines)	*Streptomyces* sp.	Mitova et al. (2008)
Caboxamycin	*Streptomyces* sp.	Hohmann et al. (2009)
Tirandamycin	*Streptomyces* sp.	Carlson et al. (2009)
Salinisporamycin	*Salinispora arenicora*	Matsuda et al. (2009)
Arenimycin	*Salinispora arenicola*	Asolkar et al. (2010)
TP-1161	*Nocardiopsis* sp.	Engelhardt et al. (2010)

Continued

TABLE 22.1 Examples of Novel Antibacterial Metabolites Produced by Marine Actinobacteria—cont'd

Compound	**Species**	**References**
Diazepinomicin	*Micromonospora*	Charan et al. (2004)
1,4-Dihydroxy-2-(3-hydroxybutyl)-9,10 anthraquinone 9,10-anthrac	*Streptomyces* sp.	Ravikumar et al. (2012)
Aureolic acid	*Streptomyces* sp.	Lu et al. (2012)
Pyridinium	*Amycolatopsis alba*	Dasari et al. (2012)
Violapyrones	*Streptomyces* sp.	Shin et al. (2014)
Curvularin-7-O-α-D-glucopyranoside	*Pseudonocardia* sp.	Ye et al. (2015)
Lagumycin-B, Micromonospora sp.	*Micromonospora* sp.	Mullowney et al. (2015)
Monacyclinone	*Streptomyces* sp.	Vicente et al. (2015)
Salinamides	*Streptomyces* sp.	Hassan et al. (2015)
Manumycins	*Streptomyces* sp.	Sattler et al. (1998), Zhang et al. (2016), and Eqler et al. (2016)
N-acetyl-*N*-demethyl-mayamycin	*Streptomyces* sp.	Liang et al. (2016)
Streptoanthraquinone	*Streptomyces* sp.	Liang et al. (2016)
Ilamycins	*Streptomyces atratus*	Ma et al. (2017)
Borrelidins C–E	*Nocardiopsis* sp.	Kim et al. (2017)

TABLE 22.2 Examples of Novel Antifungal Metabolites Produced by Marine Actinobacteria

Compound	**Species**	**References**
Bonactin	*Streptomyces* sp.	Schumacher et al. (2003)
Chandrananimycin	*Actinomadura* sp.	Maskey et al. (2003b)
Daryamides A–C	*Streptomyces* sp.	Asolkar et al. (2006)
Marinomycins	*Marinispora*	Kwon et al. (2006)
Staurosporine	*Streptomyces* sp.	Pimentel-Elardo et al. (2010)
Saadamycin	*Streptomyces* sp.	El-Gendy and El-Bondkly (2010)
N-(2-hydroxyphenyl)-2-phenazinamine (NHP)	*Nocardia dassonvillei*	Gao et al. (2012)
PM100117, PM100118	*Streptomyces caniferus*	Marta et al. (2015)
Neomaclafungins A–I	*Actinoalloteichus* sp.	Ebaa et al. (2017)

TABLE 22.3 Examples of Novel Anticancer/Antitumor Bioactive Metabolites Produced by Marine Actinobacteria

Compound	Species	References
Mitomycin C	*Streptomyces lavendulae*	Mao et al. (1999)
Salinosporamide A	*Salinispora tropica*	Feling et al. (2003)
3,6-Disubstituted indoles	*Streptomyces* sp.	López et al. (2003)
IB-00208	*Actinomadura* sp.	Rodriguez et al. (2003)
ZHD-0501	*Actinomadura* sp.	Han et al. (2003)
Caprolactones	*Streptomyces* sp.	Stritzke et al. (2004)
Aureoverticillactam	*Streptomyces aureoverticillatus*	Mitchell et al. (2004)
Trioxacarcin A, B	*Streptomyces ochraceus, S. bottropensis*	Maskey et al. (2004b)
Mechercharmycins	*Thermoactinomyces* sp.	Manam et al. (2005)
1-Hydroxy-1-norresistomycin	*Streptomyces chinaensis*	Gorajana et al. (2005) and Kock et al. (2005)
Chinikomycins	*Streptomyces* sp.	Li et al. (2005)
Chloro-dihydroquinones	*Streptomyces* sp.	Soria-Mercado et al. (2005)
Glyciapyrroles	*Streptomyces* sp.	Macherla et al. (2005)
Mechercharmycin A	*Thermoactinomyces* sp.	Kanoh et al. (2005)
Arenimycin	*Salinispora arenicola*	Asolkar et al. (2006)
1,8-Dihydroxy-2-ethyl-3-methylanthraquinone	*Streptomyces* sp.	Huang et al. (2006)
Daryamides	*Streptomyces* sp.	Asolkar et al. (2006)
Staurosporinone	*Streptomyces* sp.	Wu et al. (2006)
Streptokordin	*Streptomyces* sp.	Jeong et al. (2006)
Arenicolides	*Salinispora arenicola*	Williams et al. (2005)
Chalcomycin	*Streptomyces* sp.	Wu et al. (2007)
Lodopyridone	*Saccharomonospora* sp.	Maloney et al. (2009)
Elaiomycins B and C	*Streptomyces* sp.	Helaly et al. (2011)
Aureolic acid	*Streptomyces* sp.	Lu et al. (2012)
N-(2-hydroxyphenyl)-2-phenazinamine (NHP)	*Nocardia dassonvillei*	Gao et al. (2012)
Strepsesquitriol	*Streptomyces* sp.	Yang et al. (2013)
Curvularin-7-O-α-D-glucopyranoside	*Pseudonocardia* sp.	Ye et al. (2015)
PM100117, PM100118	*Streptomyces caniferus*	Marta et al. (2015)
Streptoanthraquinone A	*Streptomyces* sp.	Liang et al. (2016)

TABLE 22.4 Examples of Novel Cytotoxic Bioactive Metabolites Produced by Marine Actinobacteria

Compound	Species	References
Neomarinones	*Actinomycetales*	Hardt et al. (2000)
Chlorinated dihydroquinones	*Streptomyces* sp.	Soria-Mercado et al. (2005)
Salinosporamide B and C	*Salinipora tropica*	Williams et al. (2005)
Actinofuranones	*Streptomyces*	Cho et al. (2006)
Nonactin	*Streptomyces* sp.	Jeong et al. (2006)
Resistoflavine	*Streptomyces chibaensis*	Gorajana et al. (2007)
Piperazimycins	*Streptomyces* sp.	Miller et al. (2007)
Piericidins	*Streptomyces* sp.	Hayakawa et al. (2007)
Lucentamycins	*Nocardiopsis lucentensis*	Cho et al. (2007)
Piperazimycins	*Streptomyces* sp.	Miller et al. (2007)
Arenamides	*Salinipora arenicola*	Asolkar et al. (2008)
Mansouramycin C	*Streptomyces* sp.	Hawas et al. (2009)
ML-449 (macrolactam)	*Streptomyces* sp.	Jorgensen et al. (2010)
Usabamycins	*Streptomyces* sp.	Sato et al. (2011)
Pyridinium	*Amycolatopsis alba.*	Dasari et al. (2012)
Phenazines 1,2	*Streptomyces* sp.	Kondratyuk et al. (2012)
Salinamides	*Streptomyces* sp.	Hassan et al. (2015)
Monacyclinone F	*Streptomyces* sp.	Vicente et al. (2015)

22.6 FUTURE PERSPECTIVES IN ACTINOBACTERIA RESEARCH

The sea is rich of diverse unexplored ecosystem that could be considered for the isolation of novel species of *Actinobacteria*. On cultivable isolates, shorter time of cultivation is needed to achieve significant production of secondary metabolites in higher yields. The establishment of culture-dependent *Actinobacteria* resources is one of the basic requirements for the future prospect of unexplored *Actinobacteria* for the production of secondary metabolites. Based on laboratory cultivation of novel marine actinobacteria metabolite structure, novelty and biological activity can be assessed directly. This cultivation-dependent technology is revealing the biosynthetic potential of the marine bacteria, and it provides an effective choice to improve the production of microbial fermentation. Studying the biology of individual marine actinobacteria will provide tools for genetics and enzymology of secondary metabolites biosynthesis. The current approach on genome-based bioprospecting will depend on the development of efficient tools for bioinformatic analysis of the genomes of cultured and uncultured *Actinobacteria* allowing the identification of unique biosynthetic gene clusters. It may reveal the biosynthetic gene clusters with the potential to direct the production of an ample number of novel, structurally diverse bioactive compounds. In the advent of molecular genetics, next-generation genome analysis may result in the establishment of robust pipeline for actinobacterial metabolite-based drug discovery.

Acknowledgments

The authors are thankful to the authorities of the Central University of Kerala for providing facilities and Kerala State Council for Science, Technology, and Environment (KSCSTE) for the research fellowships. The authors are also thankful to UGC, DBT, and DST-SERB for the funding of the research projects.

Conflict of Interest

The authors declare no conflicts of interest.

References

Asolkar, R.N., Schroeder, D., Heckmann, R., Lang, S., Wagner-Doebler, I., Laatsch, H., 2004. Helquinoline, a new tetrahydroquinoline antibiotic from *Janibacter limosus* Hel. J. Antibiot. (Tokyo) 57, 17–23.

Asolkar, R.N., Jensen, P.R., Kauffman, C.A., Fenical, W., Daryamides, A.C., 2006. Weakly cytotoxic polyketides from a marine-derived actinomycete of the genus *Streptomyces* strain CNQ-085. J. Nat. Prod. 69, 1756–1759.

Asolkar, R.N., Freel, K.C., Jensen, P.R., Fenical, W., Kondratyuk, T.P., Park, E.J., 2008. Arenamides A-C, cytotoxic NFκB inhibitors from the marine actinomycete *Salinispora arenicola*. J. Nat. Prod. 72, 396–402.

Asolkar, R.N., Kirkland, T.N., Jensen, P.R., Fenical, W., 2010. Arenimycin, an antibiotic effective against rifampin- and methicillin-resistant *Staphylococcus aureus* from the marine actinomycete *Salinispora arenicola*. J. Antibiot. (Tokyo) 63, 37–39.

Blunt, J.W., Copp, B.R., Keyzers, R.A., Munro, M.H., Prinsep, M.R., 2016. Marine natural products. Nat. Prod. Rep. 33, 382–431.

Bruntner, C., Binder, T., Pathom-aree, W., Goodfellow, M., Bull, A.T., Potterat, O., et al., 2005. Frigocyclinone, a novel angucyclinone antibiotic produced by a *Streptomyces griseus* strain from Antarctica. J. Antibiot. (Tokyo) 58, 346–349.

Bull, A.T., Stach, J.E.M., 2007. Marine actinobacteria: new opportunities for natural product search and discovery. Trends Microbiol. 15, 491–499.

Bultel-Ponce, V.V., Debitus, C., Berge, J.P., Cerceau, C., Guyot, M., 1998. Metabolites from the sponge-associated bacterium *Micrococcus luteus*. J. Mar. Biotechnol. 6, 233–236.

Carlson, J.C., Li, S., Burr, D.A., Sherman, D.H., 2009. Isolation and characterization of tirandamycins from a marine-derived *Streptomyces* sp. J. Nat. Prod. 72, 2076–2079.

Charan, R.D., Schlingmann, G., Janso, J., Bernan, V., Feng, X., Carter, G.T., 2004. Diazepinomicin, a new antimicrobial alkaloid from a marine *Micromonospora* sp. J. Nat. Prod. 67, 1431–1433.

Cho, J.Y., Kwon, H.C., Williams, P.G., Kauffman, C.A., Jensen, P.R., Fenical, W., 2006. Actinofuranones A and B, polyketides from a marine-derived bacterium related to the genus *Streptomyces* (Actinomycetales). J. Nat. Prod. 69, 425–428.

Cho, J.Y., Williams, P.G., Kwon, H.C., Jensen, P.R., Fenical, W., Lucentamycins, A.D., 2007. Cytotoxic peptides from the marine-derived actinomycete *Nocardiopsis lucentensis*. J. Nat. Prod. 70, 1321–1328.

Dasari, V.R.R.K., Muthyala, M.K.K., Nikku, M.Y., Donthireddy, S.R.R., 2012. Novel Pyridinium compound from marine actinomycete, *Amycolatopsis alba* var. nov., DVR D4 showing antimicrobial and cytotoxic activities in vitro. Microbiol. Res. 167, 346–351.

Dhakal, D., Pokhrel, A.R., Shrestha, B., Sohng, J.K., 2017. Marine rare Actinobacteria: isolation, characterization, and strategies for harnessing bioactive compounds. Front. Microbiol. 8, 1106.

Donadio, S., Monciardini, P., Sosio, M., 2007. Polyketide synthases and nonribosomal peptide synthetases: the emerging view from bacterial genomics. Nat. Prod. Rep. 24, 1073–1109.

Ebaa, M.E.H., Cheng, C., Mostafa, M., Hamed, M.M., Ashraf Nageeb, E.S.H., Knut, O., Ute, H., Usama, R.A., 2017. Antifungal potential of marine natural products. Eur. J. Med. Chem. 126, 631–651.

El-Gendy, M.M., El-Bondkly, A.M., 2010. Production and genetic improvement of a novel antimycotic agent saadamycin, against dermatophytes and other clinical fungi from endophytic *Streptomyces* sp. Hedaya 48. J. Ind. Microbiol. Biotechnol. 37, 831–841.

El-Gendy, M.M., Shaaban, M., Shaaban, K.A., El-Bondkly, A.M., Laatsch, H., 2008. Essramycin: a first triazolopyrimidine antibiotic isolated from nature. J. Antibiot. 61, 149–157.

Engelhardt, K., Degnes, K.F., Kemmler, M., Bredholt, H., Fjaervik, E., Klinkenberg, G., et al., 2010. Production of a new thiopeptide antibiotic, TP-1161, by a marine *Nocardiopsis species*. Appl. Environ. Microbiol. 76, 4969–4976.

Eqler, J., et al., 2016. Stimulating effect of manumycin A on suicidal erythrocyte death. Cell. Physiol. Biochem. 38, 1147–1156.

Feling, R.H., Buchanan, G.O., Mincer, T.J., Kauffman, C.A., Jensen, P.R., Fenical, W., 2003. Salinosporamide A: a highly cytotoxic proteasome inhibitor from a novel microbial source, a marine bacterium of the new genus *Salinospora*. Angew. Chem. 42, 355–357.

Fiedler, H.P., Bruntner, C., Riedlinger, J., Bull, A.T., Knutsen, G., Goodfellow, M., et al., 2008. Proximicin A, B and C, novel aminofuran antibiotic and anticancer compounds isolated from marine strains of the actinomycete *Verrucosispora*. J. Antibiot. 61, 158–163.

Gao, X., Lu, Y., Xing, Y., Ma, Y., Lu, J., Bao, W., 2012. A novel anticancer and antifungus phenazine derivative from a marine actinomycete BM-17. Microbiol. Res. 167, 616–622.

Gogineni, V., Hamann, M.T., 2018. Marine natural product peptides with therapeutic potential: chemistry, biosynthesis, and pharmacology. Biochim. Biophys. Acta 1862, 81–196.

Goodfellow, M., Fiedler, H.P., 2010. A guide to successful bioprospecting: informed by actinobacterial systematics. Antonie Van Leeuwenhoek 98, 119–142.

Goodfellow, M., Kämpfer, P., Busse, H.J., Trujillo, M.E., Suzuki, K.I., Ludwig, W., et al., 2012. Bergey's Manual of Systematic Bacteriology: Vol. 5: The Actinobacteria, second ed. Springer, New York.

Gorajana, A., Kurada, B.V., Peela, S., Jangam, P., Vinjamuri, S., Poluri, E., et al., 2005. 1-Hydroxy1-norresistomycin, a new cytotoxic compound from a marine Actinomycete, *Streptomyces chibaensis* AUBN1/7. J. Antibiot. 58, 526–529.

Gorajana, A., Vinjamuri, S., Kurada, B.V., Peela, S., Jangam, P., Poluri, E., et al., 2007. Resistoflavine, cytotoxic compound from a marine actinomycete, *Streptomyces chibaensis* AUBN 1/7. Microbiol. Res. 162, 322–327.

Gurgui, C., Piel, J., 2010. Metagenomic approaches to identify and isolate bioactive natural products from microbiota of marine sponges. Methods Mol. Biol. 668, 247–264.

Han, S.K., Nedashkovskaya, O.I., Mikhailov, V.V., Kim, S.B., Bae, K.S., 2003. *Salinibacterium amurskyense* gen. nov., sp. a novel genus of the family Microbacteriaceae from the marine environment. Int. J. Syst. Evol. Microbiol. 53, 2061–2066.

Hardt, I.H., Jensen, P.R., Fenical, W., 2000. Neomarinone, and new cytotoxic marinone derivatives, produced by a marine filamentous bacterium (*Actinomycetales*). Tetrahedron Lett. 41, 2073–2076.

Hassan, H.M., Degen, D., Jang, K.H., Ebright, R.H., Fenical, W., 2015. Salinamide F new depsipeptide antibiotic and inhibitor of bacterial RNA polymerase from a marine-derived *Streptomyces sp*. J. Antibiot. 68, 206–209.

Hassan, S.S., Anjum, K., Abbas, S.Q., Akhter, N., Shagufta, B.I., Shah, S.A., Tasneem, U., 2017. Emerging biopharmaceuticals from marine actinobacteria. Environ. Toxicol. Pharmacol. 49, 34–47.

Hawas, U.W., Shaaban, M., Shaaban, K.A., Speitling, M., Maier, A., Kelter, G., et al., 2009. Mansouramycins A-D, cytotoxic isoquinolinequinones from a marine *Streptomycete*. J. Nat. Prod. 72, 2120–2124.

Hayakawa, Y., Shirasaki, S., Kawasaki, T., Matsuo, Y., Adachi, K., Shizuri, Y., 2007. Structures of new cytotoxic antibiotics, piericidins C7 and C8. J. Antibiot. (Tokyo) 60, 201–203.

Helaly, S.E., Pesic, A., Fiedler, H.P., Süssmuth, R.D., 2011. Elaiomycins B and C: alkylhydrazide antibiotics from *Streptomyces* sp. BK 190. Org. Lett. 13, 1052–1055.

Hohmann, C., Schneider, K., Bruntner, C., Irran, E., Nicholson, G., Bull, A.T., et al., 2009. Caboxamycin, a new antibiotic of the benzoxazole family produced by the deep-sea strain *Streptomyces* sp. NTK 937. J. Antibiot. 62, 99–104.

Huang, Y.F., Tian, L., Fu, H.W., Hua, H.M., Pei, Y.H., 2006. One new anthraquinone from marine *Streptomyces* sp. FX-58. Nat. Prod. Res. 20, 1207–1210.

Hughes, C.C., Prieto-Davo, A., Jensen, P.R., Fenical, W., 2008. The marinopyrroles, antibiotics of an unprecedented structure class from a marine *Streptomyces* sp. Org. Lett. 10, 629–631.

Ichikawa, N., Sasagawa, M., Yamamoto, M., Komaki, H., et al., 2013. Dobiscuit: a database of secondary metabolite biosynthetic gene clusters. Nucleic Acids Res. 41, D408–D414.

Imhoff, J.F., Labes, A., Wiese, J., 2011. Bio-mining the microbial treasures of the ocean: new natural products. Biotechnol. Adv. 29 (5), 468–482.

Jackson, S.A., Crossman, L., Almeida, E.L., Margassery, L.M., Kennedy, J., Dobson, A.D.W., 2018. Diverse and abundant secondary metabolism biosynthetic gene clusters in the genomes of marine sponge derived *Streptomyces* spp. isolates. Mar. Drugs 16. pii: E67.

Jensen, P.R., Moore, B.S., Fenical, W., 2015. The marine actinomycete genus *Salinispora*: a model organism for secondary metabolite discovery. Nat. Prod. Rep. 32, 738–751.

Jeong, S.Y., Shin, H.J., Kim, T.S., Lee, H.S., Park, S.K., Kim, H.M., 2006. Streptokordin, a new cytotoxic compound of the methylpyridine class from a marine-derived *Streptomyces* sp. KORDI-3238. J. Antibiot. 59, 234–240.

Jorgensen, H., Degnes, K.F., Dikiy, A., Fjærvik, E., Klinkenberg, G., Zotchev, S.B., 2010. Insights into the evolution of macrolactam biosynthesis through cloning and comparative analysis of the biosynthetic gene cluster for a novel macrocyclic lactam, ML-449. Appl. Environ. Microbiol. 76, 283–293.

Jose, P.A., Jha, B., 2016. New dimensions of research on actinomycetes: quest for next generation antibiotics. Front. Microbiol. 7, 1295.

Kanoh, K., Matsuo, Y., Adachi, K., Imagawa, H., Nishizawa, M., Shizuri, Y., 2005. Mechercharmycins A and B, cytotoxic substances from marine-derived *Thermoactinomyces* sp. YM3-251. J. Antibiot. 58, 289–292.

Kim, J., Daniel, S., Seong-Hwan, K., Wanki, P., Yoonho, S., Won, K.K., Sang, K.L., Ki-Bong, O., Jongheon, S., Dong-Chan, O., 2017. Borrelidins C–E: new antibacterial macrolides from a saltern-derived halophilic *Nocardiopsis* sp. Mar Drugs 15, 166.

Kock, I., Maskey, R.P., Biabani, M.A.F., Helmke, E., Laatsch, H., 2005. 1-Hydroxy-1-norresistomycin and resistoflavin methyl ether: new antibiotics from marine-derived *Streptomycetes*. J. Antibiot. 58, 530–535.

Kondratyuk, T.P., Park, E.J., Yu, R., Breemen, R.B., 2012. Novel marine phenazines as potential cancer chemopreventive and anti-inflammatory agents. Mar. Drugs 10, 451–464.

Kwon, H.C., Kauffman, C.A., Jensen, P.R., Fenical, W., Marinomycins, A.D., 2006. Antitumor antibiotics of a new structure class from a marine actinomycete of the recently discovered genus "Marinispora". J. Am. Chem. Soc. 128, 1622–1632.

Lam, K.S., 2006. Discovery of novel metabolites from marine actinomycetes. Curr. Opin. Microbiol. 9, 245–251.

Li, F., Maskey, R.P., Qin, S., Sattler, I., Fiebig, H.H., Maier, A., et al., 2005. Chinikomycins A and B: isolation, structure elucidation, and biological activity of novel antibiotics from a marine *Streptomyces* sp. isolate M045. J. Nat. Prod. 68, 349–353.

Liang, Y., Xie, X., Chen, L., Yan, S., Ye, X., Anjum, K., Zhang, Z., 2016. Bioactive polycyclic quinones from marine *Streptomyces* sp. 182SMLY. Mar. Drugs 14, 10.

López, J.M.S., Insua, M.M., Baz, J.P., Puentes, J.L.F., Hernández, L.M.C., 2003. New cytotoxic indolic metabolites from a marine *Streptomyces*. J. Nat. Prod. 66, 863–864.

Lozupone, C.A., Knight, R., 2007. Global patterns in bacterial diversity. PNAS 104, 11436–11440.

Lu, J., Ma, Y., Liang, J., Xing, Y., Xi, T., Lu, Y., 2012. Aureolic acids from a marine-derived *Streptomyces* sp. WBF16. Microbiol. Res. 167, 590–595.

Ma, J., Huang, H., Xie, Y., Liu, Z., Zhao, J., Zhang, C., Jia, Y., Zhang, Y., Zhang, H., Zhang, T., Ju, J., 2017. Biosynthesis of ilamycins featuring unusual building blocks and engineered production of enhanced anti-tuberculosis agents. Nat. Commun. 8, 391.

Macherla, V.R., Liu, J., Bellows, C., Teisan, S., Lam, K.S., Potts, B.C.M., 2005. Glaciapyrroles A, B and C, pyrrolosesquiterpenes from a *Streptomyces* sp. isolated from an Alaskan marine sediment. J. Nat. Prod. 68, 780–783.

Macherla, V.R., Liu, J., Sunga, M., White, D.J., Grodberg, J., Teisan, S., Lam, K.S., Potts, B.C., 2007. Lipoxazolidinones A, B, and C: antibacterial 4-oxazolidinones from a marine actinomycete isolated from a Guam marine sediment. J. Nat. Prod. 70, 1454–1457.

Maloney, K.N., Macmillan, J.B., Kauffman, C.A., Jensen, P.R., Dipasquale, A.G., Rheingold, A.L., Fenical, W., 2009. Lodopyridone, a structurally unprecedented alkaloid from a marine actinomycete. Org. Lett. 11, 5422–5424.

Manam, R.R., Teisan, S., White, D.J., Nicholson, B., Grodberg, J., Neuteboom, S.T.C., et al., 2005. Lajollamycin, a nitro-tetraene spiro-b-lactone-g-lactam antibiotic from the marine actinomycete *Streptomyces nodosus*. J. Nat. Prod. 68, 240–243.

Mao, Y., Varoglu, M., Sherman, D.H., 1999. Molecular characterization and analysis of the biosynthetic gene cluster for the antitumor antibiotic mitomycin C from *Streptomyces lavendulae* NRRL 2564. Chem. Biol. 6, 251–263.

Marta, P., Carmen, S., Rogelio, F., Pilar, R., Fernando, R., Paz, Z., Fernando, C., Carmen, C., 2015. PM100117 and PM100118, new antitumor macrolides produced by a marine *Streptomyces caniferus* GUA-06-05-006A. J. Antibiot. 69, 388–394.

Maskey, R.P., Helmke, E., Laatsch, H., 2003a. Himalomycin A and B: isolation and structure elucidation of new fridamycin type antibiotics from a marine *Streptomyces* isolate. J. Antibiot. 56, 942–949.

Maskey, R.P., Li, F., Qin, S., Fiebig, H.H., Laatsch, H., 2003b. Chandrananimycins A approximately C: production of novel anticancer antibiotics from a marine *Actinomadura* sp. isolate M048 by variation of medium composition and growth conditions. J. Antibiot. 56, 622–634.

Maskey, R.P., Sevvana, M., Uson, I., Helmke, E., Laatsch, H., 2004a. Gutingimycin: a highly complex metabolite from a marine streptomycete. Angew. Chem. 43, 1281–1283.

Maskey, R.P., Helmke, E., Kayser, O., Fiebig, H.H., Maier, A., Busche, A., et al., 2004b. Anti-cancer and antibacterial trioxacarcins with high anti-malaria activity from a marine *Streptomycete* and their absolute stereochemistry. J. Antibiot. 57, 771–779.

Matsuda, S., Adachi, K., Matsuo, Y., Nukina, M., Shizuri, Y., 2009. Salinisporamycin, a novel metabolite from *Salinispora arenicola*. J. Antibiot. 62, 519–526.

McArthur, K.A., Mitchell, S.S., Tsueng, G., Rheingold, A., White, D.J., Grodberg, J., 2008. Lynamicins A–E, chlorinated bisindole pyrrole antibiotics from a novel marine actinomycete. J. Nat. Prod. 71, 1732–1737.

Miller, E.D., Kauffman, C.A., Jensen, P.R., Fenical, W., 2007. Piperazimycins: cytotoxic hexadepsipeptides from a marine-derived bacterium of the genus *Streptomyces*. J. Organomet. Chem. 72, 323–330.

Mitchell, S.S., Nicholson, B., Teisan, S., Lam, K.S., Barbara, C., 2004. Aureoverticillactam, a novel 22-atom macrocyclic lactam from the marine actinomycete *Streptomyces aureoverticillatus*. J. Nat. Prod. 67, 1400–1402.

Mitova, M.I., Lang, G., Wiese, J., Imhoff, J.F., 2008. Subinhibitory concentrations of antibiotics induce phenazine production in a marine *Streptomyces sp*. J. Nat. Prod. 71, 824–827.

Mora, C., Tittensor, D.P., Adl, S., Simpson, A.G.B., Worm, B., 2011. How many species are there on earth and in the ocean? PLoS Biol. 9, e1001127.

Mullowney, M.W., Ó hAinmhire, E., Tanouye, U., Burdette, J.E., Pham, V.C., Murphy, B.T., 2015. A pimarane diterpene and cytotoxic angucyclines from a marine-derived *Micromonospora* sp. in Vietnam's East Sea. Mar. Drugs 13, 5815–5827.

Osbourn, A., 2010. Secondary metabolic gene clusters: evolutionary toolkits for chemical innovation. Trends Genet. 26, 449–457.

Peschke, U., Schmidt, H., Zhang, H.Z., Piepersberg, W., 2006. Molecular characterization of the lincomycin-production gene cluster of *Streptomyces lincolnensis*. Mol. Microbiol. 16, 1137–1156.

Pimentel-Elardo, S.M., Kozytska, S., Bugni, T.S., Ireland, C.M., Moll, H., Hentschel, U., 2010. Anti-parasitic compounds from *Streptomyces* sp. strains isolated from Mediterranean sponges. Mar. Drugs 8, 373–380.

Ravikumar, S., Gnanadesigan, M., Saravanan, A., Monisha, N., Brindha, V., Muthumari, S., 2012. Antagonistic properties of seagrass associated *Streptomyces* sp., RAUACT-1: a source for anthraquinone rich compound. Asian Pac. J. Trop. Med. 5, 887–890.

Riedlinger, J., Reicke, A., Zahner, H., Krismer, B., Bull, A.T., Maldonado, L.A., et al., 2004. Abyssomicins, inhibitors of the para-aminobenzoic acid pathway produced by the marine *Verrucosispora strain* AB-18-032. J. Antibiot. 57, 271–279.

Rodriguez, J.C., Fernandez, P.I.L., Perez, B.J., Canedo, L.M., 2003. IB-00208, a new cytotoxic polycyclic xanthone produced by a marine-derived *Actinomadura* II. Isolation, physico-chemical properties and structure determination. J. Antibiot. 56, 318–321.

Sato, S., Iwata, F., Yamada, S., Kawahara, H., Katayama, M., 2011. Usabamycins A–C: new anthramycin-type analogues from a marine-derived actinomycete. Bioorg. Med. Chem. Lett. 21, 7099–7101.

Sattler, I., et al., 1998. The manumycin-group metabolites. Nat. Prod. Rep. 15, 221–240.

Schumacher, R.W., Talmage, S.C., Miller, S.A., Sarris, K.E., Davidson, B.S., Goldberg, A., 2003. Isolation and structure determination of an antimicrobial ester from a marine sediment- derived bacterium. J. Nat. Prod. 66, 1291–1293.

Shin, H.J., Lee, H.S., Lee, J.S., Shin, J., Lee, M.A., 2014. Violapyrones H and I, new cytotoxic compounds isolated from *Streptomyces* sp. associated with the marine starfish *Acanthaster planci*. Mar. Drugs 12, 3283–3291.

Simon, C., Daniel, R., 2009. Achievements and new knowledge unraveled by metagenomic approaches. Appl. Microbiol. Biotechnol. 85, 265–276.

Socha, A.M., LaPlante, K.L., Rowley, D.C., 2006. New bisanthraquinone antibiotics and semi-synthetic derivatives with potent activity against clinical *Staphylococcus aureus* and *Enterococcus faecium* isolates. Bioorg. Med. Chem. 14, 8446–8454.

Soria-Mercado, I.E., Prieto-Davo, A., Jensen, P.R., Fenical, W., 2005. Antibiotic terpenoid chloro-dihydroquinones from a new marine actinomycete. J. Nat. Prod. 68, 904–910.

Stritzke, K., Schulz, S., Laatsch, H., Helmke, E., Beil, W., 2004. Novel caprolactones from a marine *Streptomycete*. J. Nat. Prod. 67, 395–401.

van Keulen, G., Dyson, P.J., 2014. Production of specialized metabolites by *Streptomyces coelicolor* A3(2). Adv. Appl. Microbiol. 89, 217–266.

Vicente, J., Stewart, A.K., Wagoner, R.M., Elliott, E., 2015. Monacyclinones, new angucyclinone metabolites isolated from *Streptomyces* sp. M7 15 associated with the puerto rican sponge *Scopalina ruetzleri*. Mar. Drugs 13, 4682–4700.

Ward, A.C., Bora, N., 2006. Diversity and biogeography of marine actinobacteria. Curr. Opin. Microbiol. 9, 279–286.

Williams, P.G., Buchanan, G.O., Feling, R.H., Kauffman, C.A., Jensen, P.R., Fenical, W., 2005. New cytotoxic salinosporamides from the marine actinomycete *Salinispora tropica*. J. Organomet. Chem. 70, 6196–6203.

Wu, S.J., Fotso, S., Li, F., Qin, S., Kelter, T., Fiebig, H.H., et al., 2006. 39-N carboxamidostaurosporine and selina-4(14),7(11)-diene-8,9-diol, new metabolites from a marine *Streptomyces* sp. J. Antibiot. 59, 331–337.

Wu, S.J., Fotso, S., Li, F., Qin, S., 2007. Laatsch Amorphane sesquiterpenes from a marine *Streptomyces* sp. J. Nat. Prod. 70, 304–306.

Yang, X.W., Peng, K., Liu, Z., Zhang, G.Y., 2013. Strepsesquitriol, a rearranged zizaane-type sesquiterpenoid from the deep-Sea-derived actinomycete *Streptomyces* sp. SCSIO 10355. J. Nat. Prod. 76, 2360–2363.

Ye, X., Anjum, K., Song, T., Wang, W., Yu, S., Huang, H., Lian, X.Y., Zhang, Z., 2015. A new curvularin glycoside and its cytotoxic and antibacterial analogues from marine actinomycete *Pseudonocardia* sp. HS7. Nat. Prod. Res. 30, 1156–1161.

Zhang, J., et al., 2016. Antitumor effect of manumycin on colorectal cancer cells by increasing the reactive oxygen species production and blocking PI3K-AKT pathway. OncoTargets Ther. 9, 2885–2895.

SECTION 5

METHODS IN MICROBIOLOGY

CHAPTER

23

Status and Recent Developments in Analytical Methods for the Detection of Foodborne Microorganisms

Seshadri Reddy Ankireddy, Jongsung Kim

Department of Chemical and Biological Engineering, Gachon University, Seongnam, South Korea

OUTLINE

23.1 INTRODUCTION

Owing to their social, economic, and public health importance, foodborne pathogens, such as bacteria, viruses, parasites, and prions, represent the most serious causative agents of foodborne diseases worldwide. The consumption of contaminated and unsafe food has resulted in these foodborne pathogens causing more than 200 diseases, ranging from diarrhea to cancer (Viswanath et al., 2018). According to the current World Health Organization (WHO) calculation, the number of significant disease outbreaks caused by foodborne pathogens is increasing. Approximately 1 in 10 (more than 600 million) people worldwide falls ill because of

https://doi.org/10.1016/B978-0-12-816328-3.00023-4

the consumption of unsafe food and drink. In addition, 40% of the children under 5 years of age suffer from these diseases, with 125,000 deaths occurring from these foodborne illnesses (WHO annual report, 2017). Foodborne outbreaks have occurred not only from the consumption of contaminated food but also because of climatic changes encountered by highly industrialized and developed countries, such as Europe, the United States, and Asia (Lake and Barker, 2018). The majority of the foodborne disease outbreaks are caused by bacterial species, such as *Escherichia coli*, *Salmonella enterica*, *Campylobacter*, enterohemorrhagic *E. coli*, *Listeria*, and *Vibrio cholerae*, along with viruses such as *Norovirus* and *hepatitis A virus* (Bhardwaj et al., 2017b; Torok et al., 2018). In general, outbreaks are mainly caused by the consumption of undercooked fresh vegetables, meat, fruit, poultry, other products of animal origin, eggs, and unpasteurized milk. By contrast, seafood and water are the major sources of viral pathogens. As a result, vomiting, diarrhea, abdominal pain, nausea, headache, and long-lasting liver diseases are the common symptoms for *Salmonella*, *Campylobacter*, enterohemorrhagic *E. coli*, *V. cholerae*, *Norovirus*, and *hepatitis A virus*, respectively.

Moreover, the consumption of unpasteurized dairy products can cause problems for pregnant women and newborn babies, such *Listeria* infection. Diarrhea is the most common symptom of infection with these pathogens, with more than 550 million people being affected and 230,000 deaths occurring every year (WHO annual report, 2017). Parasites and prions also cause foodborne diseases; however, they are mostly not as serious as those caused by bacteria and viruses. Certain parasites, such as fish-borne trematodes and tapeworms like *Echinococcus* spp. or *Taenia solium*, can infect through food or direct contact with animals. Moreover, soil is the main source of contamination for other parasites, such as *Cryptosporidium*, *Entamoeba histolytica*, or *Giardia*. Finally, prions cause neurodegenerative diseases and are composed entirely of proteins. Prions not only cause bovine spongiform encephalopathy (BSE, also known as "mad cow disease") in cattle but also infect with the variant Creutzfeldt-Jakob disease (vCJD) in humans via the consumption of specific risk materials, such as brain tissue (Prusiner, 1991).

23.2 THE NEED TO DEVELOP NEW ANALYTICAL METHODOLOGIES

In the early stages of research, several conventional molecular approaches were developed for the rapid, sensitive, and selective detection of foodborne pathogens, such as simple polymerase chain reaction (sPCR), multiplex PCR (mPCR), quantitative PCR (qPCR), isothermal amplification, loop-mediated isothermal amplification (LAMP), nucleic acid sequence-based amplification (NASBA), enzyme-linked immunosorbent assay (ELISA), lateral flow immunoassay, and immunomagnetic separation (IMS) assays. PCR is a simple and well-established nucleic acid amplification-based molecular technique (Lee, 2017). The basic principle of this method is that specific primers or single-stranded (ss) oligonucleotides anneal to single-stranded DNA from the denatured double-stranded DNA, and by the extension of these complementary primers, the full-length single-stranded DNAs are produced in the presence of polymerase enzyme under thermostable reaction conditions. Kim et al. (2012) established the simultaneous detection of *Salmonella* spp., *Listeria monocytogenes*, and *E. coli* O157:H7 using SYBR green-based multiplex real-time PCR using a specific set of primers. The analysis could detect 10^3 CFU/mL for each pathogen in pure culture and 1 cell per 25 g of inoculated pork sample; however, more than 30 h was required to complete the reaction. In another study, Li et al. developed a lengthy protocol for the simultaneous detection of *E. coli* O157:H7, *Salmonella*, and *Shigella* pathogens using mPCR in apple cider, cantaloupe, lettuce, tomato, watermelon, and alfalfa sprouts with a limit of detection of 8×10^{-1} CFU/g (or CFU/ml) and 8×10^{1} CFU/g (Li and Mustapha, 2004). Notably, more than 24 h is required to complete the final reaction for qPCR and NASBA. However, Lin et al. (2010) demonstrated the rapid detection of *Shigella* spp. in fresh produce using a qPCR technique with 0.12–0.74 CFU per reaction as a detection limit. As mentioned earlier, despite their advantages of high selectivity and accuracy to detect pathogens, these PCR-based techniques take a long time to complete, are expensive, and require sophisticated equipment and hygienic handling conditions. To overcome these problems, a number of nanoparticle-based biosensor strategies and methods have been developed to detect foodborne pathogens (Bozal-Palabiyik et al., 2018). These nanoparticle-based biosensors can not only detect pathogens using different kinds of biomarkers but also provide accurate results with high selectivity and sensitivity in a shorter time compared with conventional molecular approaches. Moreover, these emerging biosensor devices have the potential for cheap on-site analysis (Zhao et al., 2014).

These sensors have been developed using various nanoparticle formulations, such as quantum dots (QDs), carbon quantum dots (CQDs), graphene oxide (GO), gold nanoparticles (AuNPs), gold nanoclusters (AuNCs), silver nanoparticles (AgNPs), carbon-metal hybrid materials, and metal-organic frameworks (MOFs) in the presence of biomarkers for the specific pathogens. The biomarkers are unique for the specified pathogen factors (bioreceptors), which include various biological molecules such as proteins, ssDNA, enzymes, tissues, and carbohydrates (Bozal-Palabiyik et al., 2018; Gupta et al., 2018; Carlson et al., 2018). As a result, the biosensing signals can be recorded via interactions between corresponding biomarker and its bioreceptor. In addition, the detection and identification of these interactions (and thus, indirectly, the pathogen) are achieved by measuring a signal in the form of absorbance, fluorescence, chemiluminescence, electrochemiluminescence, photoluminescence, electrochemical signals, Raman scattering, surface plasmon resonance, or electromagnetic signals. The main purpose of this future article is to review and summarize such methods for rapid detection and identification and to discuss certain recent, novel analytic methodologies to detect toxic foodborne pathogens.

23.3 RECENTLY DEVELOPED NANOPARTICLE-BASED BIOSENSING METHODS TO DETECT FOODBORNE PATHOGENS

In recent years, nanomaterial-based analytic methods and innovations have been developed in the medical, chemical, and industrial fields by exploiting their peculiar properties, such as their natural optical, electric, and magnetic behaviors (Zhou et al., 2018; Gao et al., 2018; Cheng et al., 2017). To date, many reviews have summarized developments in analytic methods and approaches to detect foodborne pathogens (Law et al., 2015; Yeni et al., 2014). The most recent reviews have included advanced analytic methods to detect only foodborne bacterial pathogens. In addition, these reviews have discussed recent biosensing analytic developments, such as smartphone-mediated and paper-based detection, and several analytic techniques, such nuclear magnetic resonance (NMR), Fourier-transform infrared spectroscopy, Raman spectroscopy, and surface-enhanced Raman spectroscopy (SERS) along with several recent nanoparticle-mediated colorimetric, fluorometric, and electrochemical sensors to detect foodborne pathogens with high sensitivity and selectivity. In this review, we have included recent insights and developments in advanced nanoparticle-based analytic methods to detect viruses and prions, in addition to bacterial pathogens. Recent progress and developments to detect foodborne pathogens using various analytic methods are summarized in Table 23.1.

Nanoparticle-based biosensing methodologies are becoming more popular because they are less time-consuming than conventional molecular approaches. Nanoparticle-mediated analytic methods permit the rapid detection of foodborne pathogens on-site. Recently, Zeinhom et al. proposed a magnetic nanoparticle-mediated portable smartphone device for the rapid, specific, and selective detection of *E. coli* O157:H7 in yogurt and eggs (Zeinhom et al., 2018). The working principle of this sensor used the fluorescent imager on the smartphone, involving a compact laser-diode-based photosource, a long-pass (LP) thin-film interference filter, and a high-quality insert lenses as a basic component, together with an inbuilt sandwich ELISA working principle. The limit of detection was 1 and 10 CFU/mL in yogurt and egg, respectively, within less than 2 h. A smartphone-based nanoflower signal amplification-mediated nanoparticle immune sensor for the rapid detection of *Salmonella enteritidis* in milk, cheese, and water has been developed by the same group via a different strategy (Zeinhom et al., 2017). In this method, anti-*S. enteritidis* streptavidin-attached magnetic beads and biotin-labeled antibodies, as a capture platform, were tested against different concentrations of *S. enteritidis* from milk, cheese, and water. Thereafter, signal amplification was achieved by coupling the nanocomposite with the horseradish peroxidase (HRP) enzyme, which eventually produces a change in the smartphone-detectable color in the reaction well. Using this method, very low concentrations of *S. enteritidis* could be detected, with a level of detection (LOD) of 1 CFU/mL and 1 CFU/g in tap water and milk and cheese, respectively. As a result, this method cold detect lower concentrations of pathogens with high selectively and specificity than traditional methods.

In another study, an innovative, rapid, and quantitative colorimetric detection of *E. coli* O157:H7 bacteria using a magnetic nanoparticle-peptide (MNP-specific)-mediated paper-based sensor was developed by Suaifan et al. (2017). The surface-modified MNP-specific paper-based sensor has the ability to change color upon the interaction between the *E. coli* O157:H7 protease and the MNP-specific peptide-modified surface surface protein, in which the surface protein is hydrolyzed by the *E. coli* protease. Moreover, the intensity of the developed color is directly proportional to the concentration of the bacteria in the sample. For quantitative detection, the color change in the reaction can be detected using image analysis software. Based on this sensing system, very low concentrations of the pathogens could be detected, with a LOD of 12 and 30–300 CFU/mL in broth and spiked complex food matrices, respectively. The response time for the reaction is just 30 s to see the visible color change, after the addition of all reagents, which represents a rapid response. According to the obtained selectivity and sensitivity results, this sensing system could be used not only in in vitro studies but also in on-site analysis. Yang et al. introduced a novel and innovative mussel-inspired molecular imprinted polymer-mediated resonance light scattering (RLS) sensor for the highly sensitive and selective detection of *hepatitis A virus* (HAV). In this study, SiO_2 nanoparticles with a surface modification comprising polydopamine (PDA)-coated *totivirus*-imprinted

TABLE 23.1 Recent Developments of a Wide Range of Sensors for the Detection of Foodborne Pathogens

S. No.	Type of Sensor	Type of Nanoparticle	Mechanism Involved	Pathogen Detection	LOD	Reference
1	Fluorometric	Carbon quantum dot (CQD)-encapsulated organosilica nanocapsules	Antigen-antibody immune sensor	*Staphylococcus aureus*	30 CFU/mL	Yang et al. (2018)
2	Fluorometric	Nanohybrid QD complex	Complex formation between QDs and functionalized polyamidoamine dendrimers	*Listeria*	5.19×10^3 CFU/mL	Donoso et al. (2017)
3	Colorimetric	Carboxyfunctionalized cobalt-based magnetic nanoparticles	Antigen-antibody immune sensor	All bacterial foodborne pathogens	10 CFU/mL	Alamer et al. (2018)
4	Colorimetric	Gold nanoparticles	Antigen-antibody visible immune sensor	*Salmonella*	10 CFU/mL	Hahn et al. (2017)
5	Electrochemical	Chitosan, MWCNT, polypyrrole with gold nanoparticle hybrid electrode	Immunosensor	*E. coli* O157:H7	30 CFU/mL	Güner et al. (2017)
6	Electrochemical	Nanoporous Au-Cu alloy-coated glassy carbon electrode	Aptasensor	*Salmonella typhimurium*	1 CFU/mL	Ranjbar et al. (2018)
7	Fluorometric	NH_2-MIL-53 (Fe) MOF	Photoluminescence quenching phenomena	*Staphylococcus aureus*	31 CFU/mL	Bhardwaj et al. (2017a)
8	Colorimetric	Cu-MOF nanoparticles	Peroxidase-like activity that can catalyze 3,3′,5,5′-tetramethylbenzidine to produce a yellow chromogenic reaction in the presence of H_2O_2	*Staphylococcus aureus*	20 CFU/mL	Wang et al. (2017)
9	Electrochemical	Graphene-gold nanocomposite	Aptasensor	*Norovirus*	100 pM	Chand and Neethirajan (2017)
10	Electrochemical	Disposable gold electrode	DNA biosensor	Hepatitis A virus	0.65 pM	Manzano et al. (2018)
11	Photoelectrochemical	β-Cyclodextrin (β-CD) and rhodamine B (RhB)-Au-TiO_2	Aptasensor	Prion	50.9 fM	Yan et al. (2018)
12	Electrochemical	GCE-hemin/G-quadruplex DNAzyme	Immunoassay	Prion	0.38 pg/mL	Li et al. (2018)

polymers were synthesized, and the target virus was detected via an increment in the RLS signal when the target virus was perfectly captured by the imprinted polymer films. The signal enhancement was directly proportional to the concentration of the virus present in the sample. This sensing method was able to detect HAV levels as low as 8.6 pM in the potency range of 0.04–6.0 nM (Yang et al., 2017).

As mentioned earlier, NMR, FTIR, and Raman spectroscopy have also been employed to detect foodborne pathogens via several chemical interactions. Liu et al. established the detection of *E. coli* by evaluating the electrolyzed water (EW)-mediated response using ^{1}H NMR spectroscopy (Liu et al., 2017). In this study, high oxidation-reduction potential containing EW was used to induce oxidative damage to the pathogen membrane. Treatment of *E. coli* with EW produced significant changes in osmotic adjustment, reduced amino acid and nucleoside biosynthesis, and promoted fatty acid metabolism. Therefore, the surface morphological changes and metabolic rates could be monitored using ^{1}H NMR. Ultimately, EW treatment caused significant morphological and metabolic changes in *E. coli*. In another report, Nguyen et al. employed Raman and FTIR spectroscopy to detect antibiotic-induced morphological changes in *E. coli* (Nguyen et al., 2017). In this study, inhibition of different metabolic pathways in the bacteria, such as cell wall synthesis, protein synthesis, and DNA synthesis by ampicillin, tetracycline, and ciprofloxacin, respectively, and the resulting morphological changes were measured. As a result, significant changes in protein, carbohydrate, nucleic acid, and phenylalanine bands were identified using FTIR and Raman after the addition of these antibiotics.

These Raman and FTIR spectroscopies could be used not only to detect *E. coli* but also to identify other bacterial foodborne pathogens, like *Salmonella typhimurium*. Ma et al. developed a spiny gold nanoparticle (SGNP)-mediated surface-enhanced resonance spectroscopy (SERS) aptasensor to detect *S. typhimurium* (Ma et al., 2018). In this study, 4-mercaptobenzoic acid and thiolated *S. typhimurium* aptamers were functionalized on the surface of SGNPs as SERS nanoprobes. Ultimately, the SERS signal intensity increased when the biotinylated aptamer interacted with the thiolated aptamer in the presence of *S. typhimurium*. Notably, the SERS intensity band for p-MBA was observed at 1568 cm^{-1} after proper recognition between the two aptamers in the presence of the bacteria. This method could detect *S. typhimurium* at levels as low as 10^1–10^5 CFU/mL with an LOD of 4 CFU/mL. This simple strategy could detect specific bacteria rapidly, selectively, and sensitively. Using the same technique, Kearns et al. developed nanosensors as working materials for the simultaneous detection of all foodborne bacterial pathogens (Kearns et al., 2017). In this study, the SERS signal was identified by the reaction between lectin-functionalized magnetic nanoparticles and SERS-active nanoparticles functionalized with specific antibodies via optical detection by using a magnetic plug and interrogated with a laser. This method could detect each strain at a lower LOD of 10^1 CFU/mL. This multiplex nanobiosensor is capable of the rapid, selective, and sensitive detection of all bacterial foodborne pathogens.

Despite the benefits of using these innovative nanoparticle-based devices for the rapid detection of foodborne pathogens, environmentally benign fluorescent carbon quantum dots (CQDs) and graphene oxide (GO)-based materials have also become important for the rapid detection of foodborne pathogens because of their simple and rapid methodologies combined with ultraselective and sensitive detection. Fig. 23.1 shows the schematic representation of foodborne pathogen detection using nanomaterial-based biosensing systems. Weng et al. (2015) established

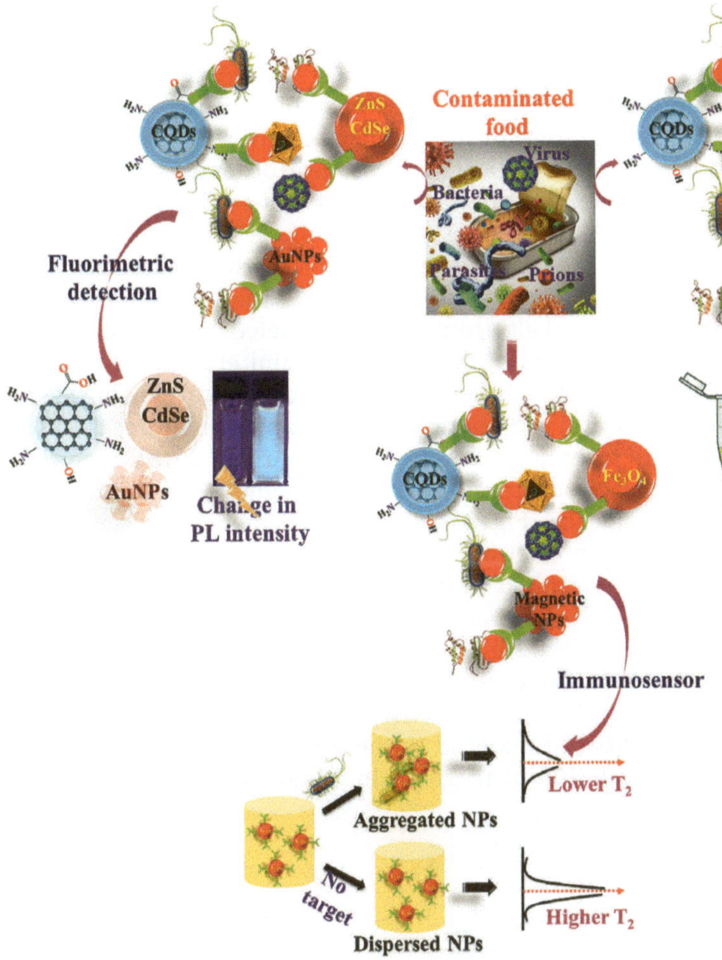

FIG. 23.1 Schematic illustration of different strategies in nanomaterial-based biosensors for the foodborne pathogen detection.

a simple and one-pot hydrothermal synthesis of mannose-modified fluorescent carbon quantum dots (Man-CQDs) from solid ammonium citrate and mannose for the label-free detection of *E. coli* via a fluorescence enhancement method. The fluorescence enhancement is possibly derived from the interaction between the FimH lectin unit in the flagella of the wild-type *E. coli.*

K12 strain and the Man-CQDs. Under optimized conditions, this method could detect *E. coli* levels as low as 450 CFU/mL. A fluorometric GO-attached truncated DNA aptamer for the detection of *Salmonella* spp. was proposed by Chinnappan et al. (2018). GO-mediated fluorescent materials can detect very low concentrations because of their large surface area and high catalytic properties. In this fluorescent displacement assay, a fluorescence/quencher pair was formed by the hybridization between the truncated aptamer with fluorescein and a quencher-labeled complementary sequence. Fluorescence increased in the presence of

S. enteritidis because of dissociation between the aptamer and the complimentary labeled oligonucleotides. This assay could detect *S. enteritidis* at an LOD of 25 CFU/mL, which was markedly lower compared with the previous methods.

Nanoparticles and nanomaterials based on electrochemical biosensors have also been used to detect foodborne pathogens because of the excellent electron-transfer capability between the working electrode and the biological sample. Electrochemical sensors work on the principle of the transfer of electrons from a region of higher electron concentration to a region of lower concentration. In the case of a working electrode and biological samples, there is a high possibility of electron transfer, because carbon materials have a graphitic nature with alternating double bonds and biological samples have many free electrons. The presence of these additional electrons makes the detection of biomolecules and pathogens easier, with very low limits of detection (Arduini et al., 2016). Carbon materials and metal-incorporated carbon materials in the electrochemical sensing system perform well as working electrodes and are abundantly available in nature; therefore, they do not cause any side effects to biological systems (Reta et al., 2018). Significant research has been carried out on electrochemical biosensing methods for the rapid detection of foodborne pathogens in samples. Fig. 23.2 shows a schematic representation of foodborne pathogen detection using an electrochemical biosensing system. Singh et al. developed carbon nanotubes (CNTs) wrapped with GO-biofunctionalized materials as a working electrode to detect bacterial cells in a polydimethylsiloxane (PDMS) microfluidic chip (Singh et al., 2018). The electrochemical signal was obtained from the interactions between *S. typhimurium* antibodies attached to GO-MWCNTs and *S. typhimurium*. In another study, Jijie et al. developed an electrochemical biosensor comprising reduced graphene oxide/polyethylenimine (rGO/PEI)-based material as the working electrode for the selective and sensitive detection of *E. coli* (Jijie et al., 2017). In this study, the detection of the

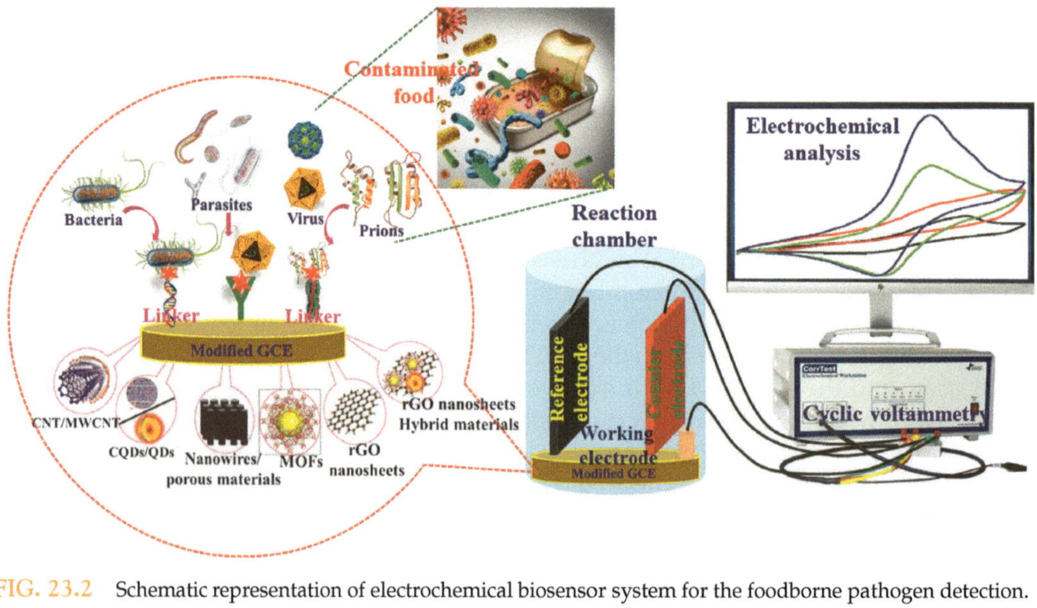

FIG. 23.2 Schematic representation of electrochemical biosensor system for the foodborne pathogen detection.

bacteria was also based on the generation of an electrochemical signal between anti-*E. coli* antibodies attached to the electrode and the bacteria. In contrast, metal oxides, MOFs, and metal clusters can be used for electrochemical biosensors along together with carbon and GO materials because of their flexibility, high conductivity, and large surface area. A large surface area can allow more biomarkers to bind, which improves the performance of the material in electrochemical biosensing applications. Zhu et al. established the selective detection of foodborne pathogens such as *L. monocytogenes* using crystalline frameworks of mesoporous tungsten oxide materials (Zhu et al., 2017). An electrochemical signal was recorded using the release of 3-hydroxy-2-butanone chemical from the bacteria via surface catalysis.

In addition, a graphene and nanoparticle binary structure-based biosensing system for the highly sensitive detection of *Norovirus*-like particles was proposed by Lee et al. (2017). In this assay, the developed gold/magnetic nanoparticle-decorated graphene electrode detected the binding of norovirus-specific antibodies via an electrochemical signal. Another group revealed an electrochemical sensing method to detect prions. A poly-β-cyclodextrin (pβ-CD)/gold nanoparticle/glassy carbon electrode was used to sensitively detect prions via an electrochemical immunosensing method (Li et al., 2017). Based on the prion antigen and antibody reactions on the electrode, an electrochemical signal was recorded. In comparison with fluorometric and colorimetric methods, severalfold higher limits of detection for foodborne pathogens have been achieved using electrochemical sensors. Based on these recent innovations, nanoparticle-mediated biosensing methods are more rapid, easier, environmentally benign, and less expensive compared with conventional molecular approaches.

23.4 LIMITATIONS AND PROSPECTS FOR NANOPARTICLE-BASED BIOSENSORS TO DETECT FOODBORNE PATHOGENS

Nanomaterial-based biosensors offer certain advantages, including improved reliability, simplicity, speed, selectivity, sensitivity, reduced cost, and environmentally friendly design compared with conventional molecular methods. In addition, the biosensing systems enable label-free, real-time or near-real-time detection. Nanomaterial-based analytic methods are promising to detect foodborne pathogens, because they require the detection of analytes that are much bigger than the molecular analytes, such as proteins, nucleic acids, enzymes, and biomolecules. However, nanomaterial-based biosensing systems may have certain limitations. First, the requirement for many washing steps to prepare a suitable catalyst preparation might result in the loss of some of the valuable chemicals. Second, fluorescent stability is very important, particularly when working with the fluorescent materials like QDs, CQDs, and dyes (e.g., fluorescein, fluorescein isothiocyanate (FITC), and cyanine), and significant results may not be obtained without proper stability-ensuring precautions. Finally, several factors can interfere with the analysis of the results (Viswanath et al., 2018; Law et al., 2015; Zhang, 2013). In addition, it may be difficult to maintain proper physical conditions, such as time, temperature, and pressure. To overcome these challenges, more advanced analytic methodologies that are stable under all environmental conditions are necessary to detect foodborne pathogens.

23.5 CONCLUSION

Serious disease outbreaks are caused by pathogens in contaminated food and water worldwide, and there is a constant demand for preventive measures and detection methods to control these serious illnesses. Conventional methods, such as molecular approaches, can detect these pathogens; however, they are not cost-effective or rapid and often require sophisticated instruments and skilled technicians. To overcome these challenges, in the last decade, nanomaterial-based biosensing methods have been developed that offer advantages in terms of speed, reliability, accuracy, selectivity, sensitivity, eco-friendliness, low cost, and importantly point-of-care analysis without the need for specialized technicians. In addition, these nanomaterial-based colorimetric, fluorometric, and electrochemical biosensors are able to detect pathogens at very low concentrations. Moreover, several nanomaterial-based biosensor products are commercially available to detect these pathogens. These advanced analytic methods play a vital role in the medical sector by providing crucial information that could ultimately mitigate serious outbreaks, with obvious benefits to human health.

Acknowledgments

This research was supported by the Basic Science Research Program through the National Research Foundation (NRF) of Korea funded by the Ministry of Education (NRF-2017R1D1A1B04035070); and the Ministry of Trade, Industry and Energy (MOTIE) of the Republic of Korea (20174030201530).

Conflict of Interest

The authors declare no conflicts of interest.

References

Alamer, S., Eissa, S., Chinnappan, R., Zourob, M., 2018. A rapid colorimetric immunoassay for the detection of pathogenic bacteria on poultry processing plants using cotton swabs and nanobeads. Microchim. Acta 185, 164.

Arduini, F., Micheli, L., Moscone, D., Palleschi, G., Piermarini, S., Ricci, F., Volpe, G., 2016. Electrochemical biosensors based on nanomodified screen-printed electrodes: recent applications in clinical analysis. Trends Anal. Chem. 79, 114–126.

Bhardwaj, N., Bhardwaj, S.K., Mehta, J., Kim, K.H., Deep, A., 2017a. MOF–bacteriophage biosensor for highly sensitive and specific detection of Staphylococcus aureus. ACS Appl. Mater. Interfaces 9, 33589–33598.

Bhardwaj, N., Bhardwaj, S.K., Nayak, M.K., Mehta, J., Kim, K.H., Deep, A., 2017b. Fluorescent nanobiosensors for the targeted detection of foodborne bacteria. Trends Anal. Chem. 97, 120–135.

Bozal-Palabiyik, B., Gumustas, A., Ozkan, S.A., Uslu, B., 2018. Biosensor-based methods for the determination of foodborne pathogens. In: Holban, A.M., Grumezescu, A.M. (Eds.), Foodborne Diseases: Handbook of Food Bioengineering. vol. 15. Elsevier, pp. 379–420.

Carlson, K., Misra, M., Mohanty, S., 2018. Developments in micro-and nanotechnology for foodborne pathogen detection. Foodborne Pathog. Dis. 15, 16–25.

Chand, R., Neethirajan, S., 2017. Microfluidic platform integrated with graphene-gold nano-composite aptasensor for one-step detection of norovirus. Biosens. Bioelectron. 98, 47–53.

Cheng, C., Li, S., Thomas, A., Kotov, N.A., Haag, R., 2017. Functional graphene nanomaterials based architectures: biointeractions, fabrications, and emerging biological applications. Chem. Rev. 117, 1826–1914.

Chinnappan, R., AlAmer, S., Eissa, S., Rahamn, A.A., Salah, K.M., Zourob, M., 2018. Fluorometric graphene oxide-based detection of Salmonella enteritis using a truncated DNA aptamer. Microchim. Acta 185, 61.

Donoso, W., Castro, R.I., Guzmán, L., López-Cabaña, Z., Nachtigall, F.M., Santos, L.S., 2017. Fast detection of Listeria monocytogenes through a nanohybrid quantum dot complex. Anal. Bioanal. Chem. 409, 5359–5371.

Gao, M., Yao, J.C., Yan, C., Li, X.F., Hu, T.J., Chen, L., Wang, Y.X., Zhang, Y.J., Liu, H.L., Liu, Y., Cao, L.H., 2018. Novel composite nanomaterials with superior thermal and pressure stability for potential LED applications. J. Alloys Compd. 734, 282–289.

Güner, A., Çevik, E., Şenel, M., Alpsoy, L., 2017. An electrochemical immunosensor for sensitive detection of *Escherichia coli* O157:H7 by using chitosan, MWCNT, polypyrrole with gold nanoparticles hybrid sensing platform. Food Chem. 229, 358–365.

Gupta, M., Summuna, B., Gupta, S., Sharma, D., 2018. Biosensor-based techniques: a reliable and primary tool for detection of foodborne pathogens. In: AFE, S., Levin, R.E., Xu, J. (Eds.), Molecular Techniques in Food Biology: Safety, Biotechnology, Authenticity and Traceability. John Wiley & Sons, p. 361.

Hahn, J., Kim, E., You, Y.S., Gunasekaran, S., Lim, S., Choi, Y.J., 2017. A switchable linker-based immunoassay for ultrasensitive visible detection of *Salmonella* in tomatoes. J. Food Sci. 82, 2321–2328.

Jijie, R., Kahlouche, K., Barras, A., Yamakawa, N., Bouckaert, J., Gharbi, T., Szunerits, S., Boukherroub, R., 2017. Reduced graphene oxide/polyethylenimine based immunosensor for the selective and sensitive electrochemical detection of uropathogenic *Escherichia coli*. Sensors Actuators B Chem. 260, 255–263.

Kearns, H., Goodacre, R., Jamieson, L.E., Graham, D., Faulds, K., 2017. SERS detection of multiple antimicrobial-resistant pathogens using nanosensors. Anal. Chem. 89, 12666–12673.

Kim, H.J., Lee, H.J., Lee, K.H., Cho, J.C., 2012. Simultaneous detection of pathogenic Vibrio species using multiplex real-time PCR. Food Control 23, 491–498.

Lake, I.R., Barker, G.C., 2018. Climate change, foodborne pathogens and illness in higher-income countries. Curr. Environ. Health Rep. 5, 1–10.

Law, J.W., Ab Mutalib, N.S., Chan, K.G., Lee, L.H., 2015. Rapid methods for the detection of foodborne bacterial pathogens: principles, applications, advantages and limitations. Front. Microbiol. 5, 770. 1–19, doi.org/10.3389/fmicb.2014.00770.

Lee, N.Y., 2017. A portable microreactor with minimal accessories for polymerase chain reaction: application to the determination of foodborne pathogens. Microchim. Acta 184, 4225–4233.

Lee, J., Takemura, K., Kato, C.N., Suzuki, T., Park, E.Y., 2017. Binary nanoparticle graphene hybrid structure-based highly sensitive biosensing platform for norovirus-like particle detection. ACS Appl. Mater. Interfaces 9, 27298–27304.

Li, Y., Mustapha, A., 2004. Simultaneous detection of *Escherichia coli* O157:H7, *Salmonella*, and *Shigella* in apple cider and produce by a multiplex PCR. J. Food Prot. 67, 27–33.

Li, X., Li, J., Liu, Y., Zhang, X., Chen, J., 2017. A sensitive electrochemical immunosensor for prion detection based on poly-β-cyclodextrin/gold nanoparticles/glassy carbon electrode. Sensors Actuators B Chem. 250, 1–7.

Li, X., Li, J., Zhu, C., Zhang, X., Chen, J., 2018. A new electrochemical immunoassay for prion protein based on hybridization chain reaction with hemin/G-quadruplex DNAzyme. Talanta 182, 292–298.

Lin, W.S., Cheng, C.M., Van, K.T., 2010. A quantitative PCR assay for rapid detection of *Shigella* species in fresh produce. J. Food Prot. 73, 221–233.

Liu, Q., Wu, J.E., Lim, Z.Y., Aggarwal, A., Yang, H., Wang, S., 2017. Evaluation of the metabolic response of *Escherichia coli* to electrolysed water by ^{1}H NMR spectroscopy. LWT Food Sci. Technol. 79, 428–436.

Ma, X., Xu, X., Xia, Y., Wang, Z., 2018. SERS aptasensor for *Salmonella typhimurium* detection based on spiny gold nanoparticles. Food Control 84, 232–237.

Manzano, M., Viezzi, S., Mazerat, S., Marks, R.S., Vidic, J., 2018. Rapid and label-free electrochemical DNA biosensor for detecting *hepatitis a virus*. Biosens. Bioelectron. 100, 89–95.

Nguyen, N.X., Sarter, S., Nguyen, N.H., Daniel, P., 2017. Detection of molecular changes induced by antibiotics in *Escherichia coli* using vibrational spectroscopy. Spectrochim. Acta A 183, 395–401.

Prusiner, S.B., 1991. Molecular biology of prion diseases. Science 252, 1515–1522.

Ranjbar, S., Shahrokhian, S., Nurmohammadi, F., 2018. Nanoporous gold as a suitable substrate for preparation of a new sensitive electrochemical aptasensor for detection of *Salmonella typhimurium*. Sensors Actuators B Chem. 255, 1536–1544.

Reta, N., Saint, C.P., Michelmore, A., Prieto-Simon, B., Voelcker, N.H., 2018. Nanostructured electrochemical biosensors for label-free detection of water-and food-borne pathogens. ACS Appl. Mater. Interfaces 10, 6055–6072.

Singh, C., Ali, M.A., Reddy, V., Singh, D., Kim, C.G., Sumana, G., Malhotra, B.D., 2018. Biofunctionalized graphene oxide wrapped carbon nanotubes enabled microfluidic immunochip for bacterial cells detection. Sensors Actuators B Chem. 255, 2495–2503.

Suaifan, G.A., Alhogail, S., Zourob, M., 2017. Paper-based magnetic nanoparticle-peptide probe for rapid and quantitative colorimetric detection of *Escherichia coli* O157: H7. Biosens. Bioelectron. 92, 702–708.

Torok, V., Hodgson, K., McLeod, C., Tan, J., Malhi, N., Turnbull, A., 2018. National survey of foodborne viruses in Australian oysters at production. Food Microbiol. 69, 196–203.

Viswanath, B., Kristine, Y.M., Kim, S., 2018. Recent trends in the development of complementary metal oxide semiconductor image sensors to detect foodborne bacterial pathogens. Trends Anal. Chem. 98, 47–57.

Wang, S., Deng, W., Yang, L., Tan, Y., Xie, Q., Yao, S., 2017. Copper-based metal–organic framework nanoparticles with peroxidase-like activity for sensitive colorimetric detection of Staphylococcus aureus. ACS Appl. Mater. Interfaces 9, 24440–24445.

Weng, C.I., Chang, H.T., Lin, C.H., Shen, Y.W., Unnikrishnan, B., Li, Y.J., Huang, C.C., 2015. One-step synthesis of biofunctional carbon quantum dots for bacterial labeling. Biosens. Bioelectron. 68, 1–6.

WHO annual report, 2017. http://www.who.int/mediacentre/factsheets/fs399/en/.

Yan, X., Li, J., Yang, R., Li, Y., Zhang, X., Chen, J., 2018. A new photoelectrochemical aptasensor for prion assay based on cyclodextrin and rhodamine B. Sensors Actuators B Chem. 255, 2187–2193.

Yang, B., Gong, H., Chen, C., Chen, X., Cai, C., 2017. A virus resonance light scattering sensor based on mussel-inspired molecularly imprinted polymers for high sensitive and high selective detection of *hepatitis A virus*. Biosens. Bioelectron. 87, 679–685.

Yang, L., Deng, W., Cheng, C., Tan, Y., Xie, Q., Yao, S., 2018. Fluorescent immunoassay for the detection of pathogenic bacteria at the single-cell level using carbon dots-encapsulated breakable organosilica nanocapsule as labels. ACS Appl. Mater. Interfaces 10, 3441–3448.

Yeni, F., Acar, S., Polat, Ö.G., Soyer, Y., Alpas, H., 2014. Rapid and standardized methods for detection of foodborne pathogens and mycotoxins on fresh produce. Food Control 40, 359–367.

Zeinhom, M.M., Wang, Y., Sheng, L., Du, D., Li, L., Zhu, M.J., Lin, Y., 2017. Smart phone based immunosensor coupled with nanoflower signal amplification for rapid detection of Salmonella Enteritidis in milk, cheese and water. Sensors Actuators B Chem. 261, 75–82.

Zeinhom, M.M., Wang, Y., Song, Y., Zhu, M.J., Lin, Y., Du, D., 2018. A portable smart-phone device for rapid and sensitive detection of *E. coli* O157: H7 in yoghurt and egg. Biosens. Bioelectron. 99, 479–485.

Zhang, G., 2013. Foodborne pathogenic bacteria detection: an evaluation of current and developing methods. The Meducator 1 (24), 28–30.

Zhao, X., Lin, C.W., Wang, J., Oh, D.H., 2014. Advances in rapid detection methods for foodborne pathogens. J. Microbiol. Biotechnol. 24, 297–312.

Zhou, H., Liu, J., Xu, J.J., Zhang, S.S., Chen, H.Y., 2018. Optical nano-biosensing interface via nucleic acid amplification strategy: construction and application. Chem. Soc. Rev. https://doi.org/10.1039/c7cs00573c.

Zhu, Y., Zhao, Y., Ma, J., Cheng, X., Xie, J., Xu, P., Liu, H., Liu, H., Zhang, H., Wu, M., Elzatahry, A.A., 2017. Mesoporous tungsten oxides with crystalline framework for highly sensitive and selective detection of foodborne pathogens. J. Am. Chem. Soc. 139, 10365–10373.

CHAPTER

24

The Impact of Bioinformatics Tools in the Development of Antimicrobial Drugs and Other Agents

Swetha Kumari Koduru

Molecular Cloning Lab, Department of Biosciences and Sericulture, Sri Padmavati Women's University, Tirupati, India

OUTLINE

Recent Developments in Applied Microbiology and Biochemistry
https://doi.org/10.1016/B978-0-12-816328-3.00024-6

24.1 INTRODUCTION

Antimicrobial research led to the discovery and progress of therapeutic antimicrobial agents. However, in long term, the development of immunity against existing antibacterial agents has occurred. This increase in the count of antibiotic resistant bacteria led to the invention of the new age antimicrobial active agents. However, few controversies regarding the effectiveness of these drugs in terms of cost and efficacy are concerned. Bacteria will always try to show different tactics to get resistance against various antibiotics. The new genetic information may lead to the alteration in the structure of protein affecting the respective antibiotic carrying capacity in the cell, enzymatic inactivation of drugs, or even cell structure interferences during interactions between drugs and bacteria. Moreover, many natural compounds are available with antimicrobial property that can be used to combat them. Such molecules called antimicrobial peptides (AMPs) are distributed in all life forms right from a bacterial cell to multicellular organisms and are meant to oppose microbial growth. Advances in bioinformatics and depth analysis of bacterial transcriptome will provide more understanding on different microbial adaptations under environmental stress that will help to develop new AMPs.

Advancements in microbiology led to the development of novel chemical drug structures called therapeutic antimicrobial agents. Bioinformatics methods are added to the fast drug discovery due to various prediction methods that are cheap, fast, and reliable. However, development of resistance against existing antibacterial agents has also been seen in microbes. This increase in occurrence of multidrug-resistant microbes triggered the research toward the development and generation of innovative and promising antibiotics. The antibacterial activities are based on broad- or narrow-spectrum mechanisms, but the challenge is to optimize the preclinical research that will lead to suitable and dependable clinical assessments. Using bioinformatics tools, this kind of job can be done perfectly by reducing and short-listing the number of lead molecules to be used as drugs and identifies the potential drugs. In addition to this, there are wide varieties of applications of bioinformatics in microbiology. This article is about different bioinformatics tools that are helpful in different fields of the microbial world (Fig. 24.1).

24.2 METAGENOMIC SHOTGUN SEQUENCING (MSS) AND TAXONOMIC DIVERSITY

The management and analysis of large-scale sequence data generated is one of the major areas of bioinformatics and is growing fast in combination with biology, mathematics, and computer science. The high-resolution taxonomic composition and the gene expression profiles of metagenome samples can be attained by the shotgun sequencing. Studies include the human microbiome composition that is associated with several diseases like type 2 diabetes (Qin et al., 2012), obesity (Le Chatelier et al., 2013), liver cirrhosis (Qin et al., 2014), and inflammatory bowel disease (Morgan et al., 2012). Almost 60,000 type 2 diabetes-related markers were identified from 345 gut metagenome samples. The study also revealed the gene functional profile showing the membrane transport of sugars, methane metabolism, xenobiotic metabolism, and sulfate reduction indicating the composition of type 2 diabetes gut microbiome. Through MSS, it was found that the gut microbiome from obese individuals had less gene richness and depleted butyrate-producing bacteria (Le Chatelier et al., 2013).

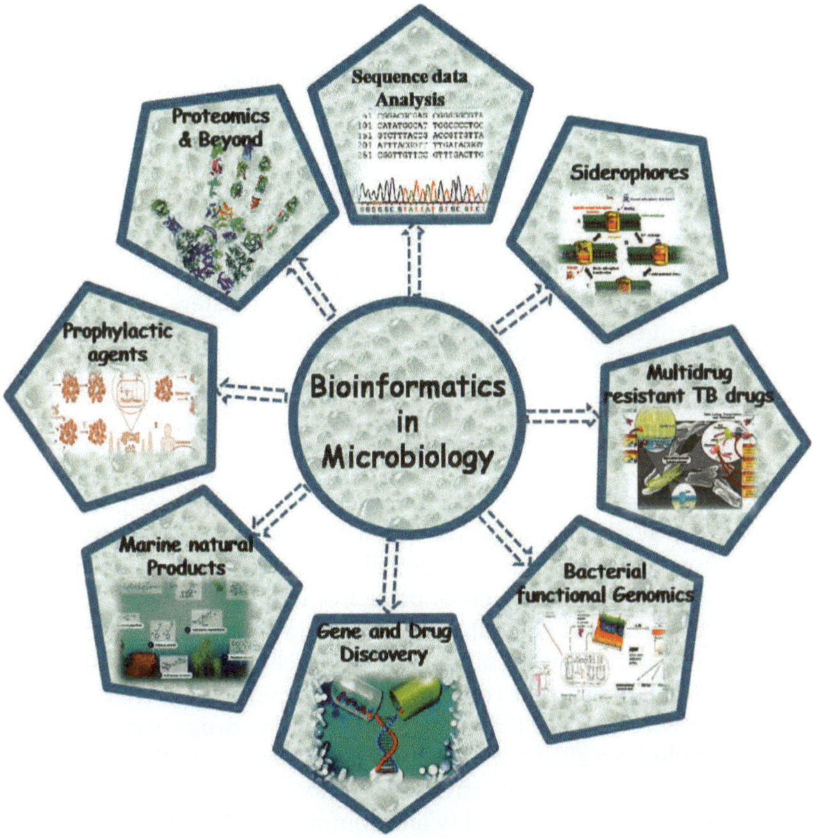

FIG. 24.1 Schematic representation of bioinformatics applications in microbiology.

Though the metagenome samples are sequenced based on 16S rRNA marker gene, the species level shows bias and lacks sensitivity. Individual bacterial isolates with identical 16S genes differ for about 15%–30% in their genomes that include antimicrobial genes and toxin production. It is possible to compute taxonomic content from MSS. The search has to be very precise as metagenomic samples contain unknown microbe species. MetaPhlAn (Segata et al., 2012) is such tool that allows rapid taxonomic profiling of metagenomic samples using a database of taxonomically informative clade-specific marker genes selected from 3000 microbial reference genomes. In addition, functional profiling of MGS data is a demanding process for several years as mapping of millions of DNA reads requires long runs and results in pseudohits to many unrelated proteins where the required sensitivity and specificity are lost. To overcome this, MG-RAST an online service by Argonne National Laboratory has an automated system for metagenomic functional profiles (Glass et al., 2010).

DIAMOND aligns large sets of DNA reads to a protein database 20,000 times faster than BLASTx on short reads with similar sensitivity (Buchfink et al., 2015), and MMseq2 used for protein-protein alignments is 400 times faster than PSI-BLAST (Steinegger and Soeding, 2017). Instead of matching translated MGS reads to a generic database of protein sequences, new approaches using fast DNA-DNA alignment with Bowtie 2 (Langmead and Salzberg, 2012) to a custom database of metagenomic gene sequences where the alignments are more specific, usually requiring 90% sequence identity produce false positive alignments.

About 2.6 million nonredundant mouse metagenomic genes were collected from 184 mice fecal MGS samples (Xiao et al., 2015). The advancements include the development of many powerful tools like HUMAnN2, MOCAT2 (Kultima et al., 2016), and MGS-Fast (Brown et al., 2017). HUMAnN2 aligns metagenomic reads to NCBI UniRef microbial genomes. MGS-Fast is a DNA alignment-based metagenomic tool, and its gene annotation is associated with IGC and KEGG. Direct sequencing of PCR-amplified 16S genes led to the creation of several operational taxonomic units (OTU) representing uncultured bacteria. The rRNA database SILVA currently contains 1,719,541 bacterial 16S rRNA sequences clustered with 99% identity level into 645,151 representative sequences (Pruesse et al., 2012).

24.3 SIDEROPHORES

Rapid development of antibiotic resistance in bacterial and fungal pathogens has drawn a novel path unlike the available antibiotics. Siderophores and their comparative study provided new insights for the development of probable targets for antibiotic discovery (Dhusia et al., 2018). The entry of an antibiotic takes place through siderophore receptors present on the pathogen cell membrane and generates a black hole via iron scarcity. Comparative study of siderophore biosynthesis pathways could allow us to achieve potential targets for inhibiting these siderophores in pathogens that control pathogenic virulence. This is also termed as a Trojan Horse Strategy, which blocks the pathogen to become drug-resistant. Medical application of siderophores includes a group of siderophore conjugates and their derivatives to blocks tRNA synthetase and start antibacterial activity. Preparation of synthetic siderophore-antibiotic conjugates includes improvised permeability and compromised cell susceptibility to overcome resistance (Holden and Bachman, 2015). Few examples that can make a microbe to commit suicide are carbacephalosporins, albomycin, enterobactin, sulfonamides, nalidixic acid, and norfloxacin. Demonstration of reduced viability of malignant melanoma cells by desferrioxamine E produced by actinobacterium species has been explained (Nakouti et al., 2013). Production of siderophore in plant-growth-promoting rhizobacteria (PGPR) is identified to provoke suppression against pathogens in plants and agricultural crops (Ahemad and Kibret, 2014). Many studies revealed that siderophores can be used both as a plant growth controller and plant pathogen biocontrol agent. They can be used in bioremediation of environmental pollutants and radioactive wastes and recycling of nuclear fuel. Phillips et al. (2015) studied on the dispersion behavior of gold nanoparticles as an optical biosensor using Fe^{3+} catecholate binding motif.

24.4 DRUG RESISTANCE

24.4.1 Antimicrobial Resistance

Antimicrobial resistance (AMR) has an evolving scenario ranging from protoresistance to untreatable clinical pathogens and hence is a universal well-being challenge. The genotype information facilitates the efficient treatment of resistant infections and new drug discovery.

Such approaches need improved microbial informatics methods. For instance, Comprehensive Antibiotic Resistance Database (CARD) originated for the prediction of resistome and resistance phenotype from genotype.

With the help of CARD, curated information on resistance mechanisms, genes, and their targets is acquired to generate a resource for algorithm development for the prediction of antibiotic resistance. However, the development of algorithms for the extraction of association and interaction information from the literature is a major area of research in the field of biocuration and is actively used to enrich a number of biological databases (Rutherford et al., 2014). Another significant task is the prediction of resistome and antibiogram. The NGS and WGS methods have revolutionized the medical microbiology making a shift from phenotype-based study to genotype-based study of resistance (Bertelli and Greub, 2013). Within the CARD, the Resistance Gene Identifier (RGI) currently provides the prediction of resistance genes, analyzes genome assemblies, and provides a detailed output of predicted antibiotic resistance genes and targeted drug classes.

24.4.2 Multidrug Resistant TB Drugs

Tuberculosis (TB) is an infectious disease of universal public health caused by *Mycobacterium tuberculosis* complex (MTC) of *M. tuberculosis* (Mtb). Development of anti-TB drugs has become a great confront in TB control. Using omics, the transformation of TB management can be fastened. A few FDA-approved drugs are being validated for repurposing, and this approach made several drugs with enhanced efficiency and capability in reducing drug tolerance or changing the host immune response (Swaminathan et al., 2016). The standardization of the database and analysis tools and the collaborations among partners, including researchers, clinicians, clinical laboratories, and commercial companies, are needed to sustain and maximize the application of next-generation sequencing technology in whole-genome sequencing of TB isolates for TB control (Faksri et al., 2016). The different tools that are available to identify the drug-resistant mutations are tabulated in Table 24.1.

24.5 BACTERIAL FUNCTIONAL GENOMICS

Since there is an urgency to develop novel classes of antibiotics for combating the threat of the development of multiple drug-resistant pathogens, the availability of the complete genome sequences of many pathogenic microbes provides huge information on every potential drug target and is an invaluable resource in the search for novel compounds. The effective confront against diseases will be done not only by structural biology, bioinformatics, and genomics but also by the collective establishment of epidemiological networks of surveillance along with enhanced measures for public health.

24.5.1 Finding Homologs

Bacterial genome size ranges from 0.6 to 8.0 megabases (Mb) and generally encodes 600–6000 proteins. In spite of abundant gene sequence data, still 30% of genes in a microbial genome are left orphaned. The key component required for such drug discovery is the presence

TABLE 24.1 Genome Analysis Tools for Drug-Resistant TB and Genotyping

Tools	Type	Input	Output	Mtb Lineages	Deposit	Geno typing
CASTB	Online	FASTQ and FASTA reads	Spoligotypes, VNTR, LSP lineages and SNP-based tree	6	Yes	4a
KvarQ	Stand-alone	FASTQ sequence reads	Spoligotype/drug resistance profile/lineages	7	No	SNP/Spol
PhyResSE	Online	FASTQ sequence reads	Drug resistance profile and lineages	12	Yes	SNP
PhytB	Online	Sequence variant data	Web portal display and text file with lineage specific SNPs and drug resistance profile	63	Yes	SNP
TB-Profiler	Online	FASTQ sequence reads	Drug resistance profile/ lineages	63	Optional	SNP
TGS-TB	Online	FASTQ reads (compressed)	Spoligotypes/VNTR results, lineage, drug-resistant pattern and tree	1	Yes	4b

of table of microbial functions that can be acquired by the identification of homologs in different organisms providing additional information on motif and superfamilies.

24.5.2 Integrated Genome Comparison and Target Prioritization

The protein function classification systems and the prediction tools together called integrated genome comparison systems are more essential for a researcher as it enables to get efficient review on the phylogenetic diversity and the metabolic mechanisms in microbes. Prioritizing targets has become the most essential components for industrial drug discovery. The detailed information on biochemical activities, knockout phenotypes and metabolic activities, and substrates and inhibitors are integrated and made into a prioritization tool (Freiberg, 2001).

24.6 BIOINFORMATICS IN GENE AND DRUG DISCOVERY

Bioinformatics has revolutionized the world of pharmacy by the advances in drug discovery and target validation.

24.6.1 Chemo-Informatics Approach in Antimicrobial Research

The merging of mathematics, statistics, computer science, information technology, and molecular biology has created an information-oriented field in biology, now known as bioinformatics. The development of effective computational tools depends on knowledge generated from these diverse disciplines (Hammami and Fliss, 2010).

24.6.2 Programs

The development of a relational database by the integration of various tools using sequence data and other associated data facilitated the ease of performing homology search, sequence alignment, evolutionary studies, and other physiochemical predications and calculations. Sequence alignment is a boon to the biologists as it paves many ways to pursue research going ahead saving time and is also cost-effective. The AntiBP server can be used to predict the antibacterial peptides in a protein sequence (Lata et al., 2010). When the structures of experimental peptides have not been elucidated, homology and comparative modeling are used in structure-based drug discovery (Cavasotto and Phatak, 2009). The accuracy of the predictions depends on the quality of the sequence alignment and template structure. High-quality structural models are produced when the target and template are closely related. BACTIBASE database has incorporated MODELER as a tool for predicting the 3-D structure of a user peptide by homology to known bacteriocins. Such a combination of relational database with molecular modeling tools should be very useful for the in silico design of novel AMPs.

24.6.3 Genomics and Target Discovery

The need for the identification of new drugs based on genome sequencing has been increasing and challenging in spite of the development of different sequencing technologies. Along with existing sequencing platforms, an improved version with GS FLX system called Titanium is released, which has the capacity to make 400–600 million bases per run with 400bp read lengths. The 454 sequencing technology that is broadly used for studies based on 16S profile of microbiomes (Bowman et al., 2012) however showed a higher rate error in repeated DNA regions. Illumina sequencing platform because of its accuracy and read lengths has shown the complete output fast. It made changes substantially for improvement that include library construction methods, sequencing chemistry, and sequencing quality estimation. HiSeq and Miseq platforms that were released in 2014 were again different using different sequencing scales producing 1000GB sequencing data in just 6days (Caporaso et al., 2012).

24.6.4 Artificial Neural Networks and Fuzzy Logic Modeling

ANNs are mathematical modeling algorithms that can offer a reliable and efficient way for in silico identification of novel AMPs (Fjell et al., 2009). ANN analysis for the prediction of the antifungal activities of pyridine derivatives of a defined structure has been successfully done by Bucinski et al. (2009). As per Mikut and Hilpert (2009), fuzzy logic with molecular descriptors for the analysis of AMPs is found able to distinguish active and inactive peptides with impressive accuracy. The application of fuzzy methods to QSAR provides useful insights for antimicrobial drug research and development.

24.6.5 Machine Learning

Machine learning algorithms are widely used in the analyses of high-dimensional gene expression datasets for gene selection (Hao and Weiss, 2016). In general, analysis methods are grouped into two groups, namely, selection of important variables and identification of subgroup structures associated with phenotypes. Taxonomic and gene function profiles are high-dimensional datasets with alike statistical parameters. The most featured generic method like random forest can be used for the high-dimensional data that has the capability to handle huge sets of data and performs accurately. Because of the generation of redundancy, the functional determination is a challenging task and is overcome by network analyses (Faust and Raes, 2012). Association networks called network interfaces are of two types namely Pairwise relationships and complex patterns that are analysed based on similarity, correlation coefficients and regression and association rule mining respectively.

24.6.6 Molecular Simulations and Dynamics

AMPs interact specifically with the bacterial membrane and kill the cell by causing leakage of its contents. The present research is focused on the bacterial membrane and AMP interactions with phosphor lipid layers. CHARMM (Brooks et al., 2009) is one such freely available program widely used to activate AMP interactions with lipid bilayers. GROMACS (Hess et al., 2008) is another freeware tool that is used in MD studies to generate trajectories.

24.7 MARINE NATURAL PRODUCTS

Other promising drug resources are marine natural products (MNPs) having many marine-derived drugs. Research studies on marine natural products would definitely assure the benefits from analysis to characterization of new chemicals using NGS techniques and advances in bioinformatics (Zhang et al., 2016).

24.8 PROPHYLACTIC AGENTS

The haphazard use of antimicrobials in the clinical, community, and agricultural settings resulted in the development of multiple drug resistance among pathogens (Hwang et al., 2016). The advantage of using commensal or probiotic cells as antimicrobial or biotherapeutic agents lies in the versatility to exert effective antimicrobial activity that can be mediated by multiple bioactive molecules, such as bacteriocins and various enzymes.

24.9 PROTEOMICS AND BEYOND

Proteome is the upcoming frontier contributing more in bioinformatics. The determination of protein structures from 2-D gel electrophoresis, mass spectrometry, and protein microarrays have already been substituted by computational methods.

24.9.1 Phylogenetic Profiling

The prime responsibility in target identification is to know the organisms in which the potential targets occur. This could be attained by phylogenetic profiling. In this strategy, the proteins are classified into families, and for each family, a set of organisms sharing the common features are determined. Such a determined set of organisms are called the phylogenetic profile of that particular family. Intelligent classification systems can help to obtain comprehensive answers easily like MurC usually seen in most pharmaceutically relevant bacteria. For instance, inorganic pyrophosphate can be seen in gram-negative bacteria and eukaryotes. But the same functioning proteins sharing different structure called analogs can be seen in gram-positive bacteria.

24.9.2 Metabolic Pathway Mapping

The phylogenetic approach that made the identification of genes indicating the functionally related ones is called metabolic pathway mapping. For instance, the identification of deoxyxylulose-5-phosphate (DOXP) pathway includes isopentyl diphosphate biosynthesis (Eisenreich et al., 2001). The proteins must be annotated with metabolic information in order to make a mapping for a metabolic pathway. Even though enzyme classification (EC) numbering system has been given, they are restricted only to the enzymatic activities of proteins but not for the cellular role of a gene. Therefore, various classification systems have to be kept for determining the genome comparison studies. Gene Ontology Consortium is aimed to explain the genes and gene products of an organism (Ashburner et al., 2000).

24.9.3 Expression Profiling

The transcriptional activity of all genes including coinduced and corepressed encoded in a microbial genome can be computed using the latest expression profiling techniques. The current research included profiling experiments on arrays having a 25-mer oligonucleotide probe for every 30bp in *Escherichia coli* genome. The initiation and termination sites are mapped using these high-resolution genome arrays more precisely than earlier. It helps in global expression profile studies against antibiotic stress. The suitable marker gene identification algorithms can be derived using expression profiling method that in turn helps in deducing more resistance mechanisms (Celis et al., 2000).

24.10 BIG DATA ANALYTICS

The big biological data have to be generated and shared for an accurate understanding of the genomic landscape of antimicrobial resistance (AMR). For instance, ribosomal RNA methyltransferases that include ArmA and Rmt enzymes are able to guard the ribosome from gentamicin action. Moreover, the approach to directly suppress the resistance mechanisms with small molecules with a clinical proved efficacy requires an understanding on the most prevalent and vital resistance genes that are said to be clinical pathogens. Furthermore, the

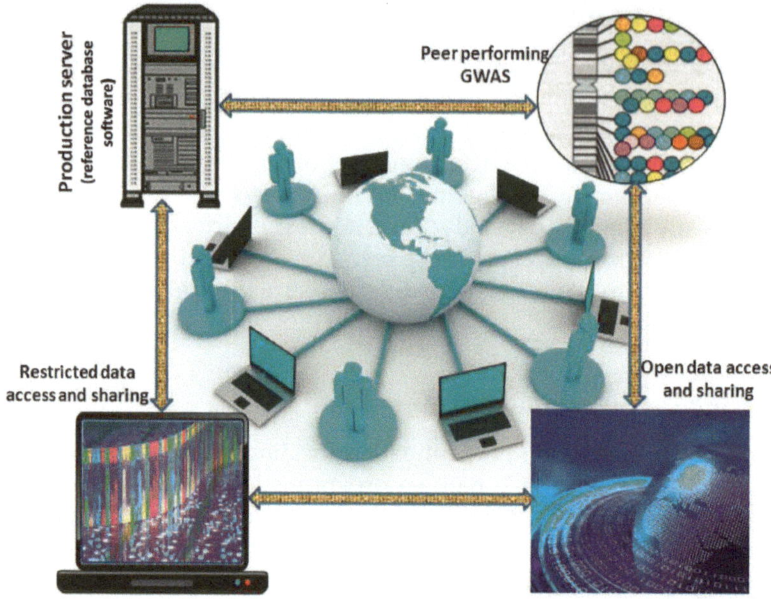

FIG. 24.2 A schematic representing global analysis of antimicrobial resistance data by a peer-to-peer approach. Example: Comprehensive Antibiotic Resistance Database (CARD) acts as the peer host, providing all downloadable software and data for resistome and antibiogram. The data are shared in both open and restricted modes like for pathogen sequences and clinical data, respectively.

advancement in the NGS platform has to be targeted to achieve novel molecular level diagnostics that is a resultant of clear understanding on resistance genotype and molecular epidemiology (Alam et al., 2014).

Comprehensive Antibiotic Resistance Database (CARD) has to develop new data sharing prototypes that help in local clinical diagnostics and research with the help of AMR computational tools. Big data sharing methods and the standards have to be generated like peer-to-peer approaches allowing the local secure data to share with customized terms and conditions and sharing agreements. The powerful AMR examining approach for targeted drug design and discovery includes computational approaches like the genome-wide association studies (GWAS) (Wang and Jia, 2016) based on global sampling (Fig. 24.2).

24.11 CONCLUSION

The advancements and emerging technologies for DNA, RNA, and protein sequencing methods have tremendously increased the knowledge and understanding on the microbes right from the level of a single cell to the complex structures of microbial consortia and from environments ranging from deep sea thermal to acidic hot springs, deserts, etc. The misuse of antimicrobials lasting recent decades has increased the spread of mutations allowing the development of multidrug-resistant microbes. The incidence of microbial infections in humans has become an important task for both medical and food sectors. There is an utmost need for the development of novel therapeutic agents. The genome sequencing technique has played its major role in determining microbial virulence and drug resistance. The current structural and functional information on AMPs created a chance to develop models for the prediction of the effectiveness of peptide candidates quantitatively as they can be used as antimicrobial drugs. Both in terms of time and cost, bioinformatics and computational analysis has provided an ultimate possibility for better understanding of antimicrobial—target relations and interactions. The combination of bioinformatics and relational databases provides the antimicrobial researcher with better tools for analyzing, linking, and comparing online search results. The computational tool generation and development is a pure interdisciplinary field that requires a thorough knowledge of mathematics, statistics, computers, information technology, and molecular biology. All these are interdependent and interlinked complementing each other for the progress and development of antimicrobial drug discoveries in the future. It is also made clear that acquisition of data is not a challenge but analysis and curation of the data is more required. To determine the direct functional activity of a microbial community, integration of multiple omic data can offer many chances to study individual microorganism or a microbiome community that generate stimuli in response to environmental and drug activities. At the outset, the longitudinal data have become more powerful and helpful for the discovery of disease-related microbiome perturbations comparing different sampling conditions. The incorporation of omic data types and datasets ultimately improved the knowledge and understanding on the functions of microbial communities in health and environmental processes that thereby lead to the enhancement in the therapeutic interventions.

Acknowledgments

The authors are indebted to all the researchers whom we cited in this review for their significant and valuable research. No funding was received to perform this review.

Conflicts of Interests

The authors declare no relevant competing financial interests to disclose.

References

Ahemad, M., Kibret, M., 2014. Mechanisms and applications of plant growth promoting rhizobacteria: current perspective. J. King Saud Univ. Sci. 26, 1–20.

Alam, M.T., Petit, R.A., Crispell, E.K., Thornton, T.A., et al., 2014. Dissecting vancomycin-intermediate resistance in Staphylococcus aureus using genome-wide association. Genome Biol. Evol. 6, 1174–1185.

Ashburner, M., Ball, C.A., Blake, J.A., Botstein, D., Butler, H., et al., 2000. Gene ontology: tool for the unification of biology. The gene ontology consortium. Nat. Genet. 25, 25–29.

Langmead, B., Salzberg, S.L., 2012. Fast gapped-read alignment with Bowtie 2. Nat. Methods 9, 357–359.

Bertelli, C., Greub, G., 2013. Rapid bacterial genome sequencing: methods and applications in clinical microbiology. Clin. Microbiol. Infect. 19, 803–813.

Bowman, J.S., Rasmussen, S., Blom, N., Deming, J.W., et al., 2012. Microbial community structure of Arctic multiyear sea ice and surface seawater by 454 sequencing of the 16S RNA gene. ISME J. 6 (1), 11–20.

Brooks, B.R., Brooks III, G.L., MacKerell Jr., A.D., Nilsson, L., Petrella, R.J., et al., 2009. CHARMM: the biomolecular simulation program. J. Comput. Chem. 30, 1545–1614.

Brown, S.M., Hao, Y., Chen, H., Laungani, B.P., et al., 2017. Fast functional annotation of metagenomic shotgun data by DNA alignment to a microbial gene catalog. bioRxiv, 120402. https://doi.org/10.1101/120402.

Buchfink, B., Xie, C., Huson, D.H., 2015. Fast and sensitive protein alignment using DIAMOND. Nat. Methods 12, 59–60.

Bucinski, A., Socha, A., Wnuk, M., Baczek, T., Nowaczyk, A., et al., 2009. Artificial neural networks in prediction of antifungal activity of a series of pyridine derivatives against *Candida albicans*. J. Microbiol. Methods 76, 25–29.

Caporaso, J.G., Lauber, C.L., Walters, W., Berg-Lyons, D., Huntley, J., et al., 2012. Ultra-high-throughput microbial community analysis on the Illumina HiSeq and MiSeq platforms. ISME J. 6, 1621–1624.

Cavasotto, C.N., Phatak, S.S., 2009. Homology modeling in drug discovery: current trends and applications. Drug Discov. Today 14, 676–683.

Celis, J.E., Kruhøffer, M., Gromova, I., Frederiksen, C., Ostergaard, M., et al., 2000. Gene expression profiling: monitoring transcription and translation products using DNA microarrays and proteomics. FEBS Lett. 480, 2–16.

Dhusia, K., Bajpai, A., Ramteke, P.W., 2018. Overcoming antibiotic resistance: is siderophore Trojan horse conjugation an answer to evolving resistance in microbial pathogens? J. Control. Release 269, 63–87.

Eisenreich, W., Rohdich, F., Bacher, A., 2001. Deoxyxylulose phosphate pathway to terpenoids. Trends Plant Sci. 6, 78–84.

Faksri, K., Tan, J.H., Chaiprasert, A., Teo, Y.Y., Ong, R.T., 2016. Bioinformatics tools and databases for whole genome sequence analysis of *Mycobacterium tuberculosis*. Infect. Genet. Evol. 45, 359–368.

Faust, K., Raes, J., 2012. Microbial interactions: from networks to models. Nat. Rev. Microbiol. 10, 538–550.

Fjell, C.D., Jenssen, H., Hilpert, K., Cheung, W.A., Panté, N., et al., 2009. Identification of novel antibacterial peptides by chemoinformatics and machine learning. J. Med. Chem. 52, 2006–2015.

Freiberg, C., 2001. Novel computational methods in anti-microbial target identification genomics supplement. Drug Discov. Today 6, 72–80.

Glass, E.M., Wilkening, J., Wilke, A., Antonopoulos, D., Meyer, F., 2010. Using the metagenomics RAST server (MG-RAST) for analyzing shotgun metagenomes. Cold Spring Harb. Protoc. https://doi.org/10.1101/pdb.prot5368.

Hammami, R., Fliss, I., 2010. Current trends in antimicrobial agent research: chemo- and bioinformatics approaches. Drug Discov. Today 15, 540–546.

Hao, Y., Weiss, G.M., 2016. In: In gene selection from microarray data for age-related macular degeneration by data mining. Proceedings of the International Conference on Data Mining (DMIN), pp. 125.

Hess, B., Kutzner, C., van der Spoel, D., Lindahl, E., 2008. GROMACS 4: algorithms for highly efficient, load-balanced, and scalable molecular simulation. J. Chem. Theory Comput. 4, 435–447.

Holden, V.I., Bachman, M.A., 2015. Diverging roles of bacterial siderophores during infection. Metallomics 7, 986–995.

Hwang, I.Y., Koh, E., Kim, H.R., Yew, W.S., Chang, M.W., 2016. Reprogrammable microbial cell-based therapeutics against antibiotic-resistant bacteria. Drug Resist. Updat. 27, 59–71.

Kultima, J.R., Coelho, L.P., Forslund, K., Huerta-Cepas, J., Li, S.S., et al., 2016. MOCAT2: a metagenomic assembly, annotation and profiling framework. Bioinformatics 32, 2520–2523.

Lata, S., Mishra, N.K., Raghava, G.P., 2010. AntiBP2: improved version of antibacterial peptide prediction. BMC Bioinform. 11 (Suppl.1), S19.

Le Chatelier, E., Nielsen, T., Qin, J., et al., 2013. Richness of human gut microbiome correlates with metabolic markers. Nature 500, 541–546.

Mikut, R., Hilpert, K., 2009. Interpretable features for the activity prediction of short antimicrobial peptides using fuzzy logic. Int. J. Pept. Res. Ther. 15, 129–137.

Morgan, X.C., Tickle, T.L., Sokol, H., et al., 2012. Dysfunction of the intestinal microbiome in inflammatory bowel disease and treatment. Genome Biol. 13, R79.

Nakouti, I., Sihanonth, P., Palaga, T., Hobbs, G., 2013. Effect of a siderophore producer on animal cell apoptosis: a possible role as anti-cancer agent. Int. J. Pharm. Med. Biol. Sci. 2, 1–5.

Phillips, D.J., Davies, G.L., Gibson, M.I., 2015. Siderophore-inspired nanoparticle-based biosensor for the selective detection of Fe^{3+}. J. Mater. Chem. B 3, 270–275.

Pruesse, E., Peplies, J., Gleockner, F.O., 2012. SINA: accurate high-throughput multiple sequence alignment of ribosomal RNA genes. Bioinformatics 28, 1823–1829.

Qin, J., Li, Y., Cai, Z., Li, S., Zhu, J., Zhang, F., et al., 2012. A metagenome-wide association study of gut microbiota in type 2 diabetes. Nature 490, 55–60.

Qin, N., Yang, F., Li, A., Prifti, E., et al., 2014. Alterations of the human gut microbiome in liver cirrhosis. Nature 513, 59–64.

Rutherford, K.M., Harris, M.A., Lock, A., Oliver, S.G., Wood, V., 2014. Canto: an online tool for community literature curation. Bioinformatics 30, 1791–1792.

Segata, N., Waldron, L., Ballarini, A., Narasimhan, V., Jousson, O., Huttenhower, C., 2012. Metagenomic microbial community profiling using unique clade-specific marker genes. Nat. Methods 9, 811–814.

Steinegger, M., Soeding, J., 2017. MMseqs2: sensitive protein sequence searching for the analysis of massive data sets. bioRxiv, 079681.

Swaminathan, S., Sundaramurthi, J.C., Palaniappan, A.N., Narayanan, S., 2016. Recent developments in genomics, bioinformatics and drug discovery to combat emerging drug-resistant tuberculosis. Tuberculosis 101, 31–40.

Wang, J., Jia, H., 2016. Metagenome-wide association studies: fine-mining the microbiome. Nat. Rev. Microbiol. 14, 508–522.

Xiao, L., Feng, Q., Liang, S., Sonne, S.B., Xia, Z., et al., 2015. A catalog of the mouse gut metagenome. Nat. Biotechnol. 33, 1103–1108.

Zhang, G., Li, J., Zhu, T., Gu, Q., Li, D., 2016. Advanced tools in marine natural drug discovery. Curr. Opin. Biotechnol. 42, 13–23.

Further Reading

Hao, Y., Pei, Z., Brown, S.M., 2017. Bioinformatics in microbiome analysis. Methods Microbiol. 44. ISSN 0580-9517.

Huynen, M., Snel, B., Lathe 3rd, W., Bork, P., 2000. Predicting protein function by genomic context: quantitative evaluation and qualitative inferences. Genome Res. 10, 1204–1210.

McArthur, A.G., Wright, G.D., 2015. Bioinformatics of antimicrobial resistance in the age of molecular epidemiology. Curr. Opin. Microbiol. 27, 45–50.

McArthur, A.G., Waglechner, N., Nizam, F., Yan, A., et al., 2013. The comprehensive antibiotic resistance database. Antimicrob. Agents Chemother. 57, 3348–3357.

CHAPTER

25

Recent Trends in the Development of Paper-Based Diagnostic Chips for the Detection of Human Viruses

*Saravanan Govindaraju**, *Min Ho Lee*[†], *Kyusik Yun**

*Department of Bionanotechnology, Gachon University, Seongnam, Republic of Korea [†]Chung Ang University, School of Integrative Engineering, Seoul, Republic of Korea

OUTLINE

25.1 INTRODUCTION

Viruses are infectious pathogens that are dangerous to human health. They can amplify infections by consuming the replication machinery of the host and using it to replicate rapidly (Liu and Gu, 2011). Annually, as many as 1 million people die due to infectious

Recent Developments in Applied Microbiology and Biochemistry
https://doi.org/10.1016/B978-0-12-816328-3.00025-8

diseases like acquired immunodeficiency syndrome (AIDS), tuberculosis (TB), malaria, Zika virus, Ebola, HIV, and influenza (Morens et al., 2004; Bissonnette and Bergeron, 2010). Increasing death rates have forced the efforts for the development of effective clinical diagnostics. Rigorous and accurate analysis diagnoses of laboratory results guarantee successful clinical management of diseases and help to control infections in the early stage. The diagnostic kits used in the laboratory depend on various performance and operational parameters such as accuracy, sensitivity, precision, and specificity (Lalkhen and McCluskey, 2008). The WHO has set up certain criteria for diagnostics in point-of-care (POC) settings such as the affordable, sensitive, specific, user-friendly, rapid and robust, equipment-free and deliverable to end users (ASSURED) criterion. Instant results and rapid recording of disease statuses are necessary to develop diagnostic devices (Souf 2016). To overcome the current requirements, paper-based sensors have gained much momentum in the sensor technology field.

The development of rapid, inexpensive point-of-care and easy-to-use diagnostic assays is urgently needed in multiple disciplines, including virus sensors, environmental monitoring, and clinical diagnostics applications. Among these, paper-based and flexible-material-based sensors seem to be the forerunners in sensor technology field (Shafiee et al., 2015). Paper-based sensors were introduced in the seventeenth century for colorimetric detection of uric acid through silver-based filter paper (Schiff, 1866). Over the following years, a few other applications including the detection of glucose in urine and cadmium were published (Bayley, 1878; Oliver, 1883). Despite these early implementations, it was only from the 1940s to 1990s that the next generation of paper-based analytic devices gained prominence with developments in chromatography and analytical chemistry (Ahmed et al., 2016). Microfluidic technology has gained precedence in developing point-of-care diagnosis through advanced sensor kits (Hawkins and Weigl, 2010). The popularity of microfluidic paper diagnostic devices increased as it was cost-effective, easy to use, and disposable. Currently, research focused on paper-based platforms, which use colorimetric, fluorometric, and electrochemical approaches, involves diagnosing through the signal received from the analyte (Shafiee et al., 2015). To enhance the sensitivity of paper-based analytic devices, nanoparticles including AuNPs, AgNPs, carbon NPs, liposome, magnetic NPs, and QDs have been used in combination with enzymes and other materials (Lopez-Marzo and Merkoci, 2016).

Molecular technology platforms are critical for the detection of viruses and other targeted pathogens. Nuclear acid testing (NAT) is prominent in molecular technology, as it promises rapid functioning, sensitivity, and selectivity over other methods. With the increase in the spread of infections, a low-cost and easy-to-perform alternative is urgently needed for diagnosis. Due to the limitations of other conventional methods, paper-based NAT was developed for the detection of viral and other diseases (Martinez et al., 2013; Akhtar et al., 2011). Integrated chip-based biosensors have emerged and seem to be highly potential in overcoming the limitations of the POC settings observed with NAT. These microfluidic devices can reduce the cost and amount of liquid, samples, and reagents used. Therefore, the cost of manufacturing devices is solely based on paper, glass, silicon, and polymers (Kumbhat et al., 2010).

In this review, we highlight the advances that the new-generation paper-based diagnostic devices developed in the recent years offer. Herein, we explain multifabrication of paper-based devices for the detection of viruses. The performance of the paper based on its shape and geometry is also discussed. Device fabrication and its methodology and configuration are discussed by considering the analytic detection, the limit of detection, and the technologies used for developing these devices. Moreover, we present an overview of the performance of paper-based devices in the research field.

25.2 IMPORTANCE OF PAPER-BASED SENSORS

In the last few years, interest in paper-based sensors has gradually increased. Paper has become one of the potential materials for fabricating bioanalytical sensors due to its wide availability, hydrophilic nature, and affordability. Beyond these well-known benefits, paper has been endowed with other possible functionalities as well, owing to its physical characteristics (Martinez et al., 2013). Moreover, its high surface-to-volume ratio, porous structure, and less volume requirement are the main reasons for the surge observed in the development of paper-based sensors (Wang et al., 2012). Analyses of blood samples can be performed within a 16 min window, in contrast to the time taken by other methods. Portable enzyme-linked immunosorbent assays provide faster results than the other current and relevant techniques. Moreover, using paper-based sensors (0.1 cents/dm^2) is less expensive than the implementation of other techniques like polyethylene terephthalate and is much cheaper than glass substrates (Nery and Kubota, 2013).

Apart from the cost of devices, the development and marketing is also easier than that of other methods. In addition, it requires less investment when introduced in the market; thus, it not only is useful in the scientific community but also can be applied in practical life (Zhao and van den Berg, 2008). Compared with other substrates like glass, silicon, and PDMS that have a low surface profile, paper provides a moderate surface profile. Moreover, paper-based devices have more flexibility than other substrates. In fluid flow reaction, the capillary action takes place in paper-based sensors. Sensitivity to moisture is also a possibility in paper-based sensors, but not in other substrates; moreover, functionalization with paper is easy to achieve than other substrates (Nery and Kubota, 2013). These are some important comparisons between paper-based sensors and other techniques and fabrication methods.

25.3 DEVELOPMENT OF PAPER-BASED SENSORS

The selection of paper is based on the fabrication steps required in developing the device and certain application areas. In the research field, filter paper and certain other paper-based devices are mainly used in the microfluidic technology and for developing the sensor. In contrast, a hydrophobic nitrocellulose membrane is suitable for nonbinding biomolecules, DNA, proteins, and other molecules (Liana et al., 2012). This membrane is smooth and has a uniform pore size, allowing for more stable results. Glossy paper-based devices are also reported as suitable sensing platforms. Due to the flexible, nondegradable, and relatively smooth surface, many nanomaterials were modified with nanomaterials for better sensing (Arena et al., 2010). Paper-based substrates and appropriate detection techniques in virus sensors are summarized in Table 25.1.

TABLE 25.1 Recent Studies on Paper-Based Sensors in Detection of Viruses

Substrate	Sensing Technique	Target	Advantages	Reference
Whatman paper-wax printer	Paper-based electrochemical sensor	Hepatitis B virus	Detection limit is 85 pM. It is user-friendly, and the cost of the device is \$0.36	Li et al. (2015)
Cellulose paper	Electric and optical sensing	Human immunodeficiency virus-1	Can detect multiple biotargets selectively, with sensitivity and repeatability	Shafiet et al. (2015)
Lateral flow test strip	Nucleic acid lateral flow assay	Human immunodeficiency virus-1	LOD is 0.1 nM	Hu et al. (2013)
Handmade-hydrophobized paper	Electrochemical detection	Influenza virus	LOD is 113 PFU mL^{-1} within 30 min	Devarakonda et al. (2017)
Neutravidin-modified paper	Fluorescence-based assay	Epstein-Barr virus (EBV) RNA	LOD is 5.0×10^5 cells. Total cost is approximately \$1.46.	(Zhang et al. (2017)

25.3.1 Fabrication and Patterning

Paper-based device fabrication requires low-cost, easy-to-use, and efficient production processes. There are several techniques and processes available for the modification and development of these devices (Li et al., 2012). Fig. 25.1 represents the different fabrication techniques involving paper-based devices. Some of these techniques reported include analog methods, etching, plasma treatment, screen printing, laser treatment, wax printing, inkjet printing, and paper cutting. The uses of these techniques are based on the material type and applications (Liana et al., 2012). Nowadays, cellulose-fiber-based materials and nitrocellulose-based paper materials are also widely used in the fabrication of point-of-care testing. The device effect and behavior efficiency are determined by the porosity and the surface chemistry of the materials. The high fluorescence-based diagnostic assay led by fluorescent molecules and selective polymeric additives was added to commercial paper (Wong and Harley, 2009).

For paper-based device fabrication, the paper device is designed to contain a hydrophilic pattern, as hydrophobic walls enable the easy flow of liquid onto the device. The main role of the pattern is to avoid the spreading of liquid in undesired areas, reaction with specific detection areas, or parallelization of multiple reactions in a single device. Photolithography is often used to make paper-based devices through an inkjet printer, a photocopying machine, and even a hand drawing based on the UV exposure. By this method, patterning and etching of gold nanoparticles on the silicon substrate do not require clean room facilities (Kane et al., 2006). Recently, printing has involved microchannel fabrication of paper-based devices by using different types of materials (Ahmed et al., 2016). Wax printing is an advanced technique in paper-based devices that prints solid wax on top of the paper and allows the wax to melt and spread on the surface of the paper (Dungchai et al., 2011). This process overtakes the hydrophobic barriers, fluid reservoirs, and reaction area from a computer design.

FIG. 25.1 Different fabrication methods involved in paper-based devices.

The fabrication and patterning of paper-based devices can be extended to 3-D microfluidic systems. The major advantages of 3-D systems include carrying out many assays, samples that penetrate through the z direction first and then enter the x and y directions and perform multiple assays in parallel. Fabrication methods also involve adsorption, covalent modification, and entrapment, while transduction systems include colorimetric transduction, plasmonic transduction, and electrochemical transductions (Ahmed et al., 2016). For a virus sensor, research is currently directed toward modification of the device by multifunctional materials, and the combination of one or two transduction methods has great potential in the field of sensing.

25.4 MICROFLUIDIC PAPER-BASED NUCLEIC ACID TEST FOR VIRUS SENSORS

Rapid nucleic acid detection is conducted by the LFA format. However, without sample preparation, these devices are not able to detect the nucleic acid in raw biological samples (Mao et al., 2009). To eliminate the risk in other fabrication methods, the paper-based nucleic acid device holds great prominence in the field. Paper-based microfluidic devices are developed to extract the DNA conjunction with direct detection of nucleic acid by using mobile phones (Fronczek et al., 2014). Extracted nucleic acid is undetectable by existing methods due to the presence of fewer samples. Therefore, nucleic acid amplification is necessary for NAT (Niemz et al., 2011; Craw and Balachandran, 2012). Reagent storage plays an important role in the amplification of the nucleic acid sensor. Porous materials are capable of storing large amounts of reagents in the dry form over a longer period (Stevens et al., 2008). Paper with large pores can capture molecules in its dried form and activated by addition of samples (Chong et al., 2013).

The recombinase polymerase amplification (RPA) technique, with paper-based devices, is a quick and simple alternative method to PCR. The RPA method depends on the enzymatically driven primer annealing process but does not rely on the temperature-primer annealing process (Lillis et al., 2014). In addition to RPA, LAMP has proved to be one of the most effective techniques for POC testing. LAMP provides a better production rate and is at least a hundred times more efficient than PCR. However, LAMP requires high heating power to attain a high level of amplification. Recently, we learned of the development of noninstrumented nucleic acid amplification (NAA) (LaBarre et al., 2010). In POC settings, HAD and SAMBA have also been identified as alternative techniques for NAA.

25.5 SENSING TECHNOLOGIES USED IN VIRUS SENSORS

Various paper-based analytic sensors, such as colorimetric, electrochemical, and chemiluminescent, have been developed to quantify virus detection. In this section, we will discuss the sensing mechanism and techniques involved and understand how these sensors were developed in this paper.

25.5.1 Colorimetric Sensor

A microfluidic paper-based analytic device has the ability to provide a semiquantitative or "yes/no" answer in colorimetric detection. Colorimetric sensors provide answers based on the interaction between the target substance and chemical reagents. The results of the colorimetric sensor can be viewed through direct observation using the naked eye and through computer software; the latter is a preferred mode of receiving data (Liana et al., 2012). Rodriguez et al. (2016) developed a paper-based colorimetric sensor for the detection of HVP 16 DNA. For this method, they used paper and pressure-sensitive adhesive sheets to extract, detect, and amplify nucleic acid from clinical samples. A similar type of sensor has been developed to detect HIV-1 and other pathogens by Shafiee et al. (2015).

The highly sensitive and specific colorimetric sandwich immunoassay model was fabricated for the detection of influenza A H1N1 and H3N2 viruses. The developed sensor offered benefits such as shorter processing time (around 1 h), fewer samples, and lower reagent volume used (5 μL) (Lei et al., 2015). The combination of RT-PCR and the PFMA-based colorimetric assay provides a simple and convenient diagnostic tool for the detection of viruses. The detection of influenza viruses, such as H1N1, H3N2, and H5N1, is with the detection limit of 10^2 copies (Kim et al., 2014). Recently, a single-stranded DNA, hybridized with gold nanoparticles, was developed for the detection of TB DNA at a concentration of 2.6nM. Due to DNA hybridization, a color change of red to blue occurred. Since this sensor operated through a mobile phone, the results of color change and accompanying data were transferred via computers (Tsai et al., 2013).

25.5.2 Electrochemical Sensor

Electrochemical detection is suitable for microfluidic PAD as it is cost-effective, user-friendly, highly sensitive and selective, and low on electrical consumption. Most electrochemical sensors were developed to detect glucose, uric acid, and molecules. The basic working principle of an electrochemical sensor is based on the electric current, which is produced from the chemical reaction, passed through the electrode (Dungchai et al., 2009; Yetisen, 2015). The hollow channel SlipPAD methodology and electrochemical-based sensor were developed for the detection of HBV virus. The obtained LOD is 85pM and costs $0.36, comparably lower than earlier detection methods. However, some problems did occur during the implementation of the system, which included reagent resolvation, time management issues in the hybridization step, and nonspecific adsorption (Li et al., 2015).

Graphene-polyaniline-modified paper-based electrochemical biosensor was developed by the inkjet printing technique for the detection of human papillomavirus (HPV). By electrochemical impendence spectroscopy, the efficiency of the sensor was measured. The results showcased a detection limit for the HPV type 16 DNA to be 2.3nM with a linear range of 10–200nM. The advantages of this sensor include cost-effectiveness, ease of use, and disposability (Teengam et al., 2017). Narang et al. (2017) fabricated an electrochemical paper-based device for the detection of herpes caused by the herpes virus 5 (HHV-5) DNA through the integration of Zn-Ag nanoblooms with a detection limit of 97 copies per mL. The device was cost-effective, user-friendly, and simple.

25.5.3 Electrochemiluminescent Sensor

Electrochemiluminescence (ECL) is the combination of chemiluminescence and electrochemical techniques that results in the generation of light. Dynamic concentration response and sensitivity, the requirement of fewer samples, and easy handling are the major advantages of ECL. More than 150 different types of immunoassays are available in the market for the detection of tumor and infectious diseases (Ge et al., 2012; Parveen et al., 2013).

ECL has gained widespread attention owing to its performance in the rapid detection of viruses. A novel paper-based ECL biosensor was developed for hepatitis B. This sensor showed high sensitivity, selectivity, and reproducibility. It performed better at a detection limit of 34.2pg/mL to 34.2ng/mL (Chen et al., 2018); however, the process is time-consuming. By using the PCR technique, the viral RNA of a dengue virus from dried blood samples was developed. We concluded that the major benefit recognized from this study was that the dengue RNA was highly stable, allowing it to be stored and transported in laboratory condition (Prado et al., 2005).

25.5.4 Enzyme-Linked Immunosorbent Assay

Microfluidic paper-based analytic devices have great potential to transform conventional immunoassays for diagnosis. Mu et al. (2014) developed a multiplex microfluidic paper-based immunoassay device for the detection of human igG antibody against HCV. They fabricated the device for multitest zones with highly flammable nitrocellulose, patterned by craft punch patterning. With this device, the LOQ of a mouse igG in chemiluminescence and colorimetry is set at 267 amol and 26.7 fmol, respectively. Pardee et al. (2016) developed a freeze-dried paper-based sensor for the detection of Zika virus by the combination of nucleic-acid-based molecular diagnostic and RNA amplification. The developed sensor employed the simple preparation of extracting the viral RNA, thus, demonstrating the robust detection of a Zika virus. In this sensor, the enzymes are combined with engineered gene circuits and freeze-dried on the paper. Further, a sensor for mRNA Ebola virus was developed, which was fabricated in 12 h and possessed a DNA input of the Ebola series, with a cost of $21 per sensor (Pardee et al., 2014). A magnetoelastic (ME) sensing system comprises a test paper for the detection of classical swine fever virus (CSFV). The anti-CSFV is immobilized on the ME surface and developed using ELISA analysis, which provided detection with a sensitivity of about 95 Hz/μg/mL and a limit of 0.6 μg/mL (Guo et al., 2016). Rohrman and Richards-Kortum (2012) used micro-PADs for the enzymatic amplification detection of the HIV DNA. This technique successfully enhanced the 10 copies of the HIV DNA within 15 min. Fig. 25.2 represents the schematic for the ELISA method.

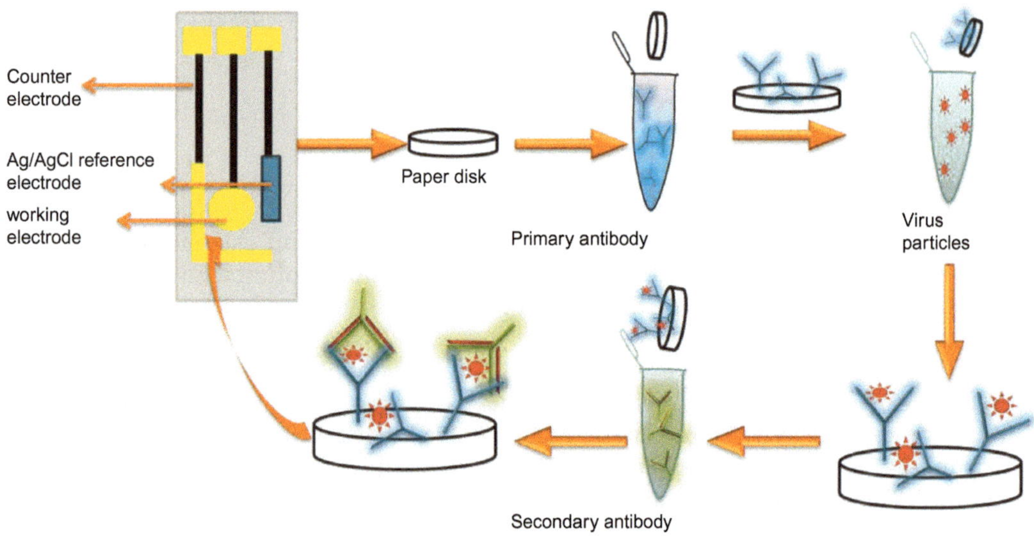

FIG. 25.2 Schematic diagram of ELISA for virus sensor.

25.5.5 Amplicon Detection in Paper-Based Diagnostics

Amplicon, including amplicon HIV-1 and HIV-1 QT, is a piece of DNA or RNA that makes it possible to analyze viruses using different techniques. Following nucleic acid amplification, a unique technique is required for the detection of amplicon. Gel electrophoresis and fluorescent detection methods are complicated, require costly equipment, and use chip-based detection techniques, creating difficulty in device fabrication (Ahmad and Hashsham, 2012; Hu et al., 2013). Single-stranded commercial DNAs are mostly used in LFA techniques. While double-stranded amplicon DNA is available and can be bound to a complementary DNA probe. A double-stranded DNA amplicon can be detected by Ag-Ab interaction. Gold nanoparticle-amplicon conjugates the binding of digoxigenin Ab on the test line, and results showed positive signals, with immobilized anti-rabbit Ab acting as a control (Kersting et al., 2014).

Metal nanoparticles are also used in the sensitivity process of LFA with thermal contrast. Upon irradiating with laser light, the metal nanoparticles released heat and helped in producing a colorimetric signal with 36 folds (Qin et al., 2012). Several fluidic control techniques are developed using paper architecture modifications and wax-printed pillars. Microfluidic paper-based analytic devices are also used as an amplicon sensor. Based on the biotin-streptavidin interaction, an instant paper-based device was fabricated using wax printing techniques for the detection of LAMP (Lo et al., 2012). The benefits of a paper-based system for NAT include cost reduction, simplicity of data, and aptness of device fabrication for POC applications. With the excellent capability possessed by paper-based platforms with advanced technology and fabrications, we expect more paper-based devices to be developed to overcome the drawbacks in the present techniques.

25.6 LIMITATIONS OF PAPER-BASED DIAGNOSTICS

Paper-based analytic devices provide cost-effective solutions in the viral sensor field. These are preliminary screening tools for health care and are useful and user-friendly (Whitesides, 2013). However, the paper-based sensor needs further improvement with respect to its cost base and clinical performance. Previously reported papers show varying ranges in the sensitivity and specificity of all its aspects (Pike et al., 2013). However, the main disadvantages observed include the following: First, different substrates and detection methods can be attached in different combinations to develop the sensor, and the cost depends on the fabrication sensor (Millipore, 2013); second, controversial readouts are reported due to subjective judgments by the operators and illumination settings, especially in the colorimetric sensor; third, environmental effects play a major role in the upkeep of the testing devices; and finally, batch-to-batch variation is a well-known challenge in POC testing (Abe et al., 2010; Li et al., 2011) and affects this method of diagnostics. Therefore, paper-based diagnostics will need to improve with respect to its sensitivity and selectivity for widescale clinical applications.

25.7 CHALLENGES AND FUTURE OUTLOOK OF PAPER-BASED CHIPS

The paper-based device has enormous potential in the sensor field. However, more development is required so that its performance match that of conventional analytic techniques. Through synergies with materials science and biomedical engineering, paper-based diagnostics are becoming simpler, more accurate, and more sensitive. The sensitivity has increased due to the distance between the sample loading in the paper and the detection zone. The analyte is spread across the paper; the concentration of the sample may decrease due to evaporation if the distance of the sensing zone and sample loading region is large. Evaporation of the sample can be addressed by modifying the paper using sticky tape or porous substrate (Apilux et al., 2010). Current research provides a high-level epidemiological monitoring of viral diseases. However, it is without question that we need to amplify public health ambitions and invest in the development of diagnostic methods, reduce cost, and bring about size-based improvements (Loman et al., 2012; Frey et al., 2014). Microfluidic technology is a prominent piece of the modern diagnostic system that includes biomedical research, DNA analysis, environmental sensor, and food safety (Sher et al., 2017). These devices can become the future of sensing kit options, even though they are currently facing some limitations such as low sensitivity and reproducibility, and the current paper-based microfluidic devices have limited capabilities in the sensor chips.

Paper-based diagnostics, mobile-phone-based optical detection, and telemedicine have gained more importance in terms of improving health-care diagnostics tools. However, limitations such as plasma separation, sample pretreatment, nucleic acid isolation, and amplification hinder their potential in POC diagnostics (Hu et al., 2014). These aspects are important in paper-based diagnostic devices with respect to sensitivity and selectivity. Lab-on-a-chip (Cao et al., 2012) is a small device that consumes small amounts of samples to perform immediate reactions with sensing devices. The reactions vary as per our analyte, detection method, and instruments. LOC devices are approved by the FDA to feature among the diagnostic devices for influenza, HIV, and HVB (Alyassin et al., 2009; Cao et al., 2012; Zhi et al., 2014). The recent developments in paper-based diagnostic devices are reconstructing clinical microbiology and the role of detection methods in reducing the prevalence of infectious diseases.

In conclusion, we described the basic concepts and developing methods and have highlighted paper-based diagnostic devices applied in the last decade in the virus-sensing field. The fundamentals, device fabrication, functionalization, and different techniques used in the paper-based devices have been critically discussed. This review also provides the advantages and disadvantages of current paper-based diagnostic devices and discusses development trends for the future. Owing to new advancements in nanotechnology, paper-based sensing rises above alternative methods for the development of rapid diagnostic sensors for food safety, environmental control, and personalized medicine.

Acknowledgments

This research was supported by the National Research Foundation of Korea (NRF) grant funded by the Korean government (MSIP) (No.2017R1A2B4004700).

This work was supported by research program of Ministry of Trade, Industry and Energy (R&D fund, 10062995).

References

Abe, K., Kotera, K., Suzuki, K., Citterio, D., 2010. Inkjet-printed paperfluidic immuno-chemical sensing device. Anal. Bioanal. Chem. 398, 885–893.

Ahmad, F., Hashsham, S.A., 2012. Miniaturized nucleic acid amplification systems for rapid and point-of-care diagnostics: a review. Anal. Chim. Acta 733, 1–15.

Ahmed, S., Bui, M.P.N., Abbas, A., 2016. Based chemical and biological sensors: engineering aspects. Biosens. Bioelectron. 77, 249–263.

Akhtar, A., Fuchs, E., Mitchison, T., Shaw, R.J., St Johnston, D., Strasser, A., Taylor, S., Walczak, C., Zerial, M., 2011. A decade of molecular cell biology: achievements and challenges. Nat. Rev. Mol. Cell Biol. 12, 669.

Alyassin, M.A., Moon, S., Keles, H.O., Manzur, F., Lin, R.L., Hæggstrom, E., Kuritzkes, D.R., Demirci, U., 2009. Rapid automated cell quantification on HIV microfluidic devices. Lab Chip 9, 3364–3369.

Apilux, A., Dungchai, W., Siangproh, W., Praphairaksit, N., Henry, C.S., Chailapakul, O., 2010. Lab-on-paper with dual electrochemical/colorimetric detection for simultaneous determination of gold and iron. Anal. Chem. 82, 1727–1732.

Arena, A., Donato, N., Saitta, G., Bonavita, A., Rizzo, G., Neri, G., 2010. Flexible ethanol sensors on glossy paper substrates operating at room temperature. Sensors Actuators B Chem. 145, 488–494.

Bayley, T., 1878. XXXVII.-On the behaviour of metallic solutions with filter paper, and on the detection of cadmium. J. Chem. Soc. 33, 304–306.

Bissonnette, L., Bergeron, M., 2010. Diagnosing infections—current and anticipated technologies for point-of-care diagnostics and home-based testing. Clin. Microbiol. Infect. 16, 1044–1053.

Cao, Q., Mahalanabis, M., Chang, J., Carey, B., Hsieh, C., Stanley, A., Odell, C.A., Mitchell, P., Feldman, J., Pollock, N.R., Klapperich, C.M., 2012. Microfluidic chip for molecular amplification of influenza a RNA in human respiratory specimens. PLoS One 7, e33176.

Chen, Y., Wang, J., Liu, Z., Wang, X., Li, X., Shan, G., 2018. A simple and versatile paper-based electrochemiluminescence biosensing platform for hepatitis B virus surface antigen detection. Biochem. Eng. J. 129, 1–6.

Chong, H., Koo, Y., Collins, B., Gomez, F.A., Yun, Y., Sankar, J., 2013. Paper-based microfluidic point-of-care diagnostic devices for monitoring drug metabolism. J. Nanomedine Biotherapeutic Discov. 3, e122.

Craw, P., Balachandran, W., 2012. Isothermal nucleic acid amplification technologies for point-of-care diagnostics: a critical review. Lab Chip 12, 2469–2486.

Devarakonda, S., Singh, R., Bhardwaj, J., Jang, J., 2017. Cost-effective and handmade paper-based immunosensing device for electrochemical detection of influenza virus. Sensors 17, 2597.

Dungchai, W., Chailapakul, O., Henry, C.S., 2009. Electrochemical detection for paper-based microfluidics. Anal. Chem. 81, 5821–5826.

Dungchai, W., Chailapakul, O., Henry, C.S., 2011. A low-cost, simple, and rapid fabrication method for paper-based microfluidics using wax screen-printing. Analyst 136, 77–82.

Frey, K.G., Herrera-Galeano, J.E., Redden, C.L., Luu, T.V., Servetas, S.L., Mateczun, A.J., Mokashi, V.P., Bishop-Lilly, K.A., 2014. Comparison of three next-generation sequencing platforms for metagenomic sequencing and identification of pathogens in blood. BMC Genomics 15, 96.

Fronczek, C.F., San Park, T., Harshman, D.K., Nicolini, A.M., Yoon, J.Y., 2014. Paper microfluidic extraction and direct smartphone-based identification of pathogenic nucleic acids from field and clinical samples. RSC Adv. 4, 11103–11110.

Ge, L., Yan, J., Song, X., Yan, M., Ge, S., Yu, J., 2012. Three-dimensional paper-based electrochemiluminescence immunodevice for multiplexed measurement of biomarkers and point-of-care testing. Biomaterials 33, 1024–1031.

Guo, X., Gao, S., Sang, S., Jian, A., Duan, Q., Ji, J., Zhang, W., 2016. Detection system based on magnetoelastic sensor for classical swine fever virus. Biosens. Bioelectron. 82, 127–131.

Hawkins, K.R., Weigl, B.H., 2010. In: Microfluidic diagnostics for low-resource settings. Proceedings of SPIE 7593, Microfluidics, BioMEMS, and Medical Microsystems VIII, 75930L, 17 February 2010.

Hu, J., Wang, L., Li, F., Han, Y.L., Lin, M., Lu, T.J., Xu, F., 2013. Oligonucleotide-linked gold nanoparticle aggregates for enhanced sensitivity in lateral flow assays. Lab Chip 13, 4352–4357.

Hu, J., Wang, S., Wang, L., Li, F., Pingguan-Murphy, B., Lu, T.J., Xu, F., 2014. Advances in paper-based point-of-care diagnostics. Biosens. Bioelectron. 54, 585–597.

Kane, R.S., Takayama, S., Ostuni, E., Ingber, D.E., Whitesides, G.M., 2006. Patterning proteins and cells using soft lithography. In: The Biomaterials: Silver Jubilee Compendium. Elsevier Science, Oxford, pp. 161–174.

Kersting, S., Rausch, V., Bier, F.F., von Nickisch-Rosenegk, M., 2014. Rapid detection of plasmodium falciparum with isothermal recombinase polymerase amplification and lateral flow analysis. Malar. J. 13, 99.
Kim, Y.T., Jung, J.H., Choi, Y.K., Seo, T.S., 2014. A packaged paper fluidic-based microdevice for detecting gene expression of influenza A virus. Biosens. Bioelectron. 61, 485–490.
Kumbhat, S., Sharma, K., Gehlot, R., Solanki, A., Joshi, V., 2010. Surface plasmon resonance based immunosensor for serological diagnosis of dengue virus infection. J. Pharm. Biomed. Anal. 52, 255–259.
LaBarre, P., Gerlach, J., Wilmoth, J., Beddoe, A., Singleton, J., Weigl, B., 2010. Non-instrumented nucleic acid amplification (NINA): instrument-free molecular malaria diagnostics for low-resource settings. EMBS 2010, 1097–1099.
Lalkhen, A.G., McCluskey, A., 2008. Clinical tests: sensitivity and specificity. Contin. Educ. Anaesth. Crit. Care Pain 8, 221–223.
Lei, K.F., Huang, C.H., Kuo, R.L., Chang, C.K., Chen, K.F., Tsao, K.C., Tsang, N.M., 2015. Based enzyme-free immunoassay for rapid detection and subtyping of influenza A H1N1 and H3N2 viruses. Anal. Chim. Acta 883, 37–44.
Li, C.Z., Vandenberg, K., Prabhulkar, S., Zhu, X., Schneper, L., Methee, K., Rosser, C.J., Almeide, E., 2011. Paper based point-of-care testing disc for multiplex whole cell bacteria analysis. Biosens. Bioelectron. 26, 4342–4348.
Li, X., Ballerini, D.R., Shen, W., 2012. A perspective on paper-based microfluidics: current status and future trends. Biomicrofluidics 6, 011301.
Li, X., Scida, K., Crooks, R.M., 2015. Detection of hepatitis B virus DNA with a paper electrochemical sensor. Anal. Chem. 87, 9009–9015.
Liana, D.D., Raguse, B., Gooding, J.J., Chow, E., 2012. Recent advances in paper-based sensors. Sensors 12, 11505–11526.
Lillis, L., Lehman, D., Singhal, M.C., Cantera, J., Singleton, J., Labarre, P., Toyama, A., Piepenburg, O., Parker, M., Wood, R., Overbaugh, J., 2014. Non-instrumented incubation of a recombinase polymerase amplification assay for the rapid and sensitive detection of proviral HIV-1 DNA. PLoS One 9, e108189.
Liu, F., Gu, J., 2011. Retinoic acid inducible gene-I, more than a virus sensor. Protein Cell 2, 351–357.
Lo, S.J., Yang, S.C., Yao, D.J., Chen, J.H., Cheng, C.M., 2012. Molecular-level dengue fever diagnostics: developing a combination of RT-LAMP and paper-based devices. IEEE Nanotechnol. Mag. 6, 26–30.
Loman, N.J., Misra, R.V., Dallman, T.J., Constantinidou, C., Gharbia, S.E., Wain, J., Pallen, M.J., 2012. Performance comparison of benchtop high-throughput sequencing platforms. Nat. Biotechnol. 30, 434.
Lopez-Marzo, A.M., Merkoci, A., 2016. Based sensors and assays: a success of the engineering design and the convergence of knowledge areas. Lab Chip 16, 3150–3176.
Mao, X., Ma, Y., Zhang, A., Zhang, L., Zeng, L., Liu, G., 2009. Disposable nucleic acid biosensors based on gold nanoparticle probes and lateral flow strip. Anal. Chem. 81, 1660–1668.
Martinez, A.W., Phillips, S.T., Whitesides, G.M., Carrilho, E., 2013. Diagnostics for the developing world: microfluidic paper-based analytical devices. Anal. Chem. 82, 3–10.
Millipore, 2013. Hi-Flow™ Plus Membranes and Surewicks Pad Materials: Quality Materials for Use in Rapid Diagnostic Lateral Flow Tests. Merck Millipore, Darmstadt, Germany.
Morens, D.M., Folkers, G.K., Fauci, A.S., 2004. The challenge of emerging and re-emerging infectious diseases. Nature 430, 242.
Mu, X., Zhang, L., Chang, S., Cui, W., Zheng, Z., 2014. Multiplex microfluidic paper-based immunoassay for the diagnosis of hepatitis C virus infection. Anal. Chem. 86, 5338–5344.
Narang, J., Singhal, C., Mathur, A., Sharma, S., Singla, V., Pundir, C.S., 2017. Portable bioactive paper based genosensor incorporated with Zn-Ag nanoblooms for herpes detection at the point-of-care. Int. J. Biol. Macromol. 107, 2559–2565.
Nery, E.W., Kubota, L.T., 2013. Sensing approaches on paper-based devices: a review. Anal. Bioanal. Chem. 405, 7573–7595.
Niemz, A., Ferguson, T.M., Boyle, D.S., 2011. Point-of-care nucleic acid testing for infectious diseases. Trends Biotechnol. 29, 240–250.
Oliver, G., 1883. On bedside urinary tests: detection of sugar in the urine by means of test papers. Lancet 121, 858–860.
Pardee, K., Green, A.A., Ferrante, T., Cameron, D.E., Daley Keyser, A., Yin, P., Collins, J.J., 2014. Based synthetic gene networks. Cell 159, 940–954.
Pardee, K., Green, A.A., Takahashi, M.K., Braff, D., Lambert, G., Lee, J.W., Ferrante, T., Ma, D., Donghia, N., Fan, M., Daringer, N.M., 2016. Rapid, low-cost detection of Zika virus using programmable biomolecular components. Cell 165, 1255–1266.
Parveen, S., Aslam, M.S., Hu, L., Xu, G., 2013. Electrogenerated Chemiluminescence: Protocols and Applications. Springer, Heidelberg, Germany.

Pike, J., Godbert, S., Johnson, S., 2013. Comparison of volunteers' experience of using, and accuracy of reading, different types of home pregnancy test formats. Expert Opin. Med. Diagn. 7, 435–441.

Prado, I., Rosario, D., Bernardo, L., Alvarez, M., Rodriguez, R., Vazquez, S., Guzman, M.G., 2005. PCR detection of dengue virus using dried whole blood spotted on filter paper. J. Virol. Methods 125, 75–81.

Qin, Z., Chan, W.C., Boulware, D.R., Akkin, T., Butler, E.K., Bischof, J.C., 2012. Significantly improved analytical sensitivity of lateral flow immunoassays by using thermal contrast. Angew. Chem. Int. Ed. 51, 4358–4361.

Rodriguez, N.M., Wong, W.S., Liu, L., Dewar, R., Klapperich, C.M., 2016. A fully integrated paperfluidic molecular diagnostic chip for the extraction, amplification, and detection of nucleic acids from clinical samples. Lab Chip 16, 753–763.

Rohrman, B.A., Richards-Kortum, R.R., 2012. A paper and plastic device for performing recombinase polymerase amplification of HIV DNA. Lab Chip 12, 3082–3088.

Schiff, H., 1866. Eine neue reihe organischer diamine. Eur. J. Org. Chem. 140, 92–137.

Shafiee, H., Asghar, W., Inci, F., Yuksekkaya, M., Jahangir, M., Zhang, M.H., Durmus, N.G., Gurkan, U.A., Kuritzkes, D.R., Demirci, U., 2015. Paper and flexible substrates as materials for biosensing platforms to detect multiple biotargets. Sci. Rep. 6, 8719.

Sher, M., Zhuang, R., Demirci, U., Asghar, W., 2017. Based analytical devices for clinical diagnosis: recent advances in the fabrication techniques and sensing mechanisms. Expert. Rev. Mol. Diagn. 17, 351–366.

Souf, S., 2016. Recent advances in diagnostic testing for viral infections. Biosci. Horiz. 9, https://doi.org/10.1093/biohorizons/hzw010.

Stevens, D.Y., Petri, C.R., Osborn, J.L., Spicar-Mihalic, P., McKenzie, K.G., Yager, P., 2008. Enabling a microfluidic immunoassay for the developing world by integration of on-card dry reagent storage. Lab Chip 8, 2038–2045.

Teengam, P., Siangproh, W., Tuantranont, A., Henry, C.S., Vilaivan, T., Chailapakul, O., 2017. Electrochemical paper-based peptide nucleic acid biosensor for detecting human papillomavirus. Anal. Chim. Acta 952, 32–40.

Tsai, T.T., Shen, S.W., Cheng, C.M., Chen, C.F., 2013. Paper-based tuberculosis diagnostic devices with colorimetric gold nanoparticles. Sci. Technol. Adv. Mater. 14, 044404. https://doi.org/10.1088/1468-6996/14/4/044404.

Wang, S., Ge, L., Song, X., Yan, M., Ge, S., Yu, J., Zeng, F., 2012. Simple and covalent fabrication of a paper device and its application in sensitive chemiluminescence immunoassay. Analyst 137, 3821–3827.

Whitesides, G.M., 2013. Cool, or simple and cheap? Why not both? Lab Chip 13, 11–13.

Wong, R.C., Harley, Y.T., 2009. Quantitative, false positive, and false negative issues for lateral flow immunoassays as exemplified by onsite drug screens. In: Lateral Flow Immunoassay. Humana Press, New York, NY, pp. 1–19.

Yetisen, A.K., 2015. Point-of-care diagnostics. In: Holographic Sensors. Springer, Cham, pp. 1–25.

Zhang, D., Broyles, D., Hunt, E.A., Dikici, E., Daunert, S., Deo, S.K., 2017. A paper-based platform for detection of viral RNA. Analyst 142, 815–823.

Zhao, W., van den Berg, A., 2008. Lab on paper. Lab Chip 8, 1988–1991.

Zhi, X., Deng, M., Yang, H., Gao, G., Wang, K., Fu, H., Zhang, Y., Chen, D., Cui, D., 2014. A novel HBV genotypes detecting system combined with microfluidic chip, loop-mediated isothermal amplification and GMR sensors. Biosens. Bioelectron. 54, 372–377.

Index

Note: Page numbers followed by *f* indicate figures and *t* indicate tables.

M

N

O

P

FSC
www.fsc.org
MIX
Papier | Fördert gute Waldnutzung
FSC® C083411

Kolibri 360 GmbH
Ferdinand-Jühlke-Straße 7
99095 Erfurt, Deutschland
produktsicherheit@kolibri360.de